Cell-Penetrating Peptides

Cell-Penetrating Peptides

Design, Development and Applications

Edited by Makoto Oba and Yosuke Demizu

WILEY-VCH

Editors

Prof. Makoto Oba
Kyoto Prefectural University of Medicine
1-5 Shimogamohangi-cho
Sakyo-ku
606-0823 Kyoto
Japan

Dr. Yosuke Demizu
National Institute of Health Sciences
Division of Organic Chemistry
3-25-26 Tonomachi
210-9501 Kawasaki
Japan

Cover: © Explode/Shutterstock

Library of Congress Card No.: applied for

British Library Cataloguing-in-Publication Data
A catalogue record for this book is available from the British Library.

Bibliographic information published by the Deutsche Nationalbibliothek
The Deutsche Nationalbibliothek lists this publication in the Deutsche Nationalbibliografie; detailed bibliographic data are available on the Internet at <http://dnb.d-nb.de>.

Print ISBN: 978-3-527-35011-7
ePDF ISBN: 978-3-527-83601-7
ePub ISBN: 978-3-527-83600-0
oBook ISBN: 978-3-527-83599-7

Cover Design: FORMGEBER, Mannheim

Typesetting: Straive, Chennai

Printing and Binding: CPI Group (UK) Ltd, Croydon, CR0 4YY

C117866_221122

Contents

Preface

The efficient and nontoxic delivery of cell membrane-impermeable bioactive molecules into cells still remains a challenging problem. Cell-penetrating peptides (CPPs) are promising candidates for this purpose. CPPs have been developed for over three decades since the first CPP, transactivator of transcription (Tat) peptide, was found in human immunodeficiency virus type 1 (HIV-1) protein. Since then, many CPP sequences have been discovered in natural peptides and proteins, as well as being artificially designed and synthesized. Their cell-penetrating mechanisms have been investigated in vitro and in vivo. Furthermore, CPPs have been used as delivery tools for drugs, peptides, proteins, nucleic acids, and nanoparticles, and several CPPs are currently under investigation in clinical trials. We wanted to create a book covering such topics.

In the first part of this book, the design of CPPs is described based on the characteristics and typical types of CPPs. In the second part, the mechanisms of CPP action both in vitro and in vivo are introduced from chemical and biological viewpoints. The third part describes CPPs as a delivery tool for drugs, peptides, proteins, nucleic acids, morpholino oligomers, polymer micelles, and lipid-based nanoparticles. The final (fourth) part examines the application of CPPs not only in medical fields but also in agricultural fields. We hope that this book will be of interest to researchers who are studying CPPs, and that it will stimulate their research and aid in the development of the next generation of experts in this field.

We sincerely thank all the contributors to this book; they graciously produced their manuscripts within their busy schedules. The publication of this book would have been impossible without the considerable effort and patience of these contributors and the staff of Wiley-VCH GmbH.

Makoto Oba
Yosuke Demizu

1

Introduction

Makoto Oba[1] *and Yosuke Demizu*[2]

[1] *Kyoto Prefectural University of Medicine, Graduate School of Medical Science, Kyoto, Japan*
[2] *National Institute of Health Sciences, Division of Organic Chemistry, Kanagawa, Japan*

Cell-penetrating peptides (CPPs) are short peptides (in general composed of 5–30 amino acids) that can be efficiently internalized into cells and have great potential for the delivery of membrane-impermeable bioactive molecules into cells. The CPP is sometimes called a protein transduction domain (PTD) or Trojan peptide. In 1988, such a peptide was first discovered in human immunodeficiency virus type 1 (HIV-1) protein transactivator of transcription (Tat). Since then, numerous sequences of CPPs have been discovered in natural peptides/proteins such as Tat peptide and CPPs have also been artificially designed. Furthermore, the cell-penetrating mechanisms of CPPs in vitro and in vivo have been investigated, and CPPs have been used as delivery tools for drugs, peptides, proteins, nucleic acids, and nanoparticles. Several CPPs are currently under investigation in clinical trials. More than 500 research papers per year have been published since 2012 with a keyword of "CPP" or "PTD," and numbers have reached slightly less than 1000 for the last few years. From these numbers, we can understand the usefulness and significance of CPPs.

This book describes the design, mechanism, delivery tools, and applications of CPPs. There are several books and journals' special issues on CPPs. However, none has previously categorized such topics. On the topic of design, the classification of CPPs is first described based on their characteristics, and then, typical types of CPPs (cationic, amphipathic, and hydrophobic peptides) and Arg-rich peptides and foldamers are introduced. The topic of mechanism deals with important factors of CPPs, such as peptide secondary structure, cellular uptake, endosomal escape, and pharmacokinetics in vivo. The topic of delivery tools is categorized based on cargos, drugs, peptides and proteins, nucleic acids, and morpholino oligomers. Furthermore, CPPs assisting in the efficient delivery of nano-sized drug delivery systems, polymeric micelles, and lipid-based nanoparticles are introduced as delivery tools. The final topic of applications covers oral delivery and intranasal delivery using CPPs, CPPs in clinical trials, and applications in plants. Table 1.1 lists the representative

Cell-Penetrating Peptides: Design, Development and Applications, First Edition. Makoto Oba and Yosuke Demizu. Published 2023 by WILEY-VCH GmbH.

CPPs introduced in this book. We hope that this book will be useful for readers studying and treating CPPs now and in the future.

Table 1.1 Lists of CPPs introduced in this book.

CPP	Sequence	Ref.
[C12-R4]	Cyclic(CXRRRR) X: (*R*,*S*)-2-amino tetradecanoic acid	[1]
[R6W3]	Cyclic(CRRWWRRWRR)	[1]
A2-17	LRKLRKRLLRLWKLRKR	[2]
A2-17KR	LRRLRRRLLRLWRLRRR	[2]
AA3H	MASIWVGHRG	[3]
Ac_5c^{NH2} peptide	$(RRX)_3$ X: Ac_5c^{NH2}, 1,3-diaminocyclopentanecarboxylic acid	[4]
Ac_5c^{Gu} peptide	$(RRX)_3$ X: Ac_5c^{Gu}, 1-amino-3-guanidinocyclopentanecarboxylic acid	[4]
Ac_6c^{NH2} peptide	$(RRX)_3$ X: Ac_6c^{NH2}, 1,4-diaminocyclohexanecarboxylic acid	[5]
Amphipathic β-peptide	[(*S*,*S*)-ACHC-β^3hArg-β^3hArg]$_3$ ACHC: *trans*-2-aminocyclohexanecarboxylic acid β^3hArg: β^3-homoarginine	[6]
Api^{C2Gu} peptide	$(RRX)_3$ X: Api^{C2Gu}, 4-aminopiperidine-4-carboxylic acid derivative	[7]
αR7W2	RRRWRRWRR	[8]
ARF (1-22)	MVRRFLVTLRIRRACGPPRVRV	[9]
ARF (19-31)	RVRVFVVHIPRLT	[9]
β-Heptaarginine	(β^3hArg)$_7$ β^3hArg: β^3-homoarginine	[10]
β-Heptalysine	(β^3hLys)$_7$ β^3hLys: β^3-homolysine	[10]
β-Tat	β^3hArg-(β^3hLys)$_2$-(β^3hArg)$_2$-β^3hGln-(β^3hArg)$_3$	[11]
$Bac_{15\text{-}24}$	RRIRPRPPRLPRPRPRPLPFPRPG	[12]
Bip2	VPTLK	[13]
Block3	LLULLULLUGGGRRRRRRRR U: Aib, 2-aminoisobutyric acid	[14]
BP100	KKLFKKKKILKYL	[15]
bPrPp (1-30)	MVKSKIGSWILVLFVAMWSDVGLCKKRPKP	[16]
Butyl-TH	AGYLLGH$_B$INLH$_B$H$_B$LAH$_B$LUH$_B$H$_B$IL U: Aib, 2-aminoisobutyric acid; H$_B$: 3-butylhistidine	[17]
C105Y	CSIPPEVKFNKPFVYLI	[18]
CADY	GLWRALWRLLRSLWRLLWRA-cysteamide	[19]
CH2R4H2C	CHHRRRRHHC	[20]

(*Continued*)

Table 1.1 (Continued)

CPP	Sequence	Ref.
cLK	Cyclic Ac-CKKLLKLLKKLLKLGGLKKLLKLLKKLLKLLK Crosslink between the side chains of C and K	[21]
CPP12	Cyclic(FXR$_4$) X: L-2-naphthylalanine	[22]
CPP2	DSLKSYWYLQKFSWR	[23]
CPP44	KRPTMRFRYTWNPMK	[23]
CPP9	Cyclic(fXRrRrQ) X: L-2-naphthylalanine	[24]
Cyclic [W(RW)$_4$]	Cyclic[W(RW)$_4$]	[25]
Cyclic R$_9$	Cyclic(CRRRRRRRRR)	[26]
Cyt c$^{77\text{-}101}$	GTKMIFVGIKKKEERADLIAYLKKA	[27]
D-Oligoarginine	r$_n$	[28]
D-Tat$^{49\text{-}57}$	Rkkrrqrrr	[28]
DPV1047	VKRGLKLRHVRPRVTRMDV	[29]
DPV3	RKKRRRESRKKRRRES	[29]
dTat-Sar-EED4	rrrqrrkkrXXXXXXGWWG X: Sar, sarcosine	[30]
FHV coat$^{35\text{-}49}$	RRRRNRTRRNRRRVR	[31]
GALA	WEAALAEALAEALAEHLAEALAEALEALAA	[32]
H5WYG	GLFHAIAHFIHGGWHGLIHGWYG	[33]
HA2	GDIMGEWGNEIFGAIAGFLG	[34]
HAad	IWLTALKFLGKAAAKAXAKQXLSKL X: L-2-aminoadipic acid	[35]
hCT$^{9\text{-}32}$-br	LGTYTQDFNK(X)FHTFPQTAIGVGAP X: PKKKRKVEDPGVGFA	[36]
HIV-1 Rev$^{34\text{-}50}$	TRQARRNRRRRWRERQR	[31]
HL	CHHHHHRRWQWRHHHHHC	[37]
HR9	CHHHHHRRRRRRRRRHHHHHC	[38]
HTLV-II Rex$^{4\text{-}16}$	TRRQRTRRARRNR	[31]
Hydrophobic MPS	VTVLAGALAGVGVG	[39]
K10H16	KKKKKKKKGHHHHHHHHHHHHHHHH	[40]
KAibA	poly(KUA) U: Aib, 2-aminoisobutyric acid	[41]
KALA	WEAKLAKALAKALAKHLAKALAKALKACEA	[42]
KLA	(KLAKLAK)$_2$	[43]
KLA10	KALKKLLAKWLAAAKALL	[44]

(Continued)

Table 1.1 (Continued)

CPP	Sequence	Ref.
L-Pro^{Gu} peptide	(RRX)$_3$ X: L-Pro^{Gu}; 4-guanidinoproline	[45]
L1-7	Cyclic FIDIIIKILLI Crosslink between the side chains of D and K	[46]
L17E	IWLTALKFLGKHAAKHEAKQQLSKL	[47]
L6	RRWQWR	[48]
LAH4	KKALLALALHHLAHLALHLALALKKA	[49]
LAH4-L1	KKALLAHALHLLALLALHLAHALKKA	[50]
LH	LHHLLHHLHHLLHH	[51]
LK	LKKLLKLLKKLLKL	[52]
LTP	RRKRRKKRRKRRKKKAC	[53]
M1	TFYGGRPKRNNFLRGIR	[54]
M918	MVTVLFRRLRIRRACGPPRVRV	[55]
MAP	KLALKLALKALKAALKLA	[56]
MAP(Aib)	KLULKLULKULKAULKLU U: Aib, 2-aminoisobutyric acid	[57]
Melittin	GIGAVLKVLTTGLPALISWIKRKRQQ	[58]
Mitoparan	INLKKLAKLUKKIL U: Aib, 2-aminoisobutyric acid	[59]
MPG	GALFLGFLGAAGSTMGAWSQPKKKRKV	[60]
MTS	AAVALLPAVLLALLAP	[61]
MTS1	AAVLLPVLLAAP	[62]
N-hxg9	(*N*-hxg)$_9$ *N*-hxg: *N*-guanidinohexylglycine	[28]
NF51	δ-(Stearyl-AGYLLG)OINLKALAALAKKIL O: Orn, L-ornithine	[63]
NF55	δ-(Stearyl-AGYLLG)OINLKALAALAKAIL O: Orn, L-ornithine	[64]
NF70	δ-(Arachidyl-HHHHYHHG)OILLKALKALAKAIL O: Orn, L-ornithine	[65]
Oligoarginine	R$_n$	[28, 31]
Oligohistidine	H$_n$	[66]
Oligolysine	K$_n$	[67]
Oligourea3	iPrNHCO-VUHUWUVUHUWURUγV γV: γ-valine	[68]
p18	LSTAADMQGVVTDGMASG	[69]

(Continued)

Table 1.1 (Continued)

CPP	Sequence	Ref.
P4	LGAQSNF	[70]
PenetraMax	KWFKIQMQIRRWKNKR	[71]
Penetratin	RQIKIWFQNRRMKWKK	[72]
Pep-1	KETWWETWWTEWSQPKKKRKV	[73]
Pep-7	SDLWEMMMVSLACQY	[74]
Pept1	PLILLRLLRGQF	[75]
PF14	Stearyl-AGYLLGKLLOOLAAAALOOLL O: Orn, L-ornithine	[76]
PF3	Stearyl-AGYLLGKINLKALAALAKKIL	[77]
PF6	Stearyl-AGYLLGK(X)INLKALAALAKKIL X: 4-trifluoromethylquinoline-based derivative	[78]
PFV	PFVYLI	[79]
PG-1	RGGRLCYCRRRFCVCVGR	[80]
pHLIC	Cyclic(EEEEWWWWWC)	[81]
pHLIP	AAEQNPIYWWARYADWLFTTPLLLLDLALLVDADEGTCG	[82]
Pip6a	RXRRBRRXRYQFLIRXRBRXRB X: aminohexanoyl; B: β-alanine	[83]
ppTG1	GLFKALLKLLKSLWKLLLKA	[84]
ppTG20	GLFRALLRLLRSLWRLLLRA	[84]
PR20	PRPRPRPRPRPRPRPRPRPRPRPRPRPRPRPRPRPRPRPR	[85]
PreS2-TLM	PLSSIFSRIGDP	[86]
PTD4	YARAAARQARA	[87]
pVEC	LLIILRRRIRKQAHAHSK	[88]
R_6/W_3	RRWWRRWRR	[89]
R_9F_2C	RRRRRRRRRFFC	[90]
RGE	RGERPPR	[91]
RICK	Kwllrwlsrllrwlarwlg	[92]
RLA	Rlarlarrlarlar	[93]
RRU peptide	$(RRU)_n$ (n = 1–6) U: Aib, 2-aminoisobutyric acid	[94]
RRX peptide	$(RRX)_3$ X: (*S*)-α-methylleucine, 1-aminocyclopentanecarboxylic acid, or (3*S*,4*S*)-1-amino-3,4-dimethoxycyclopentanecarboxylic acid	[95]
RW16	RRWRRWWRRWWRRWRR	[96]
RWRWR	RWVRVpGOWIRQ O: Orn, L-ornithine	[97]

(Continued)

Table 1.1 (Continued)

CPP	Sequence	Ref.
RXR	(R-Ahx-R)$_4$-Ahx-(β-Ala) Ahx: 6-aminohexanoic acid; β-Ala: β-alanine	[98]
SAP	VRLPPPVRLPPPVRLPPP	[99]
SAP(E)	VELPPPVELPPPVELPPP	[100]
sC18	GLRKRLRKFRNKIKEK	[101]
SLP F_4R_4	FFFFRRRR	[102]
SLP $R_2F_4R_2$	RRFFFFRR	[102]
SP	RPKPQQFGLM	[103]
Stearyl-oligoarginine	Stearyl-R_n	[104, 105]
Stitched peptide	$\underline{X_1}$RRR$\underline{X_2}$RRRRRR$\underline{X_3}$ X_1: (*S*)-2-(4-pentenyl)alanine; X_2: 2,2-bis-(4-pentenyl)glycine; X_3: (*S*)-2-(7-octenyl)alanine Crosslinks between the side chains of $\underline{X_1}$ and $\underline{X_2}$ and $\underline{X_2}$ and $\underline{X_3}$	[106]
SVS-1	KVKVKVKVpPTKVKVKVK	[107]
SynB1	RGGRLSYSRRRFSTSTGR	[108]
Tachyplesin	KWCFRVCYRGICYRRCR	[109]
$Tat^{48\text{-}60}$	GRKKRRQRRRPPQ	[110]
$Tat^{49\text{-}57}$	RKKRRQRRR	[110]
TH	AGYLLGHINLHHLAHLUHHIL U: Aib, 2-aminoisobutyric acid	[111]
TK	AGYLLGKINLKKLAKLUKKIL U: Aib, 2-aminoisobutyric acid	[112]
TLM	PLSSIFSRIGDP	[113]
TP10	AGYLLGKINLKALAALAKKIL	[114]
Transportan	GWTLNSAGYLLGKINLKALAALAKKIL	[115]
VP22	DAATATRGRSAASRPTQRPRAPARSASRPRRPVE	[116]
$Vpr^{55\text{-}82}$	TGVEALIRILQQLLFIHFRIGCRHSRIG	[117]
VT5	DPKGDPKGVTVTVTVTVTGKGDPKPD	[118]
WRAP1	LLWRLWRLLWRLWRLL	[119]
WRAP5	LLRLLRWWWRLLRLL	[119]

References

1 Amoura, M., Illien, F., Joliot, A. et al. (2019). Head to tail cyclisation of cell-penetrating peptides: impact on GAG-dependent internalization and direct translocation. *Chemical Communications* 55: 4566–4569.

2 Ohgita, T., Takeuchi-Haraya, Y., and Nadai, R. (2019). A novel amphipathic cell-penetrating peptide based on the N-terminal glycosaminoglycan binding region of human apolipoprotein E. *BBA – Biomebnranes* 1861: 541–549.

3 Kim, H.Y., Yum, S.Y., Jang, G. et al. (2015). Discovery of a non-cationic cell penetrating peptide derived from membrane-interacting human proteins and its potential as a protein delivery carrier. *Scientific Reports* 5: 11719.

4 Kato, T., Oba, M., Nishida, K. et al. (2018). Cell-penetrating peptides using cyclic α,α-disubstituted α-amino acids with basic functional groups. *ACS Biomaterials Science & Engineering* 4: 1368–1376.

5 Kato, T., Kita, Y., Iwanari, K. et al. (2021). Synthesis of six-membered carbocyclic ring α,α-disubstituted amino acids and arginine-rich peptides to investigate the effect of ring size on the properties of the peptide. *Bioorganic & Medicinal Chemistry* 38: 116111.

6 Potocky, T.B., Menon, A., and Gellman, S.H. (2005). Effects of conformational stability and geometry of guanidinium display on cell entry by β-peptides. *Journal of the American Chemical Society* 127: 3686–3687.

7 Yamashita, H., Oba, M., Misawa, T. et al. (2016). A helix-stabilized cell-penetrating peptide as an intracellular delivery tool. *ChemBioChem* 17: 137–140.

8 Walrant, A., Bauza, A., Girardet, C. et al. (1862). Ionpair-π interactions favor cell penetration of arginine/tryptophan-rich cell-penetrating peptides. *BBA – Biomembranes* 2020: 183098.

9 Johansson, H., El-Andaloussi, S., Holm, T. et al. (2008). Characterization of a novel cytotoxic cell-penetrating peptide derived from p14ARF protein. *Molecular Therapy* 16 (1): 115–123.

10 Rueping, M., Mahajan, Y., Sauer, M. et al. (2002). Cellular uptake studies with β-peptides. *ChemBioChem* 3: 257–259.

11 Umezawa, N., Gelman, M.A., Haigis, M.C. et al. (2002). Translocation of a β-peptide across cell membranes. *Journal of the American Chemical Society* 124 (3): 368–369.

12 Sadler, K., Eom, K.D., Yang, J.-L. et al. (2002). Translocating proline-rich peptides from the antimicrobial peptide bactenecin 7. *Biochemistry* 41: 14150–14157.

13 Gomez, J.A., Gama, V., Yoshida, T. et al. (2007). Bax-inhibiting peptides derived from Ku70 and cell-penetrating pentapeptides. *Biochemical Society Transactions* 35 (4): 797–801.

14 Misawa, T., Ohoka, N., Oba, M. et al. (2019). Development of 2-aminoisobutyric acid (Aib)-rich cell-penetrating foldamers for efficient siRNA delivery. *Chemical Communications* 55: 7792–7795.

15 Badosa, E., Ferre, R., Planas, M. et al. (2007). A library of linear undecapeptides with bactericidal activity against phytopathogenic bacteria. *Peptides* 28: 2276–2285.

16 Biverstahl, H., Andersson, A., Graslund, A. et al. (2004). NMR solution structure and membrane interaction of the N-terminal sequence (1-30) of the bovine prion protein. *Biochemistry* 43: 14940–14947.

17 Yao, J., Ma, Y., Zhang, W. et al. (2007). Design of new acid-activated cell-penetrating peptides for tumor drug delivery. *Peer Journal* 5: e3429.

18 Rhee, M. and Davis, P. (2006). Mechanism of uptake of C105Y, a novel cell-penetrating peptide. *The Journal of Biological Chemistry* 281 (2): 1233–1240.

19 Crombez, L., Aldrian-Herrada, G., Konate, K. et al. (2009). A new potent secondary amphipathic cell-penetrating peptide for siRNA delivery into mammalian cells. *Molecular Therapy* 17 (1): 95–103.

20 Tanaka, K., Kanazawa, T., Ogawa, T. et al. (2010). Disulfide crosslinked stearoyl carrier peptides containing histidine enhance siRNA uptake and gene silencing. *International Journal of Pharmaceutics* 398: 219.

21 Nam, S.H., Lee, Y., Ahn, J.H. et al. (2020). Augmented osteogenesis of mesenchymal stem cells using a fragmented Runx2 mixed with cell-penetrating, dimeric a-helical peptide. *European Journal of Pharmaceutical Sciences* 144: 105210.

22 Cerulli, R.A., Shehaj, L., Tosic, I. et al. (2020). Cytosolic delivery of peptidic STAT3 SH2 domain inhibitors. *Bioorganic & Medicinal Chemistry* 28: 115542.

23 Kondo, E., Saito, K., Tashiro, Y. et al. (2012). Tumour lineage-homing cell-penetrating peptides as anticancer molecular delivery systems. *Nature Communications* 3: 951.

24 Salim, H., Song, J., Sahni, A. et al. (2020). Development of a cell-permeable cyclic peptidyl inhibitor against the Krap1-Nrf2 interaction. *The Journal of Organic Chemistry* 85: 1416–1424.

25 Shirazi, N.A., Tiwari, R., Chhikara, B.S. et al. (2013). Design and biological evaluation of cell-penetrating peptide-doxorubicin conjugates as prodrugs. *Molecular Pharmaceutics* 10: 488–499.

26 Uhl, P., Grundmann, C., Sauter, M. et al. (2020). Coating of PLA-nanoparticles with cyclic, arginine-rich cell penetrating peptides enables oral delivery of liraglutide. *Nanomedicine: Nanotechnology, Biology, and Medicine* 24: 102132.

27 Howl, J. and Jones, S. (2008). Proteomimetic cell penetrating peptides. *International Journal of Peptide Research and Therapeutics* 14: 359–366.

28 Wender, P.A., Mitchell, D.J., Pattabiraman, K. et al. (2000). The design, synthesis, and evaluation of molecules that enable or enhance cellular uptake: peptoid molecular transporters. *Proceedings of the National Academy of Sciences of the United States of America* 97 (24): 13003–13008.

29 De Coupade, C., Fittipald, A., Chagnas, V. et al. (2005). Novel human-derived cell-penetrating peptides for specific subcellular delivery of therapeutic biomolecules. *Biochemical Journal* 390: 407–418.

30 Miyamoto, T., Tsuchiya, K., and Numata, K. (2020). Dual peptide-based gene delivery system for the efficient transfection of plant callus cells. *Biomacromolecules* 21: 2735–2744.

31 Futaki, S., Suzuki, T., Ohashi, W. et al. (2001). Arginine-rich peptides. *The Journal of Biological Chemistry* 276 (8): 5836–5840.

32 Subbarao, N.K., Parente, R.A., Szoka, F.C. et al. (1987). pH-Dependent bilayer destabilization by an amphipathic peptide. *Biochemistry* 26: 2964–2972.

33 Midoux, P., Kichler, A., Boutin, V. et al. (1998). Membrane permeabilization and efficient gene transfer by a peptide containing several histidines. *Bioconjugate Chemistry* 9: 260–267.

34 Wadia, J.S., Stan, R.V., and Dowdy, S.F. (2004). Transducible TAT-HA fusogenic peptide enhances escape of TAT-fusion proteins after lipid raft micropinocytosis. *Nature Medicine* 10 (3): 310–315.

35 Sakamoto, K., Michibata, J., Hirai, Y. et al. (2021). Potentiating the membrane interaction of an attenuated cationic amphipathic lytic peptide for intracellular protein delivery by anchoring with pyrene moiety. *Bioconjugate Chemistry* 32: 950–957.

36 Rennert, R., Wespe, C., Beck-Sickinger, A.G. et al. (2006). Developing novel hCT derived cell-penetrating peptides with improved metabolic stability. *Biochimica et Biophysica Acta* 1758: 347–354.

37 Lee, H.-J., Huang, Y.-W., Chiou, S.-H. et al. (2019). Polyhistidine facilitates direct membrane translocation of cell-penetrating peptides into cells. *Scientific Reports* 9: 9398.

38 Liu, M.-J., Chou, J.-C., Lee, H.-J. et al. (2013). A gene delivery method mediated by three arginine-rich cell-penetrating peptides in plant cells. *Advanced Studies in Biology* 5 (2): 71–88.

39 Zhang, L., Torgerson, T.R., Liu, X.-Y. et al. (1998). Preparation of functionally active cell-permeable peptides by single-step ligation of two peptide modules. *Proceedings of the National Academy of Sciences of the United States of America* 95: 9184–9189.

40 Iwasaki, T., Murakami, N., and Kawano, T. (2020). A polylysine-polyhistidine fusion peptide for lysosome-targeted protein delivery. *Biochemical and Biophysical Research Communications* 533: 905–912.

41 Terada, K., Gimenez-Dejoz, J., Miyagi, Y. et al. (2020). Artificial cell-penetrating peptide containing periodic α-aminoisobutyric acid with long-term internalization efficiency in human and plant cells. *ACS Biomaterials Science & Engineering* 6: 3287–3298.

42 Wyman, T.B., Nicol, F., Zelphati, O. et al. (1997). Design, synthesis, and characterization of a cationic peptide that binds to nucleic acids and permeabilizes bilayers. *Biochemistry* 36: 3008–3017.

43 Javadpour, M.M., Juban, M.M., Lo, W.-C.J. et al. (1996). De novo antimicrobial peptides with low mammalian cell toxicity. *Journal of Medicinal Chemistry* 39: 3107–3113.

44 Scheller, A., Oehlke, J., Wiesner, B. et al. (1999). Structural requirements for cellular uptake of **α**-helical amphipathic peptides. *Journal of Peptide Science* 5: 185–194.

45 Yamashita, H., Kato, T., Oba, M. et al. (2016). Development of a cell-penetrating peptide that exhibits responsive change in its secondary structure in the cellular environment. *Scientific Reports* 6: 33003.

46 Schmidt, S., Adjobo-Hermans, M.J.W., Kohze, R. et al. (2017). Identification of short hydrophobic cell-penetrating peptides for cytosolic peptide delivery by rational design. *Bioconjugate Chemistry* 28: 382–389.

47 Akishiba, M., Takeuchi, T., Kawaguchi, Y. et al. (2017). Cytosolic antibody delivery by lipid-sensitive endosomolytic peptide. *Nature Chemistry* 9: 751–761.

48 Liu, B.R., Huang, Y.-W., Aronstam, R.S. et al. (2016). Identification of α short cell-penetrating peptide from bovine lactoferricin for intracellular delivery of DNA in human A549 cells. *PLOS ONE* 11 (3): e0150439.

49 Perrone, B., Miles, A.J., Salnikov, E.S. et al. (2014). Lipid interaction of LAH4, a peptide with antimicrobial and nucleic acid transfection activities. *European Biophysics Journal* 43: 499–507.

50 Xu, Y., Liang, W., Qiu, Y. et al. (2016). Incorporation of a nuclear localization signal in pH responsive LAH4-L1 peptide enhances transfection and nuclear uptake of plasmid DNA. *Molecular Pharmaceutics* 13: 3141–3152.

51 Zhang, Y., Li, L., Chang, L. et al. (2019). Design of a new pH-activatable cell-penetrating peptide for drug delivery into tumor cells. *Chemical Biology & Drug Design* 94: 1884–1893.

52 Usui, K., Kikuchi, T., Mie, M. et al. (2013). Systematic screening of the cellular uptake of designed alpha-helix peptides. *Bioorganic & Medicinal Chemistry* 21: 2560–2567.

53 Kozhikhova, K.V., Andreev, S.M., Shilovskiy, I.P. et al. (2018). A novel peptide dendrimer LTP efficiently facilitates transfection of mammalian cells. *Organic & Biomolecular Chemistry* 16: 8181–8190.

54 Jiang, Y., Yang, W., Zhang, J. et al. (2018). Protein toxin chaperoned by LRP-1 targeted virus-mimicking vesicles induces high-efficiency glioblastoma therapy *in vivo*. *Advanced Materials* 30: 1800316.

55 El-Andaloussi, S., Johansson, H.J., Holm, T. et al. (2007). A novel cell-penetrating peptide, M918, for efficient delivery of proteins and peptide nucleic acids. *Molecular Therapy* 15 (10): 1820–1826.

56 Oehlke, J., Krause, E., Wiesner, B. et al. (1996). Nonendocytic, amphipathicity dependent cellular uptake of helical model peptides. *Protein and Peptide Letters* 3 (6): 393–398.

57 Wada, S., Tsuda, H., Okada, T. et al. (2011). Cellular uptake of Aib-containing amphipathic helix peptide. *Bioorganic & Medicinal Chemistry Letters* 21: 5688–5691.

58 Chen, J., Guan, S.-M., Sun, W. et al. (2016). Melittin, the major pain-producing substance of bee venom. *Neuroscience Bulletin* 32 (3): 265–272.

59 Jones, S. and Howl, J. (2004). Charge delocalisation and the design of novel mastoparan analogues: enhanced cytotoxicity and secretory efficacy of [Lys^5, Lys^8, Aib^{10}]MP. *Regulatory Peptides* 121: 121–128.

60 Morris, M.C., Vidal, P., Chaloin, L. et al. (1997). A new peptide vector for efficient delivery of oligonucleotides into mammalian cells. *Nucleic Acids Research* 25: 2730–2736.

61 Lin, Y.-Z., Yao, S.Y., Veach, R.A. et al. (1995). Inhibition of nuclear translocation of transcription factor NF-κB by a synthetic peptide containing a cell membrane-permeable motif and nuclear localization sequence. *The Journal of Biological Chemistry* 270 (24): 14255–14258.

62 Rojas, M., Donahue, J.P., Tan, Z. et al. (1998). Genetic engineering of proteins with cell membrane permeability. *Nature Biotechnology* 16: 370–375.

63 Arukuusk, P., Parnaste, L., Oskolkov, N. et al. (2013). New generation of efficient peptide-based vectors, NickFects, for the delivery of nucleic acids. *Biochimica et Biophysica Acta* 1828: 1365–1373.

64 Freimann, K., Arukuusk, P., Kurrikoff, K. et al. (2016). Optimization of *in vivo* DNA delivery with NickFect peptide vectors. *Journal of Controlled Release* 241: 135–143.

65 Porosk, L., Arukuusk, P., Pohako, K. et al. (2019). Enhancement of siRNA transfection by the optimization of fatty acid length and histidine content in the CPP. *Biomaterials Science* 7: 4363–4374.

66 Iwasaki, T., Tokuda, Y., Kotake, A. et al. (2015). Cellular uptake and *in vivo* distribution of polyhistidine peptides. *Journal of Controlled Release* 210: 115–124.

67 Numata, K., Horii, Y., Oikawa, K. et al. (2018). Library screening of cell-penetrating peptide for BY-2 cells, leaves of Arabidopsis, tobacco, tomato, poplar, and rice callus. *Scientific Reports* 8: 10966.

68 Douat, C., Aisenbrey, C., Antunes, S. et al. (2015). A cell-penetrating foldamer with a bioreducible linkage for intracellular delivery of DNA. *Angewandte Chemie* 127: 11285–11289.

69 Taylor, B.N., Mehta, R.R., Yamada, T. et al. (2009). Noncationic peptides obtained from azurin preferentially enter cancer cells. *Cancer Research* 69 (2): 537–546.

70 Jirka, S.M.G., Heemskerk, H., Winter, C.L.T. et al. (2014). Peptide conjugation of 2′-*O*-methyl phosphorothioate antisense oligonucleotides enhances cardiac uptake and exon skipping in *mdx* mice. *Nucleic Acid Therapeutics* 24 (1): 25–36.

71 Kamei, N., Kikushi, S., Takeda-Morishita, M. et al. (2013). Determination of the optimal cell-penetrating peptide sequence for intestinal insulin delivery based on molecular orbital analysis with self-organizing maps. *Journal of Pharmaceutical Sciences* 102 (2): 469–479.

72 Derossi, D., Joliot, A.H., Chassaing, G. et al. (1994). The third helix of the Antennapedia homeodomain translocates through biological membranes. *The Journal of Biological Chemistry* 269 (14): 10444–10450.

73 Morris, M.C., Depollier, J., Mery, J. et al. (2001). A peptide carrier for the delivery of biologically active proteins into mammalian cells. *Nature Biotechnology* 19: 1173–1176.

74 Gao, C., Mao, S., Ditzel, H.J. et al. (2002). A cell-penetrating peptide from a novel pVII-pIX phage-displayed random peptide library. *Bioorganic & Medicinal Chemistry* 10: 4057–4065.

75 Marks, J.R., Placone, J., Hristova, K. et al. (2011). Spontaneous membrane-translocating peptides by orthogonal high-throughput screening. *Journal of the American Chemical Society* 133: 8995–9004.

76 Ezzat, K., El Andaloussi, S., Zaghloul, E.M. et al. (2011). PepFect 14, a novel cell-penetrating peptide for oligonucleotide delivery in solution and as solid formulation. *Nucleic Acid Research* 39 (12): 5284–5298.

77 Mae, M., El Andaloussi, S., Lundin, P. et al. (2009). A stearylated CPP for delivery of splice correcting oligonucleotides using a non-covalent co-incubation strategy. *Journal of Controlled Release* 134: 221–227.

78 El Andaloussi, S., Lehto, T., Mager, I. et al. (2011). Design of a peptide-based vector, PepFect6, for efficient delivery of siRNA in cell culture and systemically *in vivo*. *Nucleic Acid Research* 39 (9): 3972–3987.

79 Watkins, C.L., Brennan, P., Fegan, C. et al. (2009). Cellular uptake, distribution and cytotoxicity of the hydrophobic cell penetrating peptide sequence PFVYLI linked to the proapoptotic domain peptide PAD. *Journal of Controlled Release* 140: 237–244.

80 Fahrner, R.L., Dieckmann, T., Harwig, S.S.L. et al. (1996). Solution structure of protegrin-1, a broad-spectrum antimicrobial peptide from porcine leukocytes. *Chemistry & Biology* 3: 543–550.

81 Henry, K.E., Chaney, A.M., Nagle, V.L. et al. (2020). Demarcation of sepsis-induced peripheral and central acidosis with pH (low) insertion cycle peptide. *Journal of Nuclear Medicine* 61: 1361–1368.

82 An, M., Wijesinghe, D., Andreev, O.A. et al. (2010). pH-(low)-insertion-peptide (pHLIP) translocation of membrane impermeable phalloidin toxin inhibits cancer cell proliferation. *Proceedings of the National Academy of Sciences of the United States of America* 107: 20246–20250.

83 Lehto, T., Alvarez, A.C., Gauck, S. et al. (2014). Cellular trafficking determines the exon skipping activity of Pip6a-PMO in *mdx* skeletal and cardiac muscle cells. *Nucleic Acid Research* 42 (5): 3207–3217.

84 Rittner, K., Benavente, A., Bompard-Sorlet, A. et al. (2002). New basic membrane-destabilizing peptides for plasmid-based gene delivery *in vitro* and *in vivo*. *Molecular Therapy* 5 (2): 104–114.

85 Kanekura, K., Harada, Y., Fujimoto, M. et al. (2018). Characterization of membrane penetration and cytotoxicity of C9orf72-encoding arginine-rich dipeptides. *Scientific Reports* 8: 12740.

86 Oess, S. and Hildt, E. (2000). Novel cell permeable motif derived from the PreS2-domain of hepatitis-B virus surface antigens. *Gene Therapy* 7: 750–758.

87 Ho, A., Schwarze, S.R., Mermelstein, S.J. et al. (2001). Synthetic protein transduction domains: enhanced transduction potential *in vitro* and *in vivo*. *Cancer Research* 61: 474–477.

88 Elmquist, A., Lindgren, M., Bartfai, T. et al. (2001). VE-cadherin-derived cell-penetrating peptide, *p*VEC, with carrier functions. *Experimental Cell Research* 269: 237–244.

89 Delaroche, D., Aussedat, B., Aubry, S. et al. (2007). Tracking a new cell-penetrating (W/R) nonapeptide, through an enzyme-stable mass spectrometry reporter tag. *Analytical Chemistry* 79: 1932–1938.

90 Moulton, H.M., Nelson, M.H., Hatlevig, S.A. et al. (2004). Cellular uptake of antisense morpholino oligomers conjugated to arginine-rich peptides. *Bioconjugate Chemistry* 15: 290–299.

91 Gao, L., Yu, J., Liu, Y. et al. (2018). Tumor-penetrating peptide conjugated and doxorubicin loaded T_1-T_2 dual mode MRI contrast agents nanoparticles for tumor theranostics. *Theranostics* 8 (1): 92–108.

92 Vaissiere, A., Aldrian, G., Konate, K. et al. (2017). A *retro-inverso* cell-penetrating peptide for siRNA delivery. *Journal of Nanobiotechnology* 15: 34.

93 Nakase, I., Okumura, S., Katayama, S. et al. (2012). Transformation of an antimicrobial peptide into a plasma membrane-permeable, mitochondria-targeted peptide *via* the substitution of lysine with arginine. *Chemical Communications* 48: 11097–11099.

94 Oba, M., Ito, Y., Umeno, T. et al. (2019). Plasmid DNA delivery using cell-penetrating peptide foldamers composed of Arg-Arg-Aib repeating sequences. *ACS Biomaterials Science & Engineering* 5: 5660–5668.

95 Oba, M., Nagano, Y., Kato, T. et al. (2019). Secondary structures and cell-penetrating abilities of arginine-rich peptide foldamers. *Scientific Reports* 9: 1349.

96 Jobin, M.-L., Vamparys, L., Deniau, R. et al. (2019). Biophysical insight on the membrane insertion of an arginine-rich cell-penetrating peptide. *International Journal of Molecular Sciences* 20: 4441.

97 Safa, N., Anderson, J.C., Vaithiyanathan, M. et al. (2019). CPProtectides: rapid uptake of well-folded β-hairpin peptides with enhanced resistance to intracellular degradation. *Peptide Science* 111: e24092.

98 Pinto, A., Lennarz, S., Rodrigues-Correia, R. et al. (2012). Functional detection of proteins by caged aptamers. *ACS Chemical Biology* 7: 360–366.

99 Pujals, S. and Giralt, E. (2008). Proline-rich, amphipathic cell-penetrating peptides. *Advanced Drug Delivery Review* 60: 473–484.

100 Martin, I., Teixido, M., and Giralt, E. (2011). Design, synthesis and characterization of a new anionic cell-penetrating peptide: SAP(E). *ChemBioChem* 12: 896–903.

101 Hoyer, J., Schatzschneider, U., Schulz-Siegmund, M. et al. (2012). Dimerization of a cell-penetrating peptide leads to enhanced cellular uptake and drug delivery. *Beilstein Journal of Organic Chemistry* 8: 1788–1797.

102 Mello, L.R., Aguiar, R.B., Yamada, R.Y. et al. (2020). Amphipathic design dictates self-assembly, cytotoxicity and cell uptake of arginine-rich surfactant-like peptides. *Journal of Material Chemistry B* 8: 2495–2507.

103 Lamaziere, A., Burlina, F., Wolf, C. et al. (2007). Non-metabolic membrane tubulation and permeability induced by bioactive peptides. *PLOS ONE* 2 (2): e201.

104 Futaki, S., Ohashi, W., Suzuki, T. et al. (2001). Stearylated arginine-rich peptides: a new class of transfection systems. *Bioconjugate Chemistry* 12: 1005–1011.

105 Khalil, I.A., Kogure, K., Futaki, S. et al. (2006). High density of octaarginine stimulates micropinocytosis leading to efficient intracellular trafficking for gene expression. *The Journal of Biological Chemistry* 281 (6): 3544–3551.

106 Hilinski, G.J., Kim, Y.-W., Hong, J. et al. (2014). Stitched α-helical peptides via bis ring-closing metathesis. *Journal of the American Chemical Society* 136: 12314–12322.

107 Stinthvanich, C., Veiga, A.S., Gupta, K. et al. (2012). Anticancer β-hairpin peptides: membrane-induced folding triggers activity. *Journal of the American Chemical Society* 134: 6210–6217.

108 Rousselle, C., Clair, P., Lefauconnier, J.-M. et al. (2000). New advances in the transport of doxorubicin through the blood-brain barrier by a peptide vector-mediated strategy. *Molecular Pharmacology* 57 (4): 679–686.

109 Jain, A., Yadav, B.K., and Chugh, A. (2015). Marine antimicrobial peptide tachyplesin as an efficient nanocarrier for macromolecule delivery in plant and mammalian cells. *The FEBS Journal* 282: 732–745.

110 Vives, E., Brodin, P., and Lebleu, B. (1997). A truncated HIV-1 Tat protein basic domain rapidly translocates through the plasma membrane and accumulates in the cell nucleus. *The Journal of Biological Chemistry* 272: 16010–16017.

111 Zhang, W., Song, J., Zhang, B. et al. (2011). Design of acid-activated cell penetrating peptide for delivery of active molecules into cancer cells. *Bioconjugate Chemistry* 22: 1410–1415.

112 Song, J., Kai, M., Zhang, W. et al. (2011). Cellular uptake of transportan 10 and its analogs in live cells: selectivity and structure-activity relationship studies. *Peptides* 21: 1934–1941.

113 Bleifuss, E., Kammertoens, T., Hutloff, A. et al. (2006). The translocation motif of hepatitis B virus improves protein vaccination. *Cellular and Molecular Life Sciences* 63: 627–635.

114 Soomets, U., Lindgren, M., Gallet, X. et al. (2000). Deletion analogues of transportan. *Biochimica et Biophysica Acta* 1467: 165–176.

115 Pooga, M., Hallbrink, M., Zorko, M. et al. (1998). Cell penetration of transportan. *The FASEB Journal* 12: 67–77.

116 Elliott, G. and O'Hare, P. (1997). Intercellular trafficking and protein delivery by a herpesvirus structural protein. *Cell* 88: 223–233.

117 Gross, D.A., Leborgne, C., Chappert, P. et al. (2019). Induction of tumor-specific CTL responses using the C-terminal fragment of Viral protein R as cell penetrating peptide. *Scientific Reports* 9: 3937.

118 Oehlke, J., Krause, E., Wiesner, B. et al. (1997). Extensive cellular uptake into endothelial cells of an amphipathic β-sheet forming peptide. *FEBS Letters* 415: 196–199.

119 Konate, K., Dussot, M., Aldrian, G. et al. (2019). Peptide-based nanoparticles to rapidly and efficiently "Wrap 'n Roll" siRNA into cells. *Bioconjugate Chemistry* 30: 592–603.

Part I

Design

2

Classification

Makoto Oba

Kyoto Prefectural University of Medicine, Graduate School of Medical Science, Kyoto, Japan

2.1 Introduction

Numerous cell-penetrating peptides (CPPs) have been reported to date (Table 2.1) [1–5]. These CPPs can be divided into different groups based on their characteristics. This chapter describes the classification of CPPs based on their physicochemical properties, chemical structure (linear vs. cyclic, chirality, length, N- and C-terminal structures), origin, and cargo. The percentage of each category is indicated in reference to a database of CPPs, CPPsite 2.0 (http://crdd.osdd.net/raghava/cppsite/) (Figures 2.1 and 2.2) [6, 7]. This database contains >1800 entries, which provide extensive information about CPPs including the nature of peptides, chemical modifications, experimental validation techniques, structure of peptides, and types of cargo delivered. It is difficult to cover all CPPs in a brief manner; therefore, the information in this chapter is intended to help understand the trends in CPPs and the design of novel CPPs.

2.2 Physicochemical Properties

CPPs can be mainly categorized into three classes based on their physicochemical properties: cationic, amphipathic, and hydrophobic. However, the classes to which some CPPs are assigned are ambiguous. For example, penetratin is categorized as a cationic CPP in some reports [3, 5] but an amphipathic CPP in others [8], and transportan is categorized as an amphipathic CPP in some reports [3–5, 8] but a hydrophobic CPP in others [9]. The proportions of cationic and hydrophobic amino acids in the sequences are important for the categorization of CPPs by their physicochemical properties.

Cell-Penetrating Peptides: Design, Development and Applications, First Edition. Makoto Oba and Yosuke Demizu.

Table 2.1 Classification of CPPs.

Peptide	Sequence	Physicochemical property	Origin
Tat^{48-60}	GRKKRRQRRRPPQ	Cationic	HIV-1 Tat protein
Tat^{49-57}	RKKRRQRRR	Cationic	HIV-1 Tat protein
D-Tat	rkkrrqrrr[a]	Cationic	Synthetic
Oligoarginine	R_n ($n = 6–12$)	Cationic	Synthetic
DPV1047	VKRGLKLRHVRPRVTRMDV	Cationic	Synthetic
Penetratin	RQIKIWFQNRRMKWKK	Amphipathic	Drosophila Antennapedia homeodomain protein
MPG	GALFLGFLGAAGSTMGAWSQPKKKRKV	Amphipathic	HIV glycoprotein 41/ SV40 T antigen NLS (chimeric)
Pep-1	KETWWETWWTEWSQPKKKRKV	Amphipathic	Tryptophan-rich cluster/SV40 T antigen NLS (chimeric)
pVEC	LLIILRRRIRKQAHAHSK	Amphipathic	Vascular endothelial cadherin
MAP	KLALKLALKALKAALKLA	Amphipathic	Synthetic
VT5	DPKGDPKGVTVTVTVTVTGKGDPKPD	Amphipathic	Synthetic
Transportan	GWTLNSAGYLLGKINLKALAALAKKIL	Hydrophobic	Galanin-Lys-mastoparan
Melittin	GIGAVLKVLTTGLPALISWIKRKRQQ	Hydrophobic	Natural peptide from honey bee venom
SAP(E)	$(VELPPP)_3$	Anionic	Synthetic
GALA	WEAALAEALAEALAEHLAEALAEALEALAA	Anionic	Synthetic

a) Lowercase letters indicate D-amino acids.
HIV, human immunodeficiency virus; NLS, nuclear localization signal; SV40, simian virus 40; Tat, transactivator of transcription.

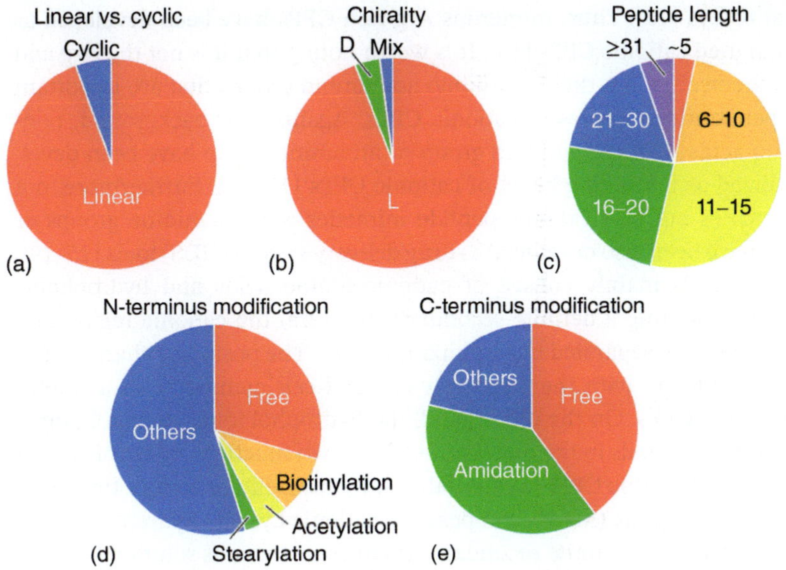

Figure 2.1 Schematic representation of the distribution of CPPs in the CPPsite 2.0 based on (a) linear and cyclic structures, (b) chirality, (c) peptide length, (d) N-terminal modification, and (e) C-terminal modification.

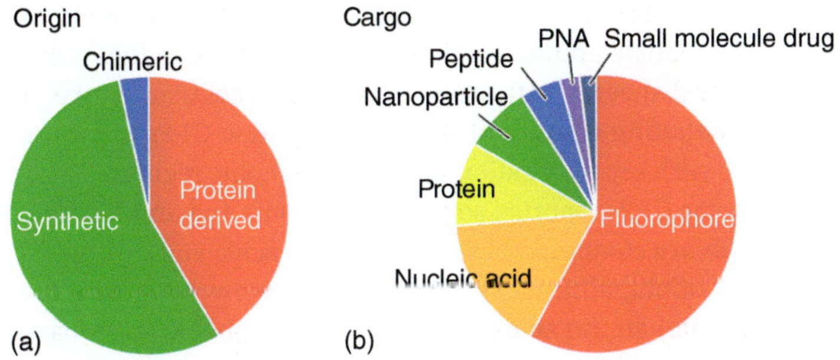

Figure 2.2 Schematic representation of the distribution of CPPs in the CPPsite 2.0 based on (a) origin and (b) delivered cargo.

The first discovered CPP, transactivator of transcription (Tat), is one of the most typical cationic CPPs [10–12]. Cationic CPPs have a high portion of cationic amino acids such as Arg, Lys, and His. Anionic functional groups, such as carboxylates, sulfates, and phosphates, are abundant on the surface of the cell membrane, and cationic functional groups in cationic CPPs can interact with these anionic functional groups by electrostatic interaction. Such interactions lead to endocytosis and/or the direct cell membrane penetration of CPPs, resulting in high cell-penetrating abilities of such CPPs. $Tat_{48\text{-}60}$ has 6 Arg and 2 Lys among 13 amino acid residues. Oligoarginine was developed based on Tat, in which Lys and hydrophobic amino acids in Tat were replaced with Arg [13, 14]. The guanidino group in the side chain of Arg can interact more strongly with anionic functional groups than the amino group in that of Lys [15]. Thus, numerous Arg-rich CPPs have been developed as artificially designed cationic CPPs [16]. It is worth noting that it is not the Arg and Lys residues themselves but their guanidino and amino groups that are important for the cell-penetrating abilities of cationic CPPs. Against this background, non-natural amino acids with a guanidino group or an amino group have been developed and utilized as a building block of cationic CPPs [17, 18]. Furthermore, not only cationic CPPs but also cationic peptide mimetics with guanidino groups or amino groups have been also developed as drug delivery system (DDS) tools [15–19].

Amphipathic CPPs mainly consist of cationic amino acids and hydrophobic amino acids. By adopting a definite secondary structure, the cationic region and hydrophobic region are separated in amphipathic CPPs. The positively charged cationic region can interact with the negatively charged cell membrane in a similar manner to cationic CPPs. On the other hand, the hydrophobic region can interact with hydrophobic lipid tails in the membrane. These two functions by two different regions enable amphipathic CPPs to internalize into intracellular compartments. A model amphipathic peptide (MAP), composed of cationic Lys and hydrophobic Leu and Ala, does not adopt a definite secondary structure in aqueous solution [20–22]. However, the MAP forms an amphipathic α-helical structure in the environment of the cell membrane. This conformational change to a helical structure gives the cell-penetrating ability to the MAP. The amphipathic CPP VT5 forms a β-sheet structure because of the presence of Val-Thr repeating units, which is a β-sheet favoring sequence [23]. The β-sheet structure is essential for the cell-penetrating ability of VT5 because substitution of a single amino acid in the VT5 sequence induces conformational changes from the β-sheet structure and decreases its cellular uptake. It is important for the cell-penetrating abilities of these amphipathic CPPs to form a specific secondary structure with separate cationic and hydrophobic regions.

There are fewer hydrophobic CPPs reported so far than cationic and amphipathic CPPs. The hydrophobic CPPs are composed of cationic amino acids and hydrophobic amino acids, similarly to amphipathic CPPs. The difference between them is the amount of charged (or cationic) amino acids. In general, hydrophobic CPPs have less than 20% of charged amino acids in their sequences [5]. In fact, transportan is a 27-amino acid peptide containing only 4 Lys residues (15% charged amino acids) [24], and melittin is a 26-amino acid peptide containing 3 Lys and 2 Arg residues (19% charged amino acids) [25]. Hydrophobic CPPs have high affinity to the hydrophobic domains of cell membranes, sometimes resulting in direct membrane penetration in an energy-independent process.

Other than the above-mentioned three types of CPPs, a few anionic peptides, which have several anionic amino acids such as Asp and Glu, have been reported to enter cells [26, 27]. Anionic peptides cannot generally associate with the negatively charged cell membrane. The GALA peptide [28, 29], which contains Glu-Ala-Leu-Ala repeating units, is negatively charged at neutral pH and adopts a random structure. However, the GALA peptide forms a membrane-fusogenic α-helical structure at acidic pH due to the protonation of carboxylate in the side chain of Glu residues. These characteristics of the GALA peptide result in endosomal/lysosomal membrane-specific disruption and migration from the endosome/lysosome to the intracellular compartment. Based on these functions, the GALA peptide has been used as an endosomal/lysosomal escape tool for nano DDS and sometimes introduced as a CPP [30].

2.3 Chemical Structure

CPPs can be also categorized based on their chemical structures. More than 90% of CPPs have a linear structure, with only approximately 5% having a cyclic structure (Figure 2.1a). In general, cyclic peptides are more stable against enzymatic degradation than linear peptides. This property is useful for CPPs to function for a long term because proteases and peptidases are abundant in the culture medium, cells, and bodies [31]. However, most protein-derived CPPs are linear, and linear peptides can be synthesized more easily than cyclic peptides. Thus, linear CPPs have been developed much more than cyclic CPPs. Cyclization has been achieved using reactions including disulfide bond formation between thiol groups in the side chain of Cys residues, an amide bond formation in the peptide backbone or in the side chain, and so on [32, 33].

The proteinogenic amino acids are L-forms, and most CPPs developed so far consist of L-amino acids. However, peptides composed of only L-amino acids are easily degraded by proteases and peptidases. One of the strategies to avoid enzymatic instability of CPPs is the use of D-amino acids [34–36]. For example, D-Tat (D-isomer of natural Tat peptide) and D-oligoarginines (composed of D-Arg) were reported to exhibit greater cellular uptake than L-Tat and L-oligoarginines, respectively, probably due to high resistance to enzymatic degradation in culture medium containing serum [13]. Another study revealed that binding of some CPPs to heparan sulfates was chirality dependent, but the efficiency of their internalization was related to chirality [37]. Furthermore, CPPs mixed with L- and D-amino acids have been developed. Despite the advantage of D isomers, as shown in Figure 2.1b, more than 90% of CPPs are L-isomers.

A wide range of peptide lengths have been reported in CPPs (~5 amino acids: 3%; 6–10 amino acids: 21%; 11–15 amino acids: 30%; 16–20 amino acids: 24%; 21–30 amino acids: 17%; ≥31~ amino acids: 5%) (Figure 2.1c). In these reports, the minimum peptide length was 3, whereas the maximum peptide length was 61. Shorter CPPs are better than longer CPPs from the viewpoints of manufacturing and handling. However, the peptide length is sometimes critical for their cell-penetrating ability. Oligoarginines from hexapeptide to dodecapeptide are usually used as CPPs. Shorter oligoarginines do not show adequate intracellular internalization, whereas the cytotoxicity of longer oligoarginines is not negligible [13, 14]. Highly positively charged polymers induce severe cytotoxicity due to their strong association with cell membrane followed by high membrane-disrupting activity [38, 39]. Therefore, appropriate peptide length with a balance between function and toxicity is very important for cationic CPPs. Peptide length is often significant for the cell-penetrating abilities of amphipathic CPPs. Furthermore, for the intracellular delivery of nucleic acids, sufficient peptide length is necessary to form stable complexes between peptides and nucleic acids [40–42]. In the case of CPPs composed of Arg-Arg-Aib repeating sequences (Aib: 2-aminoisobutyric acid), the critical repeat number was between 3 and 4 for the formation of stable complexes by CPPs and plasmid DNA and good transfection efficiencies in cultured cells [43].

CPPs are also categorized based on their N-terminal and C-terminal structures. Various molecules with a functional group that reacts with an amine can be introduced at the N-terminus of peptides as part of solid-phase synthesis. CPPs with N-terminal acetyl, biotin, and stearyl groups have been developed. Acetylation is carried out to cap the N-terminal amine; thus, no additional function is attached to CPPs. Biotinylated CPPs are used to associate with streptavidin-conjugated cargos [44]. The dissociation constant of the biotin-streptavidin interaction is very low, ~10^{-15} M, so streptavidin-conjugated cargos are easily attached to biotinylated CPPs. Stearylated CPPs are often used for the delivery of nucleic acids because a stearyl group assists the formation of stable complexes between CPPs and nucleic acids through hydrophobic interactions [45]. However, the free amine form without any chemical modification is present in the majority of CPP N-terminal structures (Figure 2.1d). The C-terminal structures of CPPs are simpler than the N-terminus and dependent on the properties of resins used in solid-phase synthesis. Approximately, 40% of C-terminal structures are –COOH (free) and $-CONH_2$ (amidation), respectively (Figure 2.1e). Because of the nature of solid-phase synthesis, it is difficult to introduce a functional molecule onto the C-terminus of peptides.

2.4 Origin

Based on their origins, CPPs can be divided into protein-derived CPPs (42%), synthetic CPPs (55%), and chimeric CPPs (3%) (Figure 2.2a). Tat [10, 11] and penetratin [46], which are first-generation CPPs, are typical protein-derived CPPs. Tat protein from human immunodeficiency virus type 1 (HIV-1) was found to have a remarkable ability to cross cell membranes, and it was revealed that the important sequence for cell membrane permeability was a just the 13-amino acid $Tat_{48\text{-}60}$. This was the first report of a CPP. Penetratin is a 16-amino acid peptide and an amphipathic helical CPP, and it is derived from the Drosophila Antennapedia homeodomain protein. The cationic CPP oligoarginines are categorized as synthetic CPP. Arg residues were found to be important for the cell-penetrating ability of Tat, and oligoarginines were designed by substitution of other amino acids in Tat with Arg and showed better membrane permeability than Tat [13, 14]. MAP is also a synthetic CPP. To form an amphipathic α-helical structure, Lys as a cationic amino acid and Leu and Ala as hydrophobic amino acids were used to design MAP, which showed good cell-penetrating ability [20–22]. Based on MAP, novel synthetic CPPs utilizing non-proteinogenic amino acids have been developed for the delivery of nucleic acids [47, 48]. Non-proteinogenic amino acids were used for the stabilization of the α-helical structure. Chimeric CPPs have two or more domains that covalently bind to each other. For example, MPG (GALFLGFLGAAGSTMGAWSQPKKKRKV) is composed of three components, a hydrophobic fusion sequence of the HIV glycoprotein 41 (GALFLGFLGAAGSTMGA), a spacer domain (WSQP), and a nuclear localization signal of simian virus 40 large T-antigen (KKKRKV) [49]. However, chimeric CPPs are only present in 3% of the total (Figure 2.2a).

2.5 Cargo

A major application of CPPs is as a DDS tool. CPPs have been used to deliver various types of cargos to date. Here, CPPs are described in terms of delivered cargos (Figure 2.2b). The cargos are sometimes conjugated to CPPs and are sometimes delivered by mixing with CPPs. More than 50% of cargos are fluorophores such as low-molecular-weight fluorescein, rhodamine, and so on. Fluorophores can be easy to conjugate to CPPs as part of solid-phase synthesis and can visualize CPP distribution both in the cells and bodies. In order to clarify the cell-penetrating abilities of CPPs, fluorophores are widely used as a cargo and are also a model for membrane-impermeable small molecule drugs. AVB-620 is a fluorescent dye (Cy5 and Cy7)-conjugated CPP (D-heptaarginine) and in clinical trials for fluorescence detection of cancer during breast cancer surgery [50, 51]. Fluorophore-conjugated CPPs can be used as a fluorescence imaging agent in patients with breast cancer. The second most common cargo is nucleic acids [52–54]. Nucleic acids, including plasmid DNA, antisense DNA, messenger RNA, small interfering RNA, and so on, have negative charges, whereas most CPPs have positive charges. Just by mixing CPPs and nucleic acids, nano-sized complexes form by electrostatic interaction, and such complexes sometimes have suitable properties as nano DDS. Proteins and peptides are also major targets for intracellular delivery by CPPs [55]. In the case of proteins, CPPs are chemically modified for amide bond formation, thiol-maleimide conjugation, disulfide bond formation, and so on. Otherwise, proteins with CPPs are produced by genetic engineering. In addition to having the same attributes as protein cargos, CPPs are easy to conjugate to peptide cargos by solid phase synthesis. CPPs are often used for assistance with the intracellular delivery of nanoparticles including liposomes and polymeric micelles [56, 57]. In order to enhance the cellular entry of less membrane-permeable nanoparticles, CPPs are attached to the surface of nanoparticles. Other cargos delivered by CPPs include peptide nucleic acids and small molecule drugs [4].

References

1 Copolovici, D.M., Langel, K., and Langel, Ü. (2014). Cell-penetrating peptides: design, synthesis, and applications. *ACS Nano* 8 (3): 1972–1994.

2 Kalafatovic, D. and Giralt, E. (2017). Cell-penetrating peptides: design strategies beyond primary structure and amphipathicity. *Molecules* 22: 1929.

3 Guidotti, G., Brambilla, L., and Rossi, D. (2017). Cell-penetrating peptides: from basic research to clinics. *Trends in Pharmacological Sciences* 38 (4): 406–424.

4 Borrelli, A., Tornesello, A., Tornesello, M.L. et al. (2018). Cell penetrating peptides as molecular carrier for anti-cancer agents. *Molecules* 23: 295.

5 Xie, J., Bi, Y., Zhang, H. et al. (2020). Cell-penetrating peptides in diagnosis and treatment of human diseases: from preclinical research to clinical application. *Frontiers in Pharmacology* 11: 697.

6 Gautam, A., Singh, H., Tyagi, A. et al. (2012). CPPsite: a curated database of cell penetrating peptides. *Database* 2012: bas015.

7 Agrawal, P., Bhalla, S., Usmani, S.S. et al. (2016). CPPsite 2.0: a repository experimentally validated cell-penetrating peptides. *Nucleic Acids Research* 44: D1098–D1103.

8 Kim, G.C., Cheon, D.H., and Lee, Y. (2021). Challenge to overcome current limitations of cell-penetrating peptides. *BBA – Proteins and Proteomics* 1869: 140604.

9 Madani, F., Lindberg, S., Langel, Ü. et al. (2011). Mechanisms of cellular uptake of cell-penetrating peptides. *Journal of Biophysics* 2011: 414729.

10 Green, M. and Loewenstein, P.M. (1988). Autonomous functional domains of chemically synthesized human immunodeficiency virus Tat *trans*-activator protein. *Cell* 55: 1179–1188.

11 Frankel, A.D. and Pabo, C.O. (1988). Cellular uptake of the Tat protein from human immunodeficiency virus. *Cell* 55: 1189–1193.

12 Rizzuti, M., Nizzardo, M., Zanetta, C. et al. (2015). Therapeutic applications of the cell-penetrating HIV-1 Tat peptide. *Drug Discovery Today* 20 (1): 76–85.

13 Wender, P.A., Mitchell, D.J., Pattabiraman, K. et al. (2000). The design, synthesis, and evaluation of molecules that enable or enhance cellular uptake: peptoid molecular transporters. *Proceedings of the National Academy of Sciences of the United States of America* 21: 13003–13008.

14 Futaki, S., Suzuki, T., Ohashi, W. et al. (2001). Arginine-rich peptides. *The Journal of Biological Chemistry* 276 (8): 5836–5840.

15 Wender, P.A., Galliher, W.C., Goun, E.A. et al. (2008). The design of guanidinium-rich transporters and their internalization mechanisms. *Advanced Drug Delivery Reviews* 60: 452–472.

16 Nakase, I., Takeuchi, T., Tanaka, G. et al. (2008). Methodological and cellular aspects that govern the internalization mechanisms of arginine-rich cell-penetrating peptides. *Advanced Drug Delivery Review* 60: 598–607.

17 Oba, M. (2019). Cell-penetrating peptide foldamers: drug-delivery tools. *ChemBioChem* 20: 2041–2045.

18 Yokoo, H., Misawa, T., and Demizu, Y. (2020). De novo design of cell-penetrating foldamers. *The Chemical Record* 20: 912–921.

19 Borneirie, M., Brion, A., Guichard, G. et al. (2021). Delivery of siRNA by tailored cell-penetrating urea-based foldamers. *Chemical Communications* 57: 1458–1461.

20 Oehlke, J., Krause, E., Wiesner, B. et al. (1996). Nonendocytic, amphipathicity dependent cellular uptake of helical model peptides. *Protein and Peptide Letters* 3 (6): 393–398.

21 Oehlke, J., Scheller, A., Wiesner, B. et al. (1998). Cellular uptake of an α-helical amphipathic model peptide with the potential to deliver polar compounds into the cell interior non-endocytically. *Biochimica et Biochysica Acta* 1414: 127–139.

22 Scheller, A., Wiesner, B., Melzig, M. et al. (2000). Evidence for an amphipathicity independent cellular uptake of amphipathic cell-penetrating peptides. *European Journal of Biochemistry* 267: 6043–6049.

23 Oehlke, J., Krause, E., Wiesner, B. et al. (1997). Extensive cellular uptake into endothelial cells of an amphipathic β-sheet forming peptide. *FEBS Letters* 415: 196–199.

24 Pooga, M., Hällbrink, M., Zorko, M. et al. (1998). Cell penetration by transportan. *The FASEB Journal* 12: 67–77.

25 Chen, J., Guan, S.-M., Sun, W. et al. (2016). Melittin, the major pain-producing substance of bee venom. *Neuroscience Bulletin* 32 (3): 265–272.

26 Marin, I., Teixido, M., and Giralt, E. (2011). Design, synthesis and characterization of a new anionic cell-penetrating peptide: SAP(E). *ChemBioChem* 12: 896–903.

27 Franz, J., Lelle, M., Peneva, K. et al. (2016). SAP(E) – a cell-penetrating polyproline helix at lipid interfaces. *Biochimica et Biophysica Acta* 1858: 2028–2034.

28 Subbarao, N.K., Parente, R.A., Szoka, F.C. et al. (1987). pH-Dependent bilayer destabilization by an amphipathic peptide. *Biochemistry* 26: 2964–2972.

29 Li, W., Nicol, F., and Szoka, F.C. (2004). GALA: a designed synthetic pH-responsive amphipathic peptide with applications in drug and gene delivery. *Advanced Drug Delivery Review* 56: 967–985.

30 Nakamura, T., Akita, H., Yamada, Y. et al. (2012). A multifunctional envelope-type nanodevice for use in nanomedicine: concept and applications. *Accounts of Chemical Research* 45 (7): 1113–1121.

31 Qian, Z., Larochelle, J.R., Jiang, B. et al. (2014). Early endosomal escape of a cyclic cell-penetrating peptide allows effective cytosolic cargo delivery. *Biochemistry* 53: 4034–4046.

32 D'Souza, C., Henriques, S.T., Wang, C.K. et al. (2014). Structural parameters modulating the cellular uptake of disulfide-rich cyclic cell-penetrating peptides: MCoTI-II and SFTI-1. *European Journal of Medicinal Chemistry* 88: 10–18.

33 Traboulsi, H., Larkin, H., Bonin, M.-A. et al. (2015). Macrocyclic cell penetrating peptides: a study of structure-penetration properties. *Bioconjugate Chemistry* 26: 405–411.

34 Elmquist, A. and Langel, Ü. (2005). in vitro uptake and stability study of pVEC and its all-D analog. *Biological Chemistry* 384 (3): 387–393.

35 Youngblood, D.S., Hatlevig, S.A., Hassinger, J.N. et al. (2007). Stability of cell-penetrating peptide–morpholino oligomer conjugates in human serum and in cells. *Bioconjugate Chemistry* 18: 50–60.

36 Pujals, S., Sabido, E., Tarrago, T. et al. (2007). All-D proline-rich cell-penetrating peptides: a preliminary in vivo internalization study. *Biochemical Society Transactions* 35: 794–796.

37 Verdurmen, W.P.R., Bovee-Geurts, P.H., Wadlhwani, P. et al. (2011). Preferential uptake of L- versus D-amino acid cell-penetrating peptides in a cell type-dependent manner. *Chemistry & Biology* 18: 1000–1010.

38 Fischer, D., Li, Y., Ahlemeyer, B. et al. (2003). in vitro cytotoxicity testing of polycations: influence of polymer structure on cell viability and hemolysis. *Biomaterials* 24: 1121–1131.

39 Moghimi, S.M., Symonds, P., Murray, J.C. et al. (2005). A two-stage poly(ethylenimine)-mediated cytotoxicity: implications for gene transfer/therapy. *Molecular Therapy* 11 (6): 990–995.

40 Katayose, S. and Kataoka, K. (1998). Remarkable increase in nuclease resistance of plasmid DNA through supramolecular assembly with poly(ethylene glycol)-poly(L-lysine) block copolymer. *Journal of Pharmaceutical Sciences* 87 (2): 160–163.

41 Kwoh, D.Y., Coffin, C.C., Lollo, C.P. et al. (1999). Stabilization of poly-L-lysine/DNA polyplexes for in vivo gene delivery to the liver. *Biochimica et Biophysica Acta* 1444: 171–190.

42 Nidome, T., Takaji, K., Urakawa, M. et al. (1999). Chain length of cationic α-helical peptide sufficient for gene delivery into cells. *Bioconjugate Chemistry* 10: 773–780.

43 Oba, M., Ito, Y., Umeno, T. et al. (2019). Plasmid DNA delivery using cell-penetrating peptide foldamers composed of Arg-Arg-Aib repeating sequences. *ACS Biomaterials Science & Engineering* 5: 5660–5668.

44 Bechara, C., Pallerla, M., Burlina, F. et al. (2015). Massive glycosaminoglycan-dependent entry of Trp-containing cell-penetrating peptides induced by exogenous sphingomyelinase or cholesterol depletion. *Cellular Molecular Life Sciences* 72: 809–820.

45 Futaki, S., Ohashi, W., Suzuki, T. et al. (2001). Stearylated arginine-rich peptides: a new class of transfection systems. *Bioconjugate Chemistry* 12: 1005–1011.

46 Derossi, D., Joliot, A.H., Chassaing, G. et al. (1994). The third helix of the Antennapedia homeodomain translocates through biological membrane. *The Journal of Biological Chemistry* 269 (14): 10444–10450.

47 Wada, S., Hashimoto, Y., Kawai, Y. et al. (2013). Effect of Ala replacement with Aib in amphipathic cell-penetrating peptide on oligonucleotide delivery into cells. *Bioorganic & Medicinal Chemistry* 21: 7669–7673.

48 Furukawa, K., Tanaka, M., and Oba, M. (2020). siRNA delivery using amphipathic cell-penetrating peptides into human hepatoma cells. *Bioorganic & Medicinal Chemistry* 28: 115402.

49 Morris, M.C., Vidal, P., Chaloin, L. et al. (1997). A new peptide vector for efficient delivery of oligonucleotides into mammalian cells. *Nucleic Acids Research* 25: 2730–2736.

50 Meampamba, M., Liu, J., Harootunian, A. et al. (2017). Sensitive in vivo visualization of breast cancer using ratiometric protease-activatable fluorescent imaging agent, AVB-620. *Theranostics* 7 (13): 3369–3386.

51 Unkart, J.T., Chen, S.L., Wapnir, I.L. et al. (2017). Intraoperative tumor detection using a ratiometric activatable fluorescent peptide: a first-in-human phase 1 sturdy. *Annals of Surgical Oncology* 24: 3167–3173.

52 Endoh, T. and Ohtsuki, T. (2009). Cellular siRNA delivery using cell-penetrating peptides modified for endosomal escape. *Advanced Drug Delivery Review* 61: 704–709.

53 Hoyer, J. and Neundorf, I. (2012). Peptide vectors for the nonviral delivery of nucleic acids. *Accounts of Chemical Research* 45 (7): 1048–1056.
54 Boisguerin, P., Dehayes, S., Gait, M.J. et al. (2015). Delivery of therapeutic oligonucleotides with cell penetrating peptides. *Advanced Drug Delivery Review* 87: 52–67.
55 Kristensen, M., Birch, D., and Nielsen, H.M. (2016). Applications and challenges for use of cell-penetrating peptides as delivery vectors for peptide and protein cargos. *International Journal of Molecular Sciences* 17: 185.
56 Yamada, Y., Sato, Y., Nakamura, T. et al. (2020). Evolution of drug delivery system from viewpoint of controlled intracellular trafficking and selective tissue targeting toward future nanomedicine. *Journal of Controlled Release* 327: 533–545.
57 Zhang, Y., Guo, P., Ma, Z. et al. (2021). Combination of cell-penetrating peptides with nanomaterials for the potential therapeutics of central nervous system disorders: a review. *Journal of Nanotechnology* 19: 255.

3

Arginine-Rich Cell-Penetrating Peptides: Recent Advances of Design and Applications for Intracellular Delivery

Ikuhiko Nakase

Osaka Metropolitan University, Graduate School of Science, 1-1, Gakuen-cho, Naka-ku, Sakai, Osaka, Japan

3.1 Introduction

The development of intracellular delivery systems based on cell-penetrating peptides (CPPs) has rapidly progressed. This includes innovative carrier peptides with useful functionalities, such as increased efficiency of cellular uptake and cell membrane penetration for the delivery of functional or therapeutic molecules. Arginine-rich CPPs, such as human immunodeficiency virus type 1(HIV-1) Tat peptide (Tat (48–60)) [1], oligoarginines [2–5], and their derivatives have been used as intracellular delivery carriers [6–8]. Various peptide derivatives have been discovered and developed including HIV-1 Rev (34–50) peptide, FHV Coat (35–49) peptide, and HTLV-II Rex (4–16) peptide [6–8]. In recent years, new CPPs that contain specialized amino acids in the peptide sequences and unique structures have been developed along with conjugation and coating technologies to achieve sophisticated delivery.

With respect to cellular uptake and cell membrane penetration mechanisms, the strong interaction of the arginine residues of the peptides with phosphate, sulfate, and carboxylate through ion-pairing and hydrogen bonding decides the overall effectiveness of cellular uptake [6–8]. Interaction of proteoglycans bearing heparan- and chondroitin-sulfate with arginine-rich peptides has been shown to be important for the efficient accumulation of peptides on the plasma membranes of targeted cells, resulting in effective uptake of the peptides into cells [6–11]. In addition, the macropinocytosis pathway, which includes actin-dependent cellular uptake mechanisms, with ruffling of the plasma membrane and engulfment of large volumes (>1 μm diameter) of outside fluids, contributes to the uptake process of arginine-rich peptides [6–9, 12]. The unique phenomena of induced multilamellar formations at the

Cell-Penetrating Peptides: Design, Development and Applications, First Edition. Makoto Oba and Yosuke Demizu.

point of membrane penetration by arginine-rich CPPs have been observed during the process of peptide-based intracellular delivery [13, 14]. Receptors (e.g. CXCR4) [15], ion channels (e.g. KCNQ5, KCNN4, and KCNK5) [16], cell membrane-related molecules (e.g. acid sphingomyelinase (ASMase) and ceramide [17], and cholesterol [18]) participate in cell membrane penetration of arginine-rich CPPs. However, the underlying mechanisms are unknown and further studies are needed.

In this chapter, the latest technology related to arginine-rich CPPs is summarized including unnatural amino acids, peptide structures, targeting cell processes, and assembly system for achieving efficient delivery of functional, diagnostic, and therapeutic molecules, especially in the cytosol following cell membrane penetration (Figure 3.1). Each technology has unique characteristics, methodologies, and mechanisms that will contribute to the development of next-generation drug delivery systems and to understanding the cellular function from the viewpoints of the chemical biology of cells for future therapy.

3.2 Arginine-Rich Peptides with Unnatural Amino Acids

For increasing the stability and cellular uptake efficacy, unnatural amino acids have been incorporated into the sequence of arginine-rich CPPs. Kato et al. developed arginine-rich CPPs with α,α-disubstituted α-amino acids (dAAs), which consisted of a six-membered carbocyclic ring and an amino group (Ac_6c^{NH2}) [19]. The peptides exhibited increased resistance against trypsin and the cellular uptake efficacy was much greater and without cytotoxicity compared with that of nonaarginine (R9) peptides in HeLa cells [19]. In addition, colocalization of the peptides (2 μM) with lysotracker following cellular uptake was also comparatively lower [19]. However, the size of the carbocyclic ring did not affect uptake efficacy or trypsin resistance in experiments using the six-membered carbocyclic ring, Ac_6c^{NH2}, and the five-membered carbocyclic ring, Ac_5c^{NH2} [19].

Oba et al. developed arginine-rich CPP foldamers using unnatural amino acids, such as aminoisobutyric acid (Aib), for plasmid DNA delivery [20]. Repeat sequences

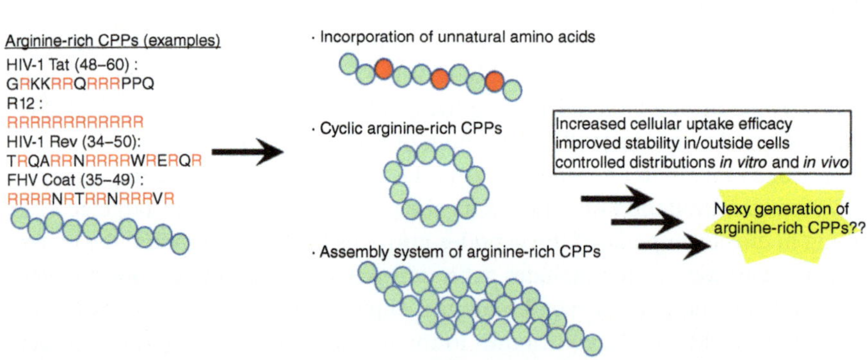

Figure 3.1 Development of arginine-rich CPPs for increasing cellular uptake efficacy, improving stability in/outside cells, and controlling distributions in vitro and in vivo.

of (Arg-Arg-Aib)$_n$ (n = 1–6) were assessed for physicochemical and pharmacological properties including secondary structures, cellular uptake, and gene delivery efficacy. Aib is a dAA that has been shown to stabilize helical structures of peptides. Oba et al. found strong induction of right-handed helix formation by incorporating the Aib residues into arginine-rich repeat sequences (n = 2–6). The highest cellular uptake efficacy of (Arg-Arg-Aib)$_5$ and (Arg-Arg-Aib)$_6$ was confirmed in comparison with other repeat sequence peptides [20]. Interestingly, (Arg-Arg-Aib)$_4$ exhibited a much higher cellular uptake efficacy compared with that of R9 peptides (Huh-7 or HeLa cells, 1 μM for 2 hours) [20]. The transfection efficiency of plasmid DNA was also assessed. The results indicated that longer peptides ($n \geq 4$) exhibited excellent gene delivery and expression (luciferase), suggesting the importance of artificial control of helical structures for arginine-rich peptides in cytosolic gene delivery [20].

Oba et al. also developed arginine-rich nonapeptide-containing dAAs in repeating units of (Arg-Arg-AA)$_3$ (AA: amino acid) such as leucine, (S)- α-methylleucine ((S)-(αMe)Leu), 1-aminocyclopentanecarboxylic acid (Ac$_5$c), and (3S,4S)-1-amino-3,4-dimethoxycyclopentanecarboxilic acid [(S,S)-Ac$_5$c^{dOM}] [21]. The stability of the helical structures was: (αMe)Leu peptide > Ac$_5$c peptide ≈ Leu peptide > Ac$_5$c^{dOM} peptide ≈ Arg peptide. Each peptide containing a dAA exhibited similar resistance against proteases [21]. In addition, by incorporating dAAs into arginine-rich peptides, the enhanced stable helical structure increased cellular uptake efficacy [21], suggesting the importance of artificially controlled secondary structures for arginine-rich peptides.

3.3 β-Hairpin Arginine Peptides for Enhanced Stabilization Under Cytosolic Condition

Safa et al. assessed the cellular uptake and stability of β-hairpin arginine peptides under cytosolic conditions [22]. β-hairpin arginine peptides, RWRWR (amino acid sequence: RWVRVpGO(FAM)WIRQ, p: D-form of proline, O: ornithine) exhibited β-hairpin structures, and the uptake efficacy was considerably lower compared with that of unstructured RWRWR peptides (approximately 5-fold at 100 minutes treatment with 10 μM into HeLa cells) or Tat peptides (approximately, 2-fold under similar experimental conditions) [22]. However, a strong inhibition of RWRWR peptide degradation in HeLa cell lysates (80% intact remaining structures even after 180 minutes treatment, $t_{1/2}$ = 423 minutes) was observed compared with unstructured RWRWR peptides (complete loss of peptide structures after 120 minutes, $t_{1/2}$ = 28 minutes) and Tat peptides (complete loss of peptide structures after 30 minutes, $t_{1/2}$ = 10 minutes) [22].

3.4 Cyclic Arginine-Rich Peptides

Peptide cyclization affects stability against proteases and cellular uptake efficacy. Uhl et al. developed cyclic-R9 peptide-decorated nanoparticles for oral delivery, which facilitated intestinal retention and mucosal penetration of loaded drugs [23]. Cyclic-R9 peptides were conjugated to polylactide acid (PLA)-nanoparticles using Michael addition chemistry with a maleimide moiety [23]. The antidiabetic peptide, liraglutide (LRT), was also encapsulated in the nanoparticles and the size was approximately 300 nm [23]. In a Wistar rat model, the oral administration of the cyclic-R9-decorated nanoparticles with encapsulation of LRT significantly improved blood levels of LRT (approximately, a 4.5-fold increase of bioavailability compared with that of free LRT or use of nanoparticles without decoration of cyclic-R9 peptide), suggesting an effective therapy and application of cyclic-arginine-rich CPPs [23].

Fujii et al. have developed artificial cyclized helix-loop-helix (HLH) peptides, which are scaffolds for grafting the p53 epitope onto the C-terminal helix and six arginine residues on the N-terminal helix for enhancing membrane penetration. This resulted in the inhibition of protein–protein interactions between the HDM2 oncoprotein and the p53 tumor suppressor in the cytosol [24]. The p53 induces cell-cycle arrest and apoptotic cell death after dissociation from HDM2. However, high expression of HDM2 in cancer cells reduces the release of p53, leading to inhibition of p53-related cancer cell death [25–28]. The cyclization of the HLH peptide resulted in increased stabilization of the secondary structure, leading to strong binding to HDM2 ($K_d = 10$ nM, approximately 40-fold increased efficacy compared with that of the linear HLH peptide) and increased proteolytic stability ($t_{1/2} > 200$ hours in the presence of cathepsin G) [24]. The arginine grafting of the HLH peptide sequence enhanced the efficacy of cellular uptake and cytosolic release of the peptide, leading to the inhibition of human colon cancer-derived HCT-116 and human prostate adenocarcinoma-derived LnCap cell proliferation [24].

To further improve peptide stability, bicyclization of arginine-rich peptides has been conducted by Wolfe et al. [29]. By using the cysteine perfluoroarylation reaction, two methods for preparing bicyclic dodecaarginine (R12) were successfully developed; (i) the coupling of multiple cysteine residues with perfluoroarene for conjugation to 1,3,5-benzenetrithiol, (ii) incorporating two perfluoroarenes in the peptide sequence, which undergoes intramolecular S_NAr to form two i, $i+7$ cycles [29]. The bicyclic R12 peptides exhibited much higher stability against trypsin compared with that of monocyclic peptides [29]. The bicyclic R12 peptides efficiently carried phosphorodiamidate morpholino oligonucleotides (PMOs) through conjugation, leading to a 14-fold higher PMO exon-skipping activity compared with that of unconjugated PMOs [29].

3.5 Cancer-Targeting Cell-Penetrating Peptides

The development of tumor-penetrating peptides (TPPs) that function in tumor-receptor targeting and membrane penetration has progressed and recently reported techniques associated with these TPPs are discussed in this section.

The motif of C-end Rule (CendR) with the consensus sequence of R/KXXR/K was able to target tumors through the neuropilin-1 (NRP-1) receptor, which contributes to angiogenesis and vascular permeability [30, 31]. The RPARPAR sequence, which is a prototypic CendR, may be internalized by tumor cells through its interaction with the loop III structure of NRP-1 receptors [31–33]. The iRGD sequence (amino acid sequence: CRGDKGPDC) can interact with $\alpha v\beta 3/5$ integrins. Cleavage of the iRGD sequence by tumor proteases resulting in the CendR motif, CRGDK, resulted in tumor cell uptake through NRP-1 receptors [34]. In addition, the LinTT1 sequence (AKRGARSTA) interacts with p32 on the tumor cell surface, resulting in cleavage of the peptide by tumor proteases and tumor cell uptake by NRP-1 receptors [35, 36]. Savier et al. prepared each NRP-1-targeted peptide fused with an interfering peptide, which can inhibit the interaction of phosphatase PP2A and oncoprotein SET. These peptides were selectively internalized by hepatocellular carcinoma cells derived from patients [37]. The cellular uptake efficiency and apoptosis induction by these peptides were correlated with the expression of integrin, p32, and NRP-1 receptors on tumor cells [37]. Although a relatively high peptide concentration (25 μM) was required to induce apoptotic cell death, this strategy of using bifunctional TPPs for developing a specific tumor-targeting system is considered useful and has potential for future applications.

Zhang et al. developed nanocarriers with R9 peptides, which are activated in acidic environments, such as in tumors [38]. As nanocarriers consisting of amphiphilic diblock polymers, the R9 peptide was conjugated with poly(ethylene glycol) (PEG) and poly(hexyl ethylene phosphate) (PHEP) [38]. In addition, the surface of the nanocarriers was electrostatically coated with PEGylated polyanionic polyphosphoester (PPC-DA) to shield the nanocarriers and avoid nonspecific targeting [38]. In the acidic environment of the tumor (~pH 6.5), PPC-DA was removed and exposure to the R9 peptides increased tumor uptake and achieved effective delivery of anticancer drugs (doxorubicin) in tumor-xenografted (4T1 breast cancer cells) mice [38].

Tumor-homing CPPs were developed by random peptide screening, which was constructed using an mRNA display technique [39]. CPP2 (amino acid sequence: DSLKSYWYLQKFSWR) and CPP44 (amino acid sequence: KRPTMRFRYTWNPMK) targeted and penetrated the cell membrane of colon adenocarcinoma and acute myelogenous leukemia, respectively [39]. The peptides fused with an antitumor, minimal inhibitory peptide sequence of p16 exhibited significant biological activity in vitro and in vivo [39].

3.6 Assembly System Using Arginine-Rich Cell-Penetrating Peptides

Artificial assembly of arginine-rich CPPs resulted in unique structures and cellular uptake mechanisms. Mello et al. designed self-assembly of surfactant-like peptides (SLPs) and studied the self-assembly and uptake efficiency of SLPs, including $(Phe)_4$-$(Arg)_4$ (F4R4) as a diblock construct and $(Arg)_2$-$(Phe)_4$-$(Arg)_2$ (R2F4R2) as a bolaamphiphile analog [40]. Interestingly, cryo-transmission electron microscopy revealed micelle/vesicle formation of F4R4 and tape-like structures of R2F4R2 [40].

The structures of the micelles/vesicles showed efficient cellular uptake and nuclear accumulation, whereas the tape-like structures resulted in low cellular uptake efficacy [40]. The F4R4 peptides exhibited anticancer effects in human skin melanoma SK-MEL-28 cells [40].

Lipopeptides may be applied for gene delivery in targeted cells. Liang et al. developed an assembly system using lipopeptides with a disulfide linkage between the lipid and arginine-rich dendritic structures [41]. Dendritic structures with a different generation of R_1LS (first-generation dendritic arginine), R_2LS (second-generation dendritic arginine), and R_3LS (third-generation dendritic arginine) were designed and synthesized using a dioleoyl-L-lysinate hydrophobic chain [41]. Assembly structures of R_2LS (average diameter: 189.1 nm) and R_3LS (average diameter: 494.1 nm) appeared as spherical and spindle structures, as well as fiber-like structures of R_1LS (average diameter: 239.3 nm) [41]. The R_2LS exhibited efficient gene transfection efficacy, even in 10% serum compared with that of lipofectamine (higher transfection efficacy was approximately 59% higher than that of lipofectamine 2000 and polyethylenimine) in HeLa cells, whereas very low transfection was observed with R_1LS or R_3LS [41]. In addition, a molecular dynamic simulation revealed flexibility of the R_2LS structure, which may affect the cell membrane interaction and uptake [41].

3.7 Gene Interaction of Arginine-Rich Cell-Penetrating Peptides

Obitz et al. developed a bioconjugate of ruthenium (II) dipyridophenazine (dppz), which shows DNA intercalation with CPPs, such as TAT-9 (amino acid sequence: RKKRRQRRR) or Xentry peptide (amino acid sequence: LCLRPVG), derived from hepatitis B virus [42]. The carboxylic acid of dppz was condensed with the N-terminus of each cell-penetrating peptide (dppz-CPP) [42]. The dppz-CPPs exhibited binding affinity with DNA (1.3×10^6 K/M^{-1} for dppz-Xentry peptide and $2.8\times10^6 K/M^{-1}$ for dppz-TAT-9) [42]. Interestingly, the dppz-Xentry peptide displayed high cellular uptake efficacy and fluorescent intensity in cells resulting from an interaction with DNA. However, dppz-TAT-9 did not show fluorescence in cells possibly because of the minimal internalization into the nucleus [42]. This experimental technique is considered useful for the detection of gene interactions in cells.

3.8 Development of Arginine-Rich Cell-Penetrating Peptides for the Delivery of Genes and Oligonucleotides

Oligonucleotide delivery using arginine-rich peptides has been developed to control gene expression in the cytosol of disease-related target cells. Wood et al. developed an antisense oligonucleotide (ASO) delivery system for myotonic dystrophy (DM1) [43]. In this study, the arginine-rich CPP, Pip6a [amino acid sequence: RXRRBRRXRYQFLIRXRBRXRB; aminohexanoyl (X), β-alanine (B)] was generated, which is a derivative of the Drosophila Antennapedia homeodomain-derived peptide with six additional arginine residues [44]. The peptide was conjugated with PMO by amide bonding at the 3′ end of the PMO [44]. The ASOs against the nucleus-retained mutant, DM1 protein kinase (DMPK) transcripts containing CUG expansions (CUGexps), were efficiently delivered by Pip6a conjugation into strained muscles in DM1 mice by intravenous injection. This resulted in specific targeting of mutant CUGexp-DMPK transcripts and the inhibition of detrimental sequestration of the MBNL1 splicing factor, which contributes to the DM1 pathophysiological mechanism [43].

Rostami et al. assessed the delivery of HIV-1 Nef DNA or protein complexed with a histidine-rich nonaarginine (HR9) (amino acid sequence: CHHHHHRRRRRRRRR HHHHHC), MPG (hydrophobic sequence-linker-nuclear localization sequence) (amino acid sequence: GALFLGFLGAAGSTMGAWSQPKKKRKV), penetratin (amino acid sequence: RQIKIWFQNRRMKWKK), and M918 (derived from C-terminal of tumor suppressor protein p14ARF) (amino acid sequence: MVTVLFRRLRIRRACG PPRVRV) for HIV-1 vaccine development [45]. Enhanced cellular uptake of the DNA complexed with HR9 or MPG and the proteins complexed with M918 or penetratin was demonstrated in HEK-293T cells. A boost regimen of heterologous small heat shock protein 20 (sHsp20 as adjuvant)-Nef DNA/MPG and recombinant Hsp20-Nef protein/M918 complexes in BALB/c mice significantly enhanced the levels of IgG2a, IgG2b, and IFN-γ after three months along with an immunological response against HIV-1 Nef antigen [45]. This technique is useful for the development of a CPP-based vaccine system using only mixing techniques.

Dendrimer structures of an arginine-rich peptide are also useful for gene delivery. Kozhikhova et al. developed an arginine-rich dendrimer, LTP with the amino acid sequence: $\{[(R_2K)_2K]_2KAC\}$. These LTP peptides could efficiently deliver pGL2 plasmid (firefly luciferase reporter gene) into 293T and HeLa HI cells compared with other linear- or lipo-arginine-rich peptides [46]. The hemolytic activity of the LTP peptides was very low [46]. In addition, intravenous injection of the LTP peptides (250 μg) into BALB/c mice resulted in a high concentration in the lung (approximately 5000 ng/g, 2 minutes after injection) [46].

3.9 Molecular Coating with Arginine-Rich Cell-Penetrating Peptides

Using molecular surface coating chemistry, the coating of macromolecules with arginine-rich CPPs is a useful technique for enhanced intracellular delivery. Bolaños et al. developed an experimental technique for coating gold nanoparticles (AuNPs), which have been used effectively for biomedical applications including disease diagnosis, control of drug release, as well as photodynamic and photothermal therapy, with bovine serum albumin (BSA) conjugated with the D-form octaarginine peptides (r8) [47]. Low cellular uptake efficacy of AuNPs limits their biomedical application and technical improvement is required. For this bioconjugate technique, BSA was reacted with iminothiolane through the amino groups of BSA, resulting in sulfhydryl groups that could be reacted with brominated r8 peptides (BSA-r8) [47]. By coating with BSA-r8, cellular uptake of AuNPs was significantly increased [47]. Interestingly, the location of the coated AuNPs in early endosomes was confirmed even after a 24 hours treatment and approximately 15 AuNPs per endosome were observed by electron microscopy [47].

In addition, direct conjugation of arginine-rich peptides to AuNPs was reported by Boussoufi et al. [48]. Direct conjugation to the surface of AuNPs by the cysteine sulfhydryl group of the arginine-rich CPP (amino acid sequence: CRRRRRRRRGDS), which has a fibronectin-derived RGD motif, resulted in cancer cell targeting [48].

Arginine-coating of objective molecules has also been studied. Gosh et al. developed arginine-terminated nanoparticles with quantum dots (QDs), which have been widely used for fluorescent imaging of cells [49]. The synthesized arginine-acrylate was used for polyacrylate coating of the QDs (R-QDs, diameter: approximately 15–30 nm) [49]. The arginine-coating enhanced the cell membrane penetration of ODs, even under conditions of low temperature (4 °C) for the inhibition of energy-dependent cellular uptake pathways, including endocytosis. This suggests direct cell membrane penetration of the R-QDs [49]. Interestingly, when the arginine moiety was replaced with ammonium, the cellular uptake of the QDs was altered to energy-dependent endocytic pathways, suggesting important functionalities of the guanidium group for direct membrane penetration [49].

3.10 Enhanced Cytosolic Release using Arginine-Rich Cell-Penetrating Peptides as Additives

Schneider et al. developed enhanced cytosolic release systems using thiol-reactive, arginine-rich peptides [50]. The addition of these peptides with cysteine-introduced cyclic or linear decaarginine (R10) peptides significantly enhanced the cytosolic release of enzymes, nanobodies, and IgG antibodies during co-treatment with cysteine-introduced peptides [50]. Using this technique, anti-TOMM20 antibody (500 nM, 1 hour, 37 °C) was targeted to mitochondria through the cell membranes of living cells (HeLa CCL-2 cells) [50].

3.11 Effective Usage of Arginine-Rich Cell-Penetrating Peptides for Extracellular Vesicles, Exosomes

Arginine-rich CPPs have been shown to induce macropinocytosis. The enhanced macropinocytotic uptake by the peptides has been applied to the intracellular delivery of various macromolecules [6–8]. Recently, we found that extracellular vesicles, exosomes, are taken up by cancer cells in pathways of macropinocytosis [51–53]. Almost all cells can secrete exosomes and encapsulate "message" molecules, such as genes (e.g. microRNAs) and enzymes. These exosomes function in cell-to-cell communication to maintain homeostasis and to participate in disease progression (e.g.

cancers) [52, 53]. Diagnostic markers for the detection of isolated exosomal contents, such as microRNAs and proteins from a patient's body fluids, have been developed [52, 53]. In addition, exosomes are anticipated to be next-generation carriers of therapeutic molecules for clinical use. They have many pharmacological advantages, such as controlling the immune response (by using self-exosomes), promoting cell-to-cell communications, brain-targeting through the blood-brain barrier, preparation of complexed proteins and gene cocktails, and expression of membrane proteins [52, 53]. Macropinocytosis induction by activating related cancer receptors, such as epidermal growth factor receptor, and expression of oncogenic K-Ras, also significantly enhance cellular exosome uptake [51–53].

Therefore, we developed a macropinocytosis-inducible exosomal system by decorating exosomal membranes with arginine-rich CPPs (Figure 3.2) [52, 54–57]. To decorate with the arginine-rich CPPs, we developed (i) peptides with a hydrophobic moiety for insertion into the exosomal membrane [54]; and (ii) covalent conjugated peptides with the side chains of exosomal membrane proteins using chemical linkers [55].

We synthesized D-formed octaarginine (r8) with a stearyl moiety at the N-terminus of the peptides and mixed stearylated peptides (stearyl-r8) with exosomes, which resulted in the decoration of the exosomal membranes with the peptides (stearyl-r8-Exo) [54]. Modification of the stearyl-r8 peptides increased the zeta-potential of the exosomes to 5.1 mV (original exosomes without the peptide modification: −12.5 mV) [54]. The stearyl-r8-Exo induced macropinocytosis in the target cells,

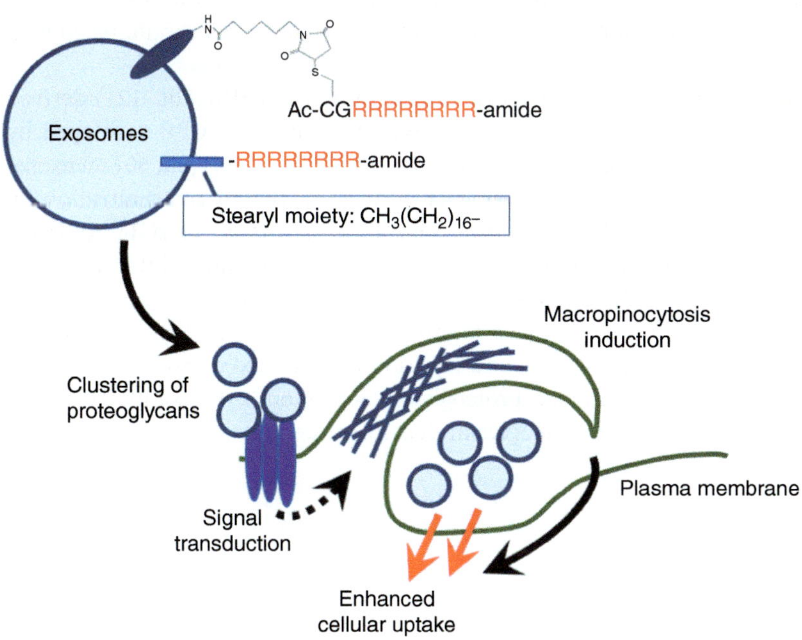

Figure 3.2 Modification of arginine-rich CPPs on exosomal membranes by covalent attachment or insertion into the membranes efficiently induces macropinocytosis, leading to enhanced cellular uptake.

leading to efficient cellular uptake of exosomes without any cytotoxicity [e.g. 33-fold uptake of exosomes (10 μg/ml) into HeLa cells after 24 hours in 10% serum-containing conditions] [54]. Clustering of syndecan-4 on the plasma membrane was also observed following treatment with stearyl-r8-Exo, suggesting effective use of exosomes as scaffolds for clustering via the peptides [54].

Covalent conjugation of oligoarginine with exosomal membrane proteins via chemical linkers was also done [55]. We synthesized oligoarginines with a cysteine residue at the C-terminus for conjugation with a maleimide moiety of sulfo-EMCS (*N*-ε-malemidocaproyl-oxysuccinimide ester), which is a crosslinker of an amine-to-sulfhydryl [55]. The peptide with a linker can covalently react with exosomal membrane proteins via a succinimidyl moiety [55]. Octa (R8)- or dodeca (R12)-arginine-conjugated exosomes exhibited a higher cellular uptake efficacy (each 29-fold cellular uptake increase) with macropinocytosis induction compared with other oligoarginines [e.g. 14-fold for tetra (R4)- and 18-fold for hexadeca (R16)-arginine modification] into CHO-K1 cells for 24 hours [55]. For assessing the delivery of therapeutic molecules, we prepared exosomes with encapsulation of saporin-toxin, which is a ribosome-inactivating protein that works in the cytosol to inhibit protein synthesis and induce cell death [54, 55, 58]. The R16-modified exosomes with saporin encapsulation exhibited the highest cytotoxicity compared with that of the other oligoarginines, although cellular uptake of the R16-modified exosomes was confirmed to be lower than that with R8- or R12-modification [55]. The results suggest enhanced cytosolic release of exosomal contents through R16 modification, balanced cellular uptake, and cytosolic release efficiencies by modifying the oligoarginines on the exosomal membranes. This should be considered to obtain robust biological activities of the exosomal contents.

In addition, modification of the antimicrobial protein, CAP18 (106–121)-derived sC18 peptides, which is an α-helical and arginine/lysine-rich CPP developed by Neundorf et al. (amino acid sequence: GLRKRLRKFRNKIKEK) [59, 60] on exosomal membranes via a stearyl moiety, significantly increased internalization by breast cancer-derived MCF-7 cells [61]. The dimer structure of sC18 [$(sC18)_2$] exhibited synergistic effects on macropinocytosis induction and cellular uptake of exosomes by modifying the peptides on the exosomes [61]. Cellular uptake of the peptide-modified exosomes was also highly dependent on proteoglycans on the plasma membrane of targeted cells, and saporin-encapsulated exosomes with modification of the peptides induced proteoglycan-dependent cytotoxicity. However, further studies to elucidate proteoglycan-dependent cellular uptake mechanisms are needed [61].

3.12 Conclusion

As summarized earlier, unique and functional designs of arginine-rich CPPs have been developed for practical use in basic biological sciences, medical sciences, and diagnostic/therapeutic technologies. Artificially designed structures of CPPs show peculiar mechanisms of cell targeting, uptake, membrane penetration, and intracellular localization. However, unknown mechanisms, especially in membrane penetration of CPPs, are issues that remain to be elucidated for further development of CPP-based intracellular delivery systems and to understand cellular functions as chemical biology. Arginine residues in CPPs exhibit unique interactions with cell membranes and complex interactions with cellular factors, signal transduction pathways, and cellular responses. Further innovative design of arginine-rich CPPs will bring together chemistry and biology to control and visualize cells for basic science and therapeutic applications.

Declaration of Competing Interest

The author declares no conflicts of interest associated with this manuscript.

Author Contribution

I.N. designed and wrote the manuscript.

Acknowledgments

This work was supported by Japan Society for the Promotion of Science (JSPS) KAKENHI (JP19H05553), Japan Science and Technology Agency (JST) CREST (JPMJCR17H3), and Takeda Science Foundation (J20ZZ00356).

References

1 Vivès, E., Brodin, P., and Lebleu, B. (1997). A truncated HIV-1 Tat protein basic domain rapidly translocates through the plasma membrane and accumulates in the cell nucleus. *Journal of Biological Chemistry* 272 (25): 16010–16017. https://doi.org/10.1074/jbc.272.25.16010.
2 Wender, P.A., Mitchell, D.J., Pattabiraman, K. et al. (2000). The design, synthesis, and evaluation of molecules that enable or enhance cellular uptake: peptoid molecular transporters. *Proceedings of the National Academy of Sciences of the United States of America* 97 (24): 13003–13008. https://doi.org/10.1073/pnas.97.24.13003.
3 Rothbard, J.B., Garlington, S., Lin, Q. et al. (2000). Conjugation of arginine oligomers to cyclosporin A facilitates topical delivery and inhibition of inflammation. *Nature Medicine* 6 (11): 1253–1257. https://doi.org/10.1038/81359.
4 Futaki, S., Suzuki, T., Ohashi, W. et al. (2001). Arginine-rich peptides. An abundant source of membrane-permeable peptides having potential as carriers for intracellular protein delivery. *Journal of Biological Chemistry* 276 (8): 5836–5840. https://doi.org/10.1074/jbc.M007540200.

5 Suzuki, T., Futaki, S., Niwa, M. et al. (2002). Possible existence of common internalization mechanisms among arginine-rich peptides. *Journal of Biological Chemistry* 277 (4): 2437–2443. https://doi.org/10.1074/jbc.M110017200.

6 Nakase, I., Takeuchi, T., Tanaka, G. et al. (2008). Methodological and cellular aspects that govern the internalization mechanisms of arginine-rich cell-penetrating peptides. *Advanced Drug Delivery Reviews* 60 (4–5): 598–607. https://doi.org/10.1016/j.addr.2007.10.006.

7 Nakase, I., Tanaka, G., and Futaki, S. (2013). Cell-penetrating peptides (CPPs) as a vector for the delivery of siRNAs into cells. *Molecular BioSystems* 9 (5): 855–861. https://doi.org/10.1039/c2mb25467k.

8 Futaki, S. and Nakase, I. (2017). Cell-surface interactions on arginine-rich cell-penetrating peptides allow for multiplex modes of internalization. *Accounts of Chemical Research* 50 (10): 2449–2456. https://doi.org/10.1021/acs.accounts.7b00221.

9 Nakase, I., Tadokoro, A., Kawabata, N. et al. (2007). Interaction of arginine-rich peptides with membrane-associated proteoglycans is crucial for induction of actin organization and macropinocytosis. *Biochemistry* 46 (2): 492–501. https://doi.org/10.1021/bi0612824.

10 Nakase, I., Osaki, K., Tanaka, G. et al. (2014). Molecular interplays involved in the cellular uptake of octaarginine on cell surfaces and the importance of syndecan-4 cytoplasmic V domain for the activation of protein kinase C**α**. *Biochemical and Biophysical Research Communications* 446 (4): 857–862. https://doi.org/10.1016/j.bbrc.2014.03.018.

11 Kawaguchi, Y., Takeuchi, T., Kuwata, K. et al. (2016). Syndecan-4 is a receptor for clathrin-mediated endocytosis of arginine-rich cell-penetrating peptides. *Bioconjugate Chemistry* 27 (4): 1119–1130. https://doi.org/10.1021/acs.bioconjchem.6b00082.

12 Nakase, I., Niwa, M., Takeuchi, T. et al. (2004). Cellular uptake of arginine-rich peptides: roles for macropinocytosis and actin rearrangement. *Molecular Therapy* 10 (6): 1011–1022. https://doi.org/10.1016/j.ymthe.2004.08.010.

13 Allolio, C., Magarkar, A., Jurkiewicz, P. et al. (2018). Arginine-rich cell-penetrating peptides induce membrane multilamellarity and subsequently enter via formation of a fusion pore. *Proceedings of the National Academy of Sciences of the United States of America* 115 (47): 11923–11928. https://doi.org/10.1073/pnas.1811520115.

14 Hirose, H., Takeuchi, T., Osakada, H. et al. (2012). Transient focal membrane deformation induced by arginine-rich peptides leads to their direct penetration into cells. *Molecular Therapy* 20 (5): 984–993. https://doi.org/10.1038/mt.2011.313.

15 Tanaka, G., Nakase, I., Fukuda, Y. et al. (2012). CXCR4 stimulates macropinocytosis: implications for cellular uptake of arginine-rich cell-penetrating peptides and HIV. *Chemistry & Biology* 19 (11): 1437–1446. https://doi.org/10.1016/j.chembiol.2012.09.011.

16 Trofimenko, E., Grasso, G., Heulot, M. et al. (2021). Genetic, cellular and structural characterization of the membrane potential-dependent cell-penetrating peptide translocation pore. *eLife* 10: e69832. https://doi.org/10.7554/eLife.69832.

17 Verdurmen, W.P., Thanos, M., Ruttekolk, I.R. et al. (2010). Cationic cell-penetrating peptides induce ceramide formation via acid sphingomyelinase: implications for

uptake. *Journal of Controlled Release* 147 (2): 171–179. https://doi.org/10.1016/j.jconrel.2010.06.030.

18 Lorents, A., Säälik, P., Langel, Ü. et al. (2018). Arginine-rich cell-penetrating peptides require nucleolin and cholesterol-poor subdomains for translocation across membranes. *Bioconjugate Chemistry* 29 (4): 1168–1177. https://doi.org/10.1021/acs.bioconjchem.7b00805.

19 Kato, T., Kita, Y., Iwanari, K. et al. (2021). Synthesis of six-membered carbocyclic ring α,α-disubstituted amino acids and arginine-rich peptides to investigate the effect of ring size on the properties of the peptide. *Bioorganic & Medicinal Chemistry* 38: 116111. https://doi.org/10.1016/j.bmc.2021.116111.

20 Oba, M., Ito, Y., Umeno, T. et al. (2019). Plasmid DNA delivery using cell-penetrating peptide foldamers composed of Arg-Arg-Aib repeating sequences. *ACS Biomaterials Science & Engineering* 5 (11): 5660–5668. https://doi.org/10.1021/acsbiomaterials.8b01451.

21 Oba, M., Nagano, Y., Kato, T. et al. (2019). Secondary structures and cell-penetrating abilities of arginine-rich peptide foldamers. *Scientific Reports* 9 (1): 1349. https://doi.org/10.1038/s41598-018-38063-8.

22 Safa, N., Anderson, J.C., Vaithiyanathan, M. et al. (2019). CPProtectides: rapid uptake of well-folded β-hairpin peptides with enhanced resistance to intracellular degradation. *Peptide Science* 111 (2): e24092. https://doi.org/10.1002/pep2.24092.

23 Uhl, P., Grundmann, C., Sauter, M. et al. (2020). Coating of PLA-nanoparticles with cyclic, arginine-rich cell penetrating peptides enables oral delivery of liraglutide. *Nanomedicine* 24: 102132. https://doi.org/10.1016/j.nano.2019.102132.

24 Fujiwara, D., Kitada, H., Oguri, M. et al. (2016). A cyclized helix-loop-helix peptide as a molecular scaffold for the design of inhibitors of intracellular protein-protein interactions by epitope and arginine grafting. *Angewandte Chemie International Edition in English* 55 (36): 10612–10615. https://doi.org/10.1002/anie.201603230.

25 Kussie, P.H., Gorina, S., Marechal, V. et al. (1996). Structure of the MDM2 oncoprotein bound to the p53 tumor suppressor transactivation domain. *Science* 274 (5289): 948–953. https://doi.org/10.1126/science.274.5289.948.

26 Perkins, J.R., Diboun, I., Dessailly, B.H. et al. (2010). Transient protein-protein interactions: structural, functional, and network properties. *Structure* 18 (10): 1233–1243. https://doi.org/10.1016/j.str.2010.08.007.

27 Vassilev, L.T., Vu, B.T., Graves, B. et al. (2004). In vivo activation of the p53 pathway by small-molecule antagonists of MDM2. *Science* 303 (5659): 844–848. https://doi.org/10.1126/science.1092472.

28 Bernal, F., Tyler, A.F., Korsmeyer, S.J. et al. (2007). Reactivation of the p53 tumor suppressor pathway by a stapled p53 peptide. *Journal of the American Chemical Society* 129 (9): 2456–2457. https://doi.org/10.1021/ja0693587.

29 Wolfe, J.M., Fadzen, C.M., Holden, R.L. et al. (2018). Perfluoroaryl bicyclic cell-penetrating peptides for delivery of antisense oligonucleotides. *Angewandte Chemie International Edition in English* 57 (17): 4756–4759. https://doi.org/10.1002/anie.201801167.

30 Teesalu, T., Sugahara, K.N., Kotamraju, V.R. et al. (2009). C-end rule peptides mediate neuropilin-1-dependent cell, vascular, and tissue penetration. *Proceedings of the National Academy of Sciences of the United States of America* 106 (38): 16157–161562. https://doi.org/10.1073/pnas.0908201106.

31 Zanuy, D., Kotla, R., Nussinov, R. et al. (2013). Sequence dependence of C-end rule peptides in binding and activation of neuropilin-1 receptor. *Journal of Structural Biology* 182 (2): 78–86. https://doi.org/10.1016/j.jsb.2013.02.006.

32 Willmore, A.M., Simón-Gracia, L., Toome, K. et al. (2016). Targeted silver nanoparticles for ratiometric cell phenotyping. *Nanoscale* 8 (17): 9096–9101. https://doi.org/10.1039/c5nr07928d.

33 Wonder, E., Simón-Gracia, L., Scodeller, P. et al. (2018). Competition of charge-mediated and specific binding by peptide-tagged cationic liposome-DNA nanoparticles in vitro and in vivo. *Biomaterials* 166: 52–63. https://doi.org/10.1016/j.biomaterials.2018.02.052.

34 Sugahara, K.N., Teesalu, T., Karmali, P.P. et al. (2009). Tissue-penetrating delivery of compounds and nanoparticles into tumors. *Cancer Cell* 16 (6): 510–520. https://doi.org/10.1016/j.ccr.2009.10.013.

35 Simón-Gracia, L., Scodeller, P., Fuentes, S.S. et al. (2018). Application of polymersomes engineered to target p32 protein for detection of small breast tumors in mice. *Oncotarget* 9 (27): 18682–18697. https://doi.org/10.18632/oncotarget.24588.

36 Paasonen, L., Sharma, S., Braun, G.B. et al. (2016). New p32/gC1qR ligands for targeted tumor drug delivery. *Chembiochem* 17 (7): 570–575. https://doi.org/10.1002/cbic.201500564.

37 Savier, E., Simon-Gracia, L., Charlotte, F. et al. (2021). Bi-functional peptides as a new therapeutic tool for hepatocellular carcinoma. *Pharmaceutics* 13 (10): 1631. https://doi.org/10.3390/pharmaceutics13101631.

38 Zhang, L., Jiang, C., Zeng, F. et al. (2020). A polymeric nanocarrier with a tumor acidity-activatable arginine-rich (R_9) peptide for enhanced drug delivery. *Biomaterials Science* 8 (8): 2255–2263. https://doi.org/10.1039/d0bm00069h.

39 Kondo, E., Saito, K., Tashiro, Y. et al. (2012). Tumour lineage-homing cell-penetrating peptides as anticancer molecular delivery systems. *Nature Communications* 3: 951. https://doi.org/10.1038/ncomms1952.

40 Mello, L.R., Aguiar, R.B., Yamada, R.Y. et al. (2020). Amphipathic design dictates self-assembly, cytotoxicity and cell uptake of arginine-rich surfactant-like peptides. *Journal of Materials Chemistry B* 8 (12): 2495–2507. https://doi.org/10.1039/c9tb02219h.

41 Liang, H., Hu, A., Chen, X. et al. (2019). Structure optimization of dendritic lipopeptide based gene vectors with the assistance from molecular dynamic simulation. *Journal of Materials Chemistry B* 7 (6): 915–926. https://doi.org/10.1039/c8tb02650e.

42 Obitz, D., Miller, R.G., and Metzler-Nolte, N. (2021). Synthesis and DNA interaction studies of Ru(II) cell penetrating peptide (CPP) bioconjugates. *Dalton Transactions* 50 (39): 13768–13777.

43 Klein, A.F., Varela, M.A., Arandel, L. et al. (2019). Peptide-conjugated oligonucleotides evoke long-lasting myotonic dystrophy correction in patient-derived cells and mice. *Journal of Clinical Investigation* 129 (11): 4739–4744. https://doi.org/10.1172/JCI128205.

44 Betts, C., Saleh, A.F., Arzumanov, A.A. et al. (2012). Pip6-PMO, a new generation of peptide-oligonucleotide conjugates with improved cardiac exon skipping activity for DMD treatment. *Molecular Therapy--Nucleic Acids* 1 (8): e38. https://doi.org/10.1038/mtna.2012.30.

45 Rostami, B., Irani, S., Bolhassani, A. et al. (2019). Gene and protein delivery using four cell penetrating peptides for HIV-1 vaccine development. *IUBMB Life* 71 (10): 1619–1633. https://doi.org/10.1002/iub.2107.

46 Kozhikhova, K.V., Andreev, S.M., Shilovskiy, I.P. et al. (2018). A novel peptide dendrimer LTP efficiently facilitates transfection of mammalian cells. *Organic & Biomolecular Chemistry* 16 (43): 8181–8190. https://doi.org/10.1039/c8ob02039f.

47 Bolaños, K., Sánchez-Navarro, M., Tapia-Arellano, A. et al. (2021). Oligoarginine peptide conjugated to BSA improves cell penetration of gold nanorods and nanoprisms for biomedical applications. *Pharmaceutics* 13 (8): 1204. https://doi.org/10.3390/pharmaceutics13081204.

48 Boussoufi, F., Gallón, S.M.N., Chang, R. et al. (2018). Synthesis and study of cell-penetrating peptide-modified gold nanoparticles. *International Journal of Nanomedicine* 13: 6199–6205. https://doi.org/10.2147/IJN.S168720.

49 Ghosh, S., Panja, P., Dalal, C. et al. (2019). Arginine-terminated, chemically designed nanoparticle for direct cell translocation. *ACS Applied Bio Materials* 2 (1): 339–348. https://doi.org/10.1021/acsabm.8b00617.

50 Schneider, A.F.L., Kithil, M., Cardoso, M.C. et al. (2021). Cellular uptake of large biomolecules enabled by cell-surface-reactive cell-penetrating peptide additives. *Nature Chemistry* 13 (6): 530–539. https://doi.org/10.1038/s41557-021-00661-x.

51 Nakase, I., Kobayashi, N.B., Takatani-Nakase, T. et al. (2015). Active macropinocytosis induction by stimulation of epidermal growth factor receptor and oncogenic Ras expression potentiates cellular uptake efficacy of exosomes. *Scientific Reports* 5: 10300. https://doi.org/10.1038/srep10300.

52 Nakase, I. (2021). Biofunctional peptide-modified extracellular vesicles enable effective intracellular delivery via the induction of macropinocytosis. *Processes* 9: 224. https://doi.org/10.3390/pr9020224.

53 Nakase, I. and Takatani-Nakase, T. (2021). Exosomes: Breast cancer-derived extracellular vesicles; recent key findings and technologies in disease progression, diagnostics, and cancer targeting. *Drug Metabolism and Pharmacokinetics* 42: 100435. https://doi.org/10.1016/j.dmpk.2021.100435.

54 Nakase, I., Noguchi, K., Fujii, I. et al. (2016). Vectorization of biomacromolecules into cells using extracellular vesicles with enhanced internalization induced by macropinocytosis. *Scientific Reports* 6: 34937. https://doi.org/10.1038/srep34937.

55 Nakase, I., Noguchi, K., Aoki, A. et al. (2017). Arginine-rich cell-penetrating peptide-modified extracellular vesicles for active macropinocytosis induction and efficient intracellular delivery. *Scientific Reports* 7 (1): 1991. https://doi.org/10.1038/s41598-017-02014-6.

56 Noguchi, K., Hirano, M., Hashimoto, T. et al. (2019). Effects of lyophilization of arginine-rich cell-penetrating peptide-modified extracellular vesicles on intracellular delivery. *Anticancer Research* 39 (12): 6701–6709. https://doi.org/10.21873/anticanres.13885.

57 Nakase, I., Ueno, N., Matsuzawa, M. et al. (2021). Environmental pH stress influences cellular secretion and uptake of extracellular vesicles. *FEBS Open Bio* 11 (3): 753–767. https://doi.org/10.1002/2211-5463.13107.

58 Fabbrini, M.S., Katayama, M., Nakase, I. et al. (2017). Plant ribosome-inactivating proteins: progesses, challenges and biotechnological applications (and a few digressions). *Toxins* 9 (10): 314. https://doi.org/10.3390/toxins9100314.

59 Hoyer, J., Schatzschneider, U., Schulz-Siegmund, M. et al. (2012). Dimerization of a cell-penetrating peptide leads to enhanced cellular uptake and drug delivery. *Beilstein Journal of Organic Chemistry* 8: 1788–1797. https://doi.org/10.3762/bjoc.8.204.

60 Gronewold, A., Horn, M., Ranđelović, I. et al. (2017). Characterization of a cell-penetrating peptide with potential anticancer activity. *ChemMedChem* 12 (1): 42–49. https://doi.org/10.1002/cmdc.201600498.

61 Noguchi, K., Obuki, M., Sumi, H. et al. (2021). Macropinocytosis-inducible extracellular vesicles modified with antimicrobial protein CAP18-derived cell-penetrating peptides for efficient intracellular delivery. *Molecular Pharmaceutics* 18 (9): 3290–3301. https://doi.org/10.1021/acs.molpharmaceut.1c00244.

4

Cationic Peptide

Takashi Iwasaki

Tottori University, Faculty of Agriculture, Tottori, Japan

4.1 Cationic Peptide

Cell-penetrating peptides (CPPs) are widely used to deliver proteins and nucleic acids into cells [1–3]. At present, over 100 CPPs are reported, with the majority being highly basic, consisting of arginine/lysine-rich sequences, such as penetratin (RQIKIWFQNRRMKWKK-NH_2) [4], HIV-Tat(48–60)(GRKKRRQRRRPPQ-NH_2) [5], and the octa-arginine peptide (RRRRRRRRR-NH_2) [6]. As these CPPs are positively charged under physiological conditions, they readily bind to negatively charged carbohydrate chains on the membrane surface and are effectively internalized into the cytoplasm via endocytosis or the direct translocation pathway [7–9]. According to these reports, positive charges are essential for ensuring that highly basic CPPs penetrate the cellular membrane. Cationic CPPs, on the other hand, perform unique functions by utilizing positive charge resources other than arginine, such as lysine or histidine (basic amino acids). This section will review lysine-rich and histidine-rich CPPs, including their design concepts.

4.1.1 Lysine-Rich CPPs

The vast majority of the lysine-rich CPPs are amphipathic cationic peptides (Table 4.1). Model amphipathic peptide (MAP) (KLALKLALKALKAALKLA) is a classical lysine-rich CPP that was designed as an amphiphilic α-helical antimicrobial peptide with hydrophobic and hydrophilic residues on opposite sides of the α-helix structure [10]. MAP transports a variety of compounds such as oligonucleotides [11], peptides [12], and small proteins (i.e. cytochrome c) [13] into mammalian cells, such as the human cervix adenocarcinoma cells (HeLa) and Chinese hamster ovary cells. Furthermore, when compared to other CPPs, MAP shows the fastest cellular uptake [14]. It has been reported that MAP(Aib), KL(Aib)LKL(Aib)LK(Aib) LKA(Aib)LKL(Aib), in which the alanine of the MAP sequence was replaced by

Cell-Penetrating Peptides: Design, Development and Applications, First Edition. Makoto Oba and Yosuke Demizu.

Table 4.1. Lysine-rich CPPs.

CPP name	Amino acid sequence	Source or design concept	Ip	Net charge (pH7.0)	Ref
MAP	KLALKLALKALKAALKLA	Analogue of synthetic amphiphilic α-helical antimicrobial peptide	10.61	4.9	10–14
MAP(Aib)	KL(Aib)LKL(Aib)LK(Aib)LKA(Aib)LKL(Aib) Aib, 2-aminoisobutyric acid	Analogue of MAP with increased membrane permeability and resistance to protease	10.61	4.9	15
Mastoparan	INLKALAALAKKIL	Natural peptide toxin from wasp venom	10.28	2.9	
Mitoparan	INLKKLAKL(Aib)KKIL	Analogue of mastoparan with increased positive charge and amphipathic helicity	10.61	4.9	16, 17
TK	AGYLLGKINLKKLAKL(Aib)KKIL-NH_2	Analogue of transportan 10 (TP10) with increased positive charge and amphipathicity	11.40	6.9	18
Cyt c^{77-101}	GTKMIFVGIKKKEERADLIAYLKKA	Partial peptide derived from human cytochrome c	9.88	3.9	20
sC18	GLRKRLRKFRNKIKEK	Highly cationic sequence derived from natural antimicrobial peptide cathelicidin and predicted as a CPP	12.02	7.9	21
K9	KKKKKKKKK	Effective CPP for plant cells	10.89	8.9	22
K18	KKKKKKKKKKKKKKKKKK	Effective CPP for plant cells	11.22	17.9	22
BP100	KKLFKKILKYL	Synthetic antimicrobial peptide with increased membrane interaction	10.30	4.9	23, 24
BP100$(KH)_9$	KKLFKKILKYLKHKHKHKHKHKHKHKHKH	Nucleic acid delivery carrier for plant cells	10.82	16.1	25–27
$(BP100)_2K_8$	KKLFKKILKYLKKLFKKILKYLKKKKKKKK	Anionic protein delivery carrier for plant cells	10.75	17.9	28

Isoelectronic points (Ip) and net charges of lysine-rich CPPs are calculated using the PROTEIN CALCULATOR v3.4 (http://protcalc.sourceforge.net/) and PepCalc.com – Peptide property calculator (https://pepcalc.com/).

2-aminoisobutyric acid (Aib) – a hydrophobic and non-coded protein functions as a helix stabilizer [15]. MAP(Aib) has enhanced helix stability and resistance to protease by concentrating Aib on one side of the helix structure, resulting in higher membrane permeability to A549 human lung carcinoma cells than the original MAP. MAP(Aib), an improved MAP, is expected to be a valuable tool for delivering hydrophilic macromolecules with resistance to protease and high cellular uptake.

Mitoparan (INLKKLAKL(Aib)KKIL) is an analog of mastoparan (INLKALAAL AKKIL), a natural peptide toxin from waspvenom. Mitoparan was designed as an artificial lysine-rich CPP with an increased positive charge and amphipathic helicity by replacing alanine in the original mastoparan with Aib and lysine [16]. Mitoparan has a high membrane permeability to U373MG human astrocytoma and ECV304 human bladder carcinoma cells due to its enhanced positive charge and amphipathic helicity [17]. Furthermore, mitoparan localizes in mitochondria after permeabilizing the cell membrane and inducing apoptosis via mitochondrial swelling. Thus, when combined with a conventional drug delivery system (DDS), mitoparan is expected to be a therapeutic candidate for selective apoptosis of target cells.

The TK peptide (AGYLLGKINLKKLAKL(Aib)KKIL-NH_2) is a lysine-rich CPP designed to increase the positive charge and amphipathic helicity by replacing alanine with Aib and lysine in the original CPP transportan 10 (AGYLLGKINL KALAALAKKIL) [18]. As TK peptide shows higher membrane permeability than the original transportan 10 in mammalian cells such as HeLa and NIH-3T3 mouse fibroblast-like cells, increasing charge and amphipathic helicity is an effective strategy for improving CPP translocation activity.

Cyt $c^{77\text{-}101}$ (GTKMIFVGIKKKEERADLIAYLKKA) is a lysine-rich CPP derived from human cytochrome c, and quantitative structure–activity relationship analysis revealed it to be a cryptic polycationic CPP motif [19]. In fact, Cyt $c^{77\text{-}101}$ is found to be a highly efficient CPP against U373MG cells, and it also plays the role of cytochrome c and apoptogenic potency in the cells [20].

sC18 (GLRKRLRKFRNKIKEK) is a highly cationic sequence corresponding to residues 106–121 of the 18 kDa cationic antimicrobial peptide cathelicidin (CAP18). Since sC18 is similar to TAT (48–60), a typical cationic CPP, it was predicted that sC18 would function as a CPP [21]. In fact, sC18 shows similar membrane permeability to TAT (48–60) in human cell lines such as HeLa, MCF-7 human breast cancer, and HEK-293 human embryonic kidney cells.

In contrast, some lysine-rich CPPs exhibit efficient cell wall and membrane permeability for plant cells. Polylysine has lower membrane permeability in mammalian cells than polyarginine [2]. However, polylysines such as K9 (KKKKKKKKK) and K18 (KKKKKKKKKKKKKKKKKK) have similar or higher membrane permeability than polyarginine, R9 (RRRRRRRRR) in plant cells [22].

BP100 (KKLFKKKKILKYL) is a synthetic peptide that was originally intended to be an antimicrobial peptide [23], but it is also a potent lysine-rich CPP for plant cells [24]. BP100(KH)$_9$ (KKLFKKILKYLKHKHKHKHKHKHKHKHKH) and (BP100)$_2$K$_8$ (KKLFK KILKYLKKLFKKILKYLKKKKHKHKH) are carriers designed to electrically bind to anionic molecules (i.e. nucleic acids and acidic proteins) by fusing BP100 with polycationic peptides [25–28].

4.1.2 Histidine-Rich CPPs

The main feature of histidine-rich CPPs is their pH-dependent selective protonation to obtain a positive charge because the side chain of histidine has a pKa value of 6.0. As a result, histidine-rich CPPs selectively increase their positive charge by protonation in a weakly acidic environment (pH < 6.0). By using this pH responsiveness, histidine-rich CPPs designed for tumor-targeting DDS or endosomal escape have been reported (Table 4.2).

The design of histidine-rich CPPs for tumor-targeting DDS takes advantage of the fact that the cancer microenvironment coincides with the weakly acidic pH region where histidine-rich CPPs have a positive charge. Solid tumor tissues develop angiogenesis that is unbalanced, and a low oxygen microenvironment known as hypoxia [29, 30]. Therefore, anerobic glycolytic systems are disorderly enhanced in the inner region of tumor tissue, and tumor cells produce acidic metabolites [31–33]. These pH selective strategies use histidine as a low pH-responsive trigger because histidine ($pK_a = 6.0$) has a positive charge in a low pH environment, but it has a neutral charge at physiological pH. Therefore, histidine-rich CPP is often used in developing strategies directed toward tumor pH targeting.

The TH peptide (AGYLLGHINLHHLAHL(Aib)HHIL-NH_2) is an excellent example of tumor pH-selective histidine-rich CPPs. TH peptide was designed to take positive charge in a weak acidity-dependent manner by replacing all the lysine in the original CPP TK peptide (AGYLLGKINLKKLAKL(Aib)KKIL-NH_2) with histidine [34]. Although TH peptide does not permeabilize the cell membrane at pH 7.4, it selectively permeabilizes and delivers antitumor agent camptothecin to tumor cells such as HeLa and MDA-MB-231 human breast cancer cells at pH 6.0. Furthermore, when compared to TK peptide, TH peptide exhibits weak cytotoxicity in vitro and in vivo, indicating high biocompatibility. The TH peptide, designed by replacing lysine with histidine, overcomes the problems associated with the original TK peptide.

The TH peptide does not permeabilize the cell membrane in the tumor microenvironment (pH 6.5–7.2), which is closer to the physiological environment, because the imidazole group of histidine is positively charged only at pH 6.0 or lower. A TH peptide analog was designed to improve the pH response of the TH peptide by introducing an electron-donating group (butyl group) into the C-2 position of histidine of the TH peptide [35]. The electron-donating group makes the N atom of the imidazole ring more alkaline and increases its ability to accept protons, thus increasing the pKa value. Even at pH 6.5, where TH peptides are not membrane-permeable, TH analogs containing electron-donating group (butyl-TH peptide) could permeabilize the membrane and deliver camptothecin to HeLa cells. Thus, by using histidine with the electron-donating group, the pH responsiveness of CPP can be improved more sensitively.

Table 4.2. Histidine-rich CPPs.

CPP name	Amino acid sequence	Design concept	Ip	Net charge (pH7.0)	Net charge (pH6.0)	Net charge (pH5.0)	Ref
TH	AGYLLGHINLHHLAHL(Aib)HHIL-NH_2 Aib, 2-aminoisobutyric acid	Acidic pH-dependent cellular uptake	9.91	1.3	4.5	5.8	34
LH	LHHLLHHLHHLLHH-NH_2	Acidic pH-dependent cellular uptake	13.80	2.8	7.1	8.8	36
H16	HHHHHHHHHHHHHHHH	Acidic pH-dependent cellular uptake	7.98	3.8	12.1	15.5	37–41
H20	HHHHHHHHHHHHHHHHHHHH	Acidic pH-dependent cellular uptake	8.08	4.7	15.2	19.4	37–41
HL6	CHHHHHRRWQWRHHHHHC	Endosomal escape by proton sponge effect	10.00	5.3	10.6	12.7	43
REDV-TAT-NLS-H12	REDVYGRKKRRQRRRPKKKRKVHHHHHHHHHHHH	Endosomal escape by proton sponge effect	12.02	14.8	21.2	24.0	44
NickFect70	Arachidyl-HHHHYHHGO*ILLKALKALAKAIL *Synthesis continued from the sidechain amino group instead of α-amino group	Endosomal escape by proton sponge effect	10.00	4.3	7.5	8.8	46
LAH4	KKALLALALHHLAHLALHLALALKKA-NH_2	Acidic pH-dependent conformational change and endosomal lysis	14.00	5.9	6.0	8.9	47–54
LAH4-L1	KKALLAHALHLLALLALHLAHALKKA-NH_2	Acidic pH-dependent conformational change and endosomal lysis	14.00	5.9	6.0	8.9	55–58

Isoelectronic points (Ip) and net charges of histidine-rich CPPs are calculated using the PROTEIN CALCULATOR v3.4 (http://protcalc.sourceforge.net/) and PepCalc.com – Peptide property calculator (https://pepcalc.com/).

LH peptide (LHHLLHHLHHLLHH-NH_2) is also a pH-activated histidine-rich CPP designed for selective membrane permeation in weakly acidic environments. LH peptide was designed to be positively charged in a weakly acidic environment by replacing all lysine on the hydrophilic side and two leucine on the hydrophobic side of the original CPP, LK peptide (LKKLLKLLKKLLKL-NH_2), with histidine [36]. In fact, LH peptide is activated and penetrates the cells at pH 6.0, whereas its membrane transduction activity could be shielded at pH 7.4. Furthermore, LH peptide can selectively transport camptothecin into tumor cells such as HeLa and MDA-MB-231 cells at pH 6.0 and exhibit pH-dependent antitumor activity. LH peptide is also a promising carrier for delivering various cargos into tumor cells in an acidic microenvironment with pH-selective activity.

On the contrary, positive charge-independent histidine-rich CPPs have also been reported. Polyhistidine peptides with more than 12 residues, all composed of histidine, have comparable cell permeability at pH 7.4, neutral, and pH 6.0, which is positively charged. Initially, polyhistidine peptides were designed and developed to be pH-responsive CPPs, but they show pH-independent, i.e. positive charge-independent, cellular uptake [37]. However, typical polyhistidine peptide (H16 peptide, HHHHHHHHHHHHHHHH-NH_2) shows 14.35- and 10.98-fold higher cell permeability than typical arginine-rich CPP (R8 peptide, RRRRRRRR-NH_2) against HT1080 human fibrosarcoma and RERF-LC human squamous lung cancer cells, respectively. The typical polyhistidine peptide (H16 peptide) also acts as an effective carrier for various cargos, such as proteins, liposomes, and organelles, into cells [38–40]. Polyhistidine peptides are coincidental but very effective CPPs. In addition, polyhistidine peptides are also effective CPPs as direct protein delivery carriers for plant cells [41].

On the other hand, histidine-rich CPPs trigger endosomal escape in transporting molecules into the cell. As histidine is protonated in a weakly acidic environment below pH 6.0, histidine-rich CPPs exert a proton sponge effect in the acidic environment of endosomes and cause endosomal membrane disruption by conformational change. In this way, histidine-rich CPPs are known to induce endosomal escape of transport molecules and release them into the cytoplasm. Histidine-rich CPPs that exert proton sponge effects include conventional CPPs fused with polyhistidine sequences.

HL6 peptide (CHHHHHRRWQWRHHHHHC) is a histidine-rich CPP designed by fusing additional pentahistidine and monocysteine residues on both sides of L6 peptide (RRWQWR), original CPP derived from bovine lactoferricin [42], to enhance the endosomal escape potency [43]. HL6 peptide forms a complex with quantum dots and is incorporated into HeLa and A549 cells more efficiently than the original L6 peptide by direct membrane translocation rather than endocytosis. Initially, pentahistidine residue fusion was used to induce endosomal escape; however, it induces efficient membrane permeabilization by direct membrane translocation.

REDV-TAT-NLS-H12 (REDVYGRKKRRQRRRPKKKRKVHHHHHHHHHHHH) is a histidine-rich CPP containing a REDV peptide that recognizes $\alpha 4\beta 1$ integrin, a nuclear localization signal (NLS), and a polyhistidine sequence (H12) added to TAT (47–57), a typical arginine-rich CPP. REDV-TAT-NLS-H12 was designed to be an efficient nucleic acid delivery carrier with endosomal escape against α4β1 integrin-expressing cells [44]. REDV-TAT-NLS-H12 achieves efficient gene delivery and gene transfection against human umbilical vein endothelial cells that highly express α4β1 integrin.

NickFect70 (Arachidyl-HHHHYHHGOILLKALKALAKAIL) is an improved CPP based on the motif of the original amphipathic CPP, NickFect55 (AGYLLGO INLKALAALAKAIL) [45]. NickFect70 was designed to be a highly functional nucleic acid transport carrier by introducing additional histidine residues to NickFect55 to make it pH-sensitive and exert a proton sponge effect in the endosomal environment [46]. Compared to the original NickFect55, NickFect70 shows the enhanced siRNA transfection and target gene knockdown both in vitro (U87 MG-luc2 human glioma cells) and in vivo (Balb/c mouse model).

Another well-known histidine-rich CPP is LAH4 (KKALLALHHLAHLALHLAL ALKKA-NH_2), a peptide that exhibits membrane permeabilization in physiological pH but induces membrane lysis in low pH environments in endosomes. LAH4 is a synthetic cationic amphipathic histidine-rich peptide originally designed as a tool to study membrane protein dynamics [47], but it undergoes an acidic pH-dependent conformational change and is oriented parallel to the membrane surface resulting in membrane lysis [48, 49]. Therefore, LAH4, as a histidine-rich CPP, is an intracellular carrier of various hydrophilic macromolecules with cell penetration and endosomal escape capabilities. In fact, LAH4 transfers a large range of various cargo, including nucleic acids [50, 51], proteins [52], quantum dots [53], or adeno-associated viruses [54], into different human cells. Furthermore, an improved LAH4, LAH4-L1 (KKALLAHALHLLALLALHLAHALKKA-NH_2), is also an effective carrier for nucleic acids with endosomal escape [55–58]. Therefore, fusing poly-histidine sequences or increasing histidine content in conventional CPPs effectively develops a delivery carrier with endosomal escape.

References

1 Bechara, C. and Sagan, S. (2013). Cell-penetrating peptides: 20 years later, where do we stand? *FEBS Letters* 587: 1693–1702. https://doi.org/10.1016/J.FEBSLET.2013.04.031.

2 Heitz, F., Morris, M.C., and Divita, G. (2009). Twenty years of cell-penetrating peptides: from molecular mechanisms to therapeutics. *British Journal of Pharmacology* 157: 195–206. https://doi.org/10.1111/J.1476-5381.2009.00057.X.

3 Koren, E. and Torchilin, V.P. (2012). Cell-penetrating peptides: breaking through to the other side. *Trends in Molecular Medicine* 18: 385–393. https://doi.org/10.1016/J.MOLMED.2012.04.012.

4 Derossit, D., Joliott, M.H., Chassaingl, G., and Prochiantztn, M. (1994). The third helix of the Antennapedia homeodomain translocates through biological membranes. *The Journal of Biological Chemistry* 269: 10444–10450. https://doi.org/10.1016/S0021-9258(17)34080-2.

5 Vivès, E., Brodin, P., and Lebleu, B. (1997). A truncated HIV-1 Tat protein basic domain rapidly translocates through the plasma membrane and accumulates in the cell nucleus. *The Journal of Biological Chemistry* 272: 16010–16017. https://doi.org/10.1074/JBC.272.25.16010.

6 Futaki, S., Suzuki, T., Ohashi, W. et al. (2001). Arginine-rich peptides. An abundant source of membrane-permeable peptides having potential as carriers for intracellular protein delivery. *The Journal of Biological Chemistry* 276: 5836–5840. https://doi.org/10.1074/JBC.M007540200.

7 Melikov, K. and Chernomordik, L.V. (2005). Arginine-rich cell penetrating peptides: from endosomal uptake to nuclear delivery. *Cellular and Molecular Life Sciences* 62: 2739–2749. https://doi.org/10.1007/S00018-005-5293-Y.

8 Futaki, S., Nakase, I., Tadokoro, A. et al. (2007). Arginine-rich peptides and their internalization mechanisms. *Biochemical Society Transactions* 35: 784–787. https://doi.org/10.1042/BST0350784.

9 Poon, G.M.K. and Gariépy, J. (2007). Cell-surface proteoglycans as molecular portals for cationic peptide and polymer entry into cells. *Biochemical Society Transactions* 35: 788–793. https://doi.org/10.1042/BST0350788.

10 Oehlke, J., Scheller, A., Wiesner, B. et al. (1998). Cellular uptake of an α-helical amphipathic model peptide with the potential to deliver polar compounds into the cell interior non-endocytically. *Biochimica et Biophysica Acta (BBA) - Biomembranes* 1414: 127–139. https://doi.org/10.1016/S0005-2736(98)00161-8.

11 Oehlke, J., Wallukat, G., Wolf, Y. et al. (2004). Enhancement of intracellular concentration and biological activity of PNA after conjugation with a cell-penetrating synthetic model peptide. *European Journal of Biochemistry* 271: 3043–3049. https://doi.org/10.1111/J.1432-1033.2004.04236.X.

12 Scheller, A., Melzig, M., and Oehlke, J. (2001). Induction of caspase-8 in human cells by the extracellular administration of peptides containing a C-terminal SLV sequence. *Letters in Peptide Science* 8: 29–34. https://doi.org/10.1023/A:1014218104739.

13 Kenien, R., Shen, W.C., and Zaro, J.L. (2012). Vesicle-to-cytosol transport of disulfide-linked cargo mediated by an amphipathic cell-penetrating peptide. *Journal of Drug Targeting* 20: 793–800. https://doi.org/10.3109/1061186X.2012.719899.

14 Hällbrink, M., Florén, A., Elmquist, A. et al. (2001). Cargo delivery kinetics of cell-penetrating peptides. *Biochimica et Biophysica Acta (BBA) - Biomembranes* 1515: 101–109. https://doi.org/10.1016/S0005-2736(01)00398-4.

15 Wada, S.I., Tsuda, H., Okada, T. et al. (2011). Cellular uptake of Aib-containing amphipathic helix peptide. *Bioorganic & Medicinal Chemistry Letters* 21: 5688–5691. https://doi.org/10.1016/J.BMCL.2011.08.030.

16 Jones, S. and Howl, J. (2004). Charge delocalisation and the design of novel mastoparan analogues: enhanced cytotoxicity and secretory efficacy of [Lys5, Lys8, Aib10]MP. *Regulatory Peptides* 121: 121–128. https://doi.org/10.1016/J.REGPEP.2004.04.015.

17 Jones, S., Martel, C., Belzacq-Casagrande et al. (2008). Mitoparan and target-selective chimeric analogues: membrane translocation and intracellular redistribution induces mitochondrial apoptosis. *Biochimica et Biophysica Acta* 1783: 849–863. https://doi.org/10.1016/J.BBAMCR.2008.01.009.

18 Song, J., Kai, M., Zhang, W. et al. (2011). Cellular uptake of transportan 10 and its analogs in live cells: selectivity and structure-activity relationship studies. *Peptides* 32: 1934–1941. https://doi.org/10.1016/J.PEPTIDES.2011.07.018.

19 Howl, J. and Jones, S. (2008). Proteomimetic cell penetrating peptides. *International Journal of Peptide Research and Therapeutics* 14: 359–366. https://doi.org/10.1007/S10989-008-9135-2.

20 Jones, S., Holm, T., Mäger, I. et al. (2010). Characterization of bioactive cell penetrating peptides from human cytochrome c: protein mimicry and the development of a novel apoptogenic agent. *Chemistry & Biology* 17: 735–744. https://doi.org/10.1016/J.CHEMBIOL.2010.05.018.

21 Neundorf, I., Rennert, R., Hoyer, J. et al. (2009). Fusion of a short HA2-derived peptide sequence to cell-penetrating peptides improves cytosolic uptake, but enhances cytotoxic activity. *Pharmaceuticals* 2: 49–65. https://doi.org/10.3390/PH2020049.

22 Numata, K., Horii, Y., Oikawa, K. et al. (2018). Library screening of cell-penetrating peptide for BY-2 cells, leaves of Arabidopsis, tobacco, tomato, poplar, and rice callus. *Scientific Reports* 8: 1–17. https://doi.org/10.1038/s41598-018-29298-6.

23 Badosa, E., Ferre, R., Planas, M. et al. (2007). A library of linear undecapeptides with bactericidal activity against phytopathogenic bacteria. *Peptides* 28: 2276–2285. https://doi.org/10.1016/J.PEPTIDES.2007.09.010.

24 Eggenberger, K., Mink, C., Wadhwani, P. et al. (2011). Using the peptide Bp100 as a cell-penetrating tool for the chemical engineering of actin filaments within living plant cells. *Chembiochem* 12: 132–137. https://doi.org/10.1002/CBIC.201000402.

25 Lakshmanan, M., Kodama, Y., Yoshizumi, T. et al. (2013). Rapid and efficient gene delivery into plant cells using designed peptide carriers. *Biomacromolecules* 14: 10–16. https://doi.org/10.1021/bm301275g.

26 Numata, K., Ohtani, M., Yoshizumi, T. et al. (2014). Local gene silencing in plants via synthetic dsRNA and carrier peptide. *Plant Biotechnology Journal* 12: 1027–1034. https://doi.org/10.1111/PBI.12208.

27 Lakshmanan, M., Yoshizumi, T., Sudesh, K. et al. (2015). Double-stranded DNA introduction into intact plants using peptide–DNA complexes. *Plant Biotechnology* 32: 39–45. https://doi.org/10.5511/PLANTBIOTECHNOLOGY.14.1210B.

28 Ng, K.K., Motoda, Y., Watanabe, S. et al. (2016). Intracellular delivery of proteins via fusion peptides in intact plants. *PLoS One* 11: e0154081. https://doi.org/10.1371/JOURNAL.PONE.0154081.

29 Tannock, I.F. and Rotin, D. (1989). Acid pH in tumors and its potential for therapeutic exploitation. *Cancer Research* 49: 4373–4384.

30 Pan, J.G. and Mak, T.W. (2007). Metabolic targeting as an anticancer strategy: dawn of a new era? *Science's STKE* 381: pe14. https://doi.org/10.1126/STKE.3812007PE14.

31 Jähde, E., Rajewsky, M.F., and Baumgärtl, H. (1982). pH distributions in transplanted neural tumors and normal tissues of BDIX rats as measured with pH microelectrodes. *Cancer Research* 42: 1498–1504.

32 Engin, K., Leeper, D.B., Cater, J.R. et al. (1995). Extracellular pH distribution in human tumours. *International Journal of Hyperthermia* 11: 211–216. https://doi.org/10.3109/02656739509022457.

33 McCoy, C.L., Parkins, C.S., Chaplin, D.J. et al. (1995). The effect of blood flow modification on intra- and extracellular pH measured by 31P magnetic resonance spectroscopy in murine tumours. *British Journal of Cancer* 72: 905–911. https://doi.org/10.1038/BJC.1995.431.

34 Zhang, W., Song, J., Zhang, B. et al. (2011). Design of acid-activated cell penetrating peptide for delivery of active molecules into cancer cells. *Bioconjugate Chemistry* 22: 1410–1415. https://doi.org/10.1021/BC200138D.

35 Yao, J., Ma, Y., Zhang, W. et al. (2017). Design of new acid-activated cell-penetrating peptides for tumor drug delivery. *PeerJ* 5: e3429. https://doi.org/10.7717/PEERJ.3429/SUPP-9.

36 Zhang, Y., Li, L., Chang, L. et al. (2019). Design of a new pH-activatable cell-penetrating peptide for drug delivery into tumor cells. *Chemical Biology & Drug Design* 94: 1884–1893. https://doi.org/10.1111/CBDD.13537.

37 Iwasaki, T., Tokuda, Y., Kotake, A. et al. (2015). Cellular uptake and in vivo distribution of polyhistidine peptides. *Journal of Controlled Release* 210: 115–124. https://doi.org/10.1016/J.JCONREL.2015.05.268.

38 Hayashi, T., Shinagawa, M., Kawano, T. et al. (2018). Drug delivery using polyhistidine peptide-modified liposomes that target endogenous lysosome. *Biochemical and Biophysical Research Communications* 501: 648–653. https://doi.org/10.1016/J.BBRC.2018.05.037.

39 Hayashi, T., Okamoto, R., Kawano, T. et al. (2019). Development of organelle replacement therapy using a stearyl-polyhistidine peptide against lysosomal storage disease cells. *Molecules* 24: 2995. https://doi.org/10.3390/MOLECULES24162995.

40 Iwasaki, T., Murakami, N., and Kawano, T. (2020). A polylysine-polyhistidine fusion peptide for lysosome-targeted protein delivery. *Biochemical and Biophysical Research Communications* 533: 905–912. https://doi.org/10.1016/J.BBRC.2020.09.087.

41 Tanaka, Y., Nanasato, Y., Omura, K. et al. (2021). Direct protein delivery into intact plant cells using polyhistidine peptides. *Bioscience, Biotechnology, and Biochemistry* 85: 1405–1414. https://doi.org/10.1093/BBB/ZBAB055.

42 Liu, B.R., Huang, Y.W., Aronstam, R.S. et al. (2016). Identification of a short cell-penetrating peptide from bovine lactoferricin for intracellular delivery of DNA in human A549 cells. *PLoS One* 11: e0150439. https://doi.org/10.1371/journal.pone.0150439.

43 Lee, H.J., Huang, Y.W., Chiou, S.H. et al. (2019). Polyhistidine facilitates direct membrane translocation of cell-penetrating peptides into cells. *Scientific Reports* 9: 1–11. https://doi.org/10.1038/s41598-019-45830-8.

44 Li, Q., Hao, X., Zaidi, S.S.A. et al. (2018). Oligohistidine and targeting peptide functionalized TAT-NLS for enhancing cellular uptake and promoting angiogenesis in vivo. *Journal of Nanobiotechnology* 16: 29. https://doi.org/10.1186/S12951-018-0358-X.

45 Freimann, K., Arukuusk, P., Kurrikoff, K. et al. (2016). Optimization of in vivo DNA delivery with NickFect peptide vectors. *Journal of Controlled Release* 241: 135–143. https://doi.org/10.1016/J.JCONREL.2016.09.022.

46 Porosk, L., Arukuusk, P., Põhako, K. et al. (2019). Enhancement of siRNA transfection by the optimization of fatty acid length and histidine content in the CPP. *Biomaterials Science* 7: 4363–4374. https://doi.org/10.1039/C9BM00688E.

47 Bechinger, B. (1996). Towards membrane protein design: pH-sensitive topology of histidine-containing polypeptides. *Journal of Molecular Biology* 263: 768–775. https://doi.org/10.1006/JMBI.1996.0614.

48 Marquette, A., Mason, A.J., and Bechinger, B. (2008). Aggregation and membrane permeabilizing properties of designed histidine-containing cationic linear peptide antibiotics. *Journal of Peptide Science* 14: 488–495. https://doi.org/10.1002/PSC.966.

49 Perrone, B., Miles, A.J., Salnikov, E.S. et al. (2014). Lipid interactions of LAH4, a peptide with antimicrobial and nucleic acid transfection activities. *European Biophysics Journal* 43: 499–507. https://doi.org/10.1007/S00249-014-0980-Y.

50 Mason, A.J., Martinez, A., Glaubitz, C. et al. (2006). The antibiotic and DNA-transfecting peptide LAH4 selectively associates with, and disorders, anionic lipids in mixed membranes. *FASEB Journal* 20: 320–322. https://doi.org/10.1096/FJ.05-4293FJE.

51 Langlet-Bertin, B., Leborgne, C., Scherman, D. et al. (2010). Design and evaluation of histidine-rich amphipathic peptides for siRNA delivery. *Pharmaceutical Research* 27: 1426–1436. https://doi.org/10.1007/S11095-010-0138-2.

52 Zhang, T.T., Kang, T.H., Ma, B. et al. (2012). LAH4 enhances CD8+ T cell immunity of protein/peptide-based vaccines. *Vaccine* 30: 784–793. https://doi.org/10.1016/J.VACCINE.2011.11.056.

53 Gemmill, K.B., Muttenthaler, M., Delehanty, J.B. et al. (2013). Evaluation of diverse peptidyl motifs for cellular delivery of semiconductor quantum dots. *Analytical and Bioanalytical Chemistry* 405: 6145–6154. https://doi.org/10.1007/S00216-013-6982-2.

54 Liu, Y., Kim, Y.J., Ji, M. et al. (2014). Enhancing gene delivery of adeno-associated viruses by cell-permeable peptides. *Molecular therapy. Methods & Clinical Development* 1: 12. https://doi.org/10.1038/MTM.2013.12.

55 Xu, Y., Liang, W., Qiu, Y. et al. (2016). Incorporation of a nuclear localization signal in pH responsive LAH4-L1 peptide enhances transfection and nuclear uptake of plasmid DNA. *Molecular Pharmaceutics* 13: 3141–3152. https://doi.org/10.1021/ACS.MOLPHARMACEUT.6B00338.

56 Liu, N., Bechinger, B., and Süss, R. (2017). The histidine-rich peptide LAH4-L1 strongly promotes PAMAM-mediated transfection at low nitrogen to phosphorus ratios in the presence of serum. *Scientific Reports* 7: 1–12. https://doi.org/10.1038/s41598-017-10049-y.

57 Wolf, J., Aisenbrey, C., and Harmouche, N. (2017). pH-dependent membrane interactions of the histidine-rich cell-penetrating peptide LAH4-L1. *Biophysical Journal* 113: 1290–1300. https://doi.org/10.1016/J.BPJ.2017.06.053.

58 Liu, J., Guo, N., and Gao, C. (2019). Effective gene silencing mediated by polypeptide nanoparticles LAH4-L1-siMDR1 in multi-drug resistant human breast cancer. *Journal of Biomedical Nanotechnology* 15: 531–543. https://doi.org/10.1166/JBN.2019.2705.

5

Amphipathic Peptide

Makoto Oba

Kyoto Prefectural University of Medicine, Graduate School of Medical Science, Kyoto, Japan

5.1 Introduction

Cell-penetrating peptides (CPPs) can be categorized into several classes based on their physicochemical properties, and amphipathic peptides are one of the major classes (Table 5.1) [1–4]. "Amphipathic" means that peptides have hydrophilic and hydrophobic moieties. Hydrophilic moieties are generally composed of cationic amino acid residues, such as Lys, Arg, and His, which can associate with the anionic surface of cell membranes or anionic cargoes such as nucleic acids and proteins. This cationic property contributes to their cell-penetrating ability and the efficient intracellular delivery of anionic biomacromolecules. A few amphipathic peptides consisting of anionic and hydrophobic amino acids have been also reported. Most amphipathic CPPs adopt a specific secondary structure, especially a helical structure. In that case, one face displays hydrophilic amino acid residues, whereas the opposite face displays hydrophobic amino acid residues. First, the sequences of amphipathic CPPs were found in natural proteins, and numerous amphipathic CPPs have recently been artificially designed. In this chapter, amphipathic CPPs are shown according to their peptide secondary structures, including α-helix, β-sheet, polyproline type II helix, and so on.

5.2 α-Helical Peptides

Penetratin (RQIKIWFQNRRMKWKK) is one of the earliest amphipathic CPPs and consists of 16 amino acids [5]. Originally, the 60-amino acid polypeptide corresponding to the sequence of the Drosophila Antennapedia homeodomain was reported to cross the cell membrane by an energy-independent mechanism [6]. Based on this finding, Prochiantz et al. also studied the internalization mechanism of the Antennapedia homeodomain and revealed that a 16-amino acid peptide,

Cell-Penetrating Peptides: Design, Development and Applications, First Edition. Makoto Oba and Yosuke Demizu.

Table 5.1 Lists of amphipathic CPPs with information on their sequences and secondary structures.

CPP	Sequence	Secondary structure
ARF (1–22)	MVRRFLVTLRIRRACGPPRVRV	None
ARF (19–31)	RVRVFVVHIPRLT	None
Bac_{15-24}	RRIRPRPPRLPRPRPRPLPFPRPG	PPII helix & α-helix
bPrPp (1–30)	MVKSKIGSWILVLFVAMWSDVGLCKKRPKP	α-helix
CADY	GLWRALWRLLRSLWRLLWRA-cysteamide	α-helix
GALA	WEAALAEALAEALAEHLAEALAEALEALAA	α-helix
KALA	WEAKLAKALAKALAKHLAKALAKALKACEA	α-helix
L17E	IWLTALKFLGKHAAKHEAKQQLSKL	α-helix
LAH4	KKALLALHHLAHLALHLALALKKA	α-helix
M918	MVTVLFRRLRIRRACGPPRVRV	None
MAP	KLALKLALKALKAALKLA	α-helix
MPG	GALFLGFLGAAGSTMGAWSQPKKKRKV	α-helix & β-sheet
p18	LSTAADMQGVVTDGMASG	α-helix & β-sheet
Penetratin	RQIKIWFQNRRMKWKK	α-helix
Pep-1	KETWWETWWTEWSQPKKKRKV	None
ppTG1	GLFKALLKLLKSLWKLLLKA	α-helix
ppTG20	GLFRALLRLLRSLWRLLLRA	α-helix
pVEC	LLIILRRRIRKQAHAHSK	None
R_6/W_3	RRWWRRWRR	α-helix
SAP	$(VRLPPP)_3$	PPII helix
SAP(E)	$(VELPPP)_3$	PPII helix
TP10	AGYLLGKINLKALAALAKKIL	α-helix
Transportan	GWTLNSAGYLLGKINLKALAALAKKIL	α-helix
VT5	DPKGDPKGVTVTVTVTVTGKGDPKPD	β-sheet

penetratin, corresponding to the third helix of the homeodomain, and lacking its N-terminal Glu translocated through the cell membrane [5]. The dominant secondary structure of penetratin in water is an α-helix, which is strongly enhanced in 30% 1,1,1,3,3,3-hexafluore-2-propanol. When penetratin adopts its α-helical structure, cationic residues such as Arg and Lys roughly congregate on one face of the helical bundle (Figure 5.1). There have been many studies on penetratin, from its internalization mechanism to the delivery of therapeutic proteins and nucleic acids [7, 8].

MAP (KLALKLALKALKAALKLA), a model amphipathic peptide, is composed of cationic Lys and hydrophobic Leu and Ala residues [9–11]. As shown in Figure 5.1, when MAP forms an α-helical structure, the hydrophilic (cationic) and hydrophobic regions are completely separated. Indeed, MAP forms an amphipathic α-helical structure in the environment of the cell membrane, although it does not adopt a

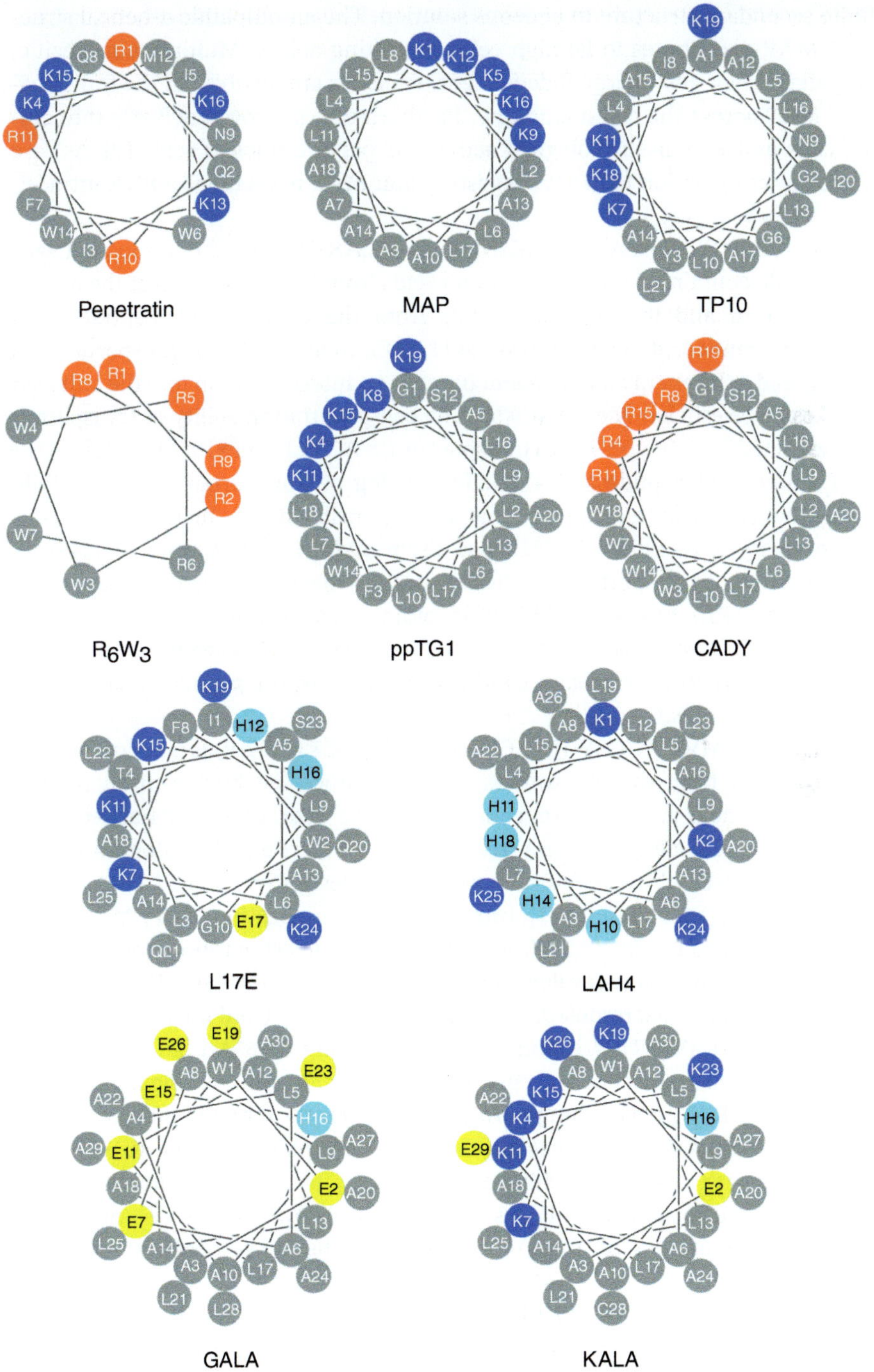

Figure 5.1 Helical wheel diagrams of amphipathic α-helical CPPs. Helical screw sense of α-helical peptides composed of L-α-amino acids is right-handed. Blue: cationic Lys (K) residues; red: cationic Arg (R) residues; light blue: cationic His (H) residues; yellow: anionic Glu (E) residues.

definite secondary structure in aqueous solution. The amphipathic α-helical structure of MAP contributes to its high cell-penetrating ability. Multiple nonspecific, energy-dependent and energy-independent processes are involved in the translocation of MAP across the cell membrane. Furthermore, improved delivery into cultured cells and enhanced biological activity of peptide nucleic acids (PNAs) are achieved by conjugation to MAP, suggesting that MAP is useful as a PNA intracellular delivery tool [12].

Transportan (GWTLNSAGYLLGKINLKALAALAKKIL) is a 27-amino acid synthetic peptide containing the first 12 amino acids from the N-terminus of the neuropeptide galanin and the 14 amino acids from the wasp venom peptide toxin mastoparan, connected via a Lys residue [13]. Cellular uptake of transportan was first proposed to be energy independent because its internalization into the cultured cells takes place efficiently even at 0 °C. The cell-penetrating ability of transportan is not restricted by cell type. TP10 (transportan 10; AGYLLGKINLKALAALAKKIL) is a 21-amino acid peptide and a deletion analog of transportan [14, 15]. While defining the regions of the sequence that are responsible for the membrane translocation property of transportan, TP10 was discovered. In TP10, 6 amino acids from the N-terminus of transportan are missing and the mastoparan-derived part is left intact. Thus, TP10 has an amphipathic moiety mastoparan part (Figure 5.1). TP10 has no effect on the activity of GTPase even at very high concentrations. These features make TP10 a promising candidate as a transport peptide with a lower potential for side effects.

bPrPp (1–30) (MVKSKIGSWILVLFVAMWSDVGLCKKRPKP) is a 30-amino acid peptide derived from the N-terminus of the unprocessed bovine prion protein (bPrPp) [16]. There are two parts consisting of the hydrophobic signal sequence (residues 1–24) and a basic domain (residues 25–30). bPrPp (1–30) adopts a largely α-helical structure, which is found between residues 8 (Ser) and 21 (Ile), in cell membrane mimetic condition [17]. Intracellular internalization of bPrPp (1–30) is mainly through macropinocytosis initiated by binding to cell-surface proteoglycans. bPrPp (1–30) can form non-covalent complexes with plasmid DNA (pDNA) and can transport an anionic macromolecule pDNA into the cytosol and nucleus.

R_6W_3 (RRWWRRRWRR) is a nonapeptide consisting of 6 Arg and 3 Trp residues [18]. R_6W_3 has partial helical content in an aqueous solution (Figure 5.1) [19]. Trp residues in R_6W_3 have a unique role in the interaction and structuring of the peptide with glycosaminoglycans, in relation to cell internalization. R_6W_3 enters cultured cells with a kinetic reaching a steady state after 30–60 minutes of incubation. This plateau is stable for 4 hours and decreased slowly afterward. Analysis of the degradation process of R_6W_3 suggested that the R_6W_3 does not traffic to lysosomes.

ppTG1 (GLFKALLKLLKSLWKLLLKA) and ppTG20 (GLFRALLRLLRSLWRLLLRA) are 20-amino acid peptides designed by replacing all Glu residues of JTS-1 (GLFEALLELLESLWELLLEA), which is an acidic peptide with membrane-destabilizing activity, by Lys or Arg residues [20]. The predominant conformations of ppTG1 and ppTG20 in an aqueous solution are α-helical structures, in which all cationic amino acids (Lys or Arg) segregate on the same side (Figure 5.1). ppTG1 and ppTG20 bind to the nucleic acid by electrostatic interaction and destabilize phospholipid liposomes at pH 5 and pH 7. Complexes of ppTG1 with pDNA showed efficient transfection in a variety of cultured cells. Moreover, intravenous injection of complexes of ppTG1 or ppTG20 with pDNA into mice achieved significant gene expression in the lung 24 hours after injection. The high gene transfer activity of these peptides is correlated with their propensity to exist in the α-helical structure.

Divita et al. designed a 20-amino acid peptide, CADY (GLWRALWRLLRSLWR LLWRA-cysteamide), based on the ppTG1 sequence as a template [21]. In order to improve both interaction of the peptide with small interfering RNA (siRNA) and its ability to interact with the cell membrane, Arg and Trp residues are included in the sequence. When CADY adopts an α-helical structure, it exposes cationic Arg residues on one side and hydrophobic Trp residues on the other (Figure 5.1). Binding to siRNA or phospholipids triggers a conformational change of CADY from an unfolded state to an α-helical structure, thereby stabilizing CADY/siRNA complexes and improving their interactions with cell membranes [22]. CADY-mediated delivery of subnanomolar concentrations of siRNA led to significant knockdown against a variety of cell lines without cytotoxicity. CADY/siRNA complexes enter cells rapidly and independently of the major endosomal pathways.

L17E (IWLTALKFLGKHAAKHEAKQQLSKL) is a 25-amino acid amphipathic α-helical peptide (Figure 5.1) [23]. Futaki et al. designed it to develop an endosomolytic peptide on the basis of M-lycotoxin, a membrane-lytic spider venom peptide. Residue 17 (Leu) in M-lycotoxin was substituted with a Glu residue. The insertion of negatively charged Glu residue into a hydrophobic face effectively attenuates the lytic activity, resulting in selectively perturbating the endosomal membrane. L17E promotes uptake by inducing macropinocytosis and efficiently delivers biomacromolecules, including antibodies, into the cytosol. The membrane-lytic activity of L17E is induced by the electrostatic interaction of the positively charged peptide with the negatively charged endosomal membranes. The cationic feature of L17E facilitates the cytosolic release of endocytosed biomacromolecules in terms of physicochemical peptide-membrane interactions.

LAH4 (KKALLALALHHLAHLALHLALALKKA) is an amphipathic α-helical peptide designed specifically to investigate the interactions that determine the membrane topology of helical peptides (Figure 5.1) [24]. LAH4 contains 4 His residues with pKa values between 5.4 and 6.0, and consequently, its charge and membrane topology in lipid bilayers are strongly pH-dependent [25]. LAH4 exhibits cell-penetration, antimicrobial, and nucleic acid transfection activities. The presence of 2 Lys residues at each end of the peptide allows for electrostatic interaction with DNA. In addition, 4 His residues have a pKa of about 6.0, the buffering, and subsequent disruption of endosomal vesicles is very important for nonviral gene vectors. The transfection efficiency of LAH4 is comparable to those of commercially available transfection agents [26].

GALA (WEAALAEALAEALAEHLAEALAEALEALAA) is a 30-amino acid peptide with a repeat sequence of Glu-Ala-Leu-Ala, and is designed as a water-soluble peptide that interacts with liposomes specifically at low pH [27]. The helical content of GALA increases when the pH is reduced from 7.6 to 5.0, and GALA forms an amphipathic structure in the helical state (Figure 5.1). As a result, GALA has a pH-sensitive fusogenic activity and facilitates endosomal escape of therapeutic drugs. Indeed, GALA has been applied for promoting the cytosol release of liposomal contents by introducing GALA into liposomal membranes [28]. Furthermore, transfection efficiency of pDNA using cationic liposomes is improved by GALA [29]. The efficient cytosolic delivery of cargo molecules is achieved by GALA.

KALA (WEAKLAKALAKALAKHLAKALAKALKACEA) is an amphipathic 30-amino acid peptide when in its α-helical structure (Figure 5.1) [30]. Szoka et al. designed KALA to combine the positive charge required for DNA binding and the amphipathic membrane-destabilizing characteristics of the GALA peptide. KALA undergoes a pH-dependent change from a random coil to an α-helical conformation as the pH increases from 5.0 to 7.5. One face displays hydrophobic Leu residues in the α-helical structure, whereas the opposite face displays hydrophilic (cationic) Lys residues. KALA binds to oligonucleotides and pDNA by electrostatic interaction, causes membrane disruption, and mediates transfection of a variety of cell lines.

5.3 β-Sheet Peptides

VT5 (DPKGDPKGVTVTVTVTVTGKGDPKPD) is a water-soluble 26-amino acid peptide with five Val-Thr repeats in the middle of sequence [31]. VT5 is unstructured in an aqueous solution at a peptide concentration <1 μM but adopts a predominantly amphipathic β-sheet structure in the presence of 15 mM sodium dodecyl sulfate (cell membrane model). Extensive internalization into cultured cells was found for VT5 compared with VT5 analogs that did not adopt a β-sheet structure [32]. The cellular uptake of VT5 is energy, temperature, and pH-dependent, which is clearly distinct from that reported for helical amphipathic peptides.

MPG (GALFLGFLGAAGSTMGAWSQPKKKRKV) is an artificially designed amphipathic CPP [33, 34]. MPG is composed of two independent domains linked by a short spacer sequence (WSQP). One is a hydrophobic N-terminal domain (GALFLGFLGAAGSTMGA) derived from the glycine-rich region of the fusion sequence of human immunodeficiency virus glycoprotein 41. This domain has functions in membrane fusion and stabilization. The other is a hydrophilic C-terminal domain (KKKRKV) derived from the nuclear localization signal of simian virus 40 large T-antigen. This domain improves the nuclear addressing of the peptide. The CD spectra of MPG showed no characteristic secondary structure in water but an α-helical structure in 20% 2,2,2-trifluoroethanol. Furthermore, MPG adopted a β-sheet structure in phosphate-buffered saline and in phosphate buffer with 1,2-dioleoyl-*sn*-glycero-3-phospho-*rac*-(1-glycerol) sodium salt. MPG can form stable non-covalent complexes with nucleic acids by electrostatic interaction and improved their intracellular delivery [34].

Azurin, a member of the cupredoxin family of copper-containing redox proteins, preferentially penetrates human cancer cells. p18 (LSTAADMQGVVTDGMASG) is an 18-amino acid peptide consisting of residues 50–67 of azurin and a minimal motif with cell-penetrating ability [35]. p18 contains residues 54–67, which is an α-helical structure of azurin as well as a partial β-sheet structure. p18 adopts an amphipathic structure when associated with phospholipid membranes. p18 is not bound by cell membrane glycosaminoglycans and penetrates cancer cells mainly via caveolae-mediated endocytosis, but partially via caveolae-independent endocytosis.

5.4 Polyproline Type II Helical Peptides

Bac 7, a 59-amino acid protein, belongs to the bactenecin family and is present in Pro-Arg-rich antimicrobials. The dual functions of cell-penetrating ability and antimicrobial activity of Bac 7 are colocalized in the N-terminal 24 residues of Bac 7, Bac_{1-24} (RRIRPRPPRLPRPRPRPLPFPRPG) [36]. Meanwhile, Bac_{15-24} (PRPLPFPRPG), a 10-amino acid pro-rich peptide with two Arg residues, translocates across the cell membrane without antimicrobial activity. The preferential conformation of Bac_{15-24} in an aqueous solution is a hybrid of the polyproline type II (PPII) helical structure and the α-helical structure. Bac_{15-24} is capable of delivering a non-covalently linked protein into cells.

Giralt et al. designed amphipathic peptides $(VXLPPP)_n$ (n = 1–3, X = His, Arg, or Lys) and examined their cell-penetrating ability [1, 37]. These sequences are designed by reference to the N-terminal repetitive domain $(VHLPPP)_8$ of maize γ-zein, which interacts with cell membranes and does not have antibiotic effects [38]. $(VRLPPP)_3$, named SAP (sweet arrow peptide), was found to be the most efficient peptide at crossing the cellular membrane. The internalization of SAP into the cells takes place via caveolae- or lipid-rafts-mediated endocytosis. SAP adopts a typical PPII helical structure in an aqueous solution (Figure 5.2) but aggregates with an increase in peptide concentration [39]. Giralt et al. also developed an amphipathic anionic CPP based on SAP, SAP(E) $(VELPPP)_3$, in which cationic Arg residues of SAP are replaced by anionic Glu residues [40, 41]. SAP(E) has a negative charge at physiological pH,

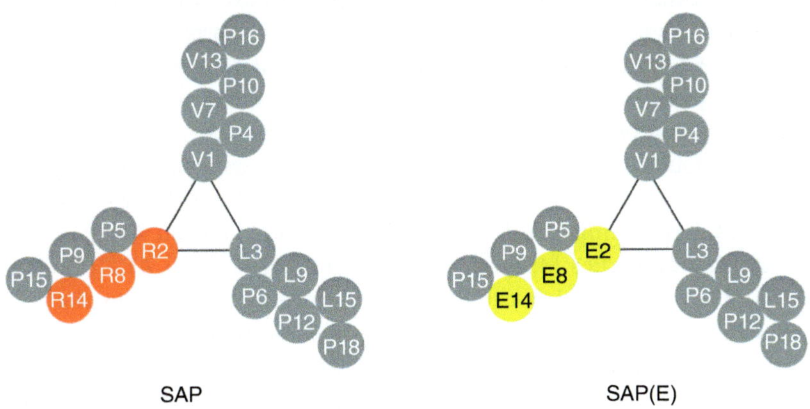

Figure 5.2 Helical wheel diagrams of amphipathic polyproline type II (PPII) helical CPPs. Helical screw sense of PPII helical peptides composed of L-α-amino acids is left-handed. Red: cationic Arg (R) residues; yellow: anionic Glu (E) residues.

presents a PPII helical structure in an aqueous solution (Figure 5.2), and is not cytotoxic even at high concentrations. Cellular uptake amounts and the proposed internalization mechanism of SAP(E) are similar to those of SAP. SAP(E) is not positively charged. So, the first electrostatic interaction with the cell membrane such as by cationic CPPs, is not necessary for its internalization.

5.5 Others

Pep-1 (KETWWETWWTEWSQPKKKRKV) is a 21-amino acid peptide consisting of three domains [42]. The first domain from the N-terminus is a hydrophobic Trp-rich sequence (KETWWETWWTEW), which can efficiently target the cell membrane and interact with proteins by hydrophobic interaction. The second domain is a spacer sequence (SQP) separating the two hydrophobic and hydrophilic domains. A Pro residue in the spacer domain improves the flexibility of both the hydrophobic and hydrophilic domains. The third domain is a hydrophilic Lys-rich domain (KKKRKV), similar to MPG, which improves the intracellular delivery and solubility of the peptide. The amphipathic CPP Pep-1 can form the complexes with anionic peptides and proteins and efficiently deliver them into the cultured cells in a fully biologically active form.

pVEC (LLIILRRRIRKQAHAHSK) is an 18-amino acid peptide derived from the murine sequence of the cell adhesion molecule vascular endothelial cadherin, amino acids 615–632 [43]. The peptide sequence contains 13 cytosolic amino acids closest to the cell membrane and 5 amino acids from the C-terminus of the transmembrane region, with an amphipathic character. pVEC shows good translocation efficiency against various cultured cells and accumulation in nuclear structures. Cellular uptake of pVEC is observed at 4 °C, suggesting an energy-independent mechanism of translocation. pVEC functions as a transporter of both hexameric PNA by covalent conjugation and a streptavidin using biotinylated pVEC. An Ala-scan study revealed that Ala substitution of the five respective N-terminal hydrophobic amino acids significantly decreased the translocation properties, whereas those of residues 6 (Arg), 8 (Arg), or 17 (Ser) enhanced it [44]. The N-terminal hydrophobicity of pVEC is crucial for its efficient cell-penetrating ability.

The tumor suppressor protein p14ARF prevents tumor transformation mainly by inhibiting human double minute 2 protein and is widely deregulated in many types of cancers. Langel et al. found that a 22-amino acid peptide (MVRRFLVTLRI RRACGPPRVRV) derived from the N-terminus of p14ARF, denoted ARF (1–22), has cell-penetrating properties [45]. Not only ARF (1–22) but also ARF (19–31) (RVRVFVVHIPRLT) showed high cellular uptake levels comparable to that of amphipathic CPP TP10. Cellular uptake of ARF (1–22) is associated with low membrane disturbance and ARF (1–22) mainly translocates cells through endocytosis. Furthermore, ARF (1–22) decreases proliferation in a dose-dependent manner and induces apoptosis.

M918 (MVTVLFRRLRIRRACGPPRVRV) is a 22-amino acid peptide derived from the tumor suppressor protein p14ARF, amino acids 1–22 with positions 3–8 inverted

[46]. M918 has a very low amphipathic moment but is more readily internalized into the cultured cells than TP10 with negligible cytotoxicity. Cellular uptake of M918 is mainly via endocytosis and, in particular, macropinocytosis, but independent of glycosaminoglycans on the cell surface. M918 can be used to transport PNAs and proteins efficiently into the cells either as a covalent conjugate or in a non-covalent complex with cargos.

References

1 Pujals, S. and Giralt, E. (2008). Proline-rich, amphipathic cell-penetrating peptides. *Advanced Drug Delivery Reviews* 60: 473–484.

2 Kalafatovic, D. and Giralt, E. (2017). Cell-penetrating peptides: design strategies beyond primary structure and amphipathicity. *Molecules* 22: 1929.

3 Guidotti, G., Brambilla, L., and Rossi, D. (2017). Cell-penetrating peptides: from basic research to clinics. *Trends in Pharmacological Sciences* 38 (4): 406–424.

4 Kim, G.C., Cheon, D.H., and Lee, Y. (2021). Challenge to overcome current limitations of cell-penetrating peptides. *BBA – Proteins and Proteomics* 1869: 140604.

5 Derossi, D., Joliot, A.H., Chassaing, G. et al. (1994). The third helix of the Antennapedia homeodomain translocates through biological membranes. *The Journal of Biological Chemistry* 269 (14): 10444–10450.

6 Joliot, A., Pernelle, C., Deagostini-Bazin, H. et al. (1991). Antennapedia homeobox peptide regulates neural morphogenesis. *Proceedings of the National Academy of Sciences of the United States of America* 88: 1864–1868.

7 Joliot, A. and Prochiantz, A. (2008). Homeoproteins as natural Penetratin cargoes with signaling properties. *Advanced Drug Delivery Reviews* 60: 608–613.

8 Dupont, E., Prochiantz, A., and Joliot, A. (2015). Penetratin story: an overview. *Methods in Molecular Biology* 1324: 29–37.

9 Oehlke, J., Krause, E., Wiesner, B. et al. (1996). Nonendocytic, amphipathicity dependent cellular uptake of helical model peptides. *Protein and Peptide Letters* 3 (6): 393–398.

10 Oehlke, J., Scheller, A., Wiesner, B. et al. (1998). Cellular uptake of an α-helical amphipathic model peptide with the potential to deliver polar compounds into the cell interior non-endocytically. *Biochimica et Biophysica Acta* 1414: 127–139.

11 Scheller, A., Wiesner, B., Melzig, M. et al. (2000). Evidence for an amphipathicity independent cellular uptake of amphipathic cell-penetrating peptides. *European Journal of Biochemistry* 267: 6043–6049.

12 Wolf, Y., Pritz, S., Abes, S. et al. (2006). Structural requirements for cellular uptake and antisense activity of peptide nucleic acids conjugated with various peptides. *Biochemistry* 45: 14944–14954.

13 Pooga, M., Hallbrink, M., Zorko, M. et al. (1998). Cell penetration of transportan. *The FASEB Journal* 12: 67–77.

14 Soomets, U., Lindgren, M., Gallet, X. et al. (2000). Deletion analogues of transportan. *Biochimica et Biophysica Acta* 1467: 165–176.

15 Yandek, L.E., Pokorny, A., Floren, A. et al. (2007). Mechanism of the cell-penetrating peptide transportan 10 permeation of lipid bilayers. *Biophysical Journal* 92: 2434–2444.
16 Magzoub, M., Sandgren, S., Lundberg, P. et al. (2006). N-terminal peptides from unprocessed prion proteins enter cells by macropinocytosis. *Biochemical and Biophysical Research Communications* 348: 379–385.
17 Biverstahl, H., Andersson, A., Graslund, A. et al. (2004). NMR solution structure and membrane interaction of the N-terminal sequence (1-30) of the bovine prion protein. *Biochemistry* 43: 14940–14947.
18 Delaroche, D., Aussedat, B., Aubry, S. et al. (2007). Tracking a new cell-penetrating (W/R) nonapeptide, through an enzyme-stable mass spectrometry reporter tag. *Analytical Chemistry* 79: 1932–1938.
19 Bechara, C., Pallerla, M., Zaltsman, Y. et al. (2013). Tryptophan within basic peptide sequences triggers glycosaminoglycan-dependent endocytosis. *The FASEB Journal* 27: 738–749.
20 Rittner, K., Benavente, A., Bompard-Sorlet, A. et al. (2002). New basic membrane-destabilizing peptides for plasmid-based gene delivery *in vitro* and *in vivo*. *Molecular Therapy* 5 (2): 104–114.
21 Crombez, L., Aldrian-Herrada, G., Konate, K. et al. (2009). A new potent secondary amphipathic cell-penetrating peptide for siRNA delivery into mammalian cells. *Molecular Therapy* 17 (1): 95–103.
22 Konate, K., Crombez, L., Dehayes, S. et al. (2010). Insight into the cellular uptake mechanism of a secondary amphipathic cell-penetrating peptide for siRNA delivery. *Biochemistry* 49: 3393–3402.
23 Akishiba, M., Takeuchi, T., Kawaguchi, Y. et al. (2017). Cytosolic antibody delivery by lipid-sensitive endosomolytic peptide. *Nature Chemistry* 9: 751–761.
24 Bechinger, B. (1996). Towards membrane protein design: pH-sensitive topology of histidine-containing polypeptides. *Journal of Molecular Biology* 263: 768–775.
25 Perrone, B., Miles, A.J., Salnikov, E.S. et al. (2014). Lipid interaction of LAH4, a peptide with antimicrobial and nucleic acid transfection activities. *European Biophysics Journal* 43: 499–507.
26 Kichler, A., Mason, A.J., and Bechinger, B. (2006). Cationic amphipathic histidine-rich peptides for gene delivery. *Biochimica et Biophysica Acta* 1758: 301–307.
27 Subbarao, N.K., Parente, R.A., Szoka, F.C. et al. (1987). pH-Dependent bilayer destabilization by an amphipathic peptide. *Biochemistry* 26: 2964–2972.
28 Kakudo, T., Chaki, S., Futaki, S. et al. (2004). Transferrin-modified liposomes equipped with a pH-sensitive fusogenic peptide: an artificial viral-like delivery system. *Biochemistry* 43: 5618–5628.
29 Nakase, I., Kogure, K., Harashima, H. et al. (2011). Application of a fusogenic peptide GALA for intracellular delivery. *Methods in Molecular Biology* 683: 525–533.
30 Wyman, T.B., Nicol, F., Zelphati, O. et al. (1997). Design, synthesis, and characterization of a cationic peptide that binds to nucleic acids and permeabilizes bilayers. *Biochemistry* 36: 3008–3017.
31 Krause, E., Beyermann, M., Fabian, H. et al. (1996). Conformation of a water-soluble β-sheet model peptide. A circular dichroism and Fourier-transform infrared

spectroscopic study of double D-amino acid replacements. *International Journal of Peptide and Protein Research* 48: 559–568.
32 Oehlke, J., Krause, E., Wiesner, B. et al. (1997). Extensive cellular uptake into endothelial cells of an amphipathic β-sheet forming peptide. *FEBS Letters* 415: 196–199.
33 Morris, M.C., Vidal, P., Chaloin, L. et al. (1997). A new peptide vector for efficient delivery of oligonucleotides into mammalian cells. *Nucleic Acids Research* 25: 2730–2736.
34 Simeoni, F., Morris, M.C., Heitz, F. et al. (2003). Insight into the mechanism of the peptide-based gene delivery system MPG: implications for delivery of siRNA into mammalian cells. *Nucleic Acids Research* 31 (11): 2717–2724.
35 Taylor, B.N., Mehta, R.R., Yamada, T. et al. (2009). Noncationic peptides obtained from azurin preferentially enter cancer cells. *Cancer Research* 69 (2): 537–546.
36 Sadler, K., Eom, K.D., Yang, J.-L. et al. (2002). Translocating proline-rich peptides from the antimicrobial peptide bactenecin 7. *Biochemistry* 41: 14150–14157.
37 Fernandez-Carneado, J., Kogan, M.J., Castel, S. et al. (2004). Potential peptide carriers: amphipathic proline-rich peptides derived from the N-terminal domain of γ-zein. *Angewandte Chemie International Edition* 43: 1811–1814.
38 Kogan, M.J., Dalcol, I., Gorostiza, P. et al. (2002). Supramolecular properties of the proline-rich γ-zein N-terminal domain. *Biophysical Journal* 83: 1194–1204.
39 Pujals, S., Fernandez-Carneado, J., Lopez-Iglesias, C. et al. (2006). Mechanistic aspects of CPP-mediated intracellular drug delivery: relevance of CPP self-assembly. *Biochimica et Biophysica Acta* 1758: 264–279.
40 Martin, I., Teixido, M., and Giralt, E. (2011). Design, synthesis and characterization of a new anionic cell-penetrating peptide: SAP(E). *Chembiochem* 12: 896–903.
41 Franz, J., Lelle, M., Peneva, K. et al. (2016). SAP(E) – A cell-penetrating polyproline helix at lipid interfaces. *Biochimica et Biophysica Acta* 1858: 2028–2034.
42 Morris, M.C., Depollier, J., Mery, J. et al. (2001). A peptide carrier for the delivery of biologically active proteins into mammalian cells. *Nature Biotechnology* 19: 1173–1176.
43 Elmquist, A., Lindgren, M., Bartfai, T. et al. (2001). VE-cadherin-derived cell-penetrating peptide, *p*VEC, with carrier functions. *Experimental Cell Research* 269: 237–244.
44 Elmquist, A., Hansen, M., and Langel, U. (2006). Structure-activity relationship study of the cell-penetrating peptide *p*VEC. *Biochimica et Biophysica Acta* 1758: 721–729.
45 Johansson, H., El-Andaloussi, S., Holm, T. et al. (2008). Characterization of a novel cytotoxic cell-penetrating peptide derived from p14ARF protein. *Molecular Therapy* 16 (1): 115–123.
46 El-Andaloussi, S., Johansson, H.J., Holm, T. et al. (2007). A novel cell-penetrating peptide, M918, for efficient delivery of proteins and peptide nucleic acids. *Molecular Therapy* 15 (10): 1820–1826.

6

Hydrophobic CPPs

Yosuke Demizu

National Institute of Health Sciences, Division of Organic Chemistry, Kasawaki, Japan

Many cell-penetrating peptides (CPPs) are composed of cationic or amphipathic (cationic and hydrophobic) amino acid residues. However, there is generally a correlation between high membrane permeability and cytotoxicity with these peptides. Therefore, it is important to develop new nontoxic CPPs that can be used to deliver cargo molecules into cells. Several hydrophobic CPPs that do not contain (or contain only a small amount of) cationic amino acids have been discovered and are being investigated for use as delivery agents. Most hydrophobic CPPs are composed of 10–20 residues, contain a high content of hydrophobic amino acids, and are usually partial sequences of proteins. Hydrophobic CPPs have a high affinity for the hydrophobic domains of cell membranes and are passively transported through the lipid membranes and migrate into the cells. Hydrophobic CPPs can be classified into bioactive natural peptides that are derived from proteins and rationally designed sequences. Table 6.1 summarizes the representative hydrophobic CPPs reported to date.

6.1 MPG [1–3]

MPG (GALFLGFLGAAGSTMGAWSQPKKKRKV), a peptide derived from the hydrophobic fusion peptide of HIV-1 gp41 and the hydrophilic nuclear localization sequence of SV40 large T antigen, can be used as a tool for the intracellular trafficking of oligonucleotides. **MPG** has low cytotoxicity, is capable of protecting against nucleic acid degradation in cell culture media by electrostatic interactions with nucleic acids, and is able to efficiently introduce plasmids into cells. **MPG** can also transport antisense nucleic acids into cells and has been shown to have the potential for gene therapy applications.

Table 6.1 Examples of hydrophobic CPPs.

Name	Sequence	Origin	Cargo	Ref
MPG	GALFLGFLGAAGSTMGAWSQPKKKRKV	Signal sequence of Ig light chain from Caiman crocodilus	Nucleic acids	[1–4]
SP	RPKPQQFGLM	Substance P and analogues	FAM	[5–7]
MTS	AAVALLPAVLLALLAP	Kaposi sarcoma fibroblast growth factor Kaposi FGF	The nuclear localization sequence of NF-κB p50	[8]
Hydrophobic MPS	VTVLAGALAGVGVG	Integrin β_3 fragment	The transcription factor NF-κB	[9, 10]
MTS1	AAVLLPVLLAAP	Grb2 (SH2 domain)	Glutathione S-transferase (GST)	[11, 12]
TP10	AGYLLGKINLKALAALAKKIL	Transportan	FITC	[13, 14]
Pep-7	SDLWEMMMVSLACQY	Phase display	FITC	[15]
TLM	PLSSIFSRIGDP	Hepatitis B virus translocation motif	Ovalbumin	[16]
hCT(9–32)-br	LGTYTQDFNK(X)FHTFPQTAIGVGAP X: PKKKRKVEDPGVGFA	Human calcitonin partial sequence 9–32	FAM	[17, 18]

(Continued)

Table 6.1 (Continued)

C105Y	CSIPPEVKFNKPFVYLI	Alpha1-antitrypsin	DNA	[19]
M918	MVTVLFRRLRIRRACGPPRVRV	Tumor suppressor protein p14ARF	PNA	[13, 20]
PFV	PFVYLI	Alpha1-antitrypsin	Proapoptotic domain peptide (PAD)	[21, 22]
Pept1	PLILLRLLRGQF	Combinatorial peptide chemistry and orthogonal high-throughput screening	TAMRA	[23]
hCT(18–32)-br	Capr-K(X)FHTFPQTAIGVGAP X:KKRKAPKKKRKFA	Human calcitonin partial sequence 9–32	siRNA	[24]
P4	LGAQSNF	Phage display	ASO	[25]
AA3H	MASIWVGHRG	Annexin	β-galactosidase	[26]
L1-7	FI<u>D</u>III<u>K</u>ILLI (cyclized)	De novo design	Model peptide sequence	[27]

Abbreviations in the table: FITC, fluoresceinisothiocyanate; FAM, carboxyfluorescein; TAMRA, tetramethylrhodamine; GST, glutathione S-transferase; PAD, Peptidyl arginine deiminase.

6.2 SP [4–6]

The substance P analog (**SP**, RPKPQQFGLM) is a membrane receptor (NK1)-specific ligand that can bind to lipid monolayers. **SP** penetrates the lipid domains of the plasma membrane and is taken up into the cell via "physical endocytosis."

6.3 MTS [7]

The hydrophobic CPP peptide, membrane-translocating sequence **MTS** (AAVALLPAVLLALLAP), derived from Kaposi sarcoma fibroblast growth factor (FGF), can efficiently transport functional cargo molecules into cells and regulate the signaling of transcription factors that mediate cellular responses, which can be used to study intracellular processes. A peptide conjugated with the hydrophobic CPP and a cargo corresponding to the nuclear localization sequence of NF-κB p50 was shown to inhibit NF-κB nuclear migration in endothelial cells and monocytic cultured cells stimulated with lipopolysaccharide, or tumor necrosis factor-α (TNF-α), in a concentration-dependent manner.

6.4 Hydrophobic MPS [8, 9]

Conjugating the membrane transport sequence of human integrin β_3, termed the hydrophobic membrane-permeable sequence (**hydrophobic MPS**; VTVLAGAL AGVGVG), to a cargo molecule results in a peptide–cargo conjugate that can be noninvasively transported into cells. For example, a chimeric peptide that combined the nuclear localization sequence of transcription factor NF-κB with **hydrophobic MPS** was shown to inhibit the nuclear translocation of transcription factor NF-κB, activator protein 1, and nuclear factor of activated T cells in agonist-stimulated Jurkat T lymphocytes.

6.5 MTS1 [10, 11]

By fusing the hydrophobic CPP, **MTS1** (AAVLLPVLLAAP), to glutathione *S*-transferase (GST), a GST-MTS1 fusion protein was constructed that could be efficiently taken up by NIH 3T3 fibroblasts and other cells. A GST-Grb2 SH2-MTS1 fusion protein (41 kDa), in which the Grb2 SH2 domain was fused to GST-MTS1, was shown to be taken up by cells and formed a complex with the phosphorylated epidermal growth factor (EGF) receptor (EGFR) resulting in the inhibition of EGF binding to EGFR-Grb2 and activation of mitogen-activated protein kinase.

6.6 TP10 [12, 13]

Transportan (GWTLNSAGYLLGKINLKALAALAKKIL), which consists of 12 amino acids from the *N*-terminus of the neuropeptide galanin conjugated with mastoparan, is a cell membrane-permeable peptide. **TP10** (AGYLLGKINLK ALAALAKKIL), which lacks six amino acids from the *N*-terminus of transportan, also has high intracellular permeability. However, deletion of the *C*-terminus or the center of the transportan sequence reduced or eliminated the cellular uptake. In addition, the length of the transportan sequence, the aromaticity, and the position of the positive charge greatly affected the orientation of the peptide in the membrane, which in turn affected the entry of the peptide into the cell.

6.7 Pep-7 [14]

Pep-7 (SDLWEMMMVSLACQY) is a hydrophobic CPP found in phage display that has been shown to transport fluorescein isothiocyanate (FITC), as a model cargo, into cells.

6.8 TLM [15]

CPPs are known to improve the antigen loading into dendritic cell vaccines. The fusion protein of the hepatitis B virus-derived cell-penetrating translocation motif (**TLM**) (PLSSIFSRIGDP) and ovalbumin (TLM-ova) has been reported to improve immunogenicity. TLM-ova vaccination improved the antitumor immune response and induced higher anti-ovalbumin IgG2a antibody titers compared with ovalbumin vaccination, indicating that CPP-protein vaccines can elicit not only a humoral immune response but also a cellular immune response.

6.9 hCT(9–32)-br [16–18]

hCT(9-32)-br (LGTYTQDFNK(X)FHTFPQTAIGVGAP, X: PKKKRKVEDPGVGFA) is a peptide in which positions 12 and 16 of hCT(9–32) and a *C*-terminally truncated CPP of human calcitonin. It had good metabolic stability and efficiently intracellularly transported the model cargo fluorescein (FAM) molecule. **hCT(9-32)-br** did not show significant cytotoxicity and is expected to be a powerful drug delivery vector.

6.10 C105Y [19]

C105Y (CSIPPEVKFNKPFVYLI) is a synthetic CPP corresponding to residues 359–374 of α1-antitrypsin. Cellular uptake of **C105Y** was localized to the nucleus and cytoplasm, and the D-isomer was readily taken up into the cell but not into the nucleus, indicating that **C105Y** may be taken up into the cell by lipid raft-mediated endocytosis. In addition, it was shown that gene expression was enhanced by encapsulating **C105Y** in DNA nanoparticles.

6.11 M918 [12, 20]

M918 (MVTVLFRRLRIRRACGPPRVRV), a partial sequence of the tumor suppressor protein p14ARF, like many other CPPs, is taken up into cells by endocytosis, particularly macropinocytosis, and is not dependent on cell surface glycosaminoglycans. In a splice modification assay using a chimeric molecule of **M918** conjugated with an antisense peptide nucleic acid (M918-PNA), a dose-dependent increase in correct splicing was observed that exceeded the effect of other CPPs.

6.12 PFV [21, 22]

The hydrophobic **PFV** (PFVYLI) peptide is taken up into cells by endocytosis. A conjugate peptide of **PFV** and the proapoptotic domain peptide PAD $(klaklak)_2$ showed cytotoxicity and induced apoptosis in mitochondria in KG1a leukemia cells.

6.13 Pept1 [23]

The hydrophobic residue-rich peptide **Pept1** (PLILLRLLRGQF), discovered using orthogonal high-throughput screening, is taken up into cells in an energy-independent manner, but not via endocytosis. Peptides that are taken up into cells in an energy-independent manner have the potential to deliver polar drugs that otherwise would not penetrate cell membranes into cells and tissues and are expected to be used as a new drug delivery system (DDS) technology.

6.14 hCT(18–32)-br [24]

hCT(18-32)-br (Capr-K(X)FHTFPQTAIGVGAP, X: KKRKAPKKKRKFA), a *C*-terminal truncated form of human calcitonin, can be used for the efficient intracellular delivery of small interfering RNA (siRNA). **hCT(18-32)-br** effectively transported siRNA targeting the human neuropeptide Y1 receptor, a member of the G protein-coupled receptor family, into HEK-293 cells, and showed knockdown effects comparable to those obtained by lipofection.

6.15 P4 [25]

The heptapeptide **P4** (LGAQSNF), identified from a phage display peptide library, can be conjugated with antisense oligonucleotides (AONs) to transport AONs into cells. Exon skipping using AONs is a promising treatment for Duchenne muscular dystrophy, and various clinical trials are currently underway. Binding of **P4** to 2′-*O*-methyl phosphorothioate AON enhanced uptake in skeletal and cardiac muscle in vivo, with significantly higher levels of exon skipping in the heart and diaphragm. Conjugation of **P4** to AONs is an interesting strategy to enhance the therapeutic effect of exon skipping for the treatment of Duchenne muscular dystrophy.

6.16 AA3H [26]

AA3H (MASIWVGHRG), which contains many hydrophobic residues, was isolated from 12 isoforms of the human membrane protein annexin, and it is taken up into cells more efficiently than the cationic CPP Tat. **AA3H** showed low cytotoxicity and long-term serum stability, and was non-immunogenic. In addition, by conjugating **AA3H** to a model protein (β-galactosidase), the protein could be introduced into mammalian cells in vitro and in vivo. **AA3H** has the potential to be used as a carrier for polymer transport.

6.17 L1-7 [27]

The hydrophobic cyclic peptide **L1-7** containing a lactam bridge, [FIDIIIKILLI (cyclized)], found by de novo screening, can achieve the transport of cargo peptides into the cytoplasm.

References

1. Morris, M.C., Chaloin, L., Mery, J. et al. (1999). A novel potent strategy for gene delivery using a single peptide vector as a carrier. *Nucleic Acids Research* 27: 3510–3517.
2. Briggs, M.S. and Gierasch, L.M. (1986). Molecular mechanism of protein secretion: the role of signal sequence. *Advances in Protein Chemistry* 38: 109–180.
3. Vives, E., Brodin, P., and Lebleu, B. (1997). A truncated HIV-1 TAT protein basic domain rapidly translocates through the plasma membrane and accumulates in the cell nucleus. *Journal of Biological Chemistry* 272: 16010–16017.
4. Lamaziere, A., Burlina, F., Wolf, C. et al. (2007). Non-metabolic membrane tubulation and permeability induced by bioactive peptides. *PLOS ONE* 2: e201.
5. Repke, A. and Bienert, M. (1987). Mast cell activation – a receptor-independent mode of substance P action. *FEBS Letters* 221: 236–240.
6. Derossi, D., Joliot, A.H., Chassaing, G. et al. (1994). The third helix of the Antennapedia homeodomain translocates through biological membranes. *Journal of Biological Chemistry* 269: 10444–10450.
7. Lin, Y.-Z., Yao, S.Y., Veach, R.A. et al. (1995). Inhibition of nuclear translocation of transcription factor NF-kappa B by a synthetic peptide containing a cell-permeable motif and nuclear localization sequence. *Journal of Biological Chemistry* 270: 14255–14258.

8 Liu, K.Y., Timmons, S., Lin, Y.-Z. et al. (1996). Identification of a functionally important sequence in the cytoplasmic tail of integrin beta 3 by using cell-permeable peptide analogs. *Proceedings of the National Academy of Sciences of the United States of America* 93: 11819–11824.
9 Zhang, L., Torgerson, T.R., Liu, X.-Y. et al. (1998). Preparation of functionally active cell-permeable peptides by single-step ligation of two peptide modules. *Proceedings of the National Academy of Sciences of the United States of America* 95: 9184–9189.
10 Rojas, M., Donahue, J.P., Tan, Z. et al. (1998). Genetic engineering of proteins with cell-membrane permeability. *Nature Biotechnology* 16: 370–375.
11 Moutal, A., François-Moutal, L., Brittain, J.M. et al. (2015). Differential neuroprotective potential of CRMP2 peptide aptamers conjugated to cationic, hydrophobic, and amphipathic cell penetrating peptides. *Frontiers in Cellular Neuroscience* 8: 471.
12 Guterstam, P., Madani, F., Hirose, H. et al. (2009). Elucidating cell-penetrating peptide mechanisms of action for membrane interaction, cellular uptake, and translocation utilizing the hydrophobic counter-anion pyrenebutyrate. *Biochimica et Biophysica Acta* 1788: 2509–2517.
13 Soomets, U., Lindgren, M., Gallet, X. et al. (2000). Deletion analogues of transportan. *Biochimica et Biophysica Acta* 1467: 165–176.
14 Gao, C., Mao, S., Ditzel, H.J. et al. (2002). A cell-penetrating peptide from a novel pVII–pIX phage-displayed random peptide library. *Bioorganic Medicinal Chemistry* 10: 4057–4065.
15 Bleifuss, E., Kammertoens, T., Hutloff, A. et al. (2006). The translocation motif of hepatitis B virus improves protein vaccination. *Cellular and Molecular Life Sciences* 63: 627–635.
16 Krauss, U., Müller, M., Stahl, M. et al. (2004). in vitro gene delivery by a novel human calcitonin (hCT)-derived carrier peptide. *Bioorganic Medicinal Chemistry Letters* 14: 51–54.
17 Rennert, R., Wespe, C., Beck-Sickinger, A.G. et al. (2006). Developing novel hCT derived cell-penetrating peptides with improved metabolic stability. *Biochimica et Biophysica Acta* 1758: 347–354.
18 Rennert, R., Neundorf, I., Jahnke, H.-G. et al. (2008). Generation of carrier peptides for the delivery of nucleic acid drugs in primary cells. *ChemMedChem* 3: 241–253.
19 Rhee, M. and Davis, P. (2006). Mechanism of uptake of C105Y, a novel cell-penetrating peptide. *Journal of Biological Chemistry* 281: 1233–1240.
20 El-Andaloussi, S., Johansson, H.J., Holm, T. et al. (2007). A novel cell-penetrating peptide, M918, for efficient delivery of proteins and peptide nucleic acids. *Molecular Therapy* 15: 1820–1826.
21 Watkins, C.L., Brennan, P., Fegan, C. et al. (2009). Cellular uptake, distribution and cytotoxicity of the hydrophobic cell penetrating peptide sequence PFVYLI linked to the proapoptotic domain peptide PAD. *Journal of Controlled Release* 140: 237–244.
22 Cai, D., Gao, W., He, B. et al. (2014). Hydrophobic penetrating peptide PFVYLI-modified stealth liposomes for doxorubicin delivery in breast cancer therapy. *Biomaterials* 35: 2283–2294.

23 Marks, J.R., Placone, J., Hristova, K. et al. (2011). Spontaneous membrane-translocating peptides by orthogonal high-throughput screening. *Journal of American Chemical Society* 133: 8995–9004.

24 Hoyer, J. and Neundorf, I. (2012). Knockdown of G protein-coupled receptor through efficient peptide-mediated siRNA delivery. *Journal of Controlled Release* 161: 826–834.

25 Jirka, S.M., Heemskerk, H., Tanganyika-deWinter, C.L. et al. (2014). Peptide conjugation of 2′-*O*-methyl phosphorothioate antisense oligonucleotides enhances cardiac uptake and exon skipping in mdx mice. *Nucleic Acid Therapeutics* 24: 25–36.

26 Kim, H.Y., Yum, S.Y., Jang, G. et al. (2015). Discovery of a non-cationic cell penetrating peptide derived from membrane-interacting human proteins and its potential as a protein delivery carrier. *Scientific Reports* 5: 11719.

27 Schmidt, S., Adjobo-Hermans, M.J.W., Kohze, R. et al. (2017). Identification of short hydrophobic cell-penetrating peptides for cytosolic peptide delivery by rational design. *Bioconjugate Chemistry* 28: 382–389.

7

Foldamer

Takuma Kato

Osaka Medical and Pharmaceutical University, Faculty of Pharmacy, Takatsuki City, Osaka, Japan

7.1 Introduction

"Foldamer" is a term proposed by Gellman in 1996 and refers to any polymer with a strong tendency to adopt new backbones with well-defined secondary structural preferences [1]. In this context, the term "well-defined" means that the conformational preference should be displayed in solution by oligomers of modest length. Specific examples of secondary structures include the α-helical, β-sheet, and double helical structures found in proteins, nucleic acids, and polysaccharides. However, the components of foldamers are not limited to amino acids, nucleic acids, and monosaccharides. Numerous other small molecules can be combined as units, and a variety of cell-penetrating foldamers have been well studied [2–5]. In this section, cell-penetrating peptide (CPP) foldamers, in which each component is linked by an amide bond, are described.

Non-proteinogenic amino acids are well-studied as constituents of peptide-type foldamers. These may be β-, γ-, or δ-amino acids or α,α-disubstituted amino acids (dAA), among others. Because these are non-proteinogenic amino acids and so are not present in naturally occurring proteins, they generally show resistance to hydrolysis by proteases. In addition, although it is difficult to control the secondary structures of peptides composed of only naturally occurring α-amino acids, which have short-chain lengths, peptide foldamers readily adopt some form of stable secondary structure. These features are particularly advantageous for CPP foldamers composed of non-proteinogenic amino acids, in part because their high stability against enzymatic degradation facilitates their actions within both cell lines and animals [6, 7]. In addition, the increased stability of the secondary structures of peptides composed of non-proteinogenic amino acids is of great merit for amphipathic helical CPPs, within which an amphipathic helical structure plays a key role. Herein, CPP foldamers are introduced by classifying them in terms of their component monomers (Figure 7.1).

Cell-Penetrating Peptides: Design, Development and Applications, First Edition. Makoto Oba and Yosuke Demizu.

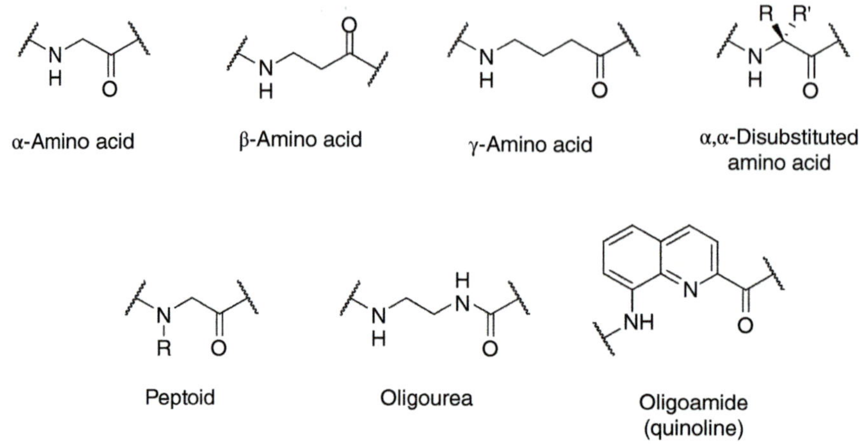

Figure 7.1 Structures of the monomers (non-proteinogenic amino acids) presented herein.

7.2 β-Amino Acids

β-Amino acids have one additional carbon atom between their amino and carboxy groups as compared to α-amino acids. Helical structures exclusively composed of β-peptides can be made much more conformationally stable than those of α-peptides, particularly at short chain lengths. As a result, β-peptides are the most studied foldamers composed of non-proteinogenic amino acids [8–10]. The α-helix, which is a typical secondary structure in peptides composed of α-amino acids, forms 13-membered rings of hydrogen bonds between backbone groups, N(i)–H···O=C(i +4) (Figure 7.2) [11].

By contrast, the representative secondary structures of β-peptides are the 14-helix (with 14-membered rings of hydrogen bonds between the backbone groups, N(i)–H···O=C(i+3)) [12–14], 12-helix (with 12-membered rings of hydrogen bonds between the backbone groups, N(i +3)–H ···O=C(i)) [15, 16], and 12/10-helix (with 12- and 10-membered rings of hydrogen bonds between backbone groups, N(i +1)–H···O=C(i) and N(i)–H···O=C(i +3)) [17–20] (Figure 7.3). The conformation-dependent dipoles of the 14- and 12-helices point in opposite directions, and there is no precedent among foldamers composed of α-amino acid for such a switch in hydrogen bond directionality.

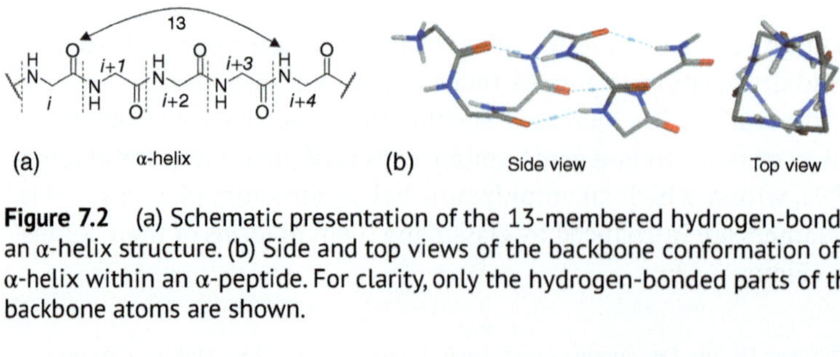

Figure 7.2 (a) Schematic presentation of the 13-membered hydrogen-bonded ring within an α-helix structure. (b) Side and top views of the backbone conformation of a right-handed α-helix within an α-peptide. For clarity, only the hydrogen-bonded parts of the peptide backbone atoms are shown.

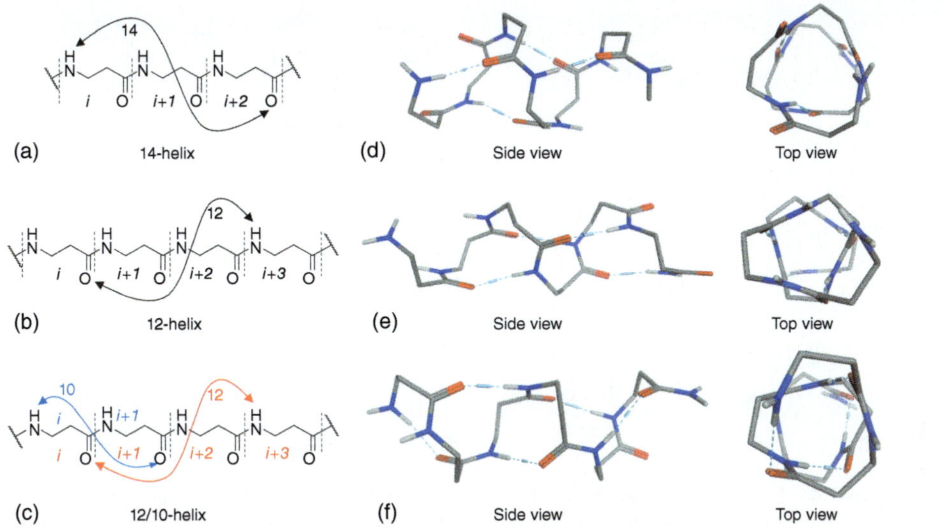

Figure 7.3 Schematic presentation of (a) the 14-membered hydrogen-bonded ring of a 14-helix structure, (b) the 12-membered hydrogen-bonded ring of a 12-helix structure, and (c) the 12- and 10-membered hydrogen-bonded rings of a 12/10-helix structure. Side and top views of the backbone conformation of a right-handed (d) 14-helix, (e) 12-helix, and (f) 12/10-helix within a β-peptide. For clarity, only the hydrogen-bonded parts of the peptide backbone atoms are shown.

In addition, peptides composed of a combination of α- and β-amino acids (α/β-peptides) are known to adopt helical structures such as the 11-helix (with 11-membered rings of hydrogen bonds between backbone groups, N(i +3)–H···O=C(i)) [21, 22], the 14/15-helix (with 14- and 15-membered rings of hydrogen bonds between the backbone groups, N(i +4)–H···O=C(i)) [23, 24] and the 11/9-helix (with 11- and 9-membered rings of hydrogen bonds between backbone groups, N(i)–H···O=C(i +1) and N(i +3)–H···O=C(i)) [25, 26] when the α:β ratio = 1:1 (Figure 7.4). Moreover, by changing the ratio of α- to β-amino acids, diverse helical structures other than those shown here may be adopted.

β/γ-Peptides composed of β- and γ-amino acids also form a wide variety of helical secondary structures. These peptides are known to adopt helical structures such as the 13-helix (with a 13-membered ring of hydrogen bonds between the backbone groups, N(i +3)–H···O=C(i)) [27, 28] (Figure 7.5). However, with combinations of various amino acid units, the naming of helical structures based on the number of atoms in the characteristic C=O···H–N hydrogen bonds becomes complicated, and it is not possible to list all of them here.

CPPs containing β-amino acids were reported at nearly the same time by the Gellman [29] and Seebach groups [30] in 2002. Gellman et al. reported β-peptides designed on the Tat (47–57) peptide sequence [29, 31], which was initially discovered as a CPP [32, 33]. A Tat peptide analogue in which all α-amino acids were replaced with β-amino acids showed cell membrane permeability comparable to that of the original Tat peptide at both 37 and 4 °C. CD spectra indicated that the

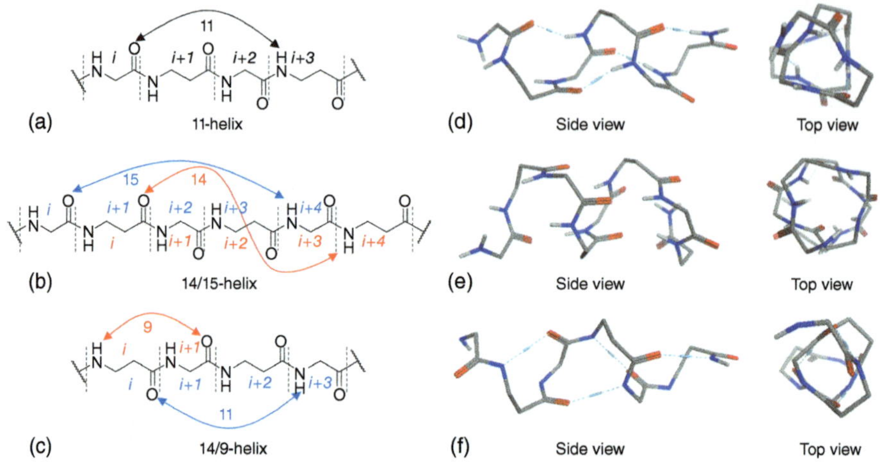

Figure 7.4 Schematic presentation of (a) the 11-membered hydrogen-bonded ring of an 11-helix structure, (b) the 14- and 15-membered hydrogen-bonded rings of a 14/15-helix structure, and (c) the 11- and 9-membered hydrogen-bonded rings of an 11/9-helix structure. Side and top views of the backbone conformation of a right-handed (d) 11-helix, (e) 14/15-helix, and (f) 11/9-helix within an α/β-peptide. For clarity, only the hydrogen-bonded parts of the peptide backbone atoms are shown.

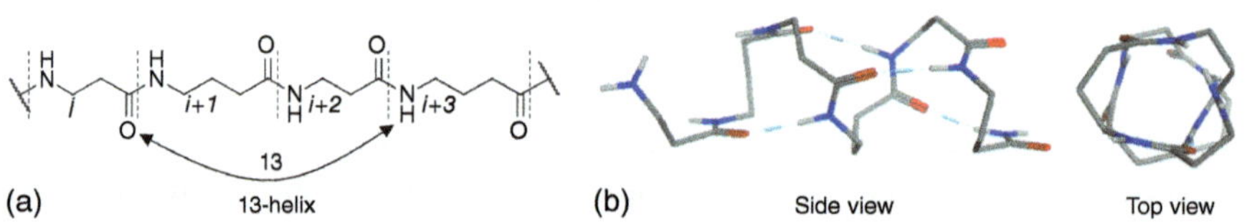

Figure 7.5 (a) Schematic presentation of the 13-membered hydrogen-bonded ring of a 13-helix structure. (b) Side and top views of the backbone conformation of a right-handed 13-helix within a β/γ-peptide. For clarity, only the hydrogen-bonded parts of the peptide backbone atoms are shown.

Tat β-peptide analogue is largely disordered in aqueous solution, like the original Tat peptide, but is highly helical in methanol. These results demonstrate that membrane translocation activity is preserved despite drastic alteration in the peptide backbone (amide group spacing). Seebach et al. designed several different types of β-peptides to determine the structural requirements for their cellular uptake [30]. In that study, they prepared β-peptides that formed a 14-helix, a 12/10-helix, or were not expected to form a helix at all. Among those, polycationic β-heptaarginine and β-heptalysine, which were not expected to form a helix, were nonetheless translocated through the membrane. The fact that β-heptaarginine was translocated through the membrane within a few minutes after its addition to the cells at a low incubation temperature indicates that its cell penetration does not depend on energy-dependent endocytosis. While this study showed that polycationic β-peptides exhibit high membrane permeability without adopting a helical structure, Gellman reported on polycationic β-peptides with a helix structure [34, 35]. They designed a peptide that formed a stable 14-helix in an aqueous solution and distributed its guanidino groups along one side of the helix structure. This β-peptide exhibited greater membrane permeability than a peptide in which the guanidino groups were distributed all around the periphery of the helix. Seebach et al. also found that β-proline oligomers permeated the cell membrane, whereas α-proline oligomers did not [36]. The rate of internalization of the β-oligoproline was much slower than that observed with the polycationic β-peptides, but oligoprolines permeate the cell membrane despite having no cationic groups. The β-oligoproline was predicted to have a secondary structure of 10_3-helix [37], and its CD spectra were consistent with that prediction. The CD spectra recorded right after dissolution and 24 hours later differed in their wavelengths and intensities of the cotton effect and may correlate with β-oligoproline's slow cell penetration. The β-oligoproline first aggregated at the surface of the cells and then crossed the membrane over time. Other groups have also reported that foldamers composed of various types of β- and α/β-peptides can be used as CPPs [38–40] and that β-amino acids are promising components of efficient CPP foldamers.

7.3 γ-Amino Acids

γ-Amino acids have two additional carbon atoms between the amino and carboxy groups as compared to α-amino acids. Like those composed of β-amino acids, peptides composed of γ-amino acids (γ-peptides) are also known to have a unique stable secondary structure. Although γ-peptides have a smaller number of potential hydrogen bonds than β-peptides for an oligomer with the same number of atoms in the main chain, they are able to adopt various stable conformations, including helices, sheets and turns [41, 42]. Regarding the helical structure, Hofmann et al. performed calculations on unsubstituted γ peptides and monosubstituted peptides with one methyl group at the α-, β-, or γ-position and showed that the 14-helix and 9-helix are the most stable conformations [43]. In fact, the representative secondary structures of γ-peptides are the 14-helix (with a 14-membered ring of hydrogen bonds between the backbone groups, N(i + 3)–H···O=C(i)) [44–46] and 9-helix (with a 9-membered ring of hydrogen bonds between the backbone groups, N(i + 2)–H···O=C(i)) [47–49] (Figure 7.6).

In addition, peptides composed of a combination of α- and γ-amino acids (α/γ-peptides) are known to form a 12-helix (with 12-membered ring hydrogen bonds between the backbone groups, N(i + 3)–H···O=C(i)) [50, 51] and a 12/10-helix (with 12- and 10-membered rings of hydrogen bonds between the backbone groups, N(i + 3)–H···O=C(i), N(i)–H···O=C(i + 2)) [52, 53] (Figure 7.7).

Royo et al. used γ-peptides composed of proline-based γ-amino acids with various functional groups in their side chains [54, 55]. Among these peptides, a γ-peptide having hydrophobic groups on the nitrogen atom of the tertiary amine exhibited efficient cellular uptake. Likewise, a γ-peptide with guanidino groups on the nitrogen atom also exhibited cellular uptake, though the uptake was not quite as efficient

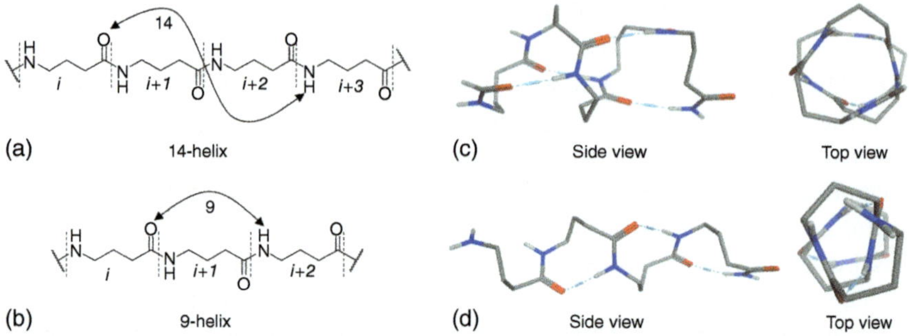

Figure 7.6 Schematic presentation of (a) the 14-membered hydrogen-bonded ring of a 14-helix structure, and (b) the 9-membered hydrogen-bonded ring of a 9-helix structure. Side and top views of the backbone conformation of a right-handed (c) 14-helix and (d) 9-helix within a γ-peptide. For clarity, only the hydrogen-bonded parts of the peptide backbone atoms are shown.

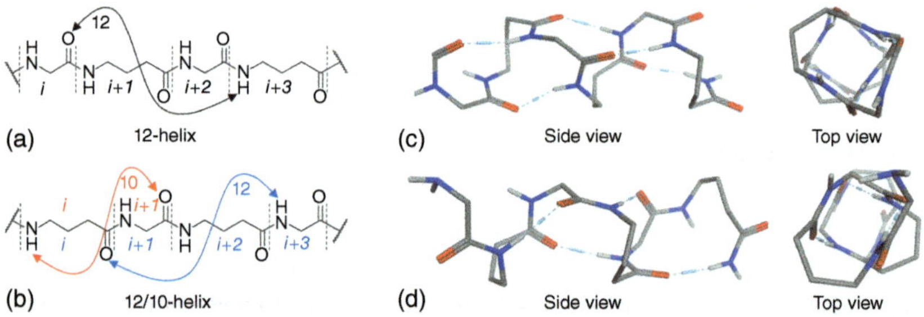

Figure 7.7 Schematic presentation of (a) the 12-membered hydrogen-bonded ring of a 12-helix structure, and (b) the 12- and 10-membered hydrogen-bonded rings of a 12/10-helix structure. Side and top views of the backbone conformation of a right-handed (c) 12-helix and (d) 12/10-helix within an α/γ-peptide. For clarity, only the hydrogen-bonded parts of the peptide backbone atoms are shown.

as with the hydrophobic peptide. These γ-peptides translocated across the cell membrane via an energy-independent mechanism and reached the nucleus. The cellular uptake of these γ-peptides was less efficient than that of the Tat peptide. However, the hydrophobic γ-peptide was resistant to proteases and both γ-peptides showed less cytotoxicity than the Tat peptide. γ-Peptide foldamers have also been applied as carriers to deliver therapeutic peptides. For example, conjugation of γ-peptides with antimicrobial peptides led to effective delivery of antimicrobial peptides into cells and enhanced anticancer activity [56]. These findings demonstrate that γ-amino acids are promising components of efficient CPP foldamers.

7.4 α,α-Disubstituted Amino Acids

In α,α-disubstituted amino acids (dAAs), the hydrogen atom at the α-position of the α-amino acid is replaced with an additional alkyl substituent [57, 58]. The introduction of dAAs into peptide sequences changes the properties of the peptides, with the conformational freedom of the secondary structure of their peptides being especially restricted. Homopeptides composed of dAAs can form a stable helical structure, even over a short sequence, and introduction of dAAs into peptides composed of α-amino acids often leads to a conformational change from a random coil to a helical structure. The representative secondary structures formed by peptides containing dAAs are the α-helix (with a 13-membered ring of hydrogen bonds between the backbone groups, N(i +4)–H···O=C(i)) (Figure 7.2) and the 3_{10}-helix (with a 10-membered ring of hydrogen bonds between the backbone groups, N(i +3)–H···O=C(i)) [59, 60] (Figure 7.8). Like the other non-proteinogenic amino acids described here, dAA-containing peptides exhibit metabolic stability against proteases and have been applied as CPPs [59].

α-Aminoisobutyric acid (Aib), which has two methyl groups at the α-position, has often been used as a building block for helical peptides and has been applied for the construction of amphipathic helical CPPs. Wada et al. reported a peptide in which alanines in a model of amphipathic peptide (MAP) were replaced with Aib for delivery of hydrophilic macromolecules [61, 62]. The Aib-containing peptide formed a more stable helical structure in an aqueous solution than the original MAP peptide composed of only α-amino acids, and it showed tolerance to proteases. This peptide also exhibited enhanced cellular uptake and was utilized as an oligonucleotide delivery carrier.

Demizu et al. described amphipathic CPP foldamers composed of Aib and Arg [63]. By utilizing L- and D-Arg, they showed that peptides with the same numbers of Aib and Arg residues but with different α-chirality adopted different secondary structures. Peptides forming stable amphipathic short helical structures showed the greatest cellular uptake. Oba et al. also investigated the impact of chain length in CPP foldamers composed of Aib and Arg when these peptides were applied to deliver plasmid DNA (pDNA) into cells [64]. Studies utilizing the properties of helical structures formed by Aib-containing peptides have also been conducted. Based on the fact that a peptide composed of L-Leu and Aib formed a stable helical structure, Misawa et al. utilized (Leu-Leu-Aib)$_n$ sequences as a helical promoter sequence to induce cationic CPPs (oligoarginine) to form a helical structure in aqueous solution [65]. These peptides adopted a helical structure in an aqueous solution without changes to the original CPP sequence and efficiently delivered small interfering RNA (siRNA) into the cells without significant cytotoxicity.

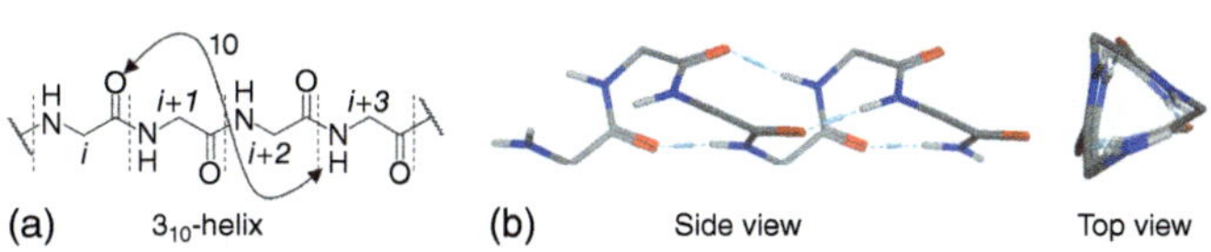

Figure 7.8 (a) Schematic presentation of the 10-membered hydrogen-bonded ring of a 3_{10}-helix structure. (b) Side and top views of the backbone conformation of a right-handed 3_{10}-helix within an α-peptide. For clarity, only the hydrogen-bonded parts of the peptide backbone atoms are shown.

The ability of dAA-containing peptides to form stable helical structures is effective not only for amphipathic CPPs but also for cationic CPPs. One study evaluated the secondary structure, enzymatic stability, and cell membrane permeability of peptides in which α-methylated dAAs and cyclic dAAs were introduced into Arg-rich peptides [66]. The secondary structures of the resultant peptides differed depending on the type of dAAs introduced, but resistance to protease digestion was greatly enhanced whether or not the peptides formed stable helical structures. Moreover, the intracellular uptake of these peptides roughly correlated with their helicity, which also affected their cell membrane permeability.

To make the best use of the properties of dAAs, cyclic dAAs with a cationic functional group have been developed, and cationic CPPs containing these cyclic dAAs have been reported [67–70]. Arg-rich CPPs containing cationic cyclic dAAs adopt a helical structure, even in an aqueous solution, and are highly membrane permeable, especially over long incubation times, due to their tolerance to protease digestion. These cationic CPPs have been used as carriers for the delivery of pDNA, as they are able to complex with pDNA via electrostatic interaction. Indeed, CPPs containing cationic cyclic dAAs delivered pDNA more effectively than a commercially available transfection reagent requiring a longer incubation time.

7.5 Others

In addition to the various non-proteinogenic amino acids described here, there are a variety of other components for foldamers. Poly-N-substituted glycine oligomers, or peptoids, are another building block for peptide foldamers [71–75]. Peptoids have an achiral backbone and typically cannot form a stable secondary structure because the tertiary backbone amides are unable to act as donors for hydrogen bonding. However, it is possible for peptoids to adopt stable secondary structures by incorporating amide side chains that restrict the backbone conformation [76–78]. For example, steric repulsion between side chains, hydrophobic interactions, and charge-charge repulsion with backbone carbonyls are predicted to play a role in peptoid folding. In addition, cell-permeable peptoids with guanidino groups have been described by Wender et al. in 2000 [79]. Although that study did not investigate the secondary structure of peptoids, their membrane permeability was reportedly better than that of oligoarginine due to their stability against enzymatic degradation. Peptoids in which basic functional groups have been introduced into their side chains in various forms have also been reported. Several studies have investigated taking advantage of their high metabolic stability to use cell-permeable peptoids to target various intracellular organelles [80–83].

The desire to stabilize secondary structures, particularly helices, has led to the development of stapled peptides in which side chains are covalently bonded [84].

Strictly speaking, stapled peptides may deviate a little from the definition of a foldamer, but I address them here because they are folded oligomers. Verdine et al. described stapled CPPs synthesized by incorporating two α-methyl α-alkenyl dAAs at defined positions followed by ring-closing olefin metathesis to insert a hydrocarbon bridge [85–87]. These peptides contained a covalent bond across one face of the helical structure, which increased the α-helical content and protease resistance, enhanced target binding affinity, and promoted cell membrane penetration. More than 200 peptides with various sequences, stapling positions, and types, and physicochemical properties were analyzed by Verdine et al. in 2015 [88]. They revealed that the stapling type and peptide charge are key physical properties correlated with membrane permeability; other parameters, including peptide molecular weight, log P and pI value, among others, do not appear to be significantly associated with it. To further investigate the properties of the membrane permeability of stapled peptides and to analyze the similarities and differences in cellular uptake between stapled peptides and the original non-stapled CPPs, Verdine et al. also designed stapled peptides based on Tat (48–60), penetratin, and octaarginine and evaluated their dose- and time-dependent cell membrane permeability and secondary structure. All stapled peptides had increased helical content, but there were differences in their cell membrane permeability. Introducing the stapling moiety into Tat peptide decreased its membrane permeability, whereas it increased the membrane permeability of penetratin and octaarginine. Although it is difficult to conclude that the incorporation of a hydrocarbon stapling into the side chain of a peptide always leads to better cellular penetration, it does appear that stapling enhances the stability of helical structures and resistance to enzymatic degradation, often resulting in good and prolonged cell-penetrating ability.

Oligoureas, which are peptide mimetics with urea bonds similar to amide bonds, also function as helical foldamers that exhibit membrane permeability [89, 90]. The representative secondary structure formed by enantiopure N,N′-linked oligoureas is a 2.5-helix (with 12- and 14-membered rings of hydrogen bonds between the backbone groups, N(i + 2)–H···O=C(i) and N(i + 3)–H···O=C(i)) that adopts a very stable helical structure due to its two intramolecular hydrogen bonds [91–93] (Figure 7.9).

Guichard et al. reported the design of 8-residue oligoureas that mimic the amphipathic α-helix host defense peptide [93]. The helicity of these oligoureas was largely independent of their primary sequence, and they formed a 2.5-helix structure in the presence of liposomes. These oligoureas also displayed potent antimicrobial activity, the mechanism of which involves permeabilization of the bacterial membrane, and they exhibited high resistance to proteolytic degradation. Focusing on oligourea's property of forming a stable 2.5-helix structure, the same group also designed oligourea foldamers in which histidine and arginine-type urea units distribute along the polar face of the helical structure [94]. Their

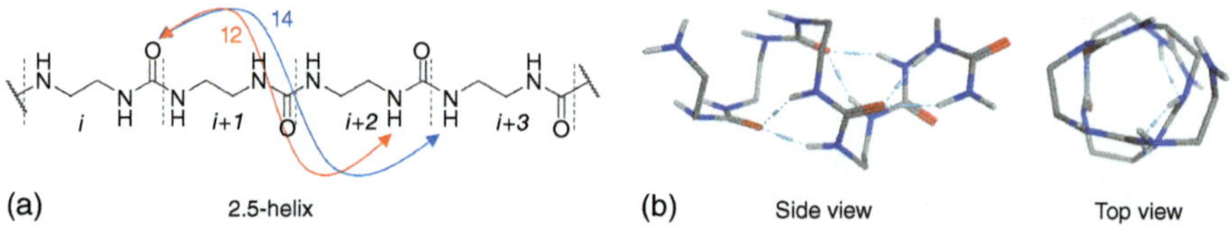

Figure 7.9 (a) Schematic presentation of the 12- and 14-membered hydrogen-bonded rings of a 2.5-helix structure. (b) Side and top views of the backbone conformation of a right-handed 2.5-helix within an oligourea foldamer. For clarity, only the hydrogen-bonded parts of the peptide backbone atoms are shown.

aim was to maximize intracellular delivery and interaction with DNA. And because this oligourea monomer showed good transfection efficiency, for even higher gene transfer efficiency they designed an amphipathic dimeric oligourea in which dimerization was mediated by disulfide bonding. The resultant dimerized oligourea foldamer maintained the same stable 2.5-helix structure as the monomer and exhibited highly efficient transfection for pDNA delivery. Moreover, this oligourea was highly resistant to protease digestion and could also be utilized for siRNA delivery [95]. Such oligourea foldamers represent potentially useful tools that overcome some of the current limitations of CPPs designed for gene delivery.

Aromatic oligoamide foldamers can also be classified as a type of foldamer having amide bonds [96–99]. Rotation about CONH—aryl bonds is not as free as rotation about CONH—alkyl bonds because of the conjugation between the amide and aromatic groups. This rotational restriction is generally strong enough to enable an accurate design of aromatic oligoamides having well-defined and predictable conformations, including helical and sheet structures. For example, when the hydrogen bonds form at the inner rim of the helix, an oligoamide having quinoline rings forms a helical structure in which only 2.5 units are necessary for one turn, while an oligoamide having pyridine rings forms a helix structure in which 4.5 units are necessary for one turn. These oligoamide foldamers form helices that are wider with a smaller helical pitch and are much more stable than those of peptides (Figure 7.10).

Because of the high hydrophobicity of their aromatic skeletons, there have been few reports evaluating the membrane permeability of oligoamide foldamers. Huc et al. reported that the introduction of cationic side chains into a quinoline-type backbone improved both the water solubility and cell membrane permeability of oligoamide foldamers [100, 101]. They revealed that a minimum length and minimum number of positive charges were required for efficient cell translocation and also demonstrated that these foldamers displayed practical advantages. In addition, Miranker et al. reported that some oligoamides in which carboxylic acids were introduced into a quinoline-type backbone formed stable helical conformations [102]. These oligoamides likely entered cells via both

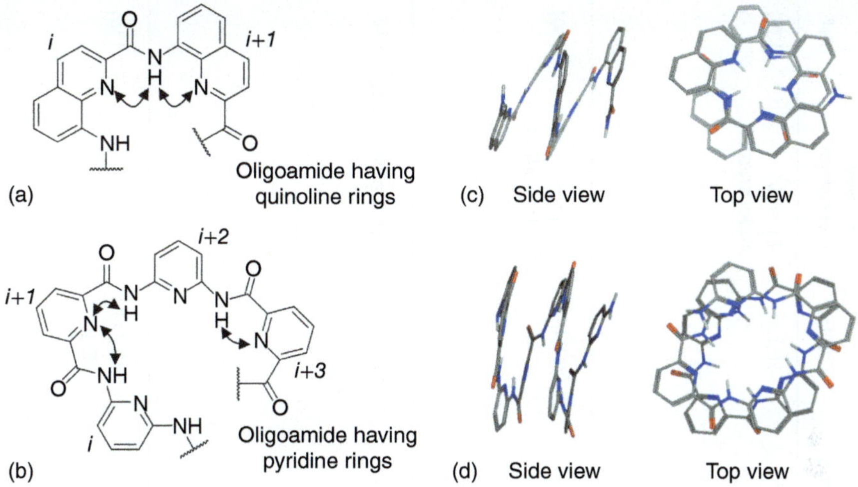

Figure 7.10 Schematic presentation of the oligoamide foldamers having (a) quinoline rings and (b) pyridine rings. Side and top views of the backbone conformation of an oligourea foldamer having (c) quinoline rings or (d) pyridine rings. For clarity, only the hydrogen-bonded parts of the peptide backbone atoms are shown.

active pathways such as endocytosis and energy-independent pathways. The latter mechanism is fundamentally different from the mechanism employed by previously reported CPPs, and the dynamic conformational changes of the molecule appear to be important for cellular uptake, even in the absence of cationic functional groups in the side chains.

7.6 Summary

In this section, various types of foldamers with amide bonds, other than those composed of naturally occurring L-a-amino acids (Table 7.1), while focusing primarily on helical foldamers, were described. The use of foldamers when designing CPPs is advantageous in that a stable secondary structure can be expected. Moreover, foldamers composed of non-proteinogenic amino acids are highly resistant to enzymatic degradation and can be expected to show high cell membrane permeability, making them potentially useful as CPPs and drug-delivery tools. Certainly, there are other polymer- and lipid-based tools for drug delivery. However, the advantages of CPP foldamers are that methods for their synthesis are well established, as for small molecules, yet molecular recognition is possible, as with high-molecular-weight compounds such as antibodies. Further development of CPP foldamers containing various building blocks is expected to contribute to advances in medicinal chemistry, drug discovery, and clinical use.

Table 7.1 A summary of studies on cell-penetrating foldamers.

β-amino acid					
Cell-penetrating foldamer	**Class**	**Cargo**	**Cell line**	**Proposed mechanism**	**Ref**
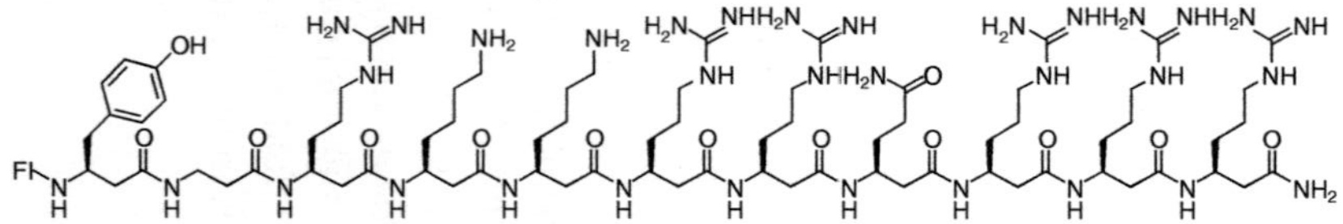 Fl-β3hTyr-β3hGly-β3hArg-β3hLys-β3hLys-β3hArg-β3hArg-β3hGln-β3hArg-β3hArg-β3hArg-NH2	Cationic	Fl[a]	HeLa	Energy independent	[29]
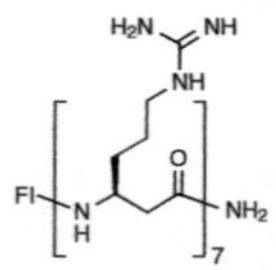 Fl-(β3hArg)7-NH2	Cationic	Fl	NIH3T3	Energy independent	[30]
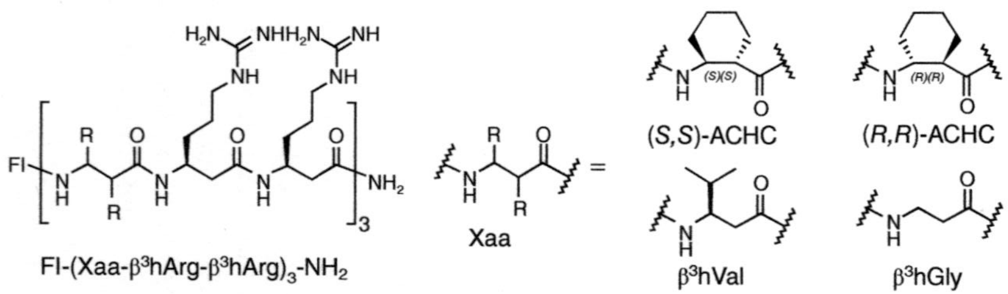 Fl-(Xaa-β3hArg-β3hArg)3-NH2	Amphipathic	Fl	HeLa	Micropinocytosis	[34, 35]
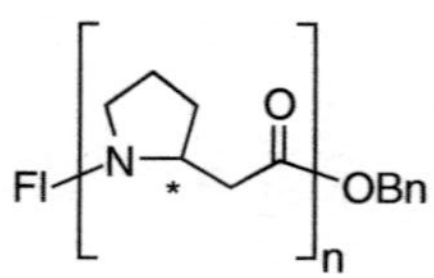 Fl-(β3hPro)n-NH2 n = 2, 6, 12 * (S) or (R)	Hydrophobic	Fl	NIT3T3, HFF	—	[36]

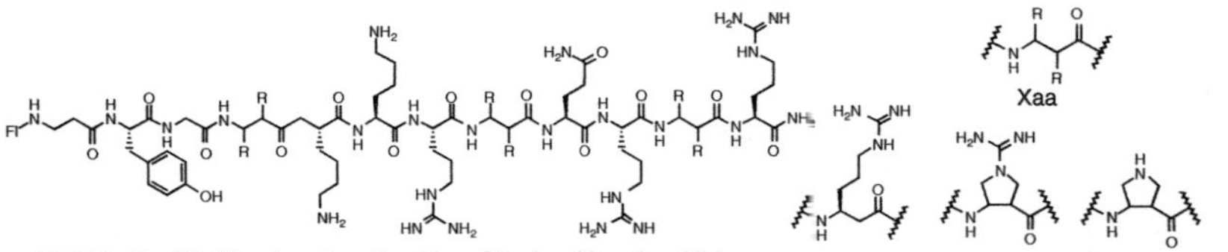 Fl-βAla-Tyr-Gly-Xaa-Lys-Lys-Arg-Xaa-Gln-Arg-Xaa-Arg-NH_2 β^{3}hArg (3*S*, 4*R*)-APC (3*S*, 4*R*)-APC^{Gu}	Cationic	Fl	HeLa	Macropinocytosis, clathrin endocytosis	[38]
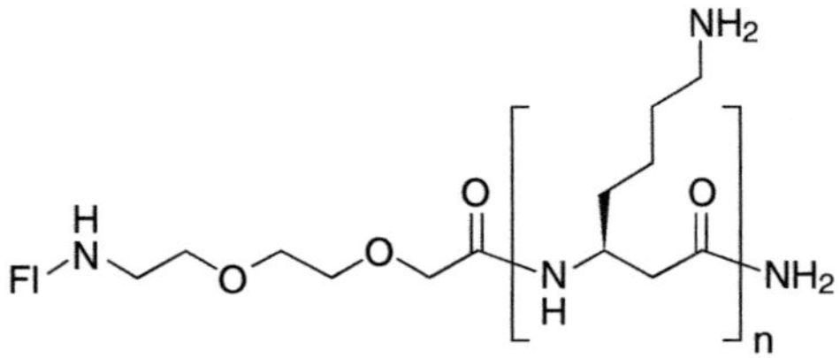 Fl-adoa-(β^{3}hLys)$_n$-NH_2 n = 5~8	Cationic	Fl	DU145, HCT15	Energy independent	[39]
γ-amino acid					
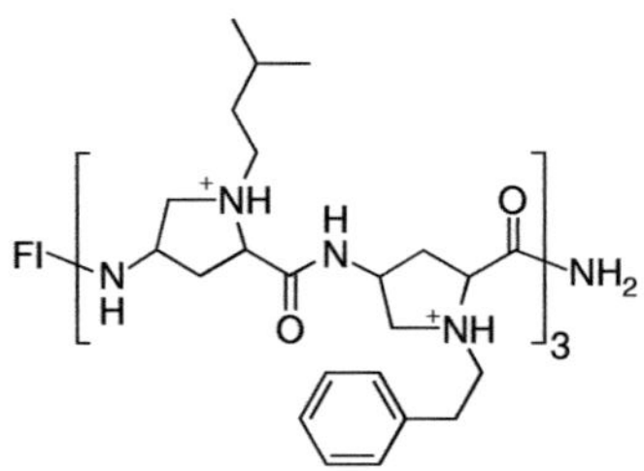 Fl-{γAmp[N^α-$CH_2CH_2CH(CH_3)_2$]-γAmp(N^α-CH_2CH_2Ph)}$_3$-NH_2 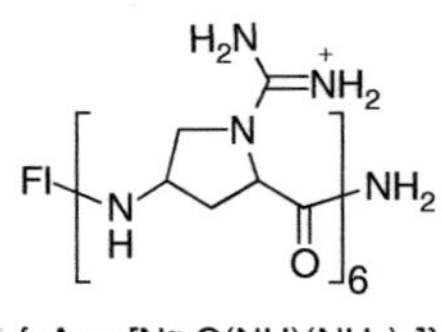 Fl-{γAmp[N^α-C(NH)$(NH_2)_2$]}$_6$-NH_2	Hydrophobic, cationic	Fl[a]	COS-1, HeLa	Endocytosis	[54, 55]

(Continued)

Table 7.1 (Continued)

Cell-penetrating foldamer	Class	Cargo	Cell line	Proposed mechanism	Ref
γ-amino acid					
R-Xaa$_1$-Lys-Leu-Phe-Lys-Lys-Ile-Leu-Lys-Xaa$_2$-Leu-PEG Xaa$_1$=Lya, Leu Xaa$_2$=Val, Phe, Tyr, Lys R=H, Ac, Ts, Bz 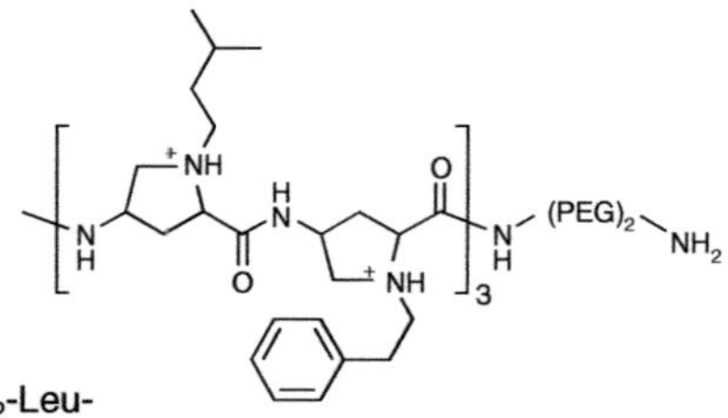 R-Xaa$_1$-Lys-Leu-Phe-Lys-Lys-Ile-Leu-Lys-Xaa$_2$-Leu--PEG-{γAmp[N$^{\alpha}$-$CH_2CH_2CH(CH_3)_2$]-γAmp(N$^{\alpha}$-CH_2CH_2Ph)}$_3$-PEG$_2$-NH_2	Cationic	Peptide	MDA-MB-231	—	[56]
α,α-disubstituted amino acid					
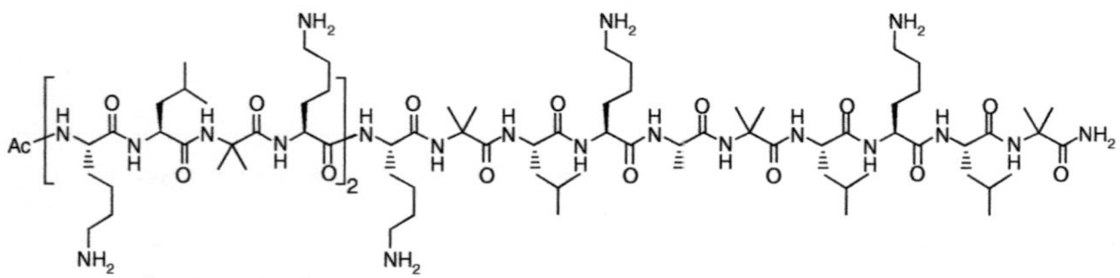 Ac-(Lys-Leu-Aib-Lys)$_2$-Lys-Aib-Leu-Lys-Ala-Aib-Leu-Lys-Leu-Aib-NH_2	Amphipathic	Oligo-nucleotide	A549	Endocytosis	[61, 62]

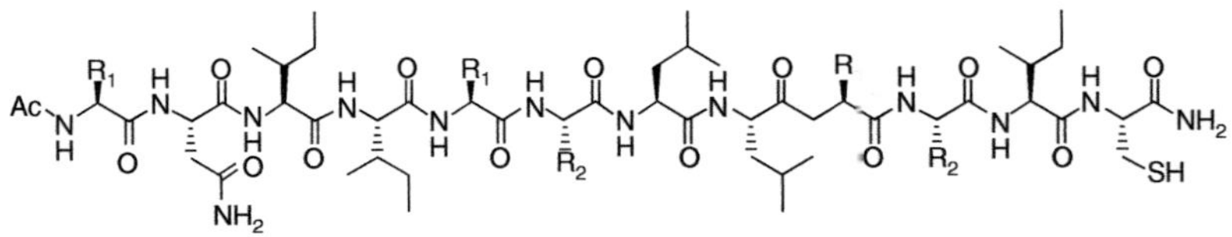 Ac-Xaa_1-Asn-Ile-Ile-Xaa_1-Xaa_2-Leu-Leu-Xaa_1-Xaa_2-Ile-Cys-NH_2 Xaa_1=Aib, Ala Xaa_2 = Pro, Aib	Hydrophobic	Oligo-nucleotide	A549, NIH3T3	Endocytosis	[62]
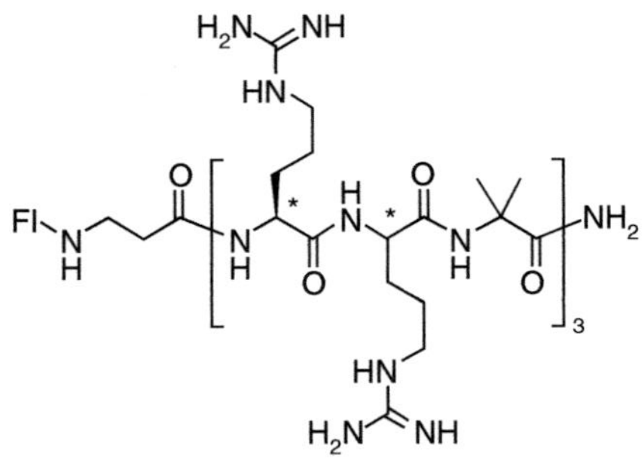 Fl-βAla-(Arg-Arg-Aib)$_3$-NH_2 *Arg = D-Arg, L-Arg	Amphipathic	Fl[a]	HeLa	Macropinocytosis	[63]
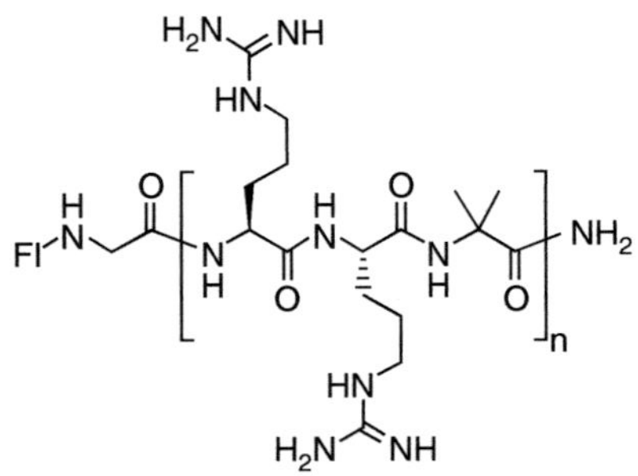 Fl-Gly-(Arg-Arg-Aib)$_n$-NH_2 n = 1~6	amphipathic	Fl, plasmid DNA	HeLa, Huh-7	Endocytosis	[64]

(Continued)

Table 7.1 (Continued)

γ-amino acid					
Cell-penetrating foldamer	**Class**	**Cargo**	**Cell line**	**Proposed mechanism**	**Ref**
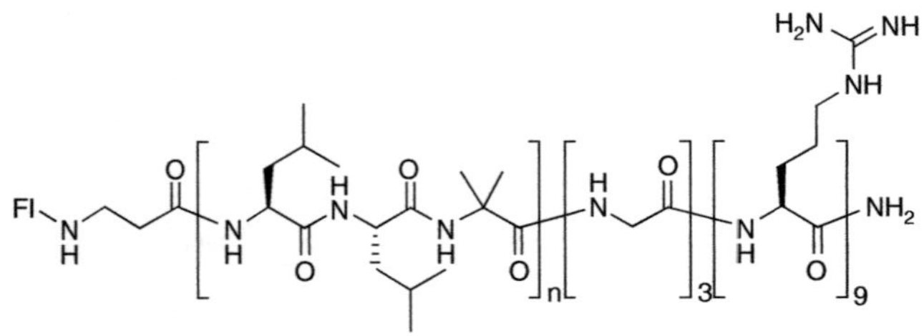 Fl-βAla-(Leu-Leu-Aib)$_n$-Gly$_3$-Arg$_9$-NH$_2$ n = 1~3	Amphipathic	siRNA	MCF-7, HeLa, NCI-H358, HepG2	Clathrin endocytosis	[65]
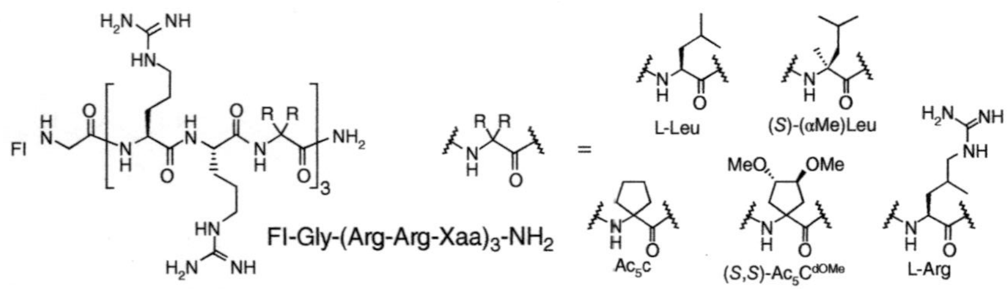	Amphipathic	Fl	HeLa, CHO-K1	Energy independent, endocytosis	[66]
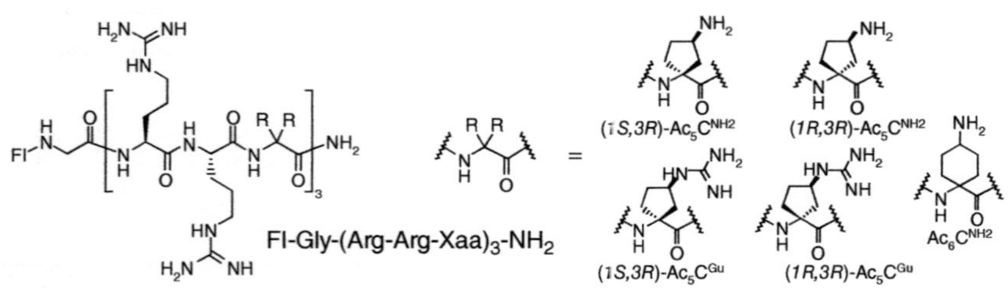	Cationic	Fl, lasmid DNA	Huh-7, HeLa, HEK293	Energy independent, endocytosis	[67, 68]

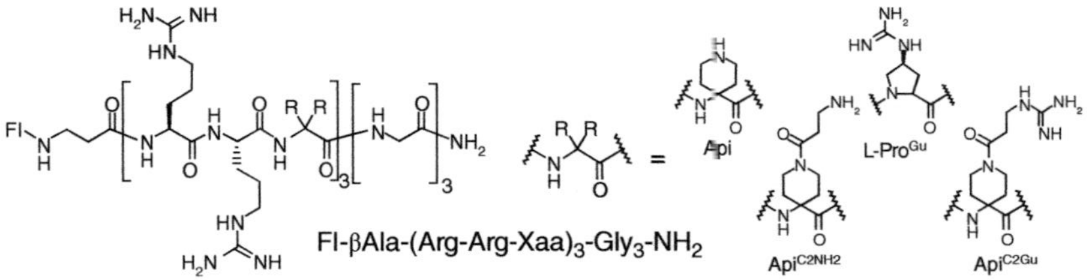	Cationic	Fl, plasmid DNA	Huh-7	Endocytosis	[69, 70]
Peptoid					
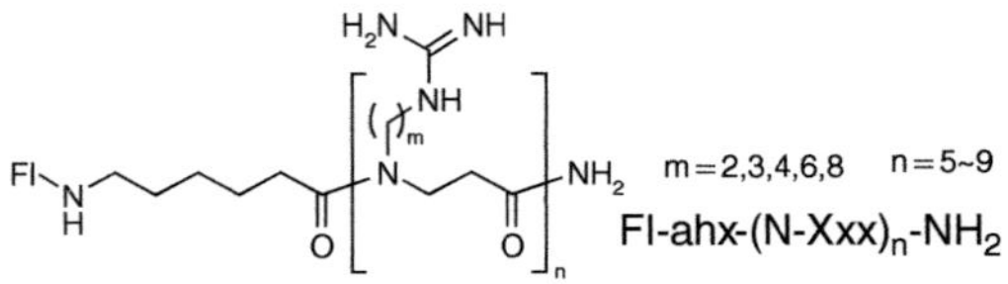	Cationic	Fl[a]	Jurkat	Energy dependent	[79]
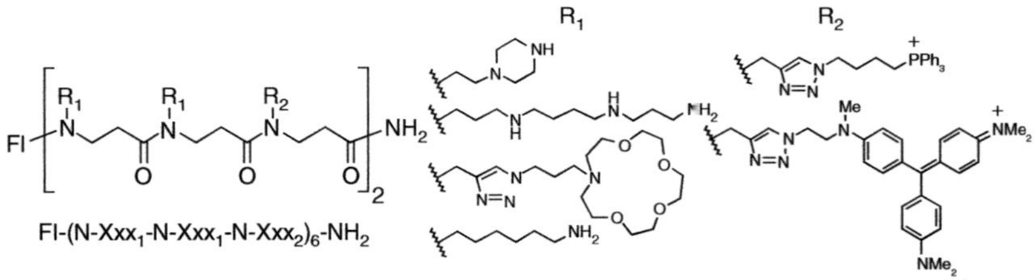	Cationic	Fl	HeLa	Endocytosis	[80]

(Continued)

Table 7.1 (Continued)

Cell-penetrating foldamer	Class	Cargo	Cell line	Proposed mechanism	Ref
Peptoid					
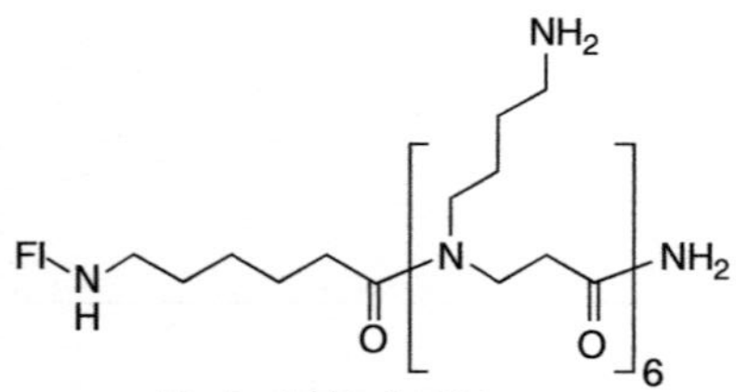 Fl-ahx-(N-Xxx)$_6$-NH$_2$ 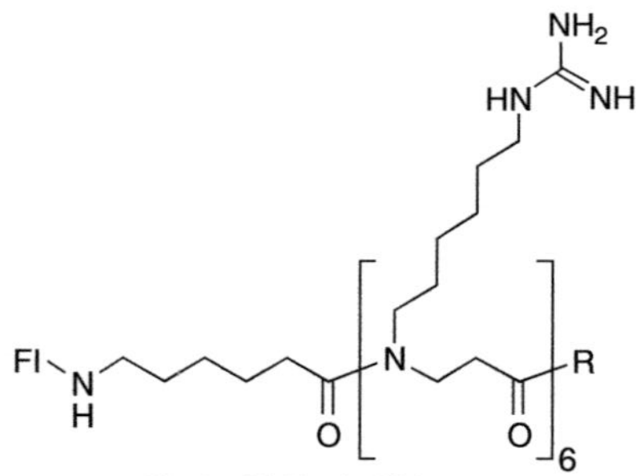 Fl-ahx-(N-Xxx)$_6$-NH$_2$	Cationic	Fl	Tobacco BY-2	Endocytosis	[81]
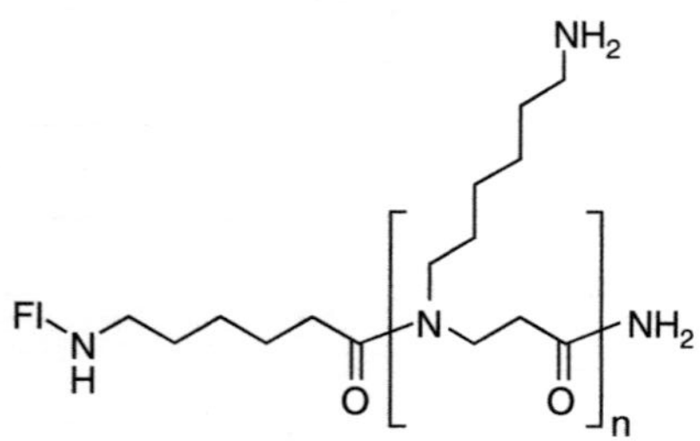 n = 3, 5, 7 Fl-ahx-(N-Xxx)$_n$-NH$_2$	Cationic	Fl	HEK293T	Energy dependent	[82, 83]

Stapled peptide					
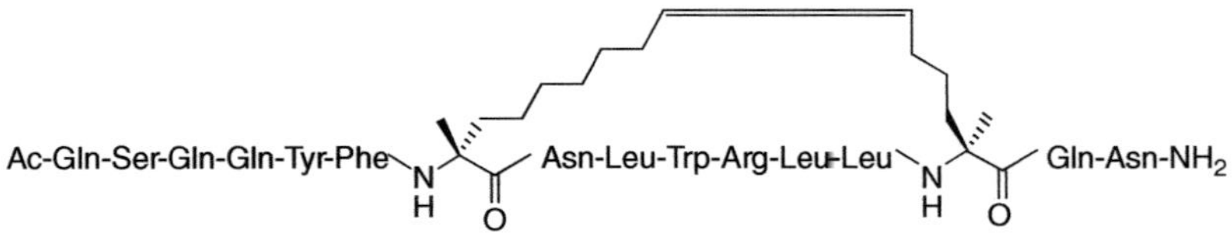 **SAH-p53-8**: Ac-Gln-Ser-Gln-Gln-Tyr-Phe-*R*$_8$-Asn-Leu-Trp-Arg-Leu-Leu-*S*$_5$-Gln-Asn-NH$_2$	—	Fl[a]	Jurkat	—	[86]
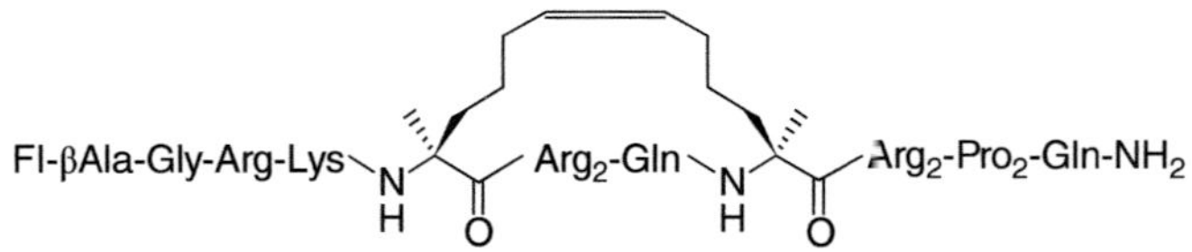 **SAH-TAT-2**: Fl-βAla-Gly-Arg-Lys-*S5*-Arg$_2$-Gln-*S5*-Arg$_2$-Pro$_2$-Gln-NH$_2$ 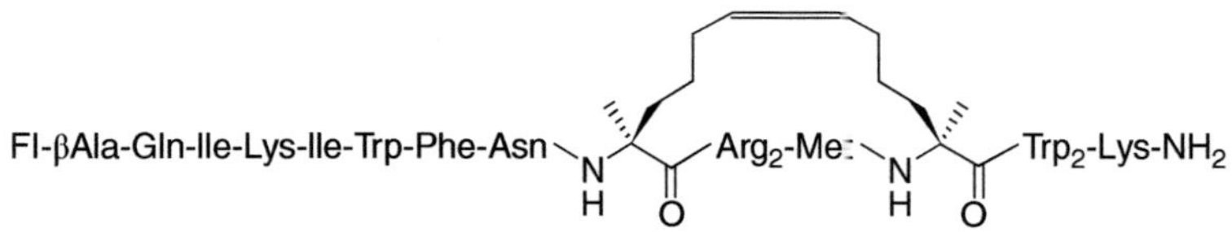 **SAH-PEN-2**: Fl-βAla-Arg-Gln-Ile-Lys-Ile-Trp-Phe-Asn-*S5*-Arg$_2$-Met-*S5*-Trp$_2$-Lys-NH$_2$ 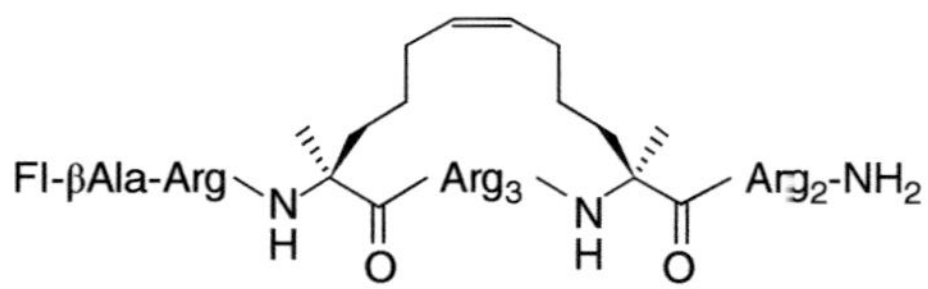 **SAH-Arg8**: Fl-βAla-Arg-*S5*-Arg$_3$-*S5*-Arg$_2$-NH$_2$	Amphipathic, cationic	Fl	U2OS, CHO	Energy dependent endocytosis	[88]

(Continued)

Table 7.1 (Continued)

oligourea					
Cell-penetrating foldamer	**Class**	**Cargo**	**Cell line**	**Proposed mechanism**	**Ref**
Oligourea					
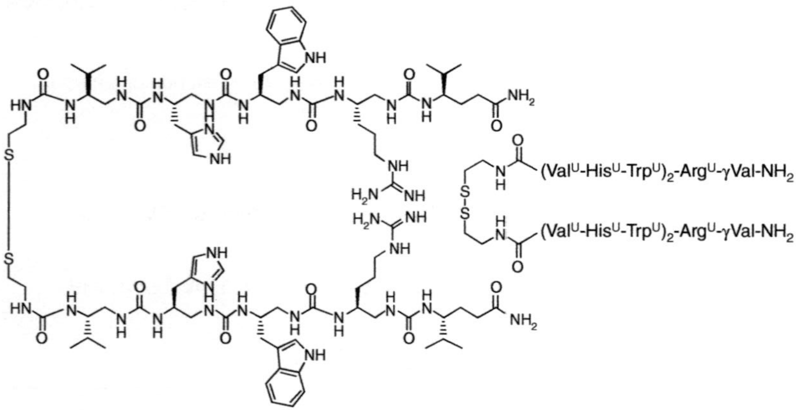	Amphipathic	Plasmid DNA	HEK293	—	[94, 95]

Oligoamide					
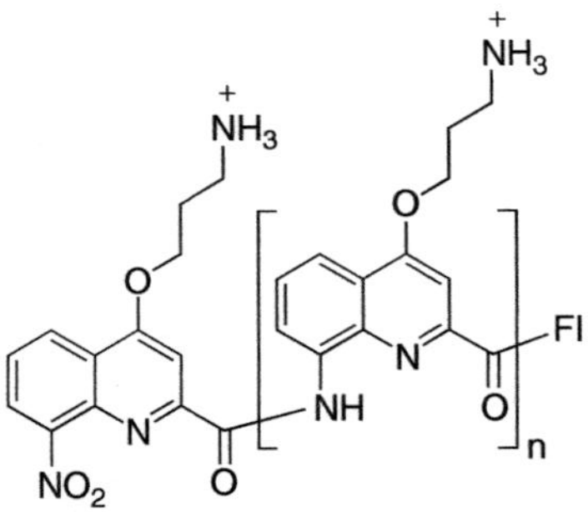 NO_2-(oligoquinoline)-Fl n = 3,5,7	Cationic	Fl[a]	HeLa, Jurkat, Huh-7	Energy dependent	[100, 101]
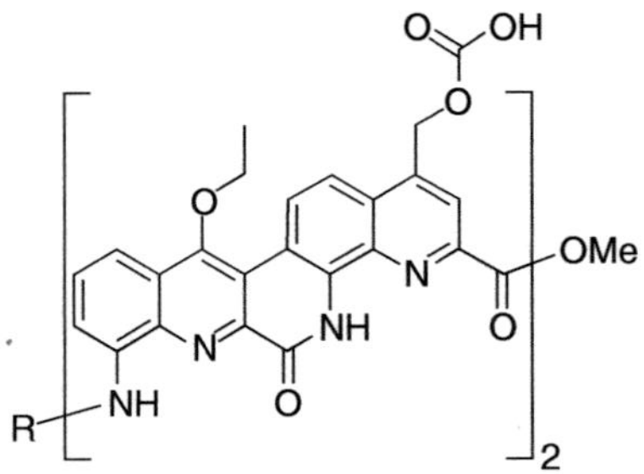 R-(oligoquinoline)-OMe R = NO_2, Fl	Anionic	Fl	INS-1, COS-1	Energy dependent	[102]

a) Fl indicates fluorescein or its derivatives.

References

1 Appella, D.H., Christianson, L.A., Karle, I.L. et al. (1996). β-Peptide foldamers: robust helix formation in a new family of β-amino acid oligomers. *Journal of the American Chemical Society* **118**: 13071–13072. https://doi.org/10.1021/ja963290l.

2 Gellman, S.H. (1998). Foldamers: a manifesto. *Accounts of Chemical Research* **31**: 173–180. https://doi.org/10.1021/ar960298r.

3 Hill, D.J., Mio, M.J., Prince, R.B. et al. (2001). A field guide to foldamers. *Chemical Reviews* **101**: 3893–4012. https://doi.org/10.1021/cr990120t.

4 Guichard, G. and Huc, I. (2011). Synthetic foldamers. *Chemical Communications* **47**: 5933–5941. https://doi.org/10.1039/c1cc11137j.

5 Gopalakrishnan, R., Frolov, A.I., Knerr, L. et al. (2016). Therapeutic potential of foldamers: from chemical biology tools to drug candidates? *Journal of Medicinal Chemistry* **59**: 9599–9621. https://doi.org/10.1021/ACS.JMEDCHEM.6B00376.

6 Pasco, M., Dolain, C., and Guichard, G. (2017). Foldamers in medicinal chemistry. *Comprehensive Supramolecular Chemistry II* **5**: 89–125. https://doi.org/10.1016/B978-0-12-409547-2.12565-X.

7 Oba, M. (2019). Cell-penetrating peptide foldamers: drug-delivery tools. *ChemBioChem* **20**: 2041–2045. https://doi.org/10.1002/CBIC.201900204.

8 Cheng, R.P., Gellman, S.H., and DeGrado, W.F. (2001). β-peptides: from structure to function. *Chemical Reviews* **101**: 3219–3232. https://doi.org/10.1021/CR000045I.

9 Seebach, D., Beck, A.K., and Bierbaum, D.J. (2004). The world of β- and γ-peptides comprised of homologated proteinogenic amino acids and other components. *Chemistry & Biodiversity* **1**: 1111–1239. https://doi.org/10.1002/CBDV.200490087.

10 Appella, D.H., Christianson, L.A., Klein, D.A. et al. (1997). Residue-based control of helix shape in β-peptide oligomers. *Nature* 387 (6631): 381–384. https://doi.org/10.1038/387381a0.

11 Crisma, M., Formaggio, F., Moretto, A., and Toniolo, C. (2006). Peptide helices based on α-amino acids. *Peptide Science* **84**: 3–12. https://doi.org/10.1002/BIP.20357.

12 Kaur, K., Sprules, T., Soliman, W. et al. (2008). Right-handed 14-helix in β3-peptides from L-aspartic acid monomers. *Biochimica et Biophysica Acta (BBA) - Proteins and Proteomics* **1784**: 658–665. https://doi.org/10.1016/J.BBAPAP.2008.01.009.

13 Hart, S.A., Bahadoor, A.B.F., Matthews, E.E. et al. (2003). Helix macrodipole control of β3-peptide 14-helix stability in water. *Journal of the American Chemical Society* **125**: 4022–4023. https://doi.org/10.1021/JA029868A.

14 Kritzer, J.A., Tirado-Rives, J., Hart, S.A. et al. (2004). Relationship between side chain structure and 14-helix stability of β3-peptides in water. *Journal of the American Chemical Society* **127**: 167–178. https://doi.org/10.1021/JA0459375.

15 Choi, S.H., Guzei, I.A., Spencer, L.C., and Gellman, S.H. (2010). Crystallographic characterization of 12-helical secondary structure in β-peptides containing side chain groups. *Journal of the American Chemical Society* **132**: 13879–13885. https://doi.org/10.1021/JA1062532.

16 Lee, H.L., Syud, F.A., Wang, X., and Gellman, S.H. (2001). Diversity in short β-peptide 12-helices: high-resolution structural analysis in aqueous solution of a hexamer containing sulfonylated pyrrolidine residues. *Journal of the American Chemical Society* **123**: 7721–7722. https://doi.org/10.1021/JA010734R.

17 Sharma, G.V.M., Reddy, N.Y., Ravi, R. et al. (2012). Synthesis of C-linked carbo-β2-amino acids and β2-peptides: design of new motifs for left-handed 12/10- and 10/12-mixed helices. *Organic & Biomolecular Chemistry* **10**: 9191–9203. https://doi.org/10.1039/C2OB26615F.

18 Mándity, I.M., Fülöp, L., Vass, E. et al. (2010). Building β-peptide H10/12 foldamer helices with six-membered cyclic side-chains: fine-tuning of folding and self-assembly. *Organic Letters* **12**: 5584–5587. https://doi.org/10.1021/OL102494M.

19 Sharma, G.V.M., Reddy, K.R., Krishna, P.R. et al. (2003). Robust mixed 10/12 helices promoted by "alternating chirality" in a new family of C-linked carbo-β-peptides. *Journal of the American Chemical Society* **125**: 13670–13671. https://doi.org/10.1021/JA035752I.

20 Shin, S., Lee, M., Guzei, I.A. et al. (2016). 12/10-helical β-peptide with dynamic folding propensity: coexistence of right- and left-handed helices in an enantiomeric foldamer. *Journal of the American Chemical Society* **138**: 13390–13395. https://doi.org/10.1021/JACS.6B08235.

21 Schmitt, M.A., Choi, S.H., Guzei, I.A., and Gellman, S.H. (2005). Residue requirements for helical folding in short α/β-peptides: crystallographic characterization of the 11-helix in an optimized sequence. *Journal of the American Chemical Society* **127**: 13130–13131. https://doi.org/10.1021/JA0536163.

22 Hayen, A., Schmitt, M.A., Ngassa, F.N. et al. (2004). Two helical conformations from a single foldamer backbone: "split personality" in short α/β-peptides. *Angewandte Chemie International Edition* **43**: 505–510. https://doi.org/10.1002/ANIE.200352125.

23 Sharma, G.V.M., Yadav, T.A., Choudhary, M., and Kunwar, A.C. (2012). Design of β-amino acid with backbone-side chain interactions: Stabilization of 14/15-helix in α/β-peptides. *Journal of Organic Chemistry* **77**: 6834–6848. https://doi.org/10.1021/JO300865D.

24 Soo, H.C., Guzei, I.A., and Gellman, S.H. (2007). Crystallographic characterization of the α/β-peptide 14/15-helix. *Journal of the American Chemical Society* **129**: 13780–13781. https://doi.org/10.1021/JA0753344.

25 Sharma, G.V.M., Nagendar, P., Jayaprakash, P. et al. (2005). 9/11 Mixed helices in α/β peptides derived from C-linked carbo-β-amino acid and L-ala repeats. *Angewandte Chemie International Edition* **44**: 5878–5882. https://doi.org/10.1002/ANIE.200501247.

26 Srinivasulu, G., Kiran Kumar, S., Sharma, G.V.M., and Kunwar, A.C. (2006). 11/9-Mixed helices in the α/β-peptides derived from alternating α- and β-amino acids with proteinogenic side chains. *Journal of Organic Chemistry* **71**: 8395–8400. https://doi.org/10.1021/JO0612980.

27 Guo, L., Almeida, A.M., Zhang, W. et al. (2010). Helix formation in preorganized β/γ-peptide foldamers: hydrogen-bond analogy to the α-helix without α-amino acid residues. *Journal of the American Chemical Society* **132**: 7868–7869. https://doi.org/10.1021/JA103233A.

28 Grison, C.M., Robin, S., and Aitken, D.J. (2016). 13-Helix folding of a β/γ-peptide manifold designed from a "minimal-constraint" blueprint. *Chemical Communications* **52**: 7802–7805. https://doi.org/10.1039/C6CC02142E.

29 Umezawa, N., Gelman, M.A., Haigis, M.C. et al. (2002). Translocation of a β-peptide across cell membranes. *Journal of the American Chemical Society* **124**: 368–369. https://doi.org/10.1021/JA017283V.

30 Rueping, M., Mahajan, Y., Sauer, M., and Seebach, D. (2002). Cellular uptake studies with β-peptides. *ChemBioChem* **3**: 257–259. https://doi.org/10.1002/1439-7633(20020301)3:2/3<257::AID-CBIC257>3.0.CO;2-S.

31 Potocky, T.B., Menon, A.K., and Gellman, S.H. (2003). Cytoplasmic and nuclear delivery of a TAT-derived peptide and a β-peptide after endocytic uptake into HeLa cells. *Journal of Biological Chemistry* **278**: 50188–50194. https://doi.org/10.1074/JBC.M308719200.

32 Frankel, A.D. and Pabo, C.O. (1988). Cellular uptake of the tat protein from human immunodeficiency virus. *Cell* **55**: 1189–1193. https://doi.org/10.1016/0092-8674(88)90263-2.

33 Green, M. and Loewenstein, P.M. (1988). Autonomous functional domains of chemically synthesized human immunodeficiency virus tat trans-activator protein. *Cell* **55**: 1179–1188. https://doi.org/10.1016/0092-8674(88)90262-0.

34 Potocky, T.B., Silvius, J., Menon, A.K., and Gellman, S.H. (2007). HeLa cell entry by guanidinium-rich β-peptides: importance of specific cation–cell surface interactions. *ChemBioChem* **8**: 917–926. https://doi.org/10.1002/CBIC.200600563.

35 Potocky, T.B., Menon, A.K., and Gellman, S.H. (2005). Effects of conformational stability and geometry of guanidinium display on cell entry by β-peptides. *Journal of the American Chemical Society* **127**: 3686–3687. https://doi.org/10.1021/JA042566J.

36 Kolesinska, B., Podwysocka, D.J., Rueping, M.A. et al. (2013). Permeation through phospholipid bilayers, skin-cell penetration, plasma stability, and CD spectra of α- and β-oligoproline derivatives. *Chemistry & Biodiversity* **10**: 1–38. https://doi.org/10.1002/CBDV.201200393.

37 Abele, S., Vögtli, K., and Seebach, D. (1999). Oligomers of β2- and of β3-homoproline: what are the secondary structures of β-peptides lacking H-bonds? *Helvetica Chimica Acta* **82**: 1539–1558. https://doi.org/10.1002/(SICI)1522-2675(19991006)82:10<1539::AID-HLCA1539>3.0.CO;2-E.

38 Demizu, Y., Oba, M., Okitsu, K. et al. (2015). A preorganized β-amino acid bearing a guanidinium side chain and its use in cell-penetrating peptides. *Organic & Biomolecular Chemistry* **13**: 5617–5620. https://doi.org/10.1039/C5OB00389J.

39 García-Echeverría, C. and Ruetz, S. (2003). β-homolysine oligomers: a new class of trojan carriers. *Bioorganic & Medicinal Chemistry Letters* **13**: 247–251. https://doi.org/10.1016/S0960-894X(02)00885-5.

40 Harker, E.A., Schepartz, A., Harker, E.A., and Schepartz, A. (2009). Cell-permeable β-peptide inhibitors of p53/hDM2 complexation. *ChemBioChem* **10**: 990–993. https://doi.org/10.1002/CBIC.200900049.

41 Bouillère, F., Thétiot-Laurent, S., Kouklovsky, C., and Alezra, V. (2011). Foldamers containing γ-amino acid residues or their analogues: Structural features and applications. *Amino Acids* **41**: 687–707. https://doi.org/10.1007/S00726-011-0893-3.

42 Legrand, B. and Maillard, L.T. (2021). α,β-unsaturated γ-peptide foldamers. *ChemPlusChem* **86**: 629–645. https://doi.org/10.1002/CPLU.202100045.

43 Baldauf, C., Günther, R., and Hofmann, H.J. (2003). Helix formation and folding in γ-peptides and their vinylogues. *Helvetica Chimica Acta* **86**: 2573–2588. https://doi.org/10.1002/HLCA.200390208.

44 Guo, L., Zhang, W., Reidenbach, A.G. et al. (2011). Characteristic structural parameters for the γ-peptide 14-helix: importance of subunit preorganization. *Angewandte Chemie International Edition* **50**: 5843–5846. https://doi.org/10.1002/ANIE.201101301.

45 Hintermann, T., Gademann, K., Jaun, B., and Seebach, D. (1998). γ-Peptides forming more stable secondary structures than α-peptides: synthesis and helical NMR solution structure of the γ-hexapeptide analog of H-(Val-Ala-Leu)2-OH. *Helvetica Chimica Acta* **81**: 983–1002. https://doi.org/10.1002/HLCA.19980810514.

46 Hanessian, S., Luo, X., Schaum, R., and Michnick, S. (1998). Design of secondary structures in unnatural peptides: stable helical γ- tetra-, hexa-, and octapeptides and consequences of α-substitution. *Journal of the American Chemical Society* **120**: 8569–8570. https://doi.org/10.1021/JA9814671.

47 Mathieu, L., Legrand, B., Deng, C. et al. (2013). Helical oligomers of thiazole-based γ-amino acids: synthesis and structural studies. *Angewandte Chemie International Edition* **52**: 6006–6010. https://doi.org/10.1002/ANIE.201302106.

48 Sharma, G.V.M., Jayaprakash, P., Narsimulu, K. et al. (2006). A left-handed 9-helix in γ-peptides: synthesis and conformational studies of oligomers with dipeptide repeats of C-linked carbo-γ4-amino acids and γ-aminobutyric acid. *Angewandte Chemie* **118**: 3010–3013. https://doi.org/10.1002/ANGE.200504380.

49 Vasudev, P.G., Shamala, N., Ananda, K. et al. (2005). C9 Helices and ribbons in γ-peptides: crystal structures of gabapentin oligomers. *Angewandte Chemie* **117**: 5052–5055. https://doi.org/10.1002/ANGE.200501055.

50 Jadhav, S.V., Bandyopadhyay, A., and Gopi, H.N. (2012). Protein secondary structure mimetics: crystal conformations of α/γ4-hybrid peptide12-helices with proteinogenic side chains and their analogy with α- and β-peptide helices. *Organic & Biomolecular Chemistry* **11**: 509–514. https://doi.org/10.1039/C2OB26805A.

51 Reddy, M.M.B., Basuroy, K., Aravinda, S., and Balaram, P. (2016). Crystallographic characterization of the α,γ C12 helix in hybrid peptide sequences. *Journal of Peptide Science* **22**: 504–510. https://doi.org/10.1002/PSC.2896.

52 Vasudev, P.G., Chatterjee, S., Ananda, K. et al. (2008). Hybrid αγ polypeptides: structural characterization of a C12/C10 helix with alternating hydrogen-bond polarity. *Angewandte Chemie International Edition* **47**: 6430–6432. https://doi.org/10.1002/ANIE.200801665.

53 Sharma, G.V.M., Jadhav, V.B., Ramakrishna, K.V.S. et al. (2006). 12/10- and 11/13-mixed helices in α/γ- and β/γ- hybrid peptides containing C-linked carbo-γ-amino acids with alternating α- and β-amino acids. *Journal of the American Chemical Society* **128**: 14657–14668. https://doi.org/10.1021/JA064875A.

54 Farrera-Sinfreu, J., Giralt, E., Castel, S. et al. (2005). Cell-penetrating cis-γ-amino-L-proline-derived peptides. *Journal of the American Chemical Society* **127**: 9459–9468. https://doi.org/10.1021/JA051648K.

55 Farrera-Sinfreu, J., Zaccaro, L., Vidal, D. et al. (2004). A new class of foldamers based on cis-γ-amino-l-proline1,2. *Journal of the American Chemical Society* **126**: 6048–6057. https://doi.org/10.1021/JA0398621.

56 Rosés, C., Carbajo, D., Sanclimens, G. et al. (2012). Cell-penetrating γ-peptide/antimicrobial undecapeptide conjugates with anticancer activity. *Tetrahedron* **68**: 4406–4412. https://doi.org/10.1016/J.TET.2012.02.003.

57 Crisma, M. and Toniolo, C. (2015). Helical screw-sense preferences of peptides based on chiral, C**α**-tetrasubstituted α-amino acids. *Peptide Science* **104**: 46–64. https://doi.org/10.1002/BIP.22581.

58 Tanaka, M. (2007). Design and synthesis of chiral α,α-disubstituted amino acids and conformational study of their oligopeptides. *Chemical and Pharmaceutical Bulletin* **55**: 349–358. https://doi.org/10.1248/CPB.55.349.

59 Millhauser, G.L. (2002). Views of helical peptides: a proposal for the position of 310-helix along the thermodynamic folding pathway. *Biochemistry* **34**: 3873–3877. https://doi.org/10.1021/BI00012A001.

60 Zhang, L. and Hermans, J. (2002). 310 helix versus .alpha.-helix: a molecular dynamics study of conformational preferences of Aib and alanine. *Journal of the American Chemical Society* **116**: 11915–11921. https://doi.org/10.1021/JA00105A034.

61 Wada, S.I., Hashimoto, Y., Kawai, Y. et al. (2013). Effect of Ala replacement with Aib in amphipathic cell-penetrating peptide on oligonucleotide delivery into cells. *Bioorganic & Medicinal Chemistry* **21**: 7669–7673. https://doi.org/10.1016/J.BMC.2013.10.029.

62 Wada, S.I., Tsuda, H., Okada, T., and Urata, H. (2011). Cellular uptake of Aib-containing amphipathic helix peptide. *Bioorganic & Medicinal Chemistry Letters* **21**: 5688–5691. https://doi.org/10.1016/J.BMCL.2011.08.030.

63 Yamashita, H., Demizu, Y., Shoda, T. et al. (2014). Amphipathic short helix-stabilized peptides with cell-membrane penetrating ability. *Bioorganic & Medicinal Chemistry* **22**: 2403–2408. https://doi.org/10.1016/J.BMC.2014.03.005.

64 Oba, M., Ito, Y., Umeno, T. et al. (2019). Plasmid DNA delivery using cell-penetrating peptide foldamers composed of Arg-Arg-Aib repeating sequences. *ACS Biomaterials Science and Engineering* **5**: 5660–5668. https://doi.org/10.1021/ACSBIOMATERIALS.8B01451/SUPPL_FILE/AB8B01451_SI_001.PDF.

65 Misawa, T., Ohoka, N., Oba, M. et al. (2019). Development of 2-aminoisobutyric acid (Aib)-rich cell-penetrating foldamers for efficient siRNA delivery. *Chemical Communications* **55**: 7792–7795. https://doi.org/10.1039/C9CC02203A.

66 Oba, M., Nagano, Y., Kato, T., and Tanaka, M. (2019). Secondary structures and cell-penetrating abilities of arginine-rich peptide foldamers. *Scientific Reports* 9 (1): 1–9. https://doi.org/10.1038/s41598-018-38063-8.

67 Kato, T., Kita, Y., Iwanari, K. et al. (2021). Synthesis of six-membered carbocyclic ring α,α-disubstituted amino acids and arginine-rich peptides to investigate the effect of ring size on the properties of the peptide. *Bioorganic & Medicinal Chemistry* **38**: 116111. https://doi.org/10.1016/J.BMC.2021.116111.

68 Kato, T., Oba, M., Nishida, K., and Tanaka, M. (2018). Cell-penetrating peptides using cyclic α,α-disubstituted α-amino acids with basic functional groups. *ACS Biomaterials Science and Engineering* **4**: 1368–1376. https://doi.org/10.1021/ACSBIOMATERIALS.8B00180.

69 Kato, T., Yamashita, H., Misawa, T. et al. (2016). Plasmid DNA delivery by arginine-rich cell-penetrating peptides containing unnatural amino acids. *Bioorganic & Medicinal Chemistry* **24**: 2681–2687. https://doi.org/10.1016/J.BMC.2016.04.031.

70 Yamashita, H., Kato, T., Oba, M. et al. (2016). Development of a cell-penetrating peptide that exhibits responsive changes in its secondary structure in the cellular environment. *Scientific Reports* 6 (1): 1–8. https://doi.org/10.1038/srep33003.

71 Patch, J.A. and Barron, A.E. (2002). Mimicry of bioactive peptides via non-natural, sequence-specific peptidomimetic oligomers. *Current Opinion in Chemical Biology* **6**: 872–877. https://doi.org/10.1016/S1367-5931(02)00385-X.

72 Simon, R.J., Kania, R.S., Zuckermann, R.N. et al. (1992). Peptoids: a modular approach to drug discovery. *Proceedings of the National Academy of Sciences* **89**: 9367–9371. https://doi.org/10.1073/PNAS.89.20.9367.

73 Zuckermann, R.N. (2011). Peptoid origins. *Peptide Science* **96**: 545–555. https://doi.org/10.1002/BIP.21573.

74 Kirshenbaum, K., Barron, A.E., Goldsmith, R.A. et al. (1998). Sequence-specific polypeptoids: a diverse family of heteropolymers with stable secondary structure. *Proceedings of the National Academy of Sciences* **95**: 4303–4308. https://doi.org/10.1073/PNAS.95.8.4303.

75 Fowler, S.A. and Blackwell, H.E. (2009). Structure–function relationships in peptoids: Recent advances toward deciphering the structural requirements for biological function. *Organic & Biomolecular Chemistry* **7**: 1508–1524. https://doi.org/10.1039/B817980H.

76 Burkoth, T.S., Beausoleil, E., Kaur, S. et al. (2002). Toward the synthesis of artificial proteins: the discovery of an amphiphilic helical peptoid assembly. *Chemistry & Biology* **9**: 647–654. https://doi.org/10.1016/S1074-5521(02)00140-0.

77 Shin, S.B.Y. and Kirshenbaum, K. (2007). Conformational rearrangements by water-soluble peptoid foldamers. *Organic Letters* **9**: 5003–5006. https://doi.org/10.1021/OL702207N.

78 Armand, P., Kirshenbaum, K., Falicov, A. et al. (1997). Chiral N-substituted glycines can form stable helical conformations. *Folding and Design* **2**: 369–375. https://doi.org/10.1016/S1359-0278(97)00051-5.

79 Wender, P.A., Mitchell, D.J., Pattabiraman, K. et al. (2000). The design, synthesis, and evaluation of molecules that enable or enhance cellular uptake: Peptoid molecular transporters. *Proceedings of the National Academy of Sciences* **97**: 13003–13008. https://doi.org/10.1073/PNAS.97.24.13003.

80 Kölmel, D.K., Hörner, A., Rönicke, F. et al. (2014). Cell-penetrating peptoids: Introduction of novel cationic side chains. *European Journal of Medicinal Chemistry* **79**: 231–243. https://doi.org/10.1016/J.EJMECH.2014.03.078.

81 Birtalan, E., Rudat, B., Kölmel, D.K. et al. (2011). Investigating rhodamine B-labeled peptoids: scopes and limitations of its applications. *Peptide Science* **96**: 694–701. https://doi.org/10.1002/BIP.21617.

82 Eggenberger, K., Birtalan, E., Schröder, T. et al. (2009). Passage of trojan peptoids into plant cells. *ChemBioChem* **10**: 2504–2512. https://doi.org/10.1002/CBIC.200900331.

83 Peretto, I., Sanchez-Martin, R.M., Wang, X.H. et al. (2003). Cell penetrable peptoid carrier vehicles: synthesis and evaluation. *Chemical Communications* **3**: 2312–2313. https://doi.org/10.1039/B306438G.

84 Verdine, G.L. and Hilinski, G.J. (2012). Stapled peptides for intracellular drug targets. *Methods in Enzymology* **503**: 3–33. https://doi.org/10.1016/B978-0-12-39696 2-0.00001-X.

85 Kim, Y.W., Grossmann, T.N., and Verdine, G.L. (2011). Synthesis of all-hydrocarbon stapled α-helical peptides by ring-closing olefin metathesis. *Nature Protocols* 6 (6): 761–771. https://doi.org/10.1038/nprot.2011.324.

86 Bernal, F., Tyler, A.F., Korsmeyer, S.J. et al. (2007). Reactivation of the p53 tumor suppressor pathway by a stapled p53 peptide. *Journal of the American Chemical Society* **129**: 2456–2457. https://doi.org/10.1021/JA0693587.

87 Schafmeister, C.E., Po, J., and Verdine, G.L. (2000). An all-hydrocarbon cross-linking system for enhancing the helicity and metabolic stability of peptides. *Journal of the American Chemical Society* **122**: 5891–5892. https://doi.org/10.1021/JA000563A.

88 Chu, Q., Moellering, R.E., Hilinski, G.J. et al. (2015). Towards understanding cell penetration by stapled peptides. *MedChemComm* **6**: 111–119. https://doi.org/10.1039/C4MD00131A.

89 Yoo, S.H., Li, B., Dolain, C. et al. (2021). Urea based foldamers. *Methods in Enzymology* **656**: 59–92. https://doi.org/10.1016/BS.MIE.2021.04.019.

90 Bécart, D., Diemer, V., Salaün, A. et al. (2017). Helical oligourea foldamers as powerful hydrogen bonding catalysts for enantioselective C-C bond-forming reactions. *Journal of the American Chemical Society* **139**: 12524–12532. https://doi.org/10.1021/JACS.7B05802.

91 Hemmerlin, C., Marraud, M., Rognan, D. et al. (2002). Helix-forming oligoureas: temperature-dependent NMR, structure determination, and circular dichroism of a nonamer with functionalized side chains. *Helvetica Chimica Acta* 85: 3692–3711. https://doi.org/10.1002/1522-2675(200211)85:11.

92 van Zeist, W.-J., Bickelhaupt, F.M., Fischer, L., and Guichard, G. (2010). Folding and self-assembly of aromatic and aliphatic urea oligomers: Towards connecting structure and function. *Organic & Biomolecular Chemistry* **8**: 3101–3117. https://doi.org/10.1039/C001090A.

93 Violette, A., Fournel, S., Lamour, K. et al. (2006). Mimicking helical antibacterial peptides with nonpeptidic folding oligomers. *Chemistry & Biology* **13**: 531–538. https://doi.org/10.1016/J.CHEMBIOL.2006.03.009.

94 Douat, C., Aisenbrey, C., Antunes, S. et al. (2015). A cell-penetrating foldamer with a bioreducible linkage for intracellular delivery of DNA. *Angewandte Chemie* **127**: 11285–11289. https://doi.org/10.1002/ANGE.201504884.

95 Bornerie, M., Brion, A., Guichard, G. et al. (2021). Delivery of siRNA by tailored cell-penetrating urea-based foldamers. *Chemical Communications* **57**: 1458–1461. https://doi.org/10.1039/D0CC06285E.

96 Dolain, C., Jiang, H., Léger, J.M. et al. (2005). Chiral induction in quinoline-derived oligoamide foldamers: Assignment of helical handedness and role of steric effects. *Journal of the American Chemical Society* **127**: 12943–12951. https://doi.org/10.1021/JA0527828.

97 Dolain, C., Grélard, A., Laguerre, M. et al. (2005). Solution structure of quinoline- and pyridine-derived oligoamide foldamers. *Chemistry – A European Journal* **11**: 6135–6144. https://doi.org/10.1002/CHEM.200500395.

98 Jiang, H., Léger, J.M., and Huc, I. (2003). Aromatic **δ**-peptides. *Journal of the American Chemical Society* **125**: 3448–3449. https://doi.org/10.1021/JA029887K.

99 Delsuc, N., Kawanami, T., Lefeuvre, J. et al. (2008). Kinetics of helix-handedness inversion: folding and unfolding in aromatic amide oligomers. *ChemPhysChem* **9**: 1882–1890. https://doi.org/10.1002/CPHC.200800310.

100 Iriondo-Alberdi, J., Laxmi-Reddy, K., Bouguerne, B. et al. (2010). Cellular internalization of water-soluble helical aromatic amide foldamers. *ChemBioChem* **11**: 1679–1685. https://doi.org/10.1002/CBIC.201000256.

101 Gillies, E.R., Deiss, F., Staedel, C. et al. (2007). Development and biological assessment of fully water-soluble helical aromatic amide foldamers. *Angewandte Chemie International Edition* **46**: 4081–4084. https://doi.org/10.1002/ANIE.200700301.

102 Kumar, S., Birol, M., Schlamadinger, D.E. et al. (2016). Foldamer-mediated manipulation of a pre-amyloid toxin. *Nature Communications* 7 (1): 1–11. https://doi.org/10.1038/ncomms11412.

Part II

Mechanism

8

Peptide Structure: Electrophysiological Analysis, Nuclear Magnetic Resonance Analysis, and Molecular Dynamics Simulation of Direct Penetration of Cell-Penetrating Peptides through Bilayer Lipid Membranes

Kayano Izumi[1], Batsaikhan Mijiddorj[2], Chihiro Saito[1], Izuru Kawamura[3], and Ryuji Kawano[1]

[1] *Department of Biotechnology and Life Science, Tokyo University of Agriculture and Technology, Koganei, Japan*
[2] *Department of Biology, School of Arts and Sciences, National University of Mongolia, Ulaanbaatar, Mongolia*
[3] *Graduate School of Engineering Science, Yokohama National University, Yokohama, Japan*

8.1 Introduction

Cell-penetrating peptides (CPPs) are short peptides consisting of 5–30 amino acids. In the last three decades, many research groups have primarily focused on the design and application of such peptides due to the extraordinary ability of CPPs to penetrate cell and tissue membranes [1–13]. To date, the CPPsite database (http://crdd.osdd.net/raghava/cppsite/), a comprehensive information resource on CPPs, includes experimentally validated information on more than 1700 CPPs [14, 15]. CPPs are classified based on origin and physical and chemical properties (Figure 8.1a) [14, 16, 17]. The amino acid composition is essential for the cellular uptake of CPPs and the further design of new agents. The arginine and lysine residues, vital for the interaction between the peptides and negatively charged head groups of membrane bilayers, were abundant in the sequences of CPPs (Figure 8.1b, stored in the CPPsite database). The CPPs are also classified depending on their physicochemical properties: cationic, amphipathic, and hydrophobic peptides (Table 8.1) [5, 18–26].

Depending on the physical and chemical properties and structures of the CPPs, the main entry mechanisms have been conferred as endocytosis and direct penetration [27]. The endocytosis takes an energy-dependent pathway. Caveolin-mediated endocytosis, clathrin-mediated endocytosis, and clathrin and caveolin-independent endocytosis are covered in the endocytosis mechanism. On the other hand, direct

Cell-Penetrating Peptides: Design, Development and Applications, First Edition. Makoto Oba and Yosuke Demizu. Published 2023 by WILEY-VCH GmbH.

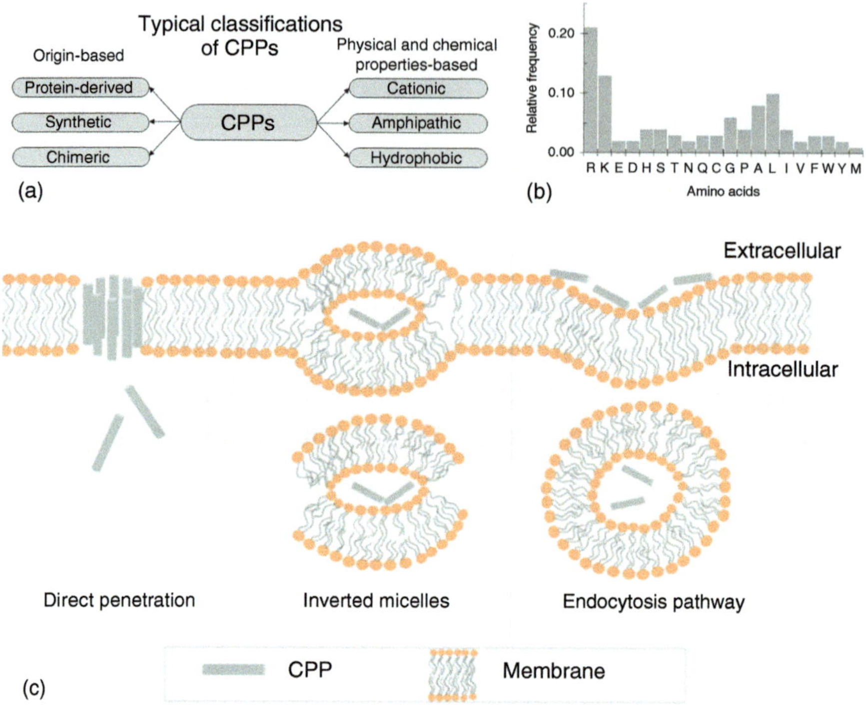

Figure 8.1 (a) Typical classifications of CPPs based on origin and physical and chemical properties. (b) Relative frequency of amino acids in nonredundant sequences of CPPs having natural residues. The sequences were retrieved from the CPPSite-2.0 database. (c) Main entry mechanisms of CPPs such as direct penetration, inverted micelles, and endocytosis pathway.

Table 8.1 Common examples of CPPs.

Peptide name	Sequence	Type	Source	Ref.
Tat	GRKKRRQRRRPPQ	Cationic	Protein-derived	[8]
Penetratin	RQIKIWFQNRRMKWKK	Cationic	Protein-derived	[10]
P4	RRRRRRRR	Cationic	Synthetic	[18]
Pep-1	KETWWETWWTEWSQPKKKRKV	Amphipathic	Synthetic	[3]
MAP	KLALKLALKALKAALKLA	Amphipathic	Chimeric	[19]
RIPL	IPLVVPLRRRRRRRRC	Amphipathic	Chimeric	[17]
Bip2 (CPP5s)	VPTLK	Hydrophobic	Protein-derived	[20]

penetration allows CPPs directly access to the cytosol through energy-independent pathways and results in highly permeable and nontoxic delivery. Direct penetration is divided into several models: toroidal, barrel-stave, and carpet-like models, and inverted micelle-mediated delivery (Figure 8.1c). This chapter summarizes the methodologies of electrophysiology, NMR, and molecular dynamics simulation in the study of direct penetration of CPPs. These methodologies are vital to understanding the interactions of CPPs with phospholipids and membrane-bound structures at the molecular and atomic levels.

8.2 Electrophysiological Analysis of the Direct Penetration of CPPs

In this session, we introduce and discuss recent advances in the electrophysiological measurements and analysis of transmembrane peptides including CPPs using a planar lipid bilayer. We initially introduce how to form the stable lipid bilayer using microdevices and, subsequently, the channel current measurements of the transmembrane peptides, such as antimicrobial peptides (AMPs) and CPPs that we have recently investigated, are introduced.

8.2.1 Stable and Reproducible Formation of Planar Bilayer Lipid Membrane using Microfabricated Devices

The planar bilayer lipid membrane (pBLM) is considered to be a simple artificial cell-membrane, and has been used for analyzing ion channel or pore-forming peptides. Traditionally, the formation of the pBLM has been performed by Montal-Mueller or painting methods [28], but these need skillful techniques, and the stability of the pBLMs is often too low to record sufficient data. Microdevices to form multiple or stable pBLMs have recently been developed for high-throughput screening (HTS) of drugs, proteomics, and nanopore DNA sequencing [29]. These devices are usually made using microfabrication technology, which has the major advantage that microfluidic channels composed in such devices can deliver the desired solution of biological materials to individual parts of the array. However, the stability of the pBLMs is still a problem for these devices. In particular, the pBLMs on microsized support pores, which are commonly used in conventional devices, are susceptible to pressure fluctuations and mechanical vibrations during solution exchange.

The droplet contact method is one of the promising methods for stable pBLM formation. In this method, pBLM is formed at the interface between droplets that are in contact and submerged in an oil/lipid mixture (Figure 8.2a, b). Such an interface has recently been referred to as a Droplet Interface Bilayer [31].

To decrease the membrane capacitance and to increase its mechanical stability, the contact area of the lipid monolayer was reduced using a poly(chloro-*p*-xylylene) (parylene) separator that was patterned with micropores [32]. The parylene film was integrated into the double-well chip to produce a parylene double-well (PDW) chip (Figure 8.2c). Therefore, the pBLMs that formed in the PDW chip were mechanically stable. Due to the high stability of the pBLMs, parallel BLMs were constructed using an injection robot; the resulting arrayed pBLM device consists of a 16-channel (16-ch) PDW chip. This improvement allows the formation of highly

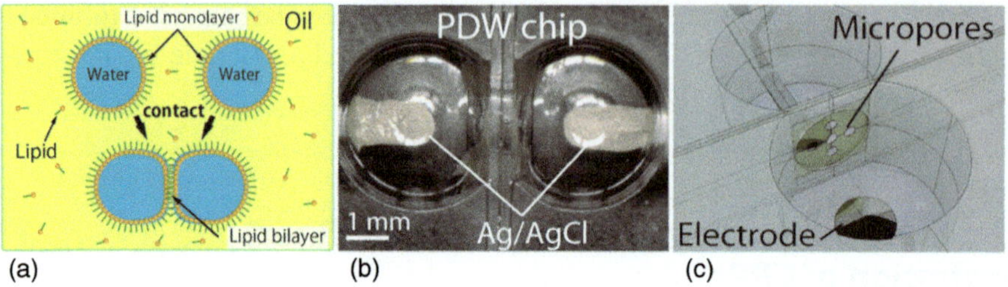

Figure 8.2 Automated, parallel ion channel-recording system using the droplet contact method. Source: Adapted with permission from Kawano et al. [30]. Copyright the author(s).

stable and reproducible pBLMs. The pBLM system has advantages over the previous system in simultaneously satisfying the following: (i) intensive data acquisition, (ii) single-channel analysis, and (iii) analysis of topologically identified targets [30].

8.2.2 Channel Current Signals and Their Classification of Transmembrane Peptides in Planer Bilayer Lipid Membranes

We have recently proposed that a channel current method would be an effective tool for the structural analysis of pore-forming peptides such as AMPs and CPPs. AMPs kill bacteria by forming pores or defects in their cell membranes, resulting in cell lysis. Pore formation can be monitored via channel-current measurement using a pBLM as the artificial cell membrane; current signals associated with various types of transmembrane peptide structures (Figure 8.3a, b).

Traditionally, channel current measurements using pBLMs require considerable operator proficiency for membrane formation, and the stability of these membranes is insufficient to acquire large datasets for stochastic analysis. Due to the high stability of the pBLM system, the channel current measurement of AMPs and CPPs and simultaneous recording of large datasets has been possible. The typical current and time traces of Magainin1 are shown in Figure 8.3a. Several shapes of the current signals were observed as previously reported. We have already proposed the assignment of these current signals into the molecular mechanism of pore or defect formation in the pBLM. We are still trying to improve this assignment, and they are updated in this study, as described below [34, 35].

1) Step signal: In this type of current signal, the current jumps up orthogonally and maintains a plateau state. This signal can apply to the barrel-stave model wherein transmembrane peptides are tightly assembled and form a rigid circular pore (Figure 8.3b).
2) Multilevel signal: In this type of current signal, the current fluctuates after jumping up and returns to the initial state. This current may indicate a toroidal model wherein transmembrane peptides form a pore with lipids. In this model, the size of the pore can change dynamically with or without the participation of the lipid between the monomers (Figure 8.3b).

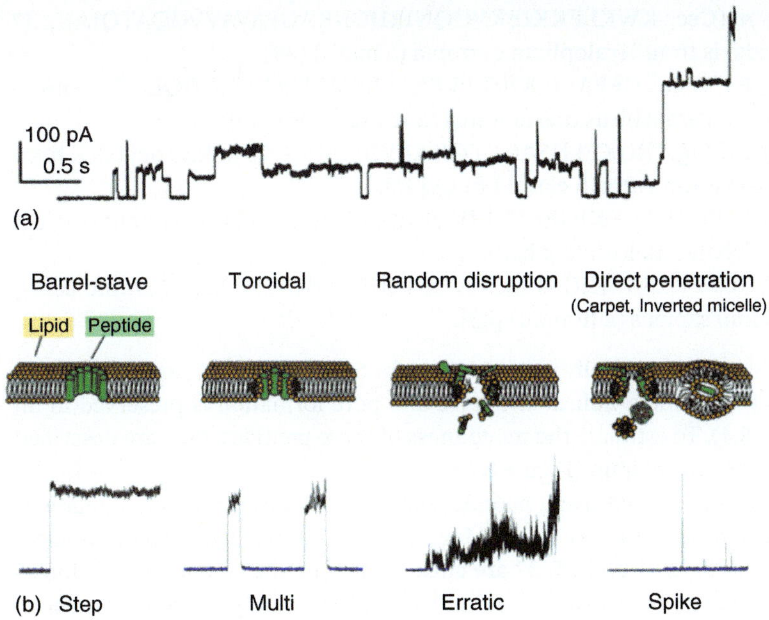

Figure 8.3 Signals of the current measurement. Source: Saigo et al. [33]/American Chemical Society/CC BY-4.0.

3) Erratic signal: In this type of current signal, the current randomly increases with fluctuation. We considered that this signal might indicate random disruption behavior [36] because there is not a suitable pore-forming model. The random current behavior of the erratic signal might be caused by the random size of the pore or cluster (Figure 8.3b).
4) Spike signal: In this type of current signal, the current suddenly rises and then returns to the baseline over a period of less than 20 ms. This signal indicates an instantaneous membrane defect.

We applied this signal to peptide permeation through the membrane. Three different models for direct penetration have been reported: pore formation, inverted micelle formation, and carpet models [37]. The pore formation was already assigned to other current signals. Therefore, we assign the spike signal as the inverted micelle and/or carpet model. The inverted micelle model suggests forming the inverted micelle structure in the lipid membrane that encapsulates the peptides and releases them into the cytosol by inversion (Figure 8.3b). The carpet model has been described as peptides that bind parallel to the lipid bilayer surface and, after reaching sufficient coverage, unwrap lipids as a peptide–lipid cluster from the membrane (Figure 8.3b).

Based on the signal classification, Saigo et al. reported the electrophysiological analysis of six different transmembrane peptides (AMPs) as follows.

1) Clavanin A (Cla, VFQFLGKIIHHVGNFVHGFSHVF, 23 amino acids) is from Styela clava (a sea squirt) [38].

2) Cecropin A (Cec, KWKLFKKIEKVGQNIRDGIIKAGPAVAVVGQATQIAK, 37 amino acids) is from Hyalophora cecropia (a moth) [39].
3) Pardaxin P5 (Par, GFFALIPKIISSPLFKTLLSAVGSALSSSGDQE, 33 amino acids) is from Pardachirus marmoratus (a red sea sole) [40].
4) Magainin 1 (Mag, GIGKFLHSAGKFGKAFVGEIMKS, 23 amino acids) is from Xenopus laevis (an African clawed frog) [41].
5) SL-37 (SLGNFFRKARKKIGEEFKRIVQRIKDFLQHLIPRTEA, 37 amino acids) is from Hylobates moloch (a gibbon) [42].
6) LL-37 (LLGDFFRKSKEKIGKEFKRIVQRIKDFLRNLVPRTES, 37 amino acids) is from Homo sapiens (a human) [43].

These peptides are genetically distant except for SL-37 and LL-37, and the antimicrobial activity with the α-helical structure and pore formation is preserved in all AMPs (Figure 8.4). To estimate the relatedness of these peptides, they are described by the molecular dendrogram (Figure 8.4a).

This diagram was created using peptide sequences because the genetic relationships among them are unknown. Par, Mag, and Cec are relatively closely related. Cla is isolated, and SL-37 and LL-37 are close, as was predicted due to their inclusion in the same gene family (cathelicidin AMP, CAMP family). These results mirror the tendency of the evolution of a biological species. These AMPs have an amphiphilic structure in the helices, commonly known as the helical wheel structure (Figure 8.4b). Based on this molecular information, the molecular basis of the pore-forming activity and its relations with antimicrobial activity was investigated using electrophysiological methods in the context of biological diversity [44].

The current signal of the six peptides is classified (Figure 8.5).

In this classification, the spike signals were eliminated to estimate the pore formation and to compare pore formation versus penetration. The step and multilevel signals reflected relatively stable pore formation in the barrel-stave and the toroidal models. In contrast, the erratic signals were assigned to the random disruption

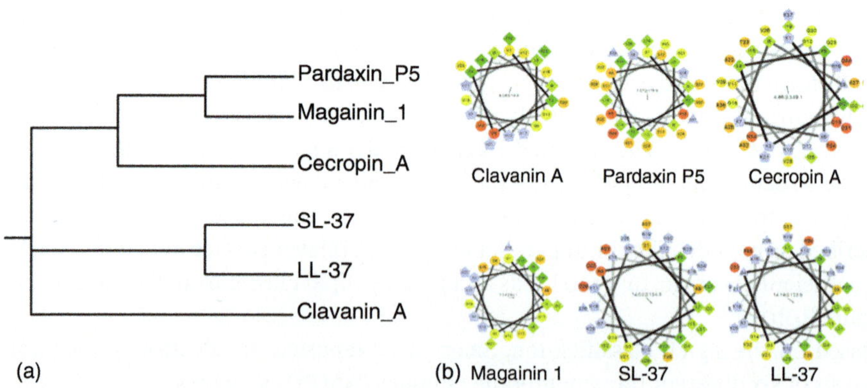

Figure 8.4 (a) Molecular dendrogram of AMPs calculated with ClustalW. (b) Helical wheel structures of AMPs. Source: Saigo et al. [33] / American Chemical Society / CC BY-4.0.

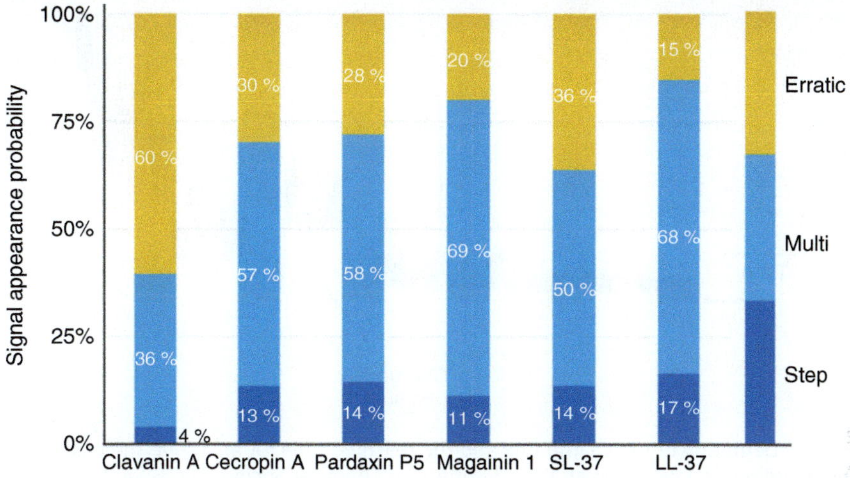

Figure 8.5 Current signal analysis and classification of six AMPs. (n = 243–681). Source: Saigo et al. [33]/ American Chemical Society / CC BY-4.0.

model that elicited detergent-like membrane disruption. Cla yielded notably unstable pore formation, whereas stable pore formation was observed in more than 60% in the other AMPs. Next, AMP pore-forming capability was analyzed by comparing stable pore formation and penetration using the spike signals, which can be assigned to the direct penetration of peptide molecules. The appearance ratio of [penetration (spike signals)]/[pore formation (step + multi + erratic signals)] was calculated to estimate the capability of pore formation. The order of the ratio was Cec > Par > LL-37 > SL-37 > Mag > Cla. These results may imply the leakage activity of AMPs. In addition, the leakage experiment is normally performed to estimate the antimicrobial activity of membrane disruption. In the current study, instead of using the leakage experiment, the focus was placed on the charge flux, which is the total amount of ion transportation through the pore [45].

The charge flux was defined as the average charge per unit time from the initial current increase to the threshold (Figure 8.6).

In this estimation, the rapid current overflow for less than 10 s that may indicate the membrane rupture was not counted. Based on the charge flux, Sekiya et al. reported the electrophysiological analysis of AMP isolated from skin secretions of the frog Bombina variegata, as all L-peptide bombinin H2 (IIGPVLGLVG-SALGGLLKKI-NH2) or its diastereomer H4 [46], which contain either L-Ile or D-allo-Ile at the second amino acid from the N-terminus. Several studies have reported that bombinin H4 showed higher antimicrobial activity than that of H2 [47]. In particular, bombinins are potent candidates as medicine for Leishmania, which constitutes an infectious disease with a global incidence of 10–12 million people infected [48]. Bombinins can directly impact Leishmania, whereas only a few other AMPs have been known to exhibit leishmanicidal activity, the majority of which act mainly on the insect stage of the parasite (promastigotes) [49]. Although the bombinins are expected to function as unique AMPs upon isomerization even against

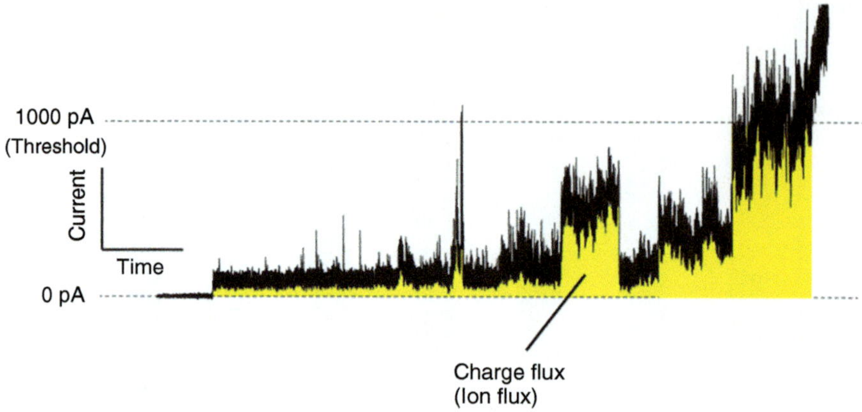

Figure 8.6 Definition of the charge flux and the reaching time for analyzing the channel current recording. Source: Saigo et al. [33]/ American Chemical Society / CC BY-4.0.

Leishmania, the molecular mechanism of their activity remains under debate. In this study, we hypothesize that the main activity of bombinins is probably due to membrane disruption as indicated from several prior experiments [50, 51]. Thus, we examined the pore-forming behavior of H2 and H4 using the pBLM as a model bacteria membrane.

Based on a systematic method, in the current study, we attempt to reveal the pore-forming activity of bombinins H2, H4, and a mixed system at the molecular basis and discuss the relationship between the pore formation behavior and antimicrobial activity. Several types of signals appeared (Table 8.2). The spike signals had the highest ratio in all three systems of H2, H4, and H2/H4 mixture, after which the order was as follows: erratic > multilevel > step signal. This tendency is reflected in the pore formability of bombinins [34]. In terms of pore formability of the four models, it is assumed that the formality of a stable pore, which means to form a more cylindrically shaped pore, is spike < erratic < multilevel < step signal. Therefore, the tendency of the ratio suggests that the capability of stable pore formation would be energetically unfavorable for these conditions.

In addition, this estimation was conducted by counting the number of signals, resulting in counting a large amount of the spike and erratic signals. The appearance probability of these two signals was relatively high, whereas the duration was shorter than other signals. Therefore, the duration time of pore formation was analyzed (Table 8.3). Results showed the duration time of each signal to be: spike < erratic < multilevel < step signal, giving the same order as of pore stability. This fact suggested that stable pore formation may strongly contribute to the membrane defect. Comparing H2 and H4, the step and multilevel signals in H4 were more than those in H2 as opposed to the erratic signals, suggesting that H4 may be likely to form a rigid pore. In addition, the mixture of H2/H4 showed a higher ratio of the step signals. This result implies the high possibility of rigid pore formation.

Table 8.2 The ratio of each signal of H2, H4, and H2/H4 mixtures.

	Spike (%)	Erratic (%)	Multilevel (%)	Step (%)
H2	57.8	36.3	5.4	0.5
H4	58.2	25.4	14.4	2.0
H2/H4	50.2	32.0	13.7	4.1

Source: Sekiya et al. [35]/with permission of American Chemical Society.

Table 8.3 Time duration of the classified signal of H2, H4, and H2/H4 mixtures.

	Spike (s)	Erratic (s)	Multilevel (s)	Step (s)
H2 average	0.0076	0.47	4.1	32
median	0.0093	0.11	2.5	27
H4 average	0.0049	0.096	0.36	12
median	0.0046	0.078	0.25	13
H2/H4 average	0.0080	0.44	0.50	48
median	0.0077	0.10	0.15	27

Source: Sekiya et al. [35]/with permission of American Chemical Society.

The charge flux of H2, H4, and H2/H4 using the threshold of 100, 500, and 1000 pA was also analyzed (Table 8.4).

The order of values was H4 ≥ H2/H4 ≥ H2 in these conditions. The charge flux should reflect the activity of ionic leakage, which corresponds to the ion permeability through the lipid membrane. In addition, the tendency of the total pore area would also be H4 ≥ H2/H4 ≥ H2 if the ability of ion permeability per unit pore area per time (permeability/ (nm^2/s)) of all bombinins is constant.

The charge flux was initially considered a suitable parameter to indicate membrane dysfunction of AMPs. However, if the pore formation is very slow, even if the pore size is large, this type of mechanism may not show higher microbial activity. Thus, examined both the dynamics and kinetics of membrane disruption were

Table 8.4 The amount of the charge flux at each threshold value.

Charge flux (pC/s)	100 pA	500 pA	1000 pA
H2	87.0	147.6	161.8
H4	100.1	190.3	251.8
H2/H4	111.7	153.7	197.7

Source: Sekiya et al. [35]/with permission of American Chemical Society.

Table 8.5 The reaching time at each threshold values.

Charge flux (pC/s)	100 pA	500 pA	1000 pA
H2	207.2	214.7	215.7
H4	159.7	163.5	164.5
H2/H4	379.7	395.5	399.0

Source: Sekiya et al. [35]/with permission of American Chemical Society.

examined. The kinetics of pore formation was analyzed by the time taken to exceed the threshold from the initial current appearance, termed "reaching time." This time indicates the kinetics for generating membrane defects by pore formation. Notably, the time in the H2/H4 mixture system was much slower than that in H2 or H4 systems. The time in H2 was slightly slower than that in the H4 system (Table 8.5).

In the previous study of the antimicrobial activities of bombinins, the activity estimated from the minimum inhibitory concentration (MIC) of H4 was higher than that of H2 to Gram-negative bacteria [47]. Several mechanisms for these activities are generally considered to exist, such as chemical enzyme inhibition in the metabolic pathway, or physical membrane disruption. According to previous experiments on bombinins, the measurements of fluorescein leakage through the lipid membrane by bombinins using living cells or liposomes show H4 > H2 [51]. These leak experiments indicate that the pore formation of bombinins into the lipid membrane constitute the important factor of antimicrobial activity and that the activity of H4 is much higher than that of H2 [51].

In electrophysiological measurements, notable results were obtained for the ion permeability from the charge flux analysis. The ion permeability had the trend: H4 > H2/H4 ≥ H2; this order is consistent with the results of the leak experiment using cells or liposomes in previous reports. It is reasonable because the permeability and leak experiments should reflect similar phenomena. The values of H2 and H2/H4 were close in this estimation.

8.2.3 Channel Current Analysis of the Direct Penetration of CPPs

CPPs have important biological functions such as mass transport through cell membranes [52, 53]. Elucidating the structure–function associations of CPPs is of considerable scientific interest for the understanding of mechanisms and for medical applications; however, the development of methodologies for structural analysis in lipid membranes remains a challenge.

LK peptides consist of a leucine (L)–lysine (K) sequence, and they are well known as CPPs [54, 55]. The single LK peptide unit is [LKKLLKL] named L4K3. Sekiya et al. reported the electrophysiological analysis of the membrane permeability of a series of LK peptides: L8K6, L12K9, and L16K12 named $[LKKLLKL]_2$, $[LKKLLKL]_3$, and $[LKKLLKL]_4$, respectively [34]. The LK peptides have an α-helical structure and have the periodic distribution of hydrophobic and hydrophilic regions in the peptide, which serve to stabilize the peptide's secondary structure [56]. Although arginine-rich peptides (e.g. R9) are also well known as CPPs [57], their secondary structure is almost a random coil. Therefore, they selected α-helical LK peptides as the membrane penetration peptides to maintain a consistent secondary structure across all the selected experimental peptides.

The signal classification of the L8K6 peptide was depicted (Figure 8.7).

The spike signal was dominant in this measurement, suggesting that the peptides may form lipid membrane defects momentarily as they translocate through the membrane, as proposed by the Shai–Matsuzaki–Huang model [58]. Although most conventional methods require fluorescent labeling to study membrane penetration, the current recording method does not require fluorescent labeling; therefore, this direct observation of the penetration phenomenon reveals the molecular mechanism of CPP directly.

Next, the spike signals of the L8K6 peptide were analyzed using three parameters: the current amplitude, which gives information on the size of the defect in the membrane; the duration time, which reflects the duration of the defect formation; and the time between spike signals, which shows the penetration event frequency (Figure 8.8a). The scatter plots were depicted for the current amplitudes and duration time of the L8K6 peptide (Figure 8.8b).

The amplitude and duration range from 10 to 100 pA and from 0.1 to 10 ms, respectively. The dependence of these three parameters on peptide length was also studied. The duration and peak amplitude did not reveal significant differences between different peptide lengths, although the values of L12K9 were slightly lower than those of all other peptides (Figure 8.8c, d). However, the time between peaks was the highest for L12K9 (Figure 8.8e).

The graph indicates that the kinetics of membrane penetration decreases with increasing molecular size up to L12K9 whereas penetration dynamics are similar regardless of size.

However, regarding the kinetics of L16K12, the event frequency increased even for the largest molecular structure. This peptide forms a strong α-helical structure in the aqueous phase, as observed from circular dichroism measurements, which may induce peptide binding on the membrane surface (Figure 8.9).

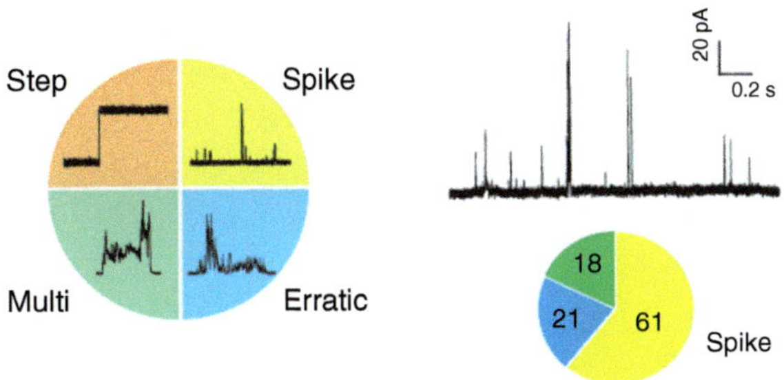

Figure 8.7 Signal classification of L8K6. The number of circular graphs indicates the percentages of each signal (n = 87–146). Source: Sekiya et al. [34]/with permission of Royal Society of Chemistry.

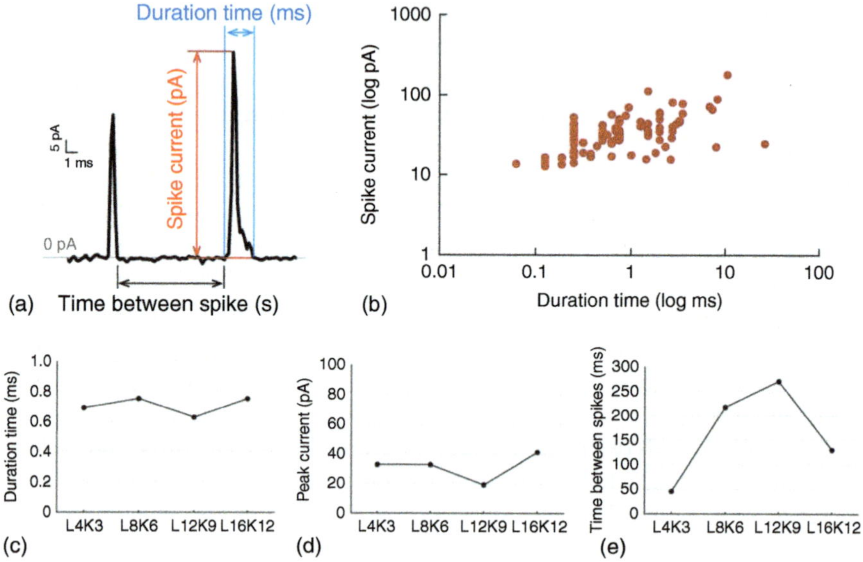

Figure 8.8 (a) Typical current–time spike signal traces of the L8K6 peptide. The spike current amplitude (red), duration time (blue), and time between spikes (black) were analyzed. (b) Scatter plots of peak current amplitudes and duration time of L8K6 peptides. The repeat length dependency of the LK peptides for (c) the duration time, (d) the peak current, and (e) time between spikes, as mean values. Source: Sekiya et al. [34]/with permission of Royal Society of Chemistry.

These peptides show relatively high event frequency, which might be owing to their concentration at the membrane surface.

In addition, the length of L16K12 (ca. 4 nm: the length of the [LKKLLKL] unit is ca. 1 nm) is almost comparable to the thickness of the lipid bilayer in this system. This might cause a higher event frequency.

The values of all data including the event rate constant are summarized (Table 8.6).

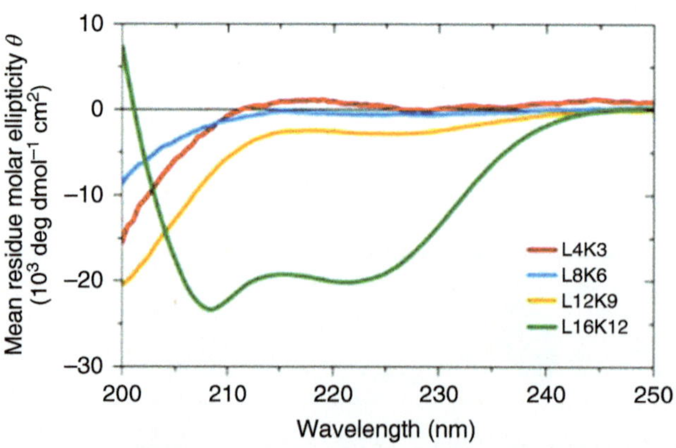

Figure 8.9 CD spectra of LK peptides [peptide] = 20 μM in 10 mM MOPS Buffer (pH 7.0). Source: Adapted with permission from Sekiya et al. [34].

Table 8.6 Several properties of LK peptides.

	L4K3	L8K6	L12K9	L16K12
Helicity, buffer (%)	0	0	1.3	58.7
Helicity, liposome (%)	0	15.5	55.8	58.5
Duration time (ms)	0.69	0.75	0.63	0.75
Peak current (pA)	32.7	32.7	19.1	41.0
Rate constant (s^{-1})	21.7	4.08	3.71	7.68

Source: Sekiya et al. [35]/with permission of American Chemical Society.

These values are much higher than the value of Tat peptide uptake in living cells ($\sim10^{-3}$ s^{-1}) [53], implying higher membrane penetration kinetics in this system. In summary, to reveal the molecular mechanism of membrane penetration by short peptides in the lipid bilayer, the current signals of three α-helical peptides were classified using the microfabricated pBLM device, and signals were assigned to the possible structural models in a lipid membrane. The step, multilevel/erratic, and spike signals are assigned as barrel stave, toroidal pore, and penetration models, respectively, using LK peptides. In addition, the spike signals assigned to the membrane penetration were analyzed, and it was concluded that the dynamics and kinetics of the membrane penetration may be estimated using the spike amplitude, duration time, and rate constant. The proposed method using channel current analysis could be a strong candidate for investigating membrane-binding peptides at the molecular level.

The next example is of arginine-rich peptides [59]. The arginine-rich peptides intrinsically exhibit various conformations depending upon the physicochemical environment due to their structural flexibility, or polymorphism [24, 60–64]. It has been suggested that the regulation of the secondary structure of peptides is a promising strategy to control the cell penetration of arginine-rich peptides [61]. Chemical technologies to restrict structural polymorphism of arginine-rich peptides with highly ordered structures have increasingly been reported [61, 65, 66]. Especially, the α-helical conformation is a primary structure, which enables amino acid residues to create an amphipathic interface at the peptide, thereby enhancing the binding to lipid membranes and subsequent cell penetration [62]. Ohgita's group previously demonstrated that the increased amphipathicity with α-helix formation enhances the cell membrane-penetrating ability of arginine-rich peptides derived from the glycosaminoglycan-binding region of human apolipoprotein E [62]. However, the relationship between the secondary structure and the membrane-penetration efficiency of arginine-rich peptides is still largely unknown.

One of the most promising approaches to restrict polymorphism of peptides is using a unique secondary structure imparted by proline residues. In particular, proline-rich peptides can adopt the polyproline II (PPII) helix structure, which is the most abundant secondary structure in proteins along with α-helix and β-sheet [67, 68]. The PPII helix is a left-handed, extended rigid, and stable helix with three residues per turn, in which all peptide bonds are in the transconfiguration due to the lack of intramolecular hydrogen bonds and the steric constraint imposed by the side-chain pyrrolidine ring [69–71]. It has been reported that proline-rich peptides, such as oligoprolines chemically modified with guanidine groups and polyproline-containing amphipathic arginine-rich peptides derived from non-lytic AMPs, exhibit efficient cell internalizations [66, 72–75]. However, the cellular internalization mechanism for such proline-rich CPPs has been only focused on endocytosis [72–75].

In Ohgita's study [59], a polyproline-containing arginine-rich peptide P9R7W (PPPPPPPPPRRRRRRRW) was designed by introducing nine proline residues into a linear R7W (RRRRRRRW) peptide, together with a non-arginine-containing P9W (PPPPPPPPPW), to investigate the effect of PPII helix on the direct cell membrane penetration of arginine-rich peptides. The secondary structure and membrane interaction of these peptides were examined in relation to the cell membrane penetration efficiencies using a variety of biophysical techniques.

To investigate the mechanism of membrane penetration of polyproline peptides without fluorescent labeling, electrophysiological measurement was performed using the pBLM. The current-time traces after the addition of R7W, P9W, and P9R7W peptides to the dioleoyl phosphatidylcholine (DOPC) bilayer membrane, exhibited "spike" and "long-lasting (duration ≥ 20 ms)" (Figure 8.10a–c).

The spike current signal can be assigned to the membrane penetration of peptides with "carpet" or "inversed micelle" models giving transient pore formation [34, 35]. The transient pore formation is also considered to be the most reasonable mechanism for the membrane penetration of arginine-rich peptides [76–78]. At 100 nM peptide concentration, the observed current signals for P9R7W and R7W included ~60–70% of spike and ~30–40% of long-lasting current signals, whereas P9W predominantly exhibited long-lasting signals (~70%) (Figure 8.10d).

At 5 μM peptide concentration, the spike event of P9R7W was still predominant (~70%), whereas the long-lasting signal ratio of R7W became close to that of P9W (~80% for R7W and ~90% for P9W, respectively). These results indicate that P9R7W exhibits the most favorable membrane penetration behavior among the peptides. Comparison of the event frequency of spike signals expressed as the mean time between spikes. (Figure 8.10e)

The frequency values of P9R7W and R7W were much larger than that of P9W at both peptide concentrations, consistent with the finding that P9R7W and R7W penetrate across the cell membranes more efficiently than P9W (Figure 8.10e and Figure 8.11b,c).

The spike frequency values of P9R7W and R7W were almost independent of peptide concentration between 100 nM and 5 μM (Figure 8.10e). It has previously been proposed that ion flux reflects the ion permeation through the membrane defects, and can be used for estimation of membrane damage [33]. Also, the ion flux was found to be correlated with the cytotoxicity of membrane-disrupting AMPs [33]. Thus, the ion flux for each peptide was determined by analyzing the area under the current in the current-time traces. The order of the ion flux was P9W > R7W > P9R7W at both peptide concentrations (Figure 8.10f). The cytotoxicity in CHO-K1

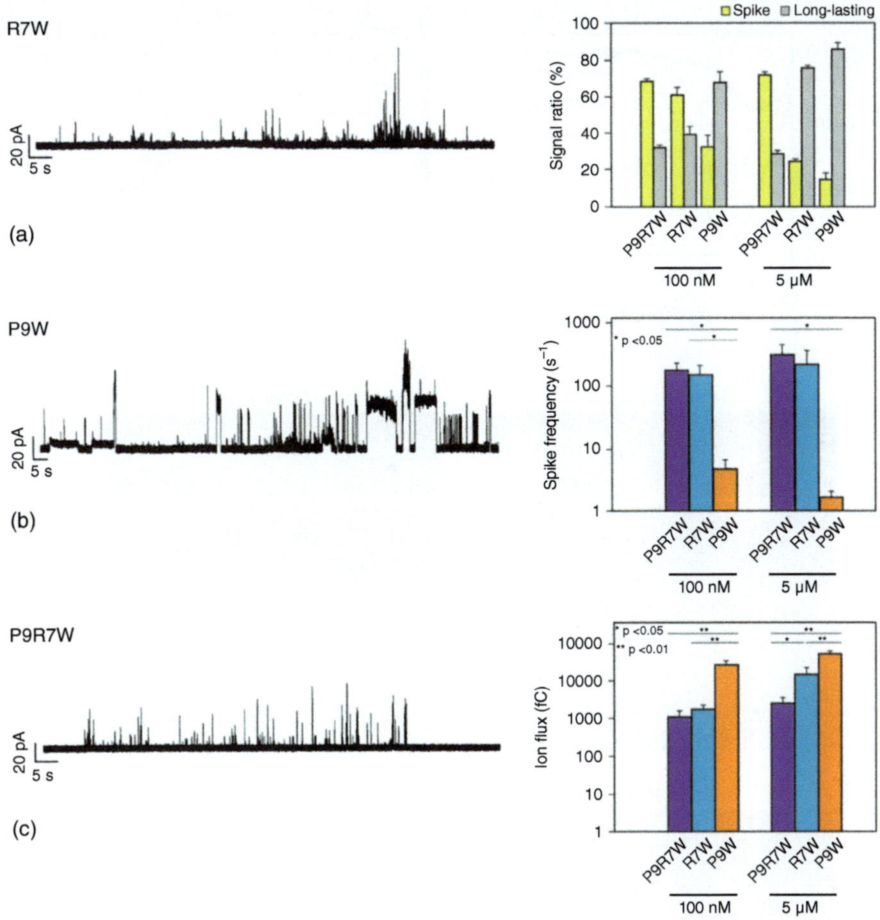

Figure 8.10 Analysis of the channel current signals of R7W, P9W and P9R7W peptides upon interaction with planar DOPC membranes. Source: Reproduced from Ohgita et al. [59]/with permission of Elsevier.

cells was not observed for all non-labeled peptides in these experimental conditions. These results indicate that the PPII helix in P9R7W does not cause perturbation of lipid membranes and cytotoxicity to cells.

The PPII helix is one of the most abundant secondary structures found in proteins along with α-helices and β-sheets [79]. A significant finding in this study is that the addition of the PPII helical motif to the arginine-rich R7W peptide significantly enhances the direct cell membrane penetration of the peptide (Figures 8.11 and 8.12).

Typical arginine-rich peptides including R7W are known to undergo a conformational transition from random coil to α-helix upon binding to GAGs or lipid membranes [25, 62, 64, 80]. In contrast, P9R7W appears to maintain its rigid PPII helix structure upon binding to GAGs or lipid membranes. Given that the secondary structure influences endocytosis and cell membrane penetration of arginine-rich

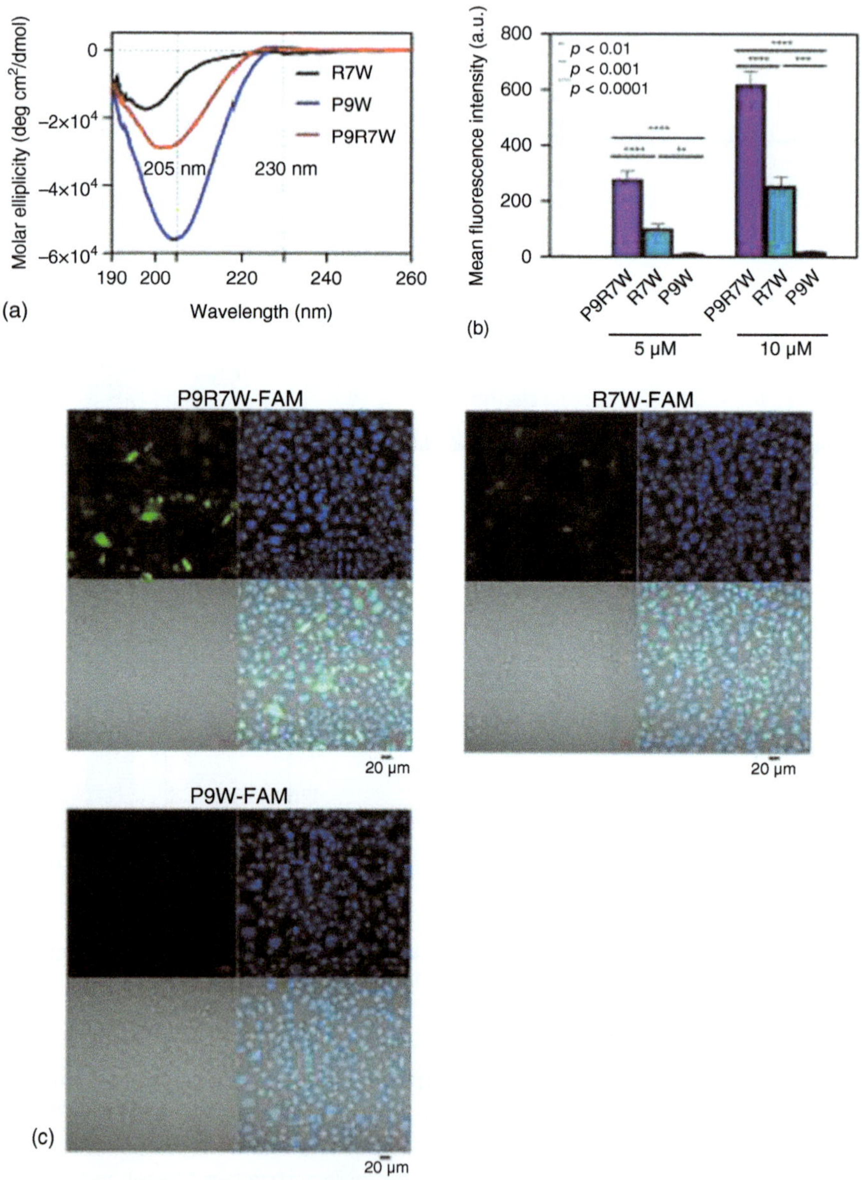

Figure 8.11 Effect of PPII helix structure on cell penetration efficiently of arginine-rich peptides. Source: Ohgita et al. [59]/with permission of Elsevier.

peptides through interactions with GAGs and lipid membranes [25, 80], the results suggest that the enhanced direct membrane penetration of P9R7W likely comes from its rigid and stable PPII helix structure. The efficient cell penetration of P9R7W also occurs at 37 °C, indicating that the PPII helix structure enhances cell penetration of peptides at which endocytic uptake is predominant [81, 82]. Similar to the case at 4 °C, no cytotoxicity was observed for all peptides at 37 °C.

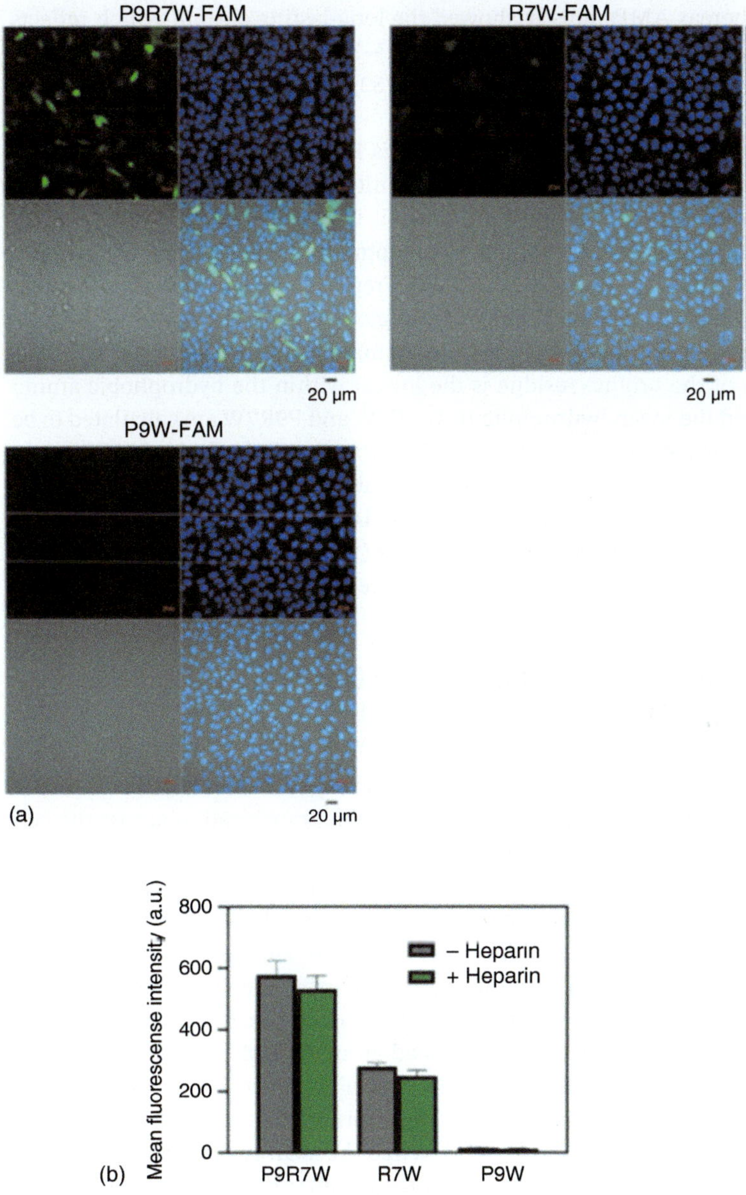

Figure 8.12 Effect of heparin washing on cell penetration assays of FAM-labeled peptides in CHO-K1 cells treated at 4 °C. Source: Ohgita et al. [59]/with permission of Elsevier.

The results of peptide-induced ion flux through stable pores in the planar bilayer membrane indicate that P9W exhibits large membrane perturbation whereas P9R7W has a negligible effect on the lipid membrane structure (Figure 8.10f). Current signals can be classified into two groups based on their previous measurements: CPPs mainly showed the spike signal which reflects their membrane penetration, whereas AMPs mostly showed the long-lasting current which reflects the pore/defect formation in lipid membranes [33–35]. The current signal pattern of P9W (Figure 8.10b) is similar to that of melittin [83], which is a representative AMP that strongly binds to lipid membranes [83].

However, the observed current signal of P9W is 10 times magnitude smaller than that of melittin, indicating that the binding affinity of P9W to lipid membranes is much smaller than that of melittin [62]. Thus, it is possible that the P9W peptide cannot cross cell membranes, but rather predominantly remains in lipid membranes, exhibiting a long-lasting signal. It was reported that in the pore-forming peptides, a peptide-induced pore in lipid membranes is stabilized by both an increase in peptide hydrophobicity and a decrease in arginine residues [78]. Given that the hydrophobicity of the proline residue is the lowest within the hydrophobic amino acids [84, 85] and the mean hydrophilicities of P9W and P9R7W are calculated to be 0.873 and 0.098, respectively, based on the Eisenberg scale [84], the PPII helix in P9R7W is unlikely to cause significant lipid membrane perturbation. The authors note that the ion flux in P9W per unit time was 14.5 ± 6.1 pC/s at 5 μM; this value was much smaller than those for cytotoxic AMPs (>100 pC/s) [33]. Indeed, all peptides did not exhibit any cytotoxicity to CHO-K1 cells.

8.3 Solid-state NMR Analysis of the Direct Penetration of CPPs

CPPs, which contain a typical Tat peptide, share an arginine-rich motif and modulate the physical properties of lipid bilayers to enable translocation across the cell membrane [86]. Although CPPs are efficient intracellular transporters, their applications are limited by the inadequate understanding of the direct permeation mechanism. Solid-state NMR spectroscopy is a valuable method for evaluating the structure and dynamics of membrane-bound proteins and peptides [87–92]. ^{31}P solid-state NMR spectroscopy is used to directly detect a lamellar or non-lamellar structure of the assembled phospholipids and dynamics of the fluid membrane containing peptides. As a phospholipid polar head group in a phospholipid molecule contains a ^{31}P nucleus with a 100% natural isotopic abundance, hence analyzing the ^{31}P static NMR signal pattern is useful for monitoring changes in the polar head group of the phospholipid bilayers [93–95]. While the axial symmetric powder pattern for the ^{31}P NMR spectrum has exhibited the lamellar phase of the phospholipids, the appearance of a signal at 4–5 ppm with a reversed ^{31}P NMR shape pattern is attributed to an inverted-hexagonal phase inside the lamellar phase (Figure 8.13) [96–98]. This behavior has been clearly observed in the NMR spectra of DOPC/1, 2-dioleoyl-*sn*-glycero-3-phosphoethanolamine (DOPE) with acylated S413-PV sharing the CPP motif, and in those of 1,2-dimyristoyl-*sn*-glycero-3-phosphocholine (DMPC) with Tat [97, 99]. Therefore, the treated ^{31}P NMR spectral patterns give us insights into the cell-penetrating mechanism, indicating that when CPPs bind to the membrane, the peptides form an inverted micelle in the lipid bilayer in a non-lytic manner.

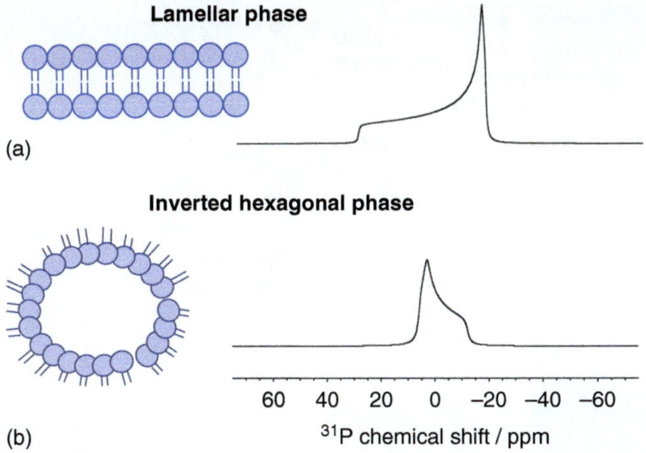

Figure 8.13 (a) A simulated ^{31}P NMR axial symmetric powder pattern of a lamellar phase. (b) A simulated spectral pattern of a hexagonal phase.

According to solid-state NMR structural analysis of Tat (48–60) in a DMPC/1,2-dimyristoyl-*sn*-glycero-3-phospho-(1′-*rac*-glycerol) (sodium salt) (DMPG) mixed membrane, the average conformation was a random coil based on the NMR chemical shift values of the isotope-labeled sites Lys4, Gln7, Arg8, and Pro11 [100]. ^{13}C-detected ^{1}H spin diffusion experiments revealed both the cross-peaks of water molecules and phospholipids with Arg8 of Tat (48–60) [100]. In ^{31}P-^{13}C rotational echo double resonance NMR experiments, the interatomic distance was estimated as 4.1 Å between the guanidium C of Arg8 and the phosphorus atom in the polar head group. These solid-state NMR results are consistent with the location of the Tat peptide in the glycerol backbone and polar head group region of the membrane as a membrane-bound model of Tat (Figure 8.14) [100, 101].

Cyclorasin 9A5, which inhibits the interaction of Ras with Raf protein, is an 11-mer cyclic amphiphilic peptide sharing the CPP-like motif Arg4-Arg5-dNal6 (D-β-naphthylalanine)-Arg7-Fpa8 (L-4-fluorophenylalanine). The cyclic peptide can permeate the cell membrane to exert its activity. NMR structural analysis revealed that the turn structure (Thr2-D-Ala3) of Cyclorasin 9A5 in water (dielectric constant, $\epsilon \approx 80$) changed to a turn/kink conformation with dimethyl sulfoxide ($\epsilon \approx 47$) as the solvent. This change allows for the formation of an amphiphilic peptide conformation inside the membrane [102]. Conformational plasticity is generally important for the cell permeabilization activities of peptides.

8.4 Molecular Dynamics Simulation of CPP-Membrane Interactions

Molecular dynamics (MD) simulation provides atomic and molecular level information on the interactions in molecular systems and gives microscopic explanations of macroscopic properties. Because of this capability, it has been applied to understand

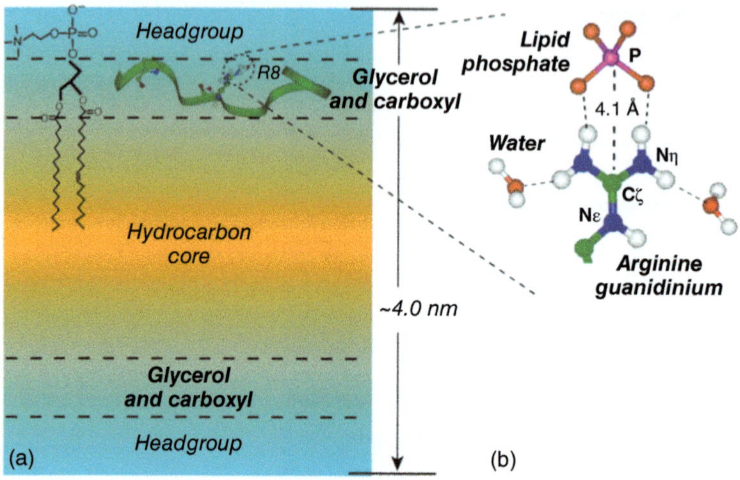

Figure 8.14 Membrane-bound model of Tat structure in DMPC/DMPG bilayers: Insertion of the Tat random coil into the membrane interface and stabilization of the Arg guanidinium group by intermolecular hydrogen bonds with the polar phosphate head group and water molecules. Source: Su et al. [100]/with permission of American Chemical Society.

membrane interaction and internalization of CPPs in many publications, and Reid and co-workers extensively reviewed the molecular dynamics approach that elucidated the molecular-level understanding of CPPs [103]. Because the translocation of CPPs can be experimentally measured on the order of a minute, classic MD simulations cannot show the penetration. Therefore, research groups mainly applied coarse-grained models [104, 105], replica-exchange simulations [106], and other nonequilibrium MD methods, including umbrella sampling [107, 108] and steered MD simulations [109].

MD simulation has been conducted on the pore formation of CPPs [76, 108]. Particularly, the Tat peptide initially aggregated on the surface of DOPC lipid bilayers due to the electrostatic interaction and hydrogen bonding between the positively charged residues and the negatively charged head groups of lipids. Notably, the positively charged residues of the CPP attracted the reverse side head groups of the membrane bilayer owing to the occurrence of membrane thinning and fluctuation. As a result, this mechanism suggested the process of pore formation in the membrane [76]. They also showed that the peptide uptake mechanism depends on the peptide concentration, and the positively charged residues are important for interaction with the head groups of phospholipids [76].

Via and co-workers investigated the insertion of R9 and R8W3 peptides in DOPC and computed insertion free-energy profiles using the coarse-grained model and metadynamics [105]. They found that a higher interfacial electrostatic potential was less affected by peptide binding to the membrane surface, and reduced the energy cost of bringing peptides to the center of the membrane bilayer [105]. Furthermore, Alaybeyoglu et al. showed the translocation mechanisms of pVEC peptide and its mutant derivatives using steered MD simulations. They identified the importance of charged amino acids for translocation by mutating them with alanine residues and further detailed atomic-level information [109]. Additionally, Gao et al. observed the correlation between the membrane translocation of CPPs and membrane potential. They set the initial membrane potential difference between intracellular and extracellular environments in these simulations. The higher potential difference displayed higher CPP translocation behavior compared to lower values [107].

Grasso et al. observed the insertion behavior of various CPPs in DOPC membranes. They found that the hydropathy of CPPs correlated with the water penetration into the membrane, with increased water penetration causing reduced order in the lipid molecules. They also showed that the physicochemical properties of the peptides are responsible for the membrane bending modulus [110]. Recently, the MD simulation technique was applied for the evaluation and ranking of machine-learning-generated sequences of CPPs. A high-ranked CPP, named Pep-MD, showed higher cell-internalization in HeLa cells than the Tat peptide did [111]. These results show that MD simulation can increase the effectiveness of designing CPPs and reduce the cost of expensive in vitro experiments [111].

References

1 Heitz, F., Morris, M.C., and Divita, G. (2009). Twenty years of cell-penetrating peptides: from molecular mechanisms to therapeutics. *British Journal of Pharmacology 157* (2): 195–206. https://doi.org/10.1111/j.1476-5381.2009.00057.x.

2 Lindgren, M., Hallbrink, M., Prochiantz, A. et al. (2000). Cell-penetrating peptides. *Trends in Pharmacological Sciences 21* (3): 99–103. https://doi.org/10.1016/s0165-6147(00)01447-4.

3 Morris, M.C., Depollier, J., Mery, J. et al. (2001). A peptide carrier for the delivery of biologically active proteins into mammalian cells. *Nature Biotechnology 19* (12): 1173–1176. https://doi.org/10.1038/nbt1201-1173.

4 Taylor, B.N., Mehta, R.R., Yamada, T. et al. (2009). Noncationic peptides obtained from azurin preferentially enter cancer cells. *Cancer Research 69* (2): 537–546. https://doi.org/10.1158/0008-5472.can-08-2932.

5 Milletti, F. (2012). Cell-penetrating peptides: classes, origin, and current landscape. *Drug Discovery Today 17* (15–16): 850–860. https://doi.org/10.1016/j.drudis.2012.03.002.

6 Frankel, A.D. and Pabo, C.O. (1988). Cellular uptake of the tat protein from human immunodeficiency virus. *Cell 55* (6): 1189–1193. https://doi.org/10.1016/0092-8674(88)90263-2.

7 Green, M. and Loewenstein, P.M. (1988). Autonomous functional domains of chemically synthesized human immunodeficiency virus tat trans-activator protein. *Cell 55* (6): 1179–1188. https://doi.org/10.1016/0092-8674(88)90262-0.

8 Vives, E., Brodin, P., and Lebleu, B. (1997). A truncated HIV-1 Tat protein basic domain rapidly translocates through the plasma membrane and accumulates in the

cell nucleus. *Journal of Biological Chemistry 272* (25): 16010–16017. https://doi.org/10.1074/jbc.272.25.16010.

9 Joliot, A., Pernelle, C., Deagostinibazin, H. et al. (1991). Antennapedia homeobox peptide regulates neural morphogenesis. *Proceedings of the National Academy of Sciences of the United States of America 88* (5): 1864–1868. https://doi.org/10.1073/pnas.88.5.1864.

10 Derossi, D., Joliot, A.H., Chassaing, G. et al. (1994). The 3rd helix of the antennapedia homeodomain translocates through biological-membranes. *Journal of Biological Chemistry 269* (14): 10444–10450. https://doi.org/10.1016/s0021-9258(17)34080-2.

11 Theodore, L., Derossi, D., Chassaing, G. et al. (1995). Intraneuronal delivery of protein-kinase-c pseudosubstrate leads to growth cone collapse. *Journal of Neuroscience 15* (11): 7158–7167. https://doi.org/10.1523/JNEUROSCI.15-11-07158.1995.

12 Elliott, G. and O'Hare, P. (1997). Intercellular trafficking and protein delivery by a herpesvirus structural protein. *Cell 88* (2): 223–233. https://doi.org/10.1016/s0092-8674(00)81843-7.

13 Phelan, A., Elliott, G., and O'Hare, P. (1998). Intercellular delivery of functional p53 by the herpesvirus protein VP22. *Nature Biotechnology 16* (5): 440–443. https://doi.org/10.1038/nbt0598-440.

14 Agrawal, P., Bhalla, S., Usmani, S.S. et al. (2016). CPPsite 2.0: a repository of experimentally validated cell-penetrating peptides. *Nucleic Acids Research 44* (D1): D1098–D1103. https://doi.org/10.1093/nar/gkv1266.

15 Kardani, K. and Bolhassani, A. (2021). Cppsite 2.0: An available database of experimentally validated cell-penetrating peptides predicting their secondary and tertiary structures. *Journal of Molecular Biology 433* (11): https://doi.org/10.1016/j.jmb.2020.11.002.

16 Chen, J.X., Xu, X.D., Yang, S. et al. (2013). Self-assembled bola-like amphiphilic peptides as viral-mimetic gene vectors for cancer cell targeted gene delivery. *Macromolecular Bioscience 13* (1): 84–92. https://doi.org/10.1002/mabi.201200283.

17 Kang, M.H., Park, M.J., Yoo, H.J. et al. (2014). RIPL peptide (IPLVVPLRRRRRRRRC)-conjugated liposomes for enhanced intracellular drug delivery to hepsin-expressing cancer cells. *European Journal of Pharmaceutics and Biopharmaceutics 87* (3): 489–499. https://doi.org/10.1016/j.ejpb.2014.03.016.

18 Tüunnemann, G., Ter-Avetisyan, G., Martin, R.M. et al. (2008). Live-cell analysis of cell penetration ability and toxicity of oligo-arginines. *Journal of Peptide Science 14* (4): 469–476. https://doi.org/10.1002/psc.968.

19 Scheller, A., Oehlke, J., Wiesner, B. et al. (1999). Structural requirements for cellular uptake of alpha-helical amphipathic peptides. *Journal of Peptide Science 5* (4): 185–194. https://doi.org/10.1002/(sici)1099-1387(199904)5:4<185::aid-psc184>3.0.co;2-9.

20 Gomez, J.A., Gama, V., Yoshida, T. et al. (2007). Bax-inhibiting peptides derived from Ku70 and cell-penetrating pentapeptides. *Biochemical Society Transactions 35*: 797–801. https://doi.org/10.1042/bst0350797.

21 Derakhshankhah, H. and Jafari, S. (2018). Cell penetrating peptides: a concise review with emphasis on biomedical applications. *Biomedicine & Pharmacotherapy 108*: 1090–1096. https://doi.org/10.1016/j.biopha.2018.09.097.

22 Jobin, M.L. and Alves, I.D. (2014). On the importance of electrostatic interactions between cell penetrating peptides and membranes: a pathway toward tumor cell selectivity? *Biochimie 107*: 154–159. https://doi.org/10.1016/j.biochi.2014.07.022.

23 Wada, S., Tsuda, H., Okada, T. et al. (2011). Cellular uptake of Aib-containing amphipathic helix peptide. *Bioorganic & Medicinal Chemistry Letters 21* (19): 5688–5691. https://doi.org/10.1016/j.bmcl.2011.08.030.

24 Deshayes, S., Decaffmeyer, M., Brasseur, R. et al. (2008). Structural polymorphism of two CPP: An important parameter of activity. *Biochimica Et Biophysica Acta-Biomembranes 1778* (5): 1197–1205. https://doi.org/10.1016/j.bbamem.2008.01.027.

25 Eiríksdóttir, E., Konate, K., Langel, U. et al. (2010). Secondary structure of cell-penetrating peptides controls membrane interaction and insertion. *Biochimica Et Biophysica Acta-Biomembranes 1798* (6): 1119–1128. https://doi.org/10.1016/j.bbamem.2010.03.005.

26 Su, Y.C., Mani, R., Doherty, T. et al. (2008). Reversible sheet-turn conformational change of a cell-penetrating peptide in lipid bilayers studied by solid-state NMR. *Journal of Molecular Biology 381* (5): 1133–1144. https://doi.org/10.1016/j.jmb.2008.06.007.

27 Ruseska, I. and Zimmer, A. (2020). Internalization mechanisms of cell-penetrating peptides. *Beilstein Journal of Nanotechnology 11*: 101–123. https://doi.org/10.3762/bjnano.11.10.

28 Montal, M. and Mueller, P. (1972). Formation of bimolecular membranes from lipid monolayers and a study of their electrical properties. *Proc. Natl. Acad. Sci. 69* (12): 3561–3566.

29 Dunlop, J., Bowlby, M., Peri, R. et al. (2008). High-throughput electrophysiology: an emerging paradigm for ion-channel screening and physiology. *Nature Reviews Drug Discovery 7* (4): 358–368. https://doi.org/10.1038/nrd2552.

30 Kawano, R., Tsuji, Y., Sato, K. et al. (2013). Automated parallel recordings of topologically identified single ion channels. *Scientific Reports 3*: 1995. https://doi.org/10.1038/srep01995.

31 Bayley, H., Cronin, B., Heron, A. et al. (2008). Droplet interface bilayers. *Molecular Biosystems 4* (12): 1191–1208. https://doi.org/10.1039/b808893d.

32 Kawano, R., Osaki, T., Sasaki, H. et al. (2011). Rapid detection of a cocaine-binding aptamer using biological nanopores on a chip. *Journal of the American Chemical Society 133* (22): 8474–8477. https://doi.org/10.1021/ja2026085.

33 Saigo, N., Izumi, K., and Kawano, R. (2019). Electrophysiological analysis of antimicrobial peptides in diverse species. *ACS Omega 4* (8): 13124–13130. https://doi.org/10.1021/acsomega.9b01033.

34 Sekiya, Y., Sakashita, S., Shimizu, K. et al. (2018). Channel current analysis estimates the pore-formation and the penetration of transmembrane peptides. *Analyst 143* (15): https://doi.org/10.1039/c8an00243f.

35 Sekiya, Y., Shimizu, K., Kitahashi, Y. et al. (2019). Electrophysiological analysis of membrane disruption by bombinin and its isomer using the lipid bilayer system. *ACS Applied Bio Materials 2* (4): 1542–1548. https://doi.org/10.1021/acsabm.8b00835.

36 Bechinger, B. and Lohner, K. (2006). Detergent-like actions of linear amphipathic cationic antimicrobial peptides. *Biochimica Et Biophysica Acta-Biomembranes 1758* (9): 1529–1539. https://doi.org/10.1016/j.bbamem.2006.07.001.

37 Guidotti, G., Brambilla, L., and Rossi, D. (2017). Cell-penetrating peptides: From basic research to clinics. *Trends in Pharmacological Sciences 38* (4): 406–424. https://doi.org/10.1016/j.tips.2017.01.003.

38 Zhao, C.Q., Liaw, L., Lee, I.H. et al. (1997). cDNA cloning of Clavanins: antimicrobial peptides of tunicate hemocytes. *Febs Letters 410* (2-3): 490–492. https://doi.org/10.1016/s0014-5793(97)00646-7.

39 Steiner, H., Hultmark, D., Engstrom, A. et al. (1981). Sequence and specificity of 2 anti-bacterial proteins involved in insect immunity. *Nature 292* (5820): 246–248. https://doi.org/10.1038/292246a0.

40 Lazarovici, P., Primor, N., and Loew, L.M. (1986). Purification and pore-forming activity of 2 hydrophobic polypeptides from the secretion of the red-sea moses sole (pardachirus-marmoratus). *Journal of Biological Chemistry 261* (35): 6704–6713.

41 Zasloff, M. (1987). Magainins, a class of antimicrobial peptides from xenopus skin - isolation, characterization of 2 active forms, and partial cdna sequence of a precursor. *Proceedings of the National Academy of Sciences of the United States of America 84* (15): 5449–5453. https://doi.org/10.1073/pnas.84.15.5449.

42 Zelezetsky, I., Pontillo, A., Puzzi, L. et al. (2006). Evolution of the primate cathelicidin – correlation between structural variations and antimicrobial activity. *Journal of Biological Chemistry 281* (29): 19861–19871. https://doi.org/10.1074/jbc.M511108200.

43 Gudmundsson, G.H., Agerberth, B., Odeberg, J. et al. (1996). The human gene FALL39 and processing of the cathelin precursor to the antibacterial peptide LL-37 in granulocytes. *European Journal of Biochemistry 238* (2): 325–332. https://doi.org/10.1111/j.1432-1033.1996.0325z.x.

44 Mylonakis, E., Podsiadlowski, L., Muhammed, M. et al. (2016). Diversity, evolution and medical applications of insect antimicrobial peptides. *Philosophical Transactions of the Royal Society B-Biological Sciences 371* (1695): https://doi.org/10.1098/rstb.2015.0290.

45 Macazo, F.C. and White, R.J. (2014). Monitoring charge flux to quantify unusual ligand-induced ion channel activity for use in biological nanopore-based sensors. *Analytical Chemistry 86* (11): 5519–5525. https://doi.org/10.1021/ac500832a.

46 Mignogna, G., Simmaco, M., Kreil, G. et al. (1993). Antibacterial and hemolytic peptides containing D-alloisoleucine from the skin of bombina-variegata. *Embo Journal 12* (12): 4829–4832. https://doi.org/10.1002/j.1460-2075.1993.tb06172.x.

47 Simmaco, M., Kreil, G., and Barra, D. (2009). Bombinins, antimicrobial peptides from Bombina species. *Biochimica Et Biophysica Acta-Biomembranes 1788* (8): 1551–1555. https://doi.org/10.1016/j.bbamem.2009.01.004.

48 Marr, A.K., McGwire, B.S., and McMaster, W.R. (2012). Modes of action of Leishmanicidal antimicrobial peptides. *Future Microbiology 7* (9): 1047–1059. https://doi.org/10.2217/fmb.12.85.

49 Mangoni, M.L., Papo, N., Saugar, J.M. et al. (2006). Effect of natural L- to D-amino acid conversion on the organization, membrane binding, and biological function of the antimicrobial peptides bombinins H. *Biochemistry 45* (13): 4266–4276. https://doi.org/10.1021/bi052150y.

50 Bozzi, A., Mangoni, M.L., Rinaldi, A.C. et al. (2008). Folding propensity and biological activity of peptides: the effect of a single stereochemical isomerization on the conformational properties of bombinins in aqueous solution. *Biopolymers 89* (9): 769–778. https://doi.org/10.1002/bip.21006.

51 Coccia, C., Rinaldi, A.C., Luca, V. et al. (2011). Membrane interaction and antibacterial properties of two mildly cationic peptide diastereomers, bombinins H2 and H4, isolated from Bombina skin. *European Biophysics Journal with Biophysics Letters 40* (4): 577–588. https://doi.org/10.1007/s00249-011-0681-8.

52 Brogden, K.A. (2005). Antimicrobial peptides: pore formers or metabolic inhibitors in bacteria? *Nature Reviews Microbiology 3* (3): 238–250. https://doi.org/10.1038/nrmicro1098.

53 Zorko, M. and Langel, U. (2005). Cell-penetrating peptides: mechanism and kinetics of cargo delivery. *Advanced Drug Delivery Reviews 57* (4): 529–545. https://doi.org/10.1016/j.addr.2004.10.010.

54 Jang, H., Arce, F.T., Ramachandran, S. et al. (2014). Disordered amyloidogenic peptides may insert into the membrane and assemble into common cyclic structural motifs. *Chemical Society Reviews 43* (19): 6750–6764. https://doi.org/10.1039/c3cs60459d.

55 Usui, K., Kikuchi, T., Mie, M. et al. (2013). Systematic screening of the cellular uptake of designed alpha-helix peptides. *Bioorganic & Medicinal Chemistry 21* (9): 2560–2567. https://doi.org/10.1016/j.bmc.2013.02.030.

56 Kim, S., Hyun, S., Lee, Y. et al. (2016). Nonhemolytic cell-penetrating peptides: Site specific introduction of glutamine and lysine residues into the alpha-helical peptide causes deletion of its direct membrane disrupting ability but retention of its cell penetrating ability. *Biomacromolecules 17* (9): 3007–3015. https://doi.org/10.1021/acs.biomac.6b00874.

57 Futaki, S., Suzuki, T., Ohashi, W. et al. (2001). Arginine-rich peptides – an abundant source of membrane-permeable peptides having potential as carriers for intracellular protein delivery. *Journal of Biological Chemistry 276* (8): 5836–5840. https://doi.org/10.1074/jbc.M007540200.

58 Zasloff, M. (2002). Antimicrobial peptides of multicellular organisms. *Nature 415* (6870): 389–395. https://doi.org/10.1038/415389a.

59 Ohgita, T., Takechi-Haraya, Y., Okada, K. et al. (2020). Enhancement of direct membrane penetration of arginine-rich peptides by polyproline II helix structure. *Biochimica et Biophysica Acta-Biomembranes 1862* (10): https://doi.org/10.1016/j.bbamem.2020.183403.

60 Deshayes, S., Konate, K., Aldrian, G. et al. (2010). Structural polymorphism of non-covalent peptide-based delivery systems: highway to cellular uptake.

Biochimica Et Biophysica Acta-Biomembranes 1798 (12): 2304–2314. https://doi.org/10.1016/j.bbamem.2010.06.005.

61 Kalafatovic, D. and Giralt, E. (2017). Cell-penetrating peptides: design strategies beyond primary structure and amphipathicity. *Molecules 22* (11): https://doi.org/10.3390/molecules22111929.

62 Ohgita, T., Takechi-Haraya, Y., Nadai, R. et al. (2019). A novel amphipathic cell-penetrating peptide based on the N-terminal glycosaminoglycan binding region of human apolipoprotein E. *Biochimica Et Biophysica Acta-Biomembranes 1861* (3): 541–549. https://doi.org/10.1016/j.bbamem.2018.12.010.

63 Oikawa, K., Islam, M.M., Horii, Y. et al. (2018). Screening of a cell-penetrating peptide library in escherichia coli: relationship between cell penetration efficiency and cytotoxicity. *ACS Omega 3* (12): 16489–16499. https://doi.org/10.1021/acsomega.8b02348.

64 Takechi-Haraya, Y., Nadai, R., Kimura, H. et al. (2016). Enthalpy-driven interactions with sulfated glycosaminoglycans promote cell membrane penetration of arginine peptides. *Biochimica Et Biophysica Acta-Biomembranes 1858* (6): 1339–1349. https://doi.org/10.1016/j.bbamem.2016.03.021.

65 Chu, Q., Moellering, R.E., Hilinski, G.J. et al. (2015). Towards understanding cell penetration by stapled peptides. *MedChemComm 6* (1): 111–119. https://doi.org/10.1039/c4md00131a.

66 Peraro, L. and Kritzer, J.A. (2018). Emerging methods and design principles for cell-penetrant peptides. *Angewandte Chemie-International Edition 57* (37): 11868–11881. https://doi.org/10.1002/anie.201801361.

67 Kakinoki, S., Hirano, Y., and Oka, M. (2005). On the stability of polyproline-I and II structures of proline oligopeptides. *Polymer Bulletin 53* (2): 109–115. https://doi.org/10.1007/s00289-004-0317-6.

68 Shi, L.Q., Holliday, A.E., Shi, H.L. et al. (2014). Characterizing intermediates along the transition from polyproline I to polyproline II using ion mobility spectrometry-mass spectrometry. *Journal of the American Chemical Society 136* (36): 12702–12711. https://doi.org/10.1021/ja505899g.

69 Horng, J.C. and Raines, R.T. (2006). Stereoelectronic effects on polyproline conformation. *Protein Science 15* (1): 74–83. https://doi.org/10.1110/ps.051779806.

70 Kelly, M.A., Chellgren, B.W., Rucker, A.L. et al. (2001). Host-guest study of left-handed polyproline II helix formation. *Biochemistry 40* (48): 14376–14383. https://doi.org/10.1021/bi011043a.

71 Pandey, A.K., Thomas, K.M., Forbes, C.R. et al. (2014). Tunable control of polyproline helix (PPII) structure via aromatic electronic effects: an electronic switch of polyproline helix. *Biochemistry 53* (32): 5307–5314. https://doi.org/10.1021/bi500696k.

72 Cardoso, M.H., Meneguetti, B.T., Costa, B.O. et al. (2019). Non-lytic antibacterial peptides that translocate through bacterial membranes to act on intracellular targets. *International Journal of Molecular Sciences 20* (19): https://doi.org/10.3390/ijms20194877.

73 Daniels, D.S. and Schepartz, A. (2007). Intrinsically cell-permeable miniature proteins based on a minimal cationic PPII motif. *Journal of the American Chemical Society 129* (47): 14578-+. https://doi.org/10.1021/ja0772445.

74 Fillon, Y.A., Anderson, J.P., and Chmielewski, J. (2005). Cell penetrating agents based on a polyproline helix scaffold. *Journal of the American Chemical Society 127* (33): 11798–11803. https://doi.org/10.1021/ja052377g.

75 Nagel, Y.A., Raschle, P.S., and Wennemers, H. (2017). Effect of preorganized charge-display on the cell-penetrating properties of cationic peptides. *Angewandte Chemie-International Edition 56* (1): 122–126. https://doi.org/10.1002/anie.201607649.

76 Herce, H.D. and Garcia, A.E. (2007). Molecular dynamics simulations suggest a mechanism for translocation of the HIV-1 TAT peptide across lipid membranes. *Proceedings of the National Academy of Sciences of the United States of America 104* (52): 20805–20810. https://doi.org/10.1073/pnas.0706574105.

77 Herce, H.D., Garcia, A.E., and Cardoso, M.C. (2014). Fundamental molecular mechanism for the cellular uptake of guanidinium-rich molecules. *Journal of the American Chemical Society 136* (50): 17459–17467. https://doi.org/10.1021/ja507790z.

78 Mishra, A., Lai, G.H., Schmidt, N.W. et al. (2011). Translocation of HIV TAT peptide and analogues induced by multiplexed membrane and cytoskeletal interactions. *Proceedings of the National Academy of Sciences of the United States of America 108* (41): 16883–16888. https://doi.org/10.1073/pnas.1108795108.

79 Adzhubei, A.A., Sternberg, M.J.E., and Makarov, A.A. (2013). Polyproline-II helix in proteins: structure and function. *Journal of Molecular Biology 425* (12): 2100–2132. https://doi.org/10.1016/j.jmb.2013.03.018.

80 Neree, A.T., Nguyen, P.T., Chatenet, D. et al. (2014). Secondary conformational conversion is involved in glycosaminoglycans-mediated cellular uptake of the cationic cell-penetrating peptide PACAP. *FEBS Letters 588* (24): 4590–4596. https://doi.org/10.1016/j.febslet.2014.10.029.

81 Bechara, C., Pallerla, M., Zaltsman, Y. et al. (2013). Tryptophan within basic peptide sequences triggers glycosaminoglycan-dependent endocytosis. *FASEB Journal 27* (2): 738–749. https://doi.org/10.1096/fj.12-216176.

82 Jiao, C.Y., Delaroche, D., Burlina, F. et al. (2009). Translocation and endocytosis for cell-penetrating peptide internalization. *Journal of Biological Chemistry 284* (49): 33957–33965. https://doi.org/10.1074/jbc.M109.056309.

83 Raghuraman, H. and Chattopadhyay, A. (2007). Melittin: a membrane-active peptide with diverse functions. *Bioscience Reports 27* (4–5): 189–223. https://doi.org/10.1007/s10540-006-9030-z.

84 Eisenberg, D., Weiss, R.M., Terwilliger, T.C. et al. (1982). Hydrophobic moments and protein-structure. *Faraday Symposia of the Chemical Society* 17: 109–120. https://doi.org/10.1039/fs9821700109.

85 Wimley, W.C. and White, S.H. (1996). Experimentally determined hydrophobicity scale for proteins at membrane interfaces. *Nature Structural Biology 3* (10): 842–848. https://doi.org/10.1038/nsb1096-842.

86 Hong, M. and Su, Y.C. (2011). Structure and dynamics of cationic membrane peptides and proteins: insights from solid-state NMR. *Protein Science 20* (4): 641–655. https://doi.org/10.1002/pro.600.

87 Bechinger, B. (1999). The structure, dynamics and orientation of antimicrobial peptides in membranes by multidimensional solid-state NMR spectroscopy. *Biochimica Et Biophysica Acta-Biomembranes 1462* (1–2): 157–183. https://doi.org/10.1016/s0005-2736(99)00205-9.

88 Kawamura, I., Norisada, K., and Naito, A. (2018). Structure determination of membrane peptides and proteins by solid-state NMR. *Experimental Approaches of NMR Spectroscopy 34*: 253–293. https://doi.org/10.1007/978-981-10-5966-7_9.

89 Molugu, T.R., Lee, S., and Brown, M.F. (2017). Concepts and methods of solid-state nmr spectroscopy applied to biomembranes. *Chemical Reviews 117* (19): 12087–12132. https://doi.org/10.1021/acs.chemrev.6b00619.

90 Naito, A. and Kawamura, I. (2007). Solid-state NMR as a method to reveal structure and membrane-interaction of amyloidogenic proteins and peptides. *Biochimica Et Biophysica Acta-Biomembranes 1768* (8): 1900–1912. https://doi.org/10.1016/j.bbamem.2007.03.025.

91 Ramamoorthy, A. (2009). Beyond NMR spectra of antimicrobial peptides: dynamical images at atomic resolution and functional insights. *Solid State Nuclear Magnetic Resonance 35* (4): 201–207. https://doi.org/10.1016/j.ssnmr.2009.03.003.

92 Shcherbarkov, A.A., Medeiros-Silva, J., Tran, N. et al. (2021). From angstroms to nanometers: Measuring interatomic distances by solid-state NMR. *Chemical Reviews* https://doi.org/10.1021/acs.chemrev.1c00662.

93 Mijiddorj, B., Kaneda, S., Sato, H. et al. (2018). The role of D-allo-isoleucine in the deposition of the anti-Leishmania peptide bombinin H4 as revealed by P-31 solid-state NMR, VCD spectroscopy, and MD simulation. *Biochimica Et Biophysica Acta-Proteins and Proteomics 1866* (7): 789–798. https://doi.org/10.1016/j.bbapap.2018.01.005.

94 Naito, A., Nagao, T., Norisada, K. et al. (2000). Conformation and dynamics of melittin bound to magnetically oriented lipid bilayers by solid-state P-31 and C-13 NMR spectroscopy. *Biophysical Journal 78* (5): 2405–2417. https://doi.org/10.1016/s0006-3495(00)76784-1.

95 Thennarasu, S., Huang, R., Lee, D.K. et al. (2010). Limiting an antimicrobial peptide to the lipid-water interface enhances its bacterial membrane selectivity: a case study of MSI-367. *Biochemistry 49* (50): 10595–10605. https://doi.org/10.1021/bi101394r.

96 Eriksson, P.O., Rilfors, L., Lindblom, G. et al. (1985). Multicomponent spectra from P-31-NMR studies of the phase-equilibria in the system dioleoylphosphatidylcholine-dioleoylphosphatidylethanolamine-water. *Chemistry and Physics of Lipids 37* (4): 357–371. https://doi.org/10.1016/0009-3084(85)90089-1.

97 Morais, C.M., Cardoso, A.M., Cunha, P.P. et al. (2018). Acylation of the S4(13)-PV cell-penetrating peptide as a means of enhancing its capacity to mediate nucleic acid delivery: relevance of peptide/lipid interactions. *Biochimica Et Biophysica Acta-Biomembranes 1860* (12): 2619–2634. https://doi.org/10.1016/j.bbamem.2018.10.002.

98 Yao, H.W. and Hong, M. (2014). Conformation and lipid interaction of the fusion peptide of the paramyxovirus PIV5 in anionic and negative-curvature membranes from solid-state NMR. *Journal of the American Chemical Society 136* (6): 2611–2624. https://doi.org/10.1021/ja4121956.

99 Afonin, S., Frey, A., Bayerl, S. et al. (2006). The cell-penetrating peptide TAT(48-60) induces a non-lamellar phase in DMPC membranes. *ChemPhysChem 7* (10): 2134–2142. https://doi.org/10.1002/cphc.200600306.

100 Su, Y.C., Waring, A.J., Ruchala, P. et al. (2010). Membrane-bound dynamic structure of an arginine-rich cell-penetrating peptide, the protein transduction domain of HIV TAT, from solid-state NMR. *Biochemistry 49* (29): 6009–6020. https://doi.org/10.1021/bi100642n.

101 Su, Y.C., Li, S.H., and Hong, M. (2013). Cationic membrane peptides: atomic-level insight of structure-activity relationships from solid-state NMR. *Amino Acids 44* (3): 821–833. https://doi.org/10.1007/s00726-012-1421-9.

102 Takeuchi, K., Misaki, I., Tokunaga, Y. et al. (2021). Conformational plasticity of cyclic ras-inhibitor peptides defines cell permeabilization activity. *Angewandte Chemie-International Edition 60* (12): 6567–6572. https://doi.org/10.1002/anie.202016647.

103 Reid, L.M., Verma, C.S., and Essex, J.W. (2019). The role of molecular simulations in understanding the mechanisms of cell-penetrating peptides. *Drug Discovery Today 24* (9): 1821–1835. https://doi.org/10.1016/j.drudis.2019.06.013.

104 Allolio, C., Magarkar, A., Jurkiewicz, P. et al. (2018). Arginine-rich cell-penetrating peptides induce membrane multilamellarity and subsequently enter via formation of a fusion pore. *Proceedings of the National Academy of Sciences of the United States of America 115* (47): 11923–11928. https://doi.org/10.1073/pnas.1811520115.

105 Via, M.A., Klug, J., Wilke, N. et al. (2018). The interfacial electrostatic potential modulates the insertion of cell-penetrating peptides into lipid bilayers. *Physical Chemistry Chemical Physics 20* (7): 5180–5189. https://doi.org/10.1039/c7cp07243k.

106 Nymeyer, H., Woolf, T.B., and Garcia, A.E. (2005). Folding is not required for bilayer insertion: Replica exchange simulations of an alpha-helical peptide with an explicit lipid bilayer. *Proteins-Structure Function and Bioinformatics 59* (4): 783–790. https://doi.org/10.1002/prot.20460.

107 Gao, X.L., Hong, S., Liu, Z.P. et al. (2019). Membrane potential drives direct translocation of cell-penetrating peptides. *Nanoscale 11* (4): 1949–1958. https://doi.org/10.1039/c8nr10447f.

108 Hu, Y., Liu, X.R., Sinha, S.K. et al. (2014). Trans location thermodynamics of linear and cyclic nonaarginine into model DPPC bilayer via coarse-grained molecular dynamics simulation: implications of pore formation and nonadditivity. *Journal of Physical Chemistry B 118* (10): 2670–2682. https://doi.org/10.1021/jp412600e.

109 Alaybeyoglu, B., Akbulut, B.S., and Ozkirimli, E. (2016). Insights into membrane translocation of the cell-penetrating peptide pVEC from molecular dynamics calculations. *Journal of Biomolecular Structure & Dynamics 34* (11): 2387–2398. https://doi.org/10.1080/07391102.2015.1117396.

110 Grasso, G., Muscat, S., Rebella, M. et al. (2018). Cell penetrating peptide modulation of membrane biomechanics by Molecular dynamics. *Journal of Biomechanics 73*: 137–144. https://doi.org/10.1016/j.jbiomech.2018.03.036.

111 Tran, D.P., Tada, S., Yumoto, A. et al. (2021). Using molecular dynamics simulations to prioritize and understand AI-generated cell penetrating peptides. *Scientific Reports 11* (1): https://doi.org/10.1038/s41598-021-90245-z.

9

Cellular Uptake Mechanisms of Arginine-Rich Cell-Penetrating Peptides

Ikuhiko Nakase

Graduate School of Science, Osaka Metropolitan University, Osaka, Japan

9.1 Introduction

Arginine-rich cell-penetrating peptides (CPPs), such as human immunodeficiency virus type 1(HIV-1) Tat peptide [Tat (48–60)] [1] and oligoarginines [2–5] are efficiently internalized by various cell types (Figure 9.1a) [6–8]. Because of the efficient uptake and membrane penetration efficiency, arginine-rich CPPs are widely and effectively used as carriers for achieving efficient intracellular delivery of bioactive molecules and therapeutics, including enzymes, peptides, genes, liposomes/micelles, magnet beads, quantum dots for regulating cell-specific functions, visualization of cellular/body systems, disease diagnostics, and therapeutics [6–8]. The arginine-rich CPPs are under development for applications in basic science and in the clinic, however, the complex cellular uptake and membrane penetration mechanisms should be further elucidated from a peptide-based chemical and biological approach. Understanding the cellular uptake mechanisms of arginine-rich CPPs will contribute to the further development of CPPs with innovative design and optimized biological/therapeutical applicability. There are many issues regarding the improvement of the biological functionalities of arginine-rich CPPs, such as the efficacy of specific cell targeting (especially for tumors), uptake, cell membrane penetration (e.g. macromolecules of biopharmaceuticals including antibodies), cytotoxicity, and regulation of the pharmacology in animals and humans.

Arginine strongly interacts with phosphate, sulfate, and carboxylate through ion-pairing and hydrogen bonding (Figure 9.1b) [6–8]. The molecular interactions on and in cells participate in the complex cell uptake and membrane penetration mechanisms associated with the uptake process [6–8]. Studies of proteoglycans have focused on understanding the accumulation of arginine-rich CPPs on plasma membranes and cellular uptake (Figure 9.2) [6–8]. "Macropinocytosis" [9, 10] refers to the

Cell-Penetrating Peptides: Design, Development and Applications, First Edition. Makoto Oba and Yosuke Demizu.

Peptide	Amino acid sequence
HIV-1 Tat (48–60)	GRKKRRQRRRPPQ
R8	RRRRRRRR
R12	RRRRRRRRRRRR
R16	RRRRRRRRRRRRRRRR
HIV-1 Rev (34–50)	TRQARRNRRRRWRERQR
FHV Coat (35–49)	RRRRNRTRRNRRRVR
aR7W2	RRRWRRWRR
RW9	RRWWRRWRR
RW16	RRWRRWWRRWWRRWRR
A2-17	LRKLRKRLLRLWKLRKR
PG-1	RGGRLCYCRRRFCVCVGR
RLA	$_{D}$(RLARLAR)$_2$

(a)

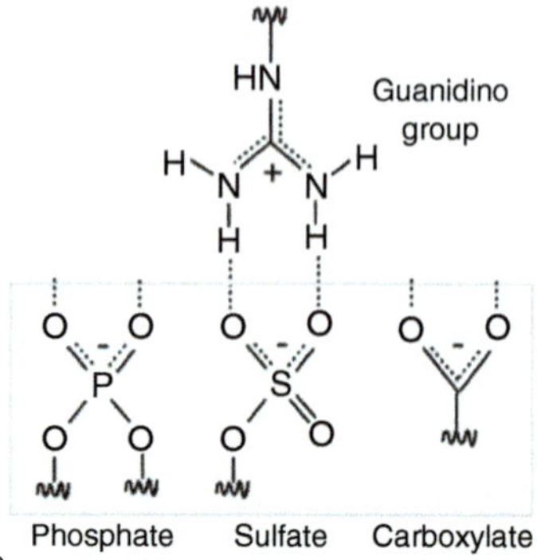

(b)

Figure 9.1 Examples of arginine-rich CPPs and hydrogen bonding of arginine residues. (a) Amino acid sequences of arginine-rich CPPs. (b) Hydrogen bonding of guanidino group with phosphate, sulfate, and carboxylate.

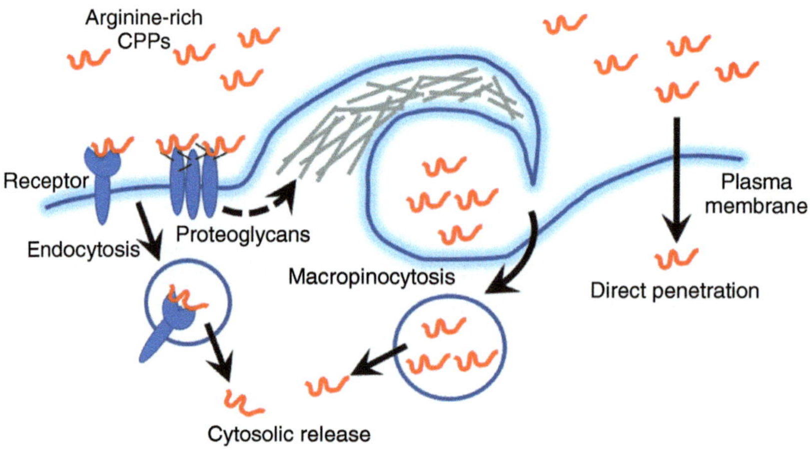

Figure 9.2 Cellular uptake pathways of arginine-rich CPPs including accumulation via proteoglycans, receptor-mediated endocytosis, macropinocytosis, and direct penetration through plasma membranes.

nonspecific uptake of extracellular material, which is important for understanding the cellular uptake mechanisms for arginine-rich CPPs (Figure 9.2) [6–8, 11]. Macropinocytotic pathways can significantly enhance the uptake of macromolecules into cells using arginine-rich peptides because of the engulfment of large volumes (>1 μm) of external fluids [12]. One of the reasons for the excellent carrier functionality of these peptides is the efficient inducibility of macropinocytosis following treatment with arginine-rich CPPs on targeted cells [6–8]. However, complicated cellular uptake mechanisms including molecular interactions with proteins inside and outside of cells through membranes, lipids, glycosaminoglycans that are involved in energy-dependent endocytic pathways, signal transduction, and direct membrane penetration, should be further examined in in vitro, in vivo, and in silico experiments.

In this chapter, the latest findings regarding our understanding of cellular uptake mechanisms of arginine-rich CPPs are summarized from the viewpoint of endocytosis and membrane penetration mechanisms, membrane curvature, toxicity, and localization after cellular uptake to obtain the desired biological activity of the cargo molecules delivered by arginine-rich CPPs.

9.2 Endocytosis Mechanisms of Arginine-Rich Cell-Penetrating Peptides

Proteoglycans participate in the cellular uptake of arginine-rich peptides (Figure 9.2) [13–17]. Our group reported the involvement of syndecan-4 in the macropinocytotic uptake of arginine-rich CPPs [16]. The syndecan family are type I transmembrane heparan sulfate proteoglycans and almost all cells express syndecan-4 on the plasma membrane with heparan sulfate chains (linear polysaccharides composed of glucuronic/iduronic acid and glucosamine residues) [18–21]. Syndecan-4 consists of an extracellular domain, transmembrane region, and conserved short C-terminal cytoplasmic domain, including C1, V, and C2 domains [18–21]. The V domain can be bound with protein kinase C-alpha (PKCα) through phosphatidylinositol (4,5) bisphosphate, which leads to the activation of PKCα [18–21]. In addition, clustering of the syndecan-4 enhances the binding of PKCα, suggesting controlled signal transduction through multimerization of the proteoglycan [18, 20–23]. Overexpression of syndecan-4 significantly enhanced macropinocytosis and cellular uptake of octaarginine (R8), and the R8 peptides increased syndecan-4 clustering on the plasma membrane [16]. Enhanced binding of PKCα to syndecan-4 following treatment with R8 peptides was also confirmed by western blot analysis, and knockdown of PKCα reduced the uptake of R8 peptides [16]. In addition, expression of the syndecan-4 lacking the V domain significantly reduced the uptake of R8 peptides; however, uptake was not affected by the deletion of the C1 or C2 domains of syndecan-4 [16]. Interestingly, tetraarginine (R4) with a lower cellular uptake efficiency did not increase syndecan-4 clustering [16].

During macropinocytotic cellular uptake of arginine-rich CPPs, actin reorganization and membrane ruffling are observed following actin staining (Figure 9.2) [13, 15, 24]. Recently, detailed dynamic membrane structures of induced lamellipodia formation by arginine-rich CPPs were observed. Membrane ruffling and cup closure induced by treatment with fluorescent-labeled R8 peptides were visualized using scanning ion-conductance microscopy and spinning-disk confocal microscopy [25]. Dynamic structural changes of the cell membrane by the peptides at low concentration (2 μM) were observed as ruffled membrane structures and cup closure resulted in macropinosome formation [25].

The incorporation of tryptophan residues into arginine-rich CPPs resulted in enhanced cellular uptake efficiency. Walrant et al. studied cell interactions and cellular uptake of various nonapeptides with arginine and tryptophan residues. The aR7W2 peptide (amino acid sequence: RRRWRRWRR) showed the highest cellular uptake compared with that of other peptides [26]. Salt bridge-π interactions participate significantly in glycosaminoglycan-dependent cellular uptake of peptides as analyzed by density functional theory, and the contribution of tryptophan residues to binding enthalpy has been shown [26].

With respect to endosomal trafficking in cells, Trofimenko et al. recently reported that the early endocytosis stage of arginine-rich CPPs, including Tat or nonaarginine (R9) peptides, showed EEA1 positive vesicles and a very low colocalization with Rab5-positive vesicles, which are typical early endosomal markers [27]. In addition, Rab14-positive vesicles were detected during the process of endosomal maturation; however, Rab7, which is a late endosomal marker, was not required for maturation [27]. Rab14-positive vesicles participate in recycling endosomes and biosynthetic (endoplasmic reticulum and Golgi) traffic [28–30]. Lamp1-positive vesicles of arginine-rich CPPs were also confirmed, suggesting that no degradation occurred [27]. In addition, the cellular uptake of transferrin and epidermal growth factor, which are internalized through clathrin-mediated endocytosis was observed in Rab5- and Rab7-positive vesicles [27]. These results suggest an independent cellular uptake process for arginine-rich CPPs in classical Rab5/Rab7 pathways, which are essential for endosomal maturation from early endosomes to lysosomes, and the existence of EEA1 and Rab14-dependent novel endocytic pathways [27].

Interestingly, endocytic uptake of arginine-rich CPPs may be applicable for the detection of intracellular pH. Xia et al. described an intracellular pH detection system using dodecaarginine peptides (R12) conjugated with 5-carboxyfluorescein (FAM) [31]. FAM is a pH-sensitive fluorescent probe and a number of arginine residues in the sequence of arginine-rich CPPs affect the pKa value of FAM after conjugation to peptides (FAM-R12) from 6.20 (only FAM) to 5.27 (FAM-R12), suggesting an application for a range of pH measurements [31]. Using FAM-R12, the pH increased following treatment with chloroquine, which is an inhibitor of pH reduction in lysosomes, was observed. The pH value of the lysosomes was calculated from 4.6 (without chloroquine treatment) to 5.7 (with chloroquine treatment) in HeLa cells following uptake of FAM-R12 (5 μM for 30 minutes) [31].

9.3 Membrane Penetration Mechanisms of Arginine-Rich Peptides

The number of arginine residues in the sequences of arginine-rich CPPs and the concentration of the peptides highly affected the efficiency of cellular uptake and cytosolic release of the peptides (Figure 9.2) [14]. Dodeca (R12)- and hexadeca (R16)-arginines exhibited a higher efficiency of membrane penetration in HeLa cells (5 μM, 30 minutes) without disturbance of the cell membrane as assessed using propidium iodide compared with R4 or R8 peptides [14]. An increase in peptide concentration enhanced the cytosolic release efficacy, which increased nonlinearly [14, 32]. Even at low temperature (4 °C), which inhibits all energy-dependent endocytosis, cell membrane penetration of the oligoarginines occurred, suggesting involvement of direct plasma membrane penetration of these peptides [14, 24]. The membrane penetration efficacy of the oligoarginines was markedly reduced in the presence of 10% bovine serum because of a strong interaction between the peptides and serum proteins (e.g. approximately 75.4% of the R12 peptides were shown to be bound to serum proteins) [14]. The contribution of proteoglycans to the cell membrane for the efficient accumulation and induction of penetration of the oligoarginines was also confirmed [14]. In addition, the V domain of syndecan-4 was demonstrated to be important for cell membrane penetration of oligoarginines (R8), which suggests the participation of syndecan-4 and PKCα-dependent signal transduction for cell membrane penetration of the peptides [16].

Temperature-dependent cell membrane penetration of arginine-rich CPPs has been demonstrated. Low temperature (4–30 °C) increased the efficiency of cytosolic release of fluorescently-labeled R8 peptides in $CD34^+$ leukemia KG1a cells (2 μM, 1 hour) [33] and HeLa cells (10 μM, 1 hour) [24], although only the endosomal signals of the peptides were observed after uptake at 37° C at similar peptide concentrations [24, 33]. In addition, cholesterol depletion from the plasma membrane following treatment with methyl-β-cyclodextrin (5 mM) enhanced cell membrane penetration of the peptides even at 37 °C and low peptide concentration (2 μM, 1 hour) [33].

Trofimenko et al. found the participation of potassium channels including KCNQ5, KCNN4, and KCNK5 in cell membrane penetration and cytosolic release of arginine-rich peptides [34]. The transmembrane potential (Vm) was an important factor in enhancing membrane penetration and decreasing Vm (~−150 mV) by increasing the positive charges on the cell membrane induced megapolarization, which lowered the free energy barrier and water pore formation (diameter of ~2 nm confirmed using dextran molecules for the cellular assay and in silico modeling: coarse-grained molecular dynamics techniques and MARTINI force field 2.2p), resulting in membrane penetration of the peptides [34]. In this study, Tat (48–57)-$RasGAP_{317\text{-}326}$ (derived from the p120 RasGap SH3 domain) [35], Tat PNA (peptide nucleic acids for splicing correction in the luciferase sequence) [36], and Tat-Cre (recombination of a loxP-RFP-STOP-loxP-GFP gene construct) [11] was used and a reduction of membrane penetration for each cargo-fused Tat peptide was confirmed in absence of the potassium channels [34].

Brock et al. reported that treatment of R9 peptides induced the translocation of acid sphingomyelinase (ASMase) and ceramide platform formation on the extracellular side of the plasma membrane [37]. The induced ceramide platform enhanced membrane penetration of R9 peptides [37]. In ASMase-deficient primary fibroblasts, the addition of ASMase from outside the cell significantly increased the efficacy of cell membrane penetration and cytosolic release of the peptides [37].

In addition, Lorents et al. found that lower concentrations of cholesterol in the cell membrane enhanced the penetration of R9 peptides as assessed using giant plasma membrane vesicles (GPMVs) [38]. Downregulation of nucleolin, which participates in cell division, decreased the GPMV membrane penetration of the R9 peptides [38].

The calcium concentration in the cytosol was also determined to be an important factor for obtaining efficient membrane penetration of arginine-rich CPPs [39].

Orai1 is a plasma membrane protein and Orai1-dependent calcium influx enhances membrane penetration and cytosolic release of arginine-rich peptides, such as the R9 peptides [39]. Calcium chelation by treatment with BAPTA-AM (*O,O′*-Bis(2-aminophenyl)ethyleneglycol-*N,N,N′,N′*-tetraacetic acid, tetraacetoxymethyl ester) significantly decreased the cytosolic release of fluorescently labeled R9 peptides (20 μM) in HeLa cells. However, the addition of ionomycin, which is a calcium carrier, increased the cytosolic release of the R9 peptides [39]. An Orai1 activator, thapsigargin, also enhanced membrane penetration of the peptides and a glycosylation site mutation in Orai1 decreased membrane penetration [39].

Using physicochemical techniques, Jobin et al. studied the membrane insertion of arginine-rich peptides [40]. The RW16 (amino acid sequence: RRWRRWWRR WWRRWRR) peptide, which has been shown to have antimicrobial and anticancer activities [41], was adopted as a model arginine-rich CPP for studying insertion into zwitterionic membranes by NMR spectroscopy, circular dichroism, fluorescence spectroscopy, and molecular dynamic simulation [40]. The tryptophan residues were calculated to be an average of 12–13 Å from the center of the membrane bilayer of dioleoylphosphatidylcholine (DOPC) LUVs and the peptides were also calculated to be located at 7 Å from the polar headgroup of the lipids [40]. In addition, the authors confirmed that the buried location of Trp3, Trp6, Trp7, and Trp10 in the hydrophobic region of the lipid bilayer was approximately 13–14 Å from the center of the membrane bilayer. In contrast, Trp11 was at the phosphate level and Trp14 was at the glycerol level, both slightly exposed to the solvent [40]. Arginine residues were shown to be located between the central glycerol atom and nitrogen atom of the choline [40]. In the design of the CPPs, detailed information regarding membrane insertion may contribute to the improvement of membrane penetration efficacy.

Saito et al. developed amphipathic CPPs derived from human apolipoprotein E [42]. The A2–17 peptide (amino acid sequence: LRKLRKRLLRLWKLRKR) was designed on the basis of a glycosaminoglycan-binding region with the α-helix bundle of apolipoprotein E. The peptides showed a strong binding efficacy to heparin as analyzed using isothermal titration calorimetry (binding enthalpy: $\Delta H° = -6.3$ kcal/mol, binding constant: $K = 3.3 \times 10^7\,M^{-1}$) [42]. The A2–17 peptides exhibited direct cell membrane penetration even at low temperatures (4 °C) for the inhibition of energy-dependent endocytosis pathways (2 μM, CHO-K1 cells) [42]. In addition, the A2–17KR peptide (LRRLRRRLLRLWRLRRR) (substitution of all lysine residues of the A2–17 peptides to arginine residues) was associated with lower cellular uptake and membrane penetration efficacies compared with that of the A2–17 peptides [42]. The distribution of the A2–17KR peptides to the cell membrane may be inhibited because of the strong binding of the peptides to glycosaminoglycans [42]. The binding balance between glycosaminoglycans and the lipid bilayer is considered to be important for determining the efficacy of cell membrane penetration of the arginine-rich peptides.

The effects of membrane disorder on the penetration of arginine-rich peptides using artificial lipid vesicles have been examined. Almedia et al. studied membrane penetration of the R9, penetratin, and RW9 containing six arginine and three tryptophan residues (amino acid sequence: RRWWRRWRR), which forms an amphipathic α-helix using large unilamellar vesicles (LUVs) [43]. In studies of egg PC composed of LUVs to mimic liquid disordered membranes, only RW9 peptides were able to penetrate through the membrane [43]. The RW9 peptides induced a more disordered cholesterol environment as analyzed using cholesterol-pyrene molecules [43]. The R9 induced diminution of cholesterol-pyrene aggregation and slightly increased the Lo membrane environmental contribution. Meanwhile, the RW9 peptides induced the Ld environment of cholesterol-pyrene even during a similar reduction of cholesterol-pyrene aggregation, suggesting the importance of redistribution of membrane cholesterol for penetration of the RW9 peptides [43]. In addition, the longer homologue, RW16 (amino acid sequence: RRWRRWWRRWWRRWRR), exhibited lipid depacking in GPMVs derived from MDCK epithelial cells [43].

9.4 Molecular Dynamic Simulation for Understanding Membrane Penetration of Arginine-Rich Cell-Penetrating Peptides

Yao et al. determined the process of membrane penetration of arginine-rich CPPs using all-atom molecular dynamics and an umbrella sampling technique [44]. The adsorption-free energy (ΔG_A) in the process for the R9 peptides from solution to the surface of the lipid bilayer was calculated as $\Delta G_A = 0.0426\Delta E_A + 36.7$, $R^2 = 0.92$ (ΔE_A: interaction energy change), suggesting a linear relationship with ΔE_A [44]. The ΔG_A (kJ/mol) of DOPC-R9, DOPC-(Lys)$_9$, and DOPC-(Glu)$_9$ were -24 ± 3, 7 ± 2, and 23 ± 3, respectively [44]. The transmembrane free energy barrier (ΔG_B) during membrane penetration of the R9 peptides was also calculated as $\Delta G_B = 0.108\Delta E_B + 135$, $R^2 = 0.73$ (ΔE_B: corresponding interaction energy change), suggesting a rough correlation with ΔE_B [44]. The accumulation and membrane penetration of peptides were considered to be enhanced by an increased negative charge of the lipid bilayer or multiple salt bridges of guanidium-PO_4 (65% of the overall interaction energy) [44]. The membrane pore length (L) was also calculated as $\Delta G_B = 46.2L–31.3$, $R^2 = 0.92$ [44].

Ramírez et al. established a molecular theory to explain membrane adsorption and insertion of arginine-rich CPPs with hydrophilic pores [45]. There was a lower number of arginine residues compared with the R9 peptides and the peptide length increased the adsorption with the membrane. Saturation was observed with higher numbers of arginine residues compared with that of the R9 peptides, because of non-display of asymptotic behavior of the negative charge density on the membrane [45]. Because of further deprotonation of acidic lipids, arginine residues increased the negative charges of the membrane [45]. During the process of peptide insertion into membrane pores, peptide chain length affects the pore size and membrane penetration of the peptides exhibits a thermal-activated process [45].

9.5 Multilamellar Formation of Membranes for Membrane Penetration of Arginine-Rich Peptides

During the process of cell membrane penetration and cytosolic release of arginine-rich CPPs, particle-like structures of the membranes at the local points where the peptides penetrate were confirmed (Figure 9.3) [46]. Treatment with Alexa488-labeled R12 peptides on cell membranes introduced "multilamellar" particle-like structures when analyzed by electron microscopy [46]. Allolio et al. also reported the multilamellar formation and fusion of cell membranes following treatment with R9 peptides [47]. The R9 peptides introduced membrane branching and layering, and fusion of the vesicle exterior and interior, which resulted in translocation of the arginine-rich CPPs with stalk formation and opening of the fusion pore as analyzed by electron microscopy and simulation [47]. Controlling the membrane dynamics is an important factor for determining the efficacy of cell membrane penetration and cytosolic release of the arginine-rich CPPs; however, the detailed mechanism of the drastic membrane behaviors must be further elucidated from physicochemical and physiochemical viewpoints.

9.6 Enhanced Membrane Penetration of Arginine-Rich Peptides by Induction of Positive Curvature

The EpN18 peptide (amino acid sequence: XSTSSLRRQXKNIVHNYS, X = norleucine), which was derived from the N-terminal sequence of epsin-1, has been shown to induce positive membrane curvature, leading to loose lipid packing [48–50]. The amphiphilic helical structure of the peptides interacts with the cell membrane and membrane insertion of the peptide's hydrophobic face is

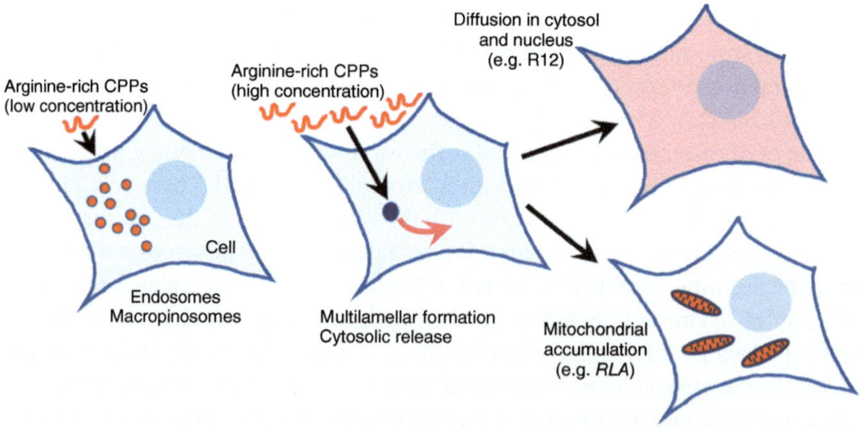

Figure 9.3 Internalization of arginine-rich CPPs by cells in experimental conditions at low or high concentrations of the peptides.

considered to expand the membrane surface to form positive curvature [48–50]. The positive curvature induction significantly increased the membrane penetration of arginine-rich CPPs [48, 50].

Conjugation of the pyrenebutyryl moiety to the N-terminus of the EpN18 peptide with a DGG linker (pBu-DGG-EpN18) further enhanced cell membrane penetration of fluorescent-labeled R8 peptides with loose cell membrane packing as analyzed and visualized using a di-4-ANEPPDHQ fluorescent probe [51]. The increased helical structure of the EpN18 in the presence of POPC/POPS (75 : 25) liposomes was also confirmed by conjugating the pyrenebutyryl moiety to the N-terminus of the EpN18 peptide [51], thus supporting the effective membrane interaction.

9.7 Internalization Mechanisms into Bacterial Cells

Understanding the uptake process in bacteria may be useful for elucidating the cytosolic internalization mechanism of arginine-rich CPPs. Frimodt-Møller et al. studied bacterial uptake of pseudopeptide DNA mimic PNAs, which target *acpP* mRNA (crucial for fatty acid synthesis), conjugated with the arginine-rich CPP, RXR [H-(R-Ahx-R)$_4$-Ahx-(βAla); Ahx, 6-aminohexanoic acid] [52]. A constitutive Cpx signaling response, which is caused by extracytoplasmic stress for maintaining the bacterial envelope, decreased the membrane potential of the cytoplasm, leading to a reduction of cellular uptake efficacy of RXR-PNA [52]. These results suggest inducible resistance against RXR-PNA by activated Cpx signaling in bacteria [52]. Meanwhile, the bacterial cellular uptake of PNA conjugated to lysine-phenylalanine-rich CPP, *L*((KFF)$_3$K), resulted in degradation of the CPP sequences in the periplasm, and the free PNA penetrated the inner bacterial membrane through SbmA, which is a transporter containing seven membrane domains (406 amino acids) [53]. Interestingly, RXR-PNA was shown to be internalized by SbmA-independent pathways [52, 53].

The mechanism of antimicrobial arginine-rich peptide membrane penetration has also been studied using molecular dynamics (MD) simulations. Lai et al. reported MD simulations of protegrin-1 (PG-1, amino acid sequence: RGGRLCYCRRRFCVCVGR), which forms a β-hairpin structure [54]. Insertion of PG-1 peptides into the membrane leads to water columns or pores in the lipid bilayer of the membrane [54]. PG-1 peptides with high cationic charges induce molecular electroporation and stabilize the pores after insertion into the membranes, thus resulting in translocation of the peptides [54].

9.8 Toxicity

Poly-proline-arginine (poly-PR), which is translated from the C9orf72 gene, is considered to be efficiently internalized by neuron cells, resulting in cell death. C9orf72-related neuronal disorders participate in amyotrophic lateral sclerosis and frontotemporal dementia (FTD) [55, 56]. Kanekura et al. reported that PR20 [amino acid sequence: (PR)$_{20}$] exhibited efficient accumulation into nucleoli

(10 μM, HEK293 cells) following cell membrane penetration, whereas other arginine-rich CPPs, such as Tat (48–60) or R12, did not [57]. In addition, PR20 inhibited the translation of proteins, resulting in cell death, because of aberrant interactions with RNAs after cellular uptake, although other arginine-rich CPPs did not [57]. Interestingly, an increase of pH (pH 8.0) around cells significantly enhanced the uptake of PR20 [57]. A low repeat number in the PR12 (amino acid sequence: $[(PR)_{12}]$) exhibited low cytotoxicity; however, fusion of GE11 peptides (amino acid sequence: YHWYGYTPQNVI) [58], which was developed to target the epidermal growth factor receptor, to the PR12 peptide significantly increased cytotoxicity [57].

Lafarga et al. reported that arginine-rich peptides with a repeat sequence of $(PR)_n$ showed displacement of RNA- and DNA-binding proteins in cells and inhibited nucleic acid processes, such as RNA transcription, translation, and DNA replication [59]. The displacement of RNA- and DNA-binding proteins likely induces the toxicity of the arginine-rich peptides [59]. Arginine-rich CPPs, including HIV-1 Tat-derived peptide and oligoarginines, have been shown to have neuroprotective effects against glutamic acid excitotoxicity and oxygen-glucose deprivation in vitro and cerebral ischemia in vivo, because of a reduction in glutamate receptor levels on neuronal cell surfaces and excitotoxic calcium influx [60].

9.9 Importance of Controlled Intracellular Localization of Arginine-Rich Cell-Penetrating Peptides After Uptake for the Delivery of Functional Moleculess

Cytosolic delivery of functional molecules through the cell membrane using arginine-rich CPPs as intracellular carriers is important for enhancing cellular uptake efficiency of targeting molecules, while avoiding degradation, especially by the lysosomes [61]. Arginine-rich peptides, including Tat (48–60) and oligoarginines, show membrane penetration and release with diffusion into the cytosol and nucleus. However, controlled locations of arginine-rich CPPs inside cells after cellular uptake should be considered to obtain the desired activity of the cargo molecules within cells. For molecules that need to function inside the mitochondria, the CPPs should efficiently carry the molecules into the mitochondria after penetrating through the cell membrane.

We previously developed plasma membrane-permeable and mitochondria-targeted arginine-rich peptides derived from an antimicrobial peptide [62]. An artificial α-helical and amphipathic peptide, KLA [amino acid sequence: $(KLAKLAK)_2$] was designed as an antimicrobial peptide by Javadpour et al., and these peptides induced low cytotoxicity toward mammalian cells [63]. In addition, conjugation of the D-form the KLA peptide (*KLA*) to the cancer-targeting RGD peptide enhanced cellular uptake and apoptosis by disrupting the mitochondrial membrane [64]. We designed a substitution of the D-lysine residues of the *KLA* peptide to D-arginine residues to enhance accumulation on cell membranes, membrane penetration, and mitochondrial accumulation [62]. The *RLA* peptides [amino acid sequence: $_D$(RLARLAR)$_2$] have a high cell membrane penetration ability and target mitochondria in a very short period of time (~10 minutes) (Figure 9.3) [62].

We tested the mitochondrial delivery of the Bcl-x_L BH4 domain peptide [65]. The BH4 domain is required for the inhibition of mitochondrial voltage-dependent anion channels, which participate in cytochrome *c* release and apoptotic cell death [65, 66]. Conjugation of the BH4 domain peptide to the *RLA* peptide (*RLA*-BH4) significantly enhanced cell membrane penetration and mitochondrial accumulation [62]. In addition, treatment with *RLA*-BH4 reduced apoptosis induction by etoposide [62]. Although the cellular uptake efficiency of R8-conjugated BH4 peptides (R8-BH4) was higher compared with that of *RLA*-BH4, treatment with R8-BH4 did not reduce apoptosis following etoposide treatment [62].

For boron neutron capture therapy (BNCT), we used *RLA* peptides for intracellular delivery of the therapeutic boron-10 (^{10}B) compound, polyhedral borane anione-based disodium mercaptoundecahydro-*closo*-dodecaborate (BSH) [67]. Thermal neutron irradiation generates alpha particles and recoiling lithium-7 (^{7}Li) nuclei from ^{10}B atoms, which results in cancer cell death. Conjugation of BSH to the *RLA* peptide increased cell membrane penetration and accumulation in the mitochondria, which resulted in caspase 3/7 activation and a reduction of ATP content in the targeted cells [67]. Conjugation of BSH to the R8 peptide increased cell membrane penetration and diffusion into the cytosol and nucleus. However, lower BNCT activity was observed [67]. Therefore, the controlled localization of bioactive molecules through the use of arginine-rich CPPs should be considered and optimized.

9.10 Conclusion

Elucidation of cellular uptake and cell membrane penetration mechanisms for arginine-rich CPPs will contribute to the development of next-generation CPPs as further functional and controllable carrier molecules (e.g. cell targeting, internalization routes, organelle targets, controlled release), which will be useful for basic science and drug development. However, the more we elucidate the pathways, the more we will understand complex cellular uptake mechanisms for arginine-rich CPPs. In addition, cargo attached to the carrier peptides may affect cellular uptake pathways and the efficacy of cell membrane penetration. With respect to chemical biology, we must elucidate the mechanisms from a physiological and physicochemical standpoint. The fusion of both disciplines will open new avenues for profoundly understanding the chemical and cellular function of arginine-rich CPPs.

Declaration of Competing Interest

The author declares no conflicts of interest associated with this manuscript.

Author Contribution

I.N. designed and wrote the manuscript.

Acknowledgments

This work was supported by Japan Society for the Promotion of Science (JSPS) KAKENHI (JP19H05553), Japan Science and Technology Agency (JST) CREST (JPMJCRI7H3), and Takeda Science Foundation (J20ZZ00356).

References

1 Vivès, E., Brodin, P., and Lebleu, B. (1997). A truncated HIV-1 Tat protein basic domain rapidly translocates through the plasma membrane and accumulates in the cell nucleus. *Journal of Biological Chemistry* 272 (25): 16010–16017. https://doi.org/10.1074/jbc.272.25.16010.
2 Wender, P.A., Mitchell, D.J., Pattabiraman, K. et al. (2000). The design, synthesis, and evaluation of molecules that enable or enhance cellular uptake: peptoid molecular transporters. *Proceedings of the National Academy of Sciences of the United States of America* 97 (24): 13003–13008. https://doi.org/10.1073/pnas.97.24.13003.
3 Rothbard, J.B., Garlington, S., Lin, Q. et al. (2000). Conjugation of arginine oligomers to cyclosporin A facilitates topical delivery and inhibition of inflammation. *Nature Medicine* 6 (11): 1253–1257. https://doi.org/10.1038/81359.
4 Futaki, S., Suzuki, T., Ohashi, W. et al. (2001). Arginine-rich peptides. An abundant source of membrane-permeable peptides having potential as carriers for intracellular protein delivery. *Journal of Biological Chemistry* 276 (8): 5836–5840. https://doi.org/10.1074/jbc.M007540200.
5 Suzuki, T., Futaki, S., Niwa, M. et al. (2002). Possible existence of common internalization mechanisms among arginine-rich peptides. *Journal of Biological Chemistry* 277 (4): 2437–2443. https://doi.org/10.1074/jbc.M110017200.
6 Nakase, I., Takeuchi, T., Tanaka, G. et al. (2008). Methodological and cellular aspects that govern the internalization mechanisms of arginine-rich cell-penetrating peptides. *Advanced Drug Delivery Reviews* 60 (4–5): 598–607. https://doi.org/10.1016/j.addr.2007.10.006.
7 Nakase, I., Tanaka, G., and Futaki, S. (2013). Cell-penetrating peptides (CPPs) as a vector for the delivery of siRNAs into cells. *Molecular BioSystems* 9 (5): 855–861. https://doi.org/10.1039/c2mb25467k.
8 Futaki, S. and Nakase, I. (2017). Cell-surface interactions on arginine-rich cell-penetrating peptides allow for multiplex modes of internalization. *Accounts of Chemical Research* 50 (10): 2449–2456. https://doi.org/10.1021/acs.accounts.7b00221.

9 Swanson, J.A. and Watts, C. (1995). Macropinocytosis. *Trends in Cell Biology* 5 (11): 424–428. https://doi.org/10.1016/s0962-8924(00)89101-1.

10 Swanson, J.A. (2008). Shaping cups into phagosomes and macropinosomes. *Nature Reviews Molecular Cell Biology* 9 (8): 639–649. https://doi.org/10.1038/nrm2447.

11 Wadia, J.S., Stan, R.V., and Dowdy, S.F. (2004). Transducible TAT-HA fusogenic peptide enhances escape of TAT-fusion proteins after lipid raft macropinocytosis. *Nature Medicine* 10 (3): 310–315. https://doi.org/10.1038/nm996.

12 Conner, S.D. and Schmid, S.L. (2003). Regulated portals of entry into the cell. *Nature* 422 (6927): 37–44. https://doi.org/10.1038/nature01451.

13 Nakase, I., Tadokoro, A., Kawabata, N. et al. (2007). Interaction of arginine-rich peptides with membrane-associated proteoglycans is crucial for induction of actin organization and macropinocytosis. *Biochemistry* 46 (2): 492–501. https://doi.org/10.1021/bi0612824.

14 Kosuge, M., Takeuchi, T., Nakase, I. et al. (2008). Cellular internalization and distribution of arginine-rich peptides as a function of extracellular peptide concentration, serum, and plasma membrane associated proteoglycans. *Bioconjugate Chemistry* 19 (3): 656–664. https://doi.org/10.1021/bc700289w.

15 Nakase, I., Hirose, H., Tanaka, G. et al. (2009). Cell-surface accumulation of flock house virus-derived peptide leads to efficient internalization via macropinocytosis. *Molecular Therapy* 17 (11): 1868–1876. https://doi.org/10.1038/mt.2009.192.

16 Nakase, I., Osaki, K., Tanaka, G. et al. (2014). Molecular interplays involved in the cellular uptake of octaarginine on cell surfaces and the importance of syndecan-4 cytoplasmic V domain for the activation of protein kinase Cα. *Biochemical and Biophysical Research Communications* 446 (4): 857–862. https://doi.org/10.1016/j.bbrc.2014.03.018.

17 Kawaguchi, Y., Takeuchi, T., Kuwata, K. et al. (2016). Syndecan-4 is a receptor for clathrin-mediated endocytosis of arginine-rich cell-penetrating peptides. *Bioconjugate Chemistry* 27 (4): 1119–1130. https://doi.org/10.1021/acs.bioconjchem.6b00082.

18 Simons, M. and Horowitz, A. (2001). Syndecan-4-mediated signalling. *Cellular Signalling* 13 (12): 855–862. https://doi.org/10.1016/s0898-6568(01)00190-5.

19 Lambaerts, K., Wilcox-Adelman, S.A., and Zimmermann, P. (2009). The signaling mechanisms of syndecan heparan sulfate proteoglycans. *Current Opinion in Cell Biology* 21 (5): 662–669. https://doi.org/10.1016/j.ceb.2009.05.002.

20 Okina, E., Manon-Jensen, T., Whiteford, J.R. et al. (2009). Syndecan proteoglycan contributions to cytoskeletal organization and contractility. *Scandinavian Journal of Medicine & Science in Sports* 19 (4): 479–489. https://doi.org/10.1111/j.1600-0838.2009.00941.x.

21 Couchman, J.R. (2003). Syndecans: proteoglycan regulators of cell-surface microdomains? *Nature Reviews Molecular Cell Biology* 4 (12): 926–937. https://doi.org/10.1038/nrm1257.

22 Oh, E.S., Woods, A., and Couchman, J.R. (1997). Multimerization of the cytoplasmic domain of syndecan-4 is required for its ability to activate protein kinase C. *Journal of Biological Chemistry* 272 (18): 11805–11811. https://doi.org/10.1074/jbc.272.18.11805.

23 Oh, E.S., Woods, A., and Couchman, J.R. (1997). Syndecan-4 proteoglycan regulates the distribution and activity of protein kinase C. *Journal of Biological Chemistry* 272 (13): 8133–8136. https://doi.org/10.1074/jbc.272.13.8133.

24 Nakase, I., Niwa, M., Takeuchi, T. et al. (2004). Cellular uptake of arginine-rich peptides: roles for macropinocytosis and actin rearrangement. *Molecular Therapy* 10 (6): 1011–1022. https://doi.org/10.1016/j.ymthe.2004.08.010.

25 Ida, H., Takahashi, Y., Kumatani, A. et al. (2021). Nanoscale visualization of morphological alteration of live-cell membranes by the interaction with oligoarginine cell-penetrating peptides. *Analytical Chemistry* 93 (13): 5383–5393. https://doi.org/10.1021/acs.analchem.0c04097.

26 Walrant, A., Bauzá, A., Girardet, C. et al. (2020). Ionpair-π interactions favor cell penetration of arginine/tryptophan-rich cell-penetrating peptides. *Biochimica et Biophysica Acta, Biomembranes* 1862 2: 183098. https://doi.org/10.1016/j.bbamem.2019.183098.

27 Trofimenko, E., Homma, Y., Fukuda, M. et al. (2021). The endocytic pathway taken by cationic substances requires Rab14 but not Rab5 and Rab7. *Cell Reports* 37 (5): 109945. https://doi.org/10.1016/j.celrep.2021.109945.

28 Junutula, J.R., De Maziére, A.M., Peden, A.A. et al. (2004). Rab14 is involved in membrane trafficking between the Golgi complex and endosomes. *Molecular Biology of the Cell* 15 (5): 2218–2229. https://doi.org/10.1091/mbc.e03-10-0777.

29 Proikas-Cezanne, T., Gaugel, A., Frickey, T. et al. (2006). Rab14 is part of the early endosomal clathrin-coated TGN microdomain. *FEBS Letters* 580 (22): 5241–5246. https://doi.org/10.1016/j.febslet.2006.08.053.

30 Linford, A., Yoshimura, S., Nunes Bastos, R. et al. (2012). Rab14 and its exchange factor FAM116 link endocytic recycling and adherens junction stability in migrating cells. *Developmental Cell* 22 (5): 952–966. https://doi.org/10.1016/j.devcel.2012.04.010.

31 Xia, M.C., Cai, L., Yang, Y. et al. (2019). Tuning the pK_a of carboxyfluorescein with arginine-rich cell-penetrating peptides for intracellular pH imaging. *Analytical Chemistry* 91 (14): 9168–9173. https://doi.org/10.1021/acs.analchem.9b01864.

32 Tünnemann, G., Ter-Avetisyan, G., Martin, R.M. et al. (2008). Live-cell analysis of cell penetration ability and toxicity of oligo-arginines. *Journal of Peptide Science* 14 (4): 469–476. https://doi.org/10.1002/psc.968.

33 Fretz, M.M., Penning, N.A., Al-Taei, S. et al. (2007). Temperature-, concentration- and cholesterol-dependent translocation of L- and D-octa-arginine across the plasma and nuclear membrane of CD34+ leukaemia cells. *Biochemical Journal* 403 (2): 335–342. https://doi.org/10.1042/BJ20061808.

34 Trofimenko, E., Grasso, G., Heulot, M. et al. (2021). Genetic, cellular and structural characterization of the membrane potential-dependent cell-penetrating peptide translocation pore. *eLife* 10: e69832. https://doi.org/10.7554/eLife.69832.

35 Michod, D., Yang, J.Y., Chen, J. et al. (2004). A RasGAP-derived cell permeable peptide potently enhances genotoxin-induced cytotoxicity in tumor cells. *Oncogene* 23 (55): 8971–8978. https://doi.org/10.1038/sj.onc.1207999.

36 Abes, S., Turner, J.J., Ivanova, G.D. et al. (2007). Efficient splicing correction by PNA conjugation to an R6-Penetratin delivery peptide. *Nucleic Acids Research* 35 (13): 4495–4502. https://doi.org/10.1093/nar/gkm418.

37 Verdurmen, W.P., Thanos, M., Ruttekolk, I.R. et al. (2010). Cationic cell-penetrating peptides induce ceramide formation via acid sphingomyelinase: implications for uptake. *Journal of Controlled Release* 147 (2): 171–179. https://doi.org/10.1016/j.jconrel.2010.06.030.

38 Lorents, A., Säälik, P., Langel, Ü. et al. (2018). Arginine-rich cell-penetrating peptides require nucleolin and cholesterol-poor subdomains for translocation across membranes. *Bioconjugate Chemistry* 29 (4): 1168–1177. https://doi.org/10.1021/acs.bioconjchem.7b00805.

39 Zuconelli, C.R., Schmidt, S., Wallbrecher, R. et al. (2020). Modulation of Orai1 by cationic peptides triggers their direct cytosolic uptake. *Biochimica et Biophysica Acta, Biomembranes* 1862 3: 183155. https://doi.org/10.1016/j.bbamem.2019.183155.

40 Jobin, M.L., Vamparys, L., Deniau, R. et al. (2019). Biophysical insight on the membrane insertion of an arginine-rich cell-penetrating peptide. *International Journal of Molecular Sciences* 20 (18): 4441. https://doi.org/10.3390/ijms20184441.

41 Jobin, M.L., Bonnafous, P., Temsamani, H. et al. (2013). The enhanced membrane interaction and perturbation of a cell penetrating peptide in the presence of anionic lipids: toward an understanding of its selectivity for cancer cells. *Biochimica et Biophysica Acta* 1828 6: 1457–1470. https://doi.org/10.1016/j.bbamem.2013.02.008.

42 Ohgita, T., Takechi-Haraya, Y., Nadai, R. et al. (2019). A novel amphipathic cell-penetrating peptide based on the N-terminal glycosaminoglycan binding region of human apolipoprotein E. *Biochimica et Biophysica Acta, Biomembranes* 1861 3: 541–549. https://doi.org/10.1016/j.bbamem.2018.12.010.

43 Almeida, C., Maniti, O., Di Pisa, M. et al. (2019). Cholesterol re-organisation and lipid de-packing by arginine-rich cell penetrating peptides: Role in membrane translocation. *PLoS One* 14 (1): e0210985. https://doi.org/10.1371/journal.pone.0210985.

44 Yao, C., Kang, Z., Yu, B. et al. (2019). All-factor analysis and correlations on the transmembrane process for arginine-rich cell-penetrating peptides. *Langmuir* 35 (28): 9286–9296. https://doi.org/10.1021/acs.langmuir.9b01169.

45 Ramírez, P.G., Del Pópolo, M.G., Vila, J.A. et al. (2019). Adsorption and insertion of polyarginine peptides into membrane pores: the trade-off between electrostatics, acid-base chemistry and pore formation energy. *Journal of Colloid and Interface Science* 552: 701–711. https://doi.org/10.1016/j.jcis.2019.05.087.

46 Hirose, H., Takeuchi, T., Osakada, H. et al. (2012). Transient focal membrane deformation induced by arginine-rich peptides leads to their direct penetration into cells. *Molecular Therapy* 20 (5): 984–993. https://doi.org/10.1038/mt.2011.313.

47 Allolio, C., Magarkar, A., Jurkiewicz, P. et al. (2018). Arginine-rich cell-penetrating peptides induce membrane multilamellarity and subsequently enter via formation of a fusion pore. *Proceedings of the National Academy of Sciences of the United States of America* 115 (47): 11923–11928. https://doi.org/10.1073/pnas.1811520115.

48 Pujals, S., Miyamae, H., Afonin, S. et al. (2013). Curvature engineering: positive membrane curvature induced by epsin N-terminal peptide boosts internalization of octaarginine. *ACS Chemical Biology* 8 (9): 1894–1899. https://doi.org/10.1021/cb4002987.

49 Sakai, T., Kawano, K., Iino, M. et al. (2019). Loosening of lipid packing by cell-surface recruitment of amphiphilic peptides by coiled-coil tethering. *Chembiochem* 20 (16): 2151–2159. https://doi.org/10.1002/cbic.201900347.

50 Hsu, W.Y., Masuda, T., Afonin, S. et al. (2020). Enhancing the activity of membrane remodeling epsin-peptide by trimerization. *Bioorganic & Medicinal Chemistry Letters* 30 (12): 127190. https://doi.org/10.1016/j.bmcl.2020.127190.

51 Kuroki, K., Sakai, T., Masuda, T. et al. (2021). Membrane anchoring of a curvature-inducing peptide, EpN18, promotes membrane translocation of octaarginine. *Bioorganic & Medicinal Chemistry Letters* 43: 128103. https://doi.org/10.1016/j.bmcl.2021.128103.

52 Frimodt-Møller, J., Koulouktsis, A., Charbon, G. et al. (2021). Activating the Cpx response induces tolerance to antisense PNA delivered by an arginine-rich peptide in *Escherichia coli*. *Molecular Therapy--Nucleic Acids* 25: 444–454. https://doi.org/10.1016/j.omtn.2021.06.009.

53 Ghosal, A., Vitali, A., Stach, J.E. et al. (2013). Role of SbmA in the uptake of peptide nucleic acid (PNA)-peptide conjugates in *E. coli*. *ACS Chemical Biology* 8 (2): 360–367. https://doi.org/10.1021/cb300434e.

54 Lai, P.K. and Kaznessis, Y.N. (2018). Insights into membrane translocation of protegrin antimicrobial peptides by multistep molecular dynamics simulations. *ACS Omega* 3 (6): 6056–6065. https://doi.org/10.1021/acsomega.8b00483.

55 Mizielinska, S., Grönke, S., Niccoli, T. et al. (2014). C9orf72 repeat expansions cause neurodegeneration in Drosophila through arginine-rich proteins. *Science* 345 (6201): 1192–1194. https://doi.org/10.1126/science.1256800.

56 Kwon, I., Xiang, S., Kato, M. et al. (2014). Poly-dipeptides encoded by the C9orf72 repeats bind nucleoli, impede RNA biogenesis, and kill cells. *Science* 345 (6201): 1139–1145. https://doi.org/10.1126/science.1254917.

57 Kanekura, K., Harada, Y., Fujimoto, M. et al. (2018). Characterization of membrane penetration and cytotoxicity of C9orf72-encoding arginine-rich dipeptides. *Scientific Reports* 8 (1): 12740. https://doi.org/10.1038/s41598-018-31096-z.

58 Li, Z., Zhao, R., Wu, X. et al. (2005). Identification and characterization of a novel peptide ligand of epidermal growth factor receptor for targeted delivery of therapeutics. *The FASEB Journal* 19 (14): 1978–1985. https://doi.org/10.1096/fj.05-4058com.

59 Lafarga, V., Sirozh, O., Díaz-López, I. et al. (2021). Widespread displacement of DNA- and RNA-binding factors underlies toxicity of arginine-rich cell-penetrating peptides. *The EMBO Journal* 40 (13): e103311. https://doi.org/10.15252/embj.2019103311.

60 Edwards, A.B., Anderton, R.S., Knuckey, N.W. et al. (2018). Perinatal hypoxic-ischemic encephalopathy and neuroprotective peptide therapies: a case for cationic arginine-rich peptides (CARPs). *Brain Sciences* 8 (8): 147. https://doi.org/10.3390/brainsci8080147.

61 Nakase, I., Kobayashi, S., and Futaki, S. (2010). Endosome-disruptive peptides for improving cytosolic delivery of bioactive macromolecules. *Biopolymers* 94 (6): 763–770.

62 Nakase, I., Okumura, S., Katayama, S. et al. (2012). Transformation of an antimicrobial peptide into a plasma membrane-permeable, mitochondria-targeted peptide via the substitution of lysine with arginine. *Chemical Communications* 48 (90): 11097–11099. https://doi.org/10.1039/c2cc35872g.

63 Javadpour, M.M., Juban, M.M., Lo, W.C. et al. (1996). De novo antimicrobial peptides with low mammalian cell toxicity. *Journal of Medicinal Chemistry* 39 (16): 3107–3113. https://doi.org/10.1021/jm9509410.

64 Ellerby, H.M., Arap, W., Ellerby, L.M. et al. (1999). Anti-cancer activity of targeted pro-apoptotic peptides. *Nature Medicine* 5 (9): 1032–1038. https://doi.org/10.1038/12469.

65 Shimizu, S., Konishi, A., Kodama, T. et al. (2000). BH4 domain of antiapoptotic Bcl-2 family members closes voltage-dependent anion channel and inhibits apoptotic mitochondrial changes and cell death. *Proceedings of the National Academy of Sciences of the United States of America* 97 (7): 3100–3105. https://doi.org/10.1073/pnas.97.7.3100.

66 Shimizu, S., Narita, M., and Tsujimoto, Y. (1999). Bcl-2 family proteins regulate the release of apoptogenic cytochrome c by the mitochondrial channel VDAC. *Nature* 399 (6735): 483–487. https://doi.org/10.1038/20959.

67 Nakase, I., Katayama, M., Hattori, Y. et al. (2019). Intracellular target delivery of cell-penetrating peptide-conjugated dodecaborate for boron neutron capture therapy (BNCT). *Chemical Communications* 55 (93): 13955–13958. https://doi.org/10.1039/c9cc03924d.

10

Endosomal Escape

Austin R. Prater and Jean-Philippe Pellois

Texas A&M University, Department of Biochemistry and Biophysics, Department of Chemistry, College Station, TX 77843, USA

10.1 Endosomal Escape

10.1.1 Introduction

Cell-penetrating peptides (CPPs) are peptides that can enter the cytoplasm of cells. Through direct conjugation or coincubation methods, they can allow the cytosolic delivery of a variety of molecular payloads. While the process of cellular delivery has been extensively studied, there still lies some ambiguity about where/how CPPs get into the cell. Two main pathways have been proposed: direct plasma membrane translocation and endosomal escape after endocytosis. Direct plasma membrane translocation seems to occur when CPP concentration is relatively high or the CPP is amphipathic in nature [1]. Several models of direct membrane translocation have been proposed, including pore formation, inverted micelles, and the "carpet" model [2–4]. Additionally, the oxidation state of the plasma membrane as well as the formation of extracellular CPP-lipid clusters appear to facilitate this process [5–7].

For many CPPs and experimental settings, endocytosis is a predominant route of cellular entry [8–10]. Endocytic uptake, which is possible via a number of mechanisms, leads to endosomal accumulation. CPPs and payloads trapped in endosomes are trafficked to lysosomes where their degradation will likely occur. Interestingly, some CPPs are able to escape endosomes and reach the cytosolic space of cells. This endosomal escape is typically necessary for the payload to reach its intracellular target and exert its biological function. Endosomal escape achieved by current CPPs has unfortunately been relatively inefficient with estimates suggesting 99% of the CPP/cargo remain trapped in endosomes [11]. Several questions arise when considering this route of cell entry for CPPs: How does endosomal escape work? Is it advantageous and can it be improved? Is it harmful to the cell? How does one know CPPs are escaping endosomes? These questions are worth answering because understanding the mechanisms CPPs use to enter cells allows for improved design of both CPPs and payloads to be delivered. This chapter will aim to answer these questions and provide an in-depth analysis of how CPPs utilize endosomal escape to penetrate cells.

10.1.2 Endocytosis as a Route of Cell Entry

In order for CPPs to reach the cytosol via an endocytic route, two events must occur: endocytosis of the peptide followed by an escape from the endosome. Endocytosis is the process in which cells take up foreign material from their surrounding environment. The material that is endocytosed is enveloped by part of the cellular membrane to form a vesicle. These vesicles, and their luminal contents, are then trafficked to different organelles. The endocytic pathway involves multiple organelles interconnected by fusion events.

10.1.2.1 Types of Endocytosis

There are several types of endocytosis performed by the cell including clathrin-mediated endocytosis, caveolae-mediated endocytosis, and macropinocytosis. Clathrin-mediated endocytosis, or receptor-mediated endocytosis, is the process where molecules bind to receptors on the cell surface and become internalized via clathrin-coated pits in the plasma membrane [12]. These pits bud off the plasma membrane as clathrin-coated vesicles. The clathrin coat is then shed and recycled by the cell, and the vesicles fuse into early endosomes. Caveolae are small invaginations of the plasma membrane which can bud off from the plasma membrane to form vesicles similar to clathrin-mediated endocytosis [13]. This budding is augmented by members of the caveolin family of proteins. Macropinocytosis, or cell drinking, is the process in which the plasma membrane becomes ruffled and engulfs the surrounding extracellular fluid. This membrane ruffling is caused by large actin protrusions that extend from the membrane [14, 15]. While these processes differ in their mechanisms, they each result in outside material being brought into the cell where the material can be trafficked along the endocytic pathway.

10.1.2.2 The Endocytic Pathway

The endocytic pathway consists of a variety of organelles distinguished by morphology, intraluminal pH, lipid composition, and protein markers. These organelles include early endosomes, multivesicular bodies, late endosomes, endolysosomes, and lysosomes (Figure 10.1).

Endocytic vesicles are first trafficked to early endosomes where they undergo homotypic fusion with other vesicles [16]. Early endosomes are marked by thin cisternal protrusions and a large main body ranging from 300–400 nm in diameter [16]. Additionally, early endosomes can encapsulate smaller vesicles known as intraluminal vesicles. In addition to morphological differences, early endosomes have unique protein components as well. Rab GTPases, proteins involved in vesicle budding, trafficking, fusion, and movement, give unique signatures to each organelle along the endocytic pathway [17]. Specifically, Rab5 is associated with early endosomes [17]. From here, the large spherical body of endosomes separates from

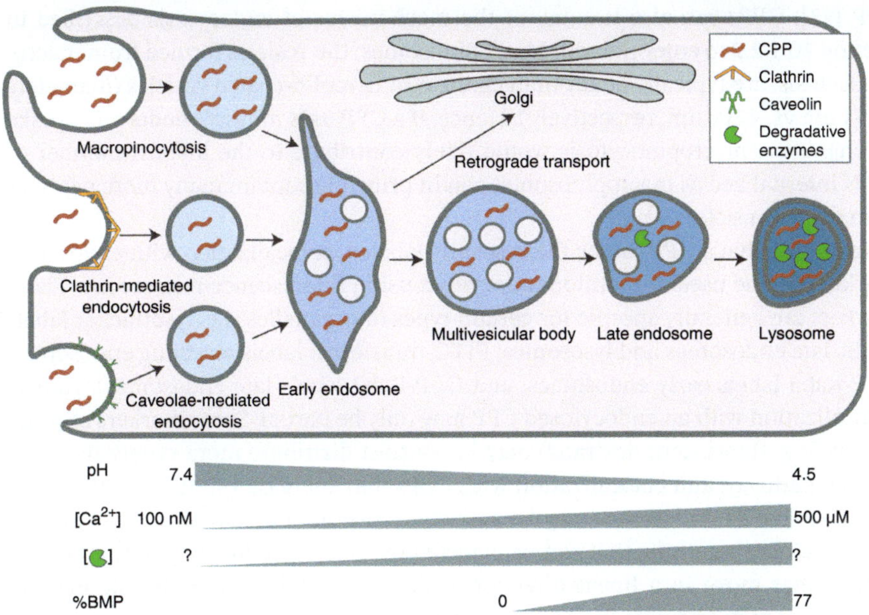

Figure 10.1 Endocytosis of CPPs: CPPs can get endocytosed via macropinocytosis, clathrin-mediated endocytosis, or caveolae-mediated endocytosis. The vesicles generated from macropinocytosis can reach sizes up to 1 μm while vesicles from clathrin or caveolae endocytosis are generally less than 100 nm which likely allows for macropinosomes to contain more extracellular material. From here, vesicles merge into early endosomes where they can be trafficked to the Golgi through retrograde transport or continue through the endocytic pathway to multivesicular bodies. Multivesicular bodies mature into late endosomes followed by maturation into lysosomes. As vesicles travel through the endocytic pathway, the pH of the vesicles drops, $[Ca^{2+}]$ increases, the concentration of degradative enzymes increases, and BMP lipid content increases.

the cisternal regions and matures into a multivesicular body, which is characterized by the presence of intraluminal vesicles and has a diameter of 400–500 nm [16]. A gradual progression then takes place resulting in maturation into late endosomes. Late endosomes have a pH of approximately 5.5 and are characterized by the presence of Rab7 and lysosomal-associated membrane protein 1 (LAMP1) and the absence of Rab5 [17, 18]. Additionally, late endosomes are highly enriched in the negatively charged lipid bis(monoacylglycerol)phosphate (BMP) with levels around 77% of the total lipid content [19–21]. Lastly, late endosomes fuse with lysosomes where their contents can be degraded. Lysosomes are highly acidic and contain a plethora of degradative enzymes that can break down most proteins, nucleic acids, and other molecules [16]. Endocytosis of CPPs has been demonstrated using a variety of tools and will be discussed in the next section.

10.1.2.3 Endocytosis of CPPs

On the cell surface, CPPs may bind to different receptors, including heparan sulfate proteoglycans [22]. Heparan sulfate is negatively charged and can interact electrostatically with cationic CPPs, and this interaction can cause endocytosis of the

CPP [23]. CPPs can also use any of the mechanisms of endocytosis described in Section 10.1.2.2 to enter the cell. Macropinosomes, the vesicles formed from macropinocytosis, are typically larger than clathrin or caveolin-coated vesicles (diameters of >1 μm vs. <100 nm, respectively). Hence, if a CPP uses multiple endocytic uptake mechanisms, macropinocytosis would likely contribute to the highest number of CPPs internalized as macropinosomes can in principle contain many more peptides than coated vesicles.

Assuming that a CPP can be fluorescently labeled, colocalization with endosomal markers can be used to monitor endocytosis using fluorescence microscopy. These markers are generally specific for certain types of organelles (i.e. lysotracker labels acidic late endosomes and lysosomes, FITC-transferrin labels recycling endosomes, GFP-Rab5 labels early endosomes, and GFP-Rab7 labels late endosomes). Hence, colocalization with an endocytosed CPP may only be partial. Some markers of pinocytosis (e.g. fluorescent dextrans) may in contrast distribute more evenly throughout the pathway, and colocalization with CPPs may increase [8]. It is worth noting that CPP-loaded endosomes may be actively transported by motor proteins on the cytoskeleton, something that can be detected by time-lapse imaging in the form of puncta that move in a linear direction inside cells [24]. In contrast, fluorescent puncta that may originate from CPPs aggregating outside cells would be expected to appear static or moving according to Brownian motions.

In addition to visualizing the localization of CPPs along the endocytic pathway to confirm endocytosis, blocking endocytosis can also confirm the route of cellular uptake and this can be done at several points along the endocytic pathway. First, endocytosis can be generally inhibited by incubating cells at 4 °C or with sodium azide as both approaches rely on the idea of blocking ATP-consuming processes [25]. Alternatively, several drugs are able to inhibit various mechanisms of endocytic uptake. Cytochalasin D and amiloride, inhibitors of macropinocytosis, work by either disrupting actin polymerization or inhibiting the epithelial Na^+ channel, respectively [26]. The presence of either one of these inhibitors has been shown to drastically reduce the endocytic uptake of Tat and Tat-conjugates [27, 28]. Additionally, chlorpromazine or methyl-β-cyclodextrin, inhibitors of clathrin-mediated or caveolae-mediated endocytosis, respectively, can also be used to monitor the endocytosis of CPPs [29, 30].

10.1.3 Mechanisms of Endosomal Escape

Once a CPP is endocytosed, there are several possible exit points CPPs can take throughout the endocytic pathway to enter the cytosol of cells. Several mechanisms have been proposed as to how CPPs can disrupt endosomes to promote leakage, and these include leaky fusion of membranes, the proton-sponge effect, and membrane destabilization (Figure 10.2).

10.1.3.1 Leaky Fusion

Leaky fusion of membranes is the process in which CPPs cause membranes to fuse with each other, and upon fusion, the luminal contents are released via pores formed

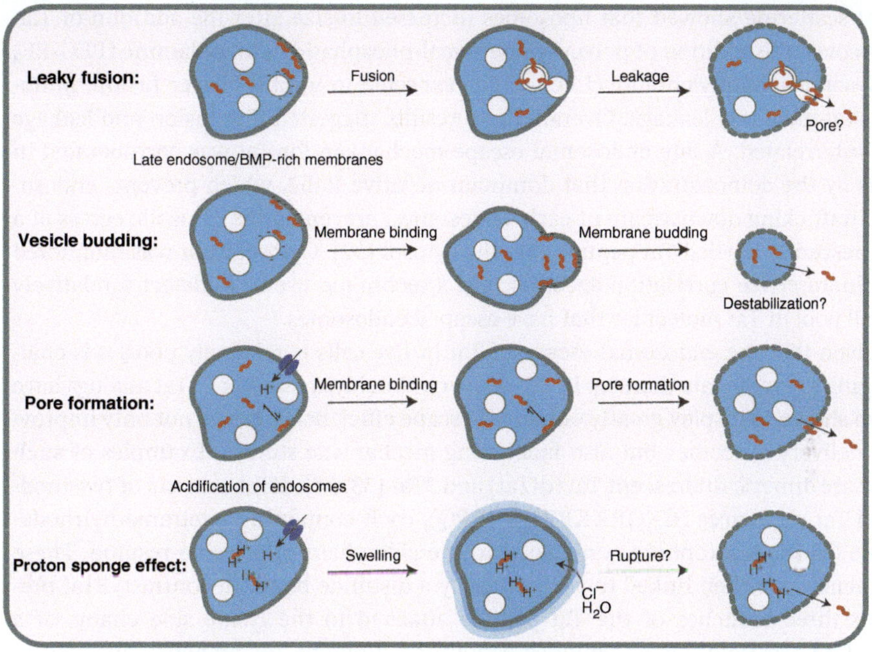

Figure 10.2 Mechanisms of endosomal escape: There are multiple mechanisms CPPs use to escape endosomes which include: leaky fusion, membrane destabilization by pore formation or vesicle budding, and the proton sponge effect. Leaky fusion is the process in which CPPs bind to endosomal membranes and cause fusion between intraluminal vesicles and parent endosomal membranes. This binding is facilitated by the cationic nature of the CPP and the increased amount of anionic BMP. This fusion event likely leads to pore formation causing leakage of the luminal contents. pH-sensitive CPPs cause leakage by forming pores in the endosomal membranes. Upon acidification of the endosome, protonation of the CPP results in conformational changes allowing the CPP to insert into the membrane and cause pores that release the luminal contents. Vesicle budding results from cyclized, cationic CPPs binding to negatively charged lipids in endosomes and causing budding of smaller vesicles. These vesicles destabilize allowing the cytosolic release of their contents. The proton sponge effect is the idea that membranes swell and rupture due to changes in the osmotic pressure resulting from the "soaking up" of protons by a CPP in an endosome.

by the CPP or by destabilization of the membrane. This phenomenon was described in vitro by Yang et al. using Tat as a model [31]. The authors postulated that since Tat is highly positively charged, it may interact with anionic BMP and potentially induce the leakage of BMP-rich late endosomes. To test this idea, liposomes mimicking the lipid composition of late endosomes and of the plasma membrane (lacking BMP) were produced with a fluorescently labeled dextran encapsulated in their lumen. This dextran was used as a probe to monitor leakage of the lipid bilayer. Tat was able to induce leakage of the dextrans in the late endosome liposomes but not the plasma membrane liposomes, suggesting that BMP is necessary for leakage to take place. Leakage was also observed to be more efficient at pH 5.5 compared to pH 7, consistent with the pH levels in late endosomes. Concomitant to leakage, dynamic light scattering showed that liposomes increased in size after the addition of Tat. Moreover, the addition of polyethylene glycol-phosphatidylethanolamine (PEG-PE) or lysophosphatidylcholine (LPC), lipids expected to inhibit bilayer fusion, inhibited Tat-mediated leakage. Overall, these results suggested that fusion and leakage are interrelated. A late endosomal escape mechanism for Tat was corroborated in cells by the demonstration that dominant-negative Rab7, which prevents endosomal trafficking downstream of early endosomes, prevented the cytosolic egress of a fluorescently labeled Tat peptide into the cytosol [32]. Cytosolic Tat was monitored by fluorescence correlation spectroscopy, a technique useful to detect a relatively small pool of Tat molecules that have escaped endosomes.

Given that the endosomal escape of Tat in live cells is relatively poor, it is challenging to investigate this step in the delivery. Notably, a number of Tat analogs have been shown to display greatly enhanced escape efficiencies, hence not only improving delivery outcomes but also facilitating mechanistic studies. Examples of such CPP are dimeric fluorescent Tat (dfTat) and 3Tat [33, 34]. dfTat consists of two modified Tat sequences (CKGRKKRRQRRRPQ), each containing a tetramethylrhodamine (TMR) fluorophore conjugated to the side chain of a lysine residue. These sequences are then linked to each other by a disulfide bond. In contrast, 3Tat presents three branches of the Tat peptide attached to the lysine side chains of a KGKGKG scaffold. Other peptides containing different Tat branches attached to various scaffold have been reported [35, 36]. Like Tat, dfTat and 3Tat induce leakage of BMP containing liposomes that mimicked the lipid content of late endosomes in vitro [37, 38]. The efficiency of leakage is dependent on the peptide to lipid ratio, the percentage of BMP present in the bilayer, and whether liposomes can contact each other. Indeed, df Tat or 3Tat do not mediate the leakage of liposomes that are restricted from interacting with one another. Conversely, leakage correlates positively with peptide-induced fusion and flocculation. In particular, without peptide, BMP-rich liposomes repulse one another because of low negative zeta potential. Upon peptide binding, the zeta potential becomes neutral, electrostatic repulsion diminishes, liposomes contact one another, and leakage occurs. Interestingly, at high peptide/lipid ratios, the zeta potential becomes positive because of the excess cationic peptide bound to the liposome surface. In turn, liposomes repulse one another again, and leakage ceases. dfTat or 3Tat-mediated liposome fusion can be achieved with negatively charged lipids other than BMP, including phosphatidylglycerol (PG), a regioisomer of BMP. However, PG-rich liposomes do not leak in the presence of dfTat or 3Tat. Fusion is therefore necessary in the process of leaky fusion, but not sufficient. In particular, electron microscopy examination has established that PG and BMP lipid bilayers do not contact in the same manner when treated with dfTat or 3Tat [38]. Specifically, CPPs-treated PG bilayers maintain an interstitial space between them post-fusion. In comparison, BMP bilayers appear to be in direct contact and appear to encapsulate the CPPs more closely, as suggested by molecular dynamics simulations. A possible explanation for this effect relates to the conical shape of the BMP lipid, which may allow the lipid to form inverted micelle-like structures around the CPP sandwiched between bilayers. Leakage may occur if these BMP/CPP structures were to coalesce and form pores. This has however not been observed directly. It is worth noting that the unique properties of BMP have been implicated in the delivery activity of cationic lipids, tools commonly used for the delivery of nucleic acids in cells [39]. There could therefore be some mechanistic similarities between polycationic lipoplexes and CPPs.

The leaky-fusion model of endosomal escape for dfTat or 3Tat is corroborated in live cells by several pieces of evidence. First, dfTat has been shown to escape from late endosomes specifically. This was determined by inhibition of endosomal trafficking (see Section 10.1.2.3 for general strategies used). Electron microscopy of cells treated with dfTat also showed a marked reduction in the number of late endosomes, and an increase in organelles consistent with endolysosomal morphology, highlighting that the former become the latter upon cell penetration. It is however worth noting that endolysosomes are not necessarily the organelles formed upon endosomal escape, but instead the organelles present after endosomal membrane repair pathways are engaged. Consistent with the importance of BMP established in vitro, incubation of cells with an anti-BMP monoclonal antibody inhibits dfTat-mediated endosomal escape. In contrast, incubation with antibodies that do not recognize BMP as an antigen does not interfere with cell penetration. in vitro, anti-BMP blocks liposomal contact and fusion without inhibiting CPP binding suggesting that it is possible anti-BMP blocks the contact between intraluminal vesicles and the limiting membrane of late endosomes in a live cell and subsequently prevents the membrane fusion induced by the CPP. Overall, molecular and cellular insights remain needed to further understand the mechanisms involved in endosomal leakage. Given that dfTat has been successfully used to deliver proteins (e.g. antibodies, transcription factors, enzymes) and nanoparticles 100 nm in diameter, one can presume that the leakage permits the translocation of relatively large species via pores that would need to be at least equally large in size. The requirement for bilayer contacts in this leaky fusion model would also imply that large payloads directly attached to dfTat or 3Tat could interfere with encapsulation of the CPP by BMP by simply introducing bulk hindrance. Although dfTat/3Tat have been used in a co-incubation format where CPP and payload are not physically interacting, they may behave differently once directly conjugated to a payload. This has not, to our knowledge, been tested yet. Finally, it is interesting to note that based on his model, the selectivity of permeabilization of the late endosomes is likely due to both the presence of BMP and the multivesicular nature of theses organelles as this feature possibly allows for membrane contacts.

10.1.3.2 Membrane Destabilization/Permeabilization

Some CPPs destabilize the membrane of endosomes via pore formation, but without the requirement for fusion highlighted in the previous section. Various sequences derived from viruses have been used to enhance CPPs' ability to enter cells, most notably, HA2 [40]. The HA2 peptide is a fusogenic peptide derived from the hemagglutinin protein of influenza. HA2 and HA2 analogs (GLFEAIEGFIENGWEGM IDGWYG) have been conjugated to Tat to improve the endosomal escape of this CPP [28, 41]. Upon acidification of the endosome, acidic residues in the HA2 sequence (glutamates and aspartates) are protonated, leading to a loss in charge and an increase in hydrophobicity. The hydrophobic peptide adopts an α-helical secondary structure and inserts into the hydrophobic environment of the membrane [42]. The inserted peptides are then thought to form pores with a variety of pore-forming mechanisms being possible in this context. Hemolysis assays with red blood cells have shown a clear pH dependence on this process (red blood cells do not perform endocytosis and permeabilization can be easily detected by the release of hemoglobin) [43]. Indeed, at a given concentration, peptides like HA2-Tat permeabilize red blood cells more extensively at acidic pH than at neutral pH. Therefore, HA2-Tat may permeabilize early endosomes or late endosomes upon acidification of the endocytic pathway without permeabilizing the plasma membrane of the treated cells. However, assuming an equilibrium between a non-protonated and protonated form of HA2-Tat modulated by pH in a manner related to the Henderson-Hasselbalch equation, a fraction of the pore-forming peptide may exist at pH 7. It is therefore not surprising that HA2-like peptides become rather toxic at high concentrations with the membrane-permeabilizing form of the peptide likely acting upon the plasma membrane of cells. Moreover, it should be noted that HA2-like peptides appear to remain associated with endosomal membranes, even after mediating leakage [43]. Hence, the egress of payloads attached to these peptides to other parts of the cell may remain hindered even though endosomal escape is technically achieved. This issue was shown to be potentially resolved by the attachment of linkers that are cleaved by the reducing environment of the cytosol [43]. The pH dependence of HA2 is a general strategy applied to various other pore-forming peptides, including peptides from other viruses, antimicrobial peptides, and venom-derived peptides [44–47].

In addition to pore formation, cyclic peptides have been synthesized that are able to cause membrane budding followed by rupture of the budded vesicles [48]. In this model, as the CPPs accumulate in the endocytic vesicles, they bind to the luminal side of the membrane inducing membrane curvature and subsequent budding of CPP enriched lipids as small vesicles. Because of the cationic nature of these cyclic CPPs, they are believed to bind more strongly to lipids present further down the endocytic pathway due to the accumulation of negatively charged lipids. After these CPP enriched vesicles bud off of the parent endosome, it is speculated that these vesicles burst following rapid dissipation of the pH gradient across their membrane due to the lack of proton pumps associated with these small vesicles.

10.1.3.3 Proton Sponge Effect

The proton sponge effect is based on the idea that when weakly basic compounds are present in the endosome, they are able to "soak up" protons like a sponge [49]. It is postulated that, when this occurs, chloride ions rush into the endosome to restore membrane potential equilibrium. This results in an increase of the osmotic pressure of the endosome causing swelling and membrane rupture. Chloride ions do in fact accumulate in endosomes in the presence of buffering polymers such as polyethyleneimine (PEI) [50]. In addition, bafilomycin A1, an inhibitor of the V-ATPase pump which prevents endosomal acidification, hindered the accumulation of chloride ions in endosomes [50]. Through light microscopy, it was demonstrated that endosomes containing PEI polyplexes increased in volume by 140% [51].

Won et al. calculated the osmotic pressure increase in endosomal vesicles upon a decrease in pH from 7.4 to 5.0 and found that this will expand vesicle membranes by roughly 2.3% [52]. However, lipid vesicles can withstand expansion up to 5%, suggesting that this increase in osmotic pressure is unlikely to burst the endocytic vesicles [52].

Even though the details of the proton sponge effect are not fully understood, this mechanism has been used as the basis for the rational design of new endosomal escape agents. PEI, which is extensively used for the transfection of nucleic acids into cells, has also been incorporated into Tat peptides leading to an increase in endosomal escape abilities [53]. Moreover, histidine residues have been introduced into CPP sequences to possibly promote this proton sponge effect. The imidazolium side chain of histidine has a pKa of 6.0, therefore, histidine residues should be protonated in endosomes. Thus, peptides containing histidines can potentially activate the proton sponge effect. Polyhistidine sequences have been added to polyarginine peptides in order to enhance the endosomal escape of peptide/siRNA complexes [54]. The HA2 peptide, as presented in Section 10.1.3.2, has also been mutated to introduce histidines [55]. This analog, H5WYG, contains five histidine residues in place of five glutamates or aspartates. In principle, H5WYG is noticeably different from other HA2 analogs in that H5WYG becomes positively charged, and not neutral, at acidic pH. Unlike HA2, H5WYG is therefore more likely to be hydrophobic outside cells, and more hydrophilic within endosomes. Protonated H5WYG may therefore not readily form pores like HA2 does, and instead promote endosomal escape via a proton sponge mechanism. Consistent with this idea, H5WYG enhances the endosomal escape of lactosylated polylysine DNA complexes and this process is inhibited by bafilomycin [56].

Overall, the mechanistic details involved in CPP-mediated endosomal escape remain to be fully elucidated. The insights gained so far will likely inspire the development of reagents with improved efficiencies. This pursuit however raises a number of questions: if CPPs make endosomes leaky, is it something that is potentially deleterious to cells? Does this not damage endocytic organelles and negatively impact the function of the endocytic pathway? Moreover, does this not become more and more problematic as the endosomal escape efficiency of CPPs increases? Isn't toxicity an inherent consequence of using CPPs that would cause the leakage of membranes efficiently?

10.1.4 Is Endosomal Escape Detrimental to Cells?

The membranes of cells, including that of endocytic organelles, are crucial to numerous processes essential to life. It would be logical that if its membranes were disrupted, the cell would become stressed and potentially die. For instance, ion concentration gradients outside the cell are maintained across biological membranes. Therefore, disruption of these concentration gradients, which could happen in the presence of CPPs, can initiate signaling cascades that lead to cell death. Among various ions, calcium is important in this respect as calcium is used by the cell to detect membrane damage [57]. Ca^{2+} concentrations reach millimolar levels outside of the cell and are kept at approximately 100 nM inside cells [57]. Disruption of membranes can lead to a drastic and sudden increase in intracellular Ca^{2+} levels which is lethal to cells. Endosomes contain high concentrations of calcium, making endosomal escape and leakage potentially toxic [16]. Additionally, endosomal membranes are a barrier to proteases required for the degradation of endocytosed material. The release of these proteases from the lumen of endosomes to the cytosol can potentially lead to the degradation of cytoplasmic components and be detrimental to a cell [16]. With this in mind, cellular delivery of CPPs and cargoes leads to an interesting conundrum; how can one disrupt an endosomal membrane to achieve the efficient cellular delivery of various payloads without killing the cell? What other cellular events are triggered upon membrane damage?

10.1.4.1 Evidence That Endosomal Escape is Toxic

CPPs can potentially exert a toxic effect at various locations: outside cells while interacting with the plasma membrane, within the endocytic pathway while potentially permeabilizing endosomal membranes, or inside cells after cell penetration is achieved. These possibilities are non-exclusive and, if or when a CPP is toxic, it is likely challenging to dissect what the triggers of cell killing are. Herein, we will highlight some lines of evidence that suggest that the event of endosomal escape itself, on its own, can be cytotoxic. The first line of evidence relates to the fact that lysosome permeabilization has been shown to initiate cell death. Permeabilization can in this case be induced by reactive oxygen species, endogenous cell components, or lysosomotropic agents. An example of lysosomotropic agent is L-leucyl-L-leucine methyl ester (LLOME) [58]. LLOME is endocytosed by cells and subsequently converted into a polymer by endosomal cathepsins [58, 59]. Lysosomal membrane damage causes cathepsins to escape from lysosomes and enter the cytosol. One downstream effect is the activation of caspases and subsequent initiation of apoptosis [58]. Depending on the specific circumstances and severity of lysosomal membrane permeabilization, the cathepsins released can also initiate necrosis, lysosome-dependent cell death, or pyroptosis [60]. Additionally, the release of iron from the lysosomal pool, along with reactive species, leads to ferroptosis [61]. In principle, the permeabilization of endocytic organelles upstream of lysosomes in the endocytic pathway can yield similar effects. This is because they also contain cathepsins, iron, and ROS, albeit at lower concentrations.

Another line of evidence highlighting that CPP-mediated endosomal escape is potentially toxic is illustrated with the example of photolytic CPPs. In particular, CPPs labeled with fluorophores or photosensitizers can be used to lyse endosomes upon light irradiation so as to mediate the cytosolic delivery of payloads with temporal and spatial control [62–65]. This general approach is referred to as photochemical internalization (PCI, PCI not requiring the use of CPPs, but that of photosensitizers). Fluorescent-CPPs such as (TMR)-Tat can be observed to be localized within endosomes (a distinct punctate distribution) by fluorescence microscopy [65]. The light used to excite the fluorophore leads to the formation of reactive oxygen species in a manner dependent on the photophysical properties of the fluorophore and on the light irradiance. In this process, the CPP, which brings the fluorophore into the vicinity of endosomal membranes, promotes the formation of ROS near lipids, subsequent lipid oxidation, and endosomal membrane rupture. Notably, when using high light irradiances, this phenomenon can be achieved within seconds with many endosomes "bursting" simultaneously. Under such circumstances, a relatively large pool of endosomal calcium is released into the cytosol of cells. In turn, this calcium induces mitochondrial pore formation, rapid plasma membrane blebbing, plasma membrane permeabilization, and cell death [66].

10.1.4.2 Evidence That Endosomal Escape is Nontoxic

Several animal viruses permeabilize the membrane of endosomes without apparent deleterious effects on cells. Of course, conditions of infections, e.g. a single viral particle permeabilizing a single endosome within a cell containing hundreds of endocytic organelles, may not do enough damage to fully reveal the effects of endosomal membrane disruption. Likewise, CPPs that are relatively inefficient at entering cells are likely not representative of how deleterious endosomal escape can be. In contrast, several of the CPPs described in Section 10.1.3 achieve high levels of endosomal escape while not killing cells. We will use dfTat as an example. After incubation with cells for a period of one hour, dfTat permeabilizes most late endosomes in a cell [33]. This is established by counting the number of late endosomes converted to endolysosomes by electron microscopy. More than 90% of a payload initially trapped in endosomes was also shown to reach the nucleus post-endosomal escape. Moreover, the amount of a protein payload delivered into the cytosol was also estimated to reach μM concentrations, indicating a high efficiency of cell entry. Overall, this indicates that endosomal escape occurs at a very high level, with close to all late endosomes leaking almost all of their content. Yet, this delivery approach does not kill cells. It also does not reduce cell proliferation rates, indicating that cells are metabolically healthy. More surprisingly, mRNA expression is minimally impacted by treatment of cells with dfTat for one hour (conditions used to induce endosomal escape) [33]. Consistent with this result, the endosomal membrane permeabilization induced by dfTat does not lead to activation of Transcription Factor EB (TFEB), a transcription factor that promotes lysosomal biogenesis and that is otherwise activated by lysosomotropic agents such as LLOME [67]. Of note, dfTat-mediated endosomal leakage increases endosomal pH, decreases the number of macromolecules internalized by macropinocytosis post dfTat treatment, and decreases the rate at which these macromolecules are degraded within the endocytic pathway [67]. These processes remain disrupted from 6 to 24 hours and depend on the cell line used. When combined with the notion that late endosomes convert to endolysosomes upon permeabilization (see Section 10.1.3.1), it appears that late endosomes mostly disappear from cells and, with TFEB not being engaged, these organelles may not be replaced for several hours. In turn, the proper function and dynamics of the endocytic pathway are disrupted, with a possible feedback mechanism that leads to the reduction of endocytosis. This may certainly further impact different aspects of the biology of the cell. Yet, it does not appear to be cytotoxic, and cells have the ability to recover.

10.1.4.3 Factors That May Dictate Whether Endosomal Escape Is Toxic or Not

In the absence of fully resolved mechanistic details about how endosomal escape takes place, it is challenging to explain why endosomal escape may be toxic in some cases and not in other. It is interesting to note that, while the PCI approach can elicit efficient endosomal escape in a few seconds, the endosomal escape of dfTat appears to take place progressively in the span of 30–60 minutes [33]. This time frame likely includes the time it takes for dfTat to reach late endosomes and the time it takes to accumulate enough peptides to induce permeabilization. Therefore, it is maybe not the efficiency of endosomal escape that dictates toxicity per se, but instead may be the speed of the process. In particular, it is likely that cells may be able to cope with a low level of calcium or proteases leaking into their cytosol over a period of time. Calcium pumps, for instance, could transport calcium out of the cell before cytotoxic responses are induced, thereby maintaining homeostasis. In contrast, a sudden burst of intracellular calcium may overwhelm the cell, triggering intracellular death cascades.

It is likely that cellular responses induced by endosomal escape may depend on the endocytic organelles disrupted. For instance, the leakage of early endosomes may release less calcium, iron, acid, or hydrolases compared to the leakage of lysosomes. It is also likely that the duration for which cells are exposed to an endosomal escape agent will determine toxicity. In particular, dfTat is relatively nontoxic when used for one hour incubation. Extending this incubation time could cause cells to have continually damaged late endosomes which may not be something cells can survive. Notably, plastic microparticles that continually damage endosomal membranes lead to high levels of cellular stress and toxicity [68]. It is also important to note that cells have the ability to detect endosomal membrane damage (Figure 10.3). For instance, charged multivesicular body proteins (CHMP) recruit components of the endosomal sorting complexes required for transport (ESCRT) complex to endosomes that are leaky [69, 70]. This leads to apparent membrane repair and restoration of membrane integrity. Alternatively, galectins can also be recruited to endosomes upon membrane disruption and exposure to intraluminal galactosides [71]. Galectin recruitment is triggered by membrane damage events that are severe and likely beyond what cells can repair. Consistent with this idea, this process initiates the degradation of damaged organelles by autophagy. Overall, it is likely that these repair and degradation pathways protect cells from the deleterious effect of endosomal escape. As with the example given for calcium leakage, it is also possible that cytotoxicity may arise if the repair processes of the cell become overwhelmed by how many organelles become damaged and how quickly or concomitantly this happens.

10.1.5 How to Determine If a CPP Escapes Endosomes?

10.1.5.1 Testing for Endosomal Escape by Inhibiting the Endocytic Pathway

As described in Section 10.1.2.3, there are a variety of ways to monitor the endocytosis of CPPs by inhibition of the endocytic pathway. If any of these agents impact

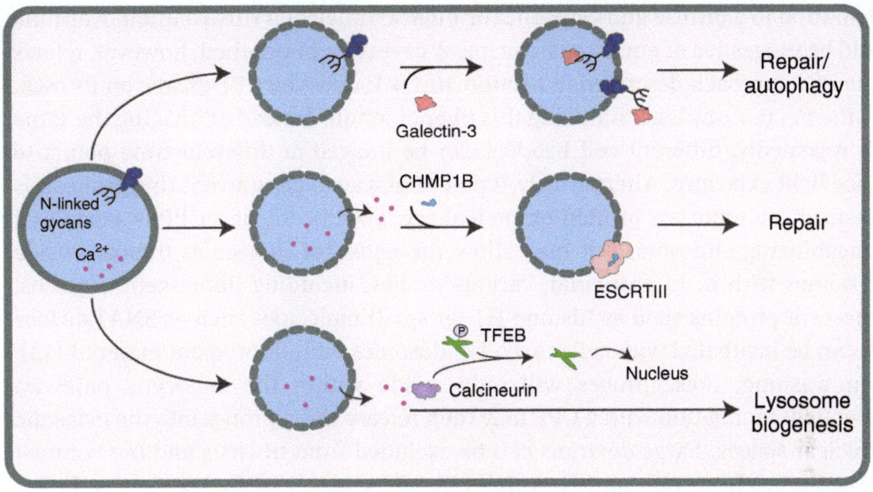

Figure 10.3 Membrane repair: Upon membrane damage, the cell can undergo several responses to repair the damage. N-linked glycans are normally sequestered into the lumen of vesicles, but upon membrane damage, can become exposed to the cytosol. Cytosolic galectin-3 can then bind to these glycans and promote signaling of repair or autophagy of the damaged vesicles. Ca^{2+} leakage can also occur after membrane damage. Free Ca^{2+} leads to the activation of calcineurin which then dephosphorylates TFEB. TFEB can then move to the nucleus where it signals lysosome biogenesis. In addition, CHMP proteins recognize cytosolic Ca^{2+} and recruit the ESCRTIII complex to the damaged endosome. These repair mechanisms are not mutually exclusive and can happen simultaneously. If the damage is too great, the cell will likely undergo apoptosis.

the cytosolic delivery of CPPs and their payloads, one may conclude that endocytic uptake and endosomal maturation are involved in the delivery process. However, there are several caveats associated with these approaches that can confound interpretation. For instance, inhibitors often lack specificity and have pleiotropic effects. Inhibition of endocytosis and maturation of the endocytic pathway may also impact the plasma membrane indirectly by either decreasing or promoting direct plasma membrane translocation. Likewise, a requirement of endosomal trafficking of a CPP for its cytosolic delivery does not necessarily imply endosomal escape. Instead, endosomal trafficking could lead to access to the Golgi or endoplasmic reticulum via retrograde transport routes. In such a scenario, which exists for several toxins, a CPP may enter the cytosol through escape from these organelles, not from endosomes [72]. Inhibitors of retrograde transport can be used to test this possibility.

10.1.5.2 Testing for Endosomal Escape Using Reporters of Membrane Leakage

Several strategies can be used to detect endosomal escape more directly than through inhibitory approaches. A straightforward strategy consists of directly observing a fluorescent CPP trapped inside endosomes (potentially labeled) by fluorescence microscopy. A CPP that appears to redistribute from a punctate distribution to a diffuse and cytosolic (or nuclear/nucleolar) distribution over time would be suggestive of endosomal escape. A caveat to this method, however, relates to the PCI approach described in Section 10.1.4.1: does the CPP escape on its own, or is the microscopy itself inducing this phenomenon. Instead of imaging the same cells repeatedly, different cell batches can be imaged at different time points to reduce light exposure. Alternatively, a potential way to circumvent this problem is to instead use reporters of membrane leakage. Specifically, if a CPP is capable of permeabilizing endosomes, it may allow the egress of molecules trapped inside endosomes with it. In particular, various probes, including fluorescent dextrans, fluorescent proteins such as histone H1, or small molecules such as SNAPsurface 488, can be incubated with cells to load endosomes with fluorescent material [73]. Upon washing, these probes will only reside within the endocytic pathway. Subsequent incubation with a CPP may then release these probes into the cytosolic or nuclear spaces. Large dextrans can be excluded from nucleus and the contrast between a bright cytoplasm (assuming efficient endosomal escape) and a darker nucleus can be used as an indication that the fluorescence detected is inside the cell, as opposed to out-of-focus fluorescence from outside the cells. Both SNAPsurface488 (if used in cells transfected with SNAP-H2) and H1 accumulate in nuclei upon endosomal escape. In turn, this can be useful to observe the endosomes that have not lost their content. Overall, the CPP-redistribution of these probes from endosomes to other intracellular localization is highly suggestive of endosomal escape. While useful, this approach does suffer from several caveats. First, fluorescence microscopy is not readily able to detect low levels of endosomal escape, and fluorescence microscopy-based approaches may only be convenient for relatively active CPPs. This could potentially be circumvented by preloading endosomes with more sensitive probes and using sensitive detection techniques where possible. Second, preloading endosomes with leakage markers only work if the CPP, externally administered after preloading, can traffic to endosomal organelles that contain these markers. This approach may therefore not work if the markers become depleted in the endosomes from which the CPP escape. For instance, these markers could potentially accumulate in lysosomes and be degraded during the course of the experiment.

A possible way to assess for endosomal escape is to detect endosomal membrane damage. This approach, which relies on Chmp1 or Gal3 probes described in Section 10.1.4.2, has been used to screen and identify small molecules that can deliver macromolecules into cells. However, when tested with CPPs, it was reported that endosomal damage and endosomal leakage are not necessarily synonymous [67]. In particular, endosomal damage can be detected in the absence of endosomal leakage, and vice versa.

10.1.6 How Can Endosomal Escape Be Improved?

While there are a variety of CPPs that have been reported to escape endosomes, the majority do so at relatively low efficiency. Finding ways to improve the endosomal escape ability of CPPs is crucial to advancing the field. Enhancing the endosomal escape properties of CPPs has generally been done in two ways: direct modification of the CPP or coincubation with endosomolytic small molecules.

10.1.6.1 Direct Modification of CPPs

A general strategy to improve the endosomal escape of a given CPP is to conjugate this CPP to an endosomolytic sequence. As previously described, the addition of the HA2 peptide to Tat has improved the overall cytosolic delivery of the CPP [28]. In such constructs, Tat improves the endocytic uptake of HA2, while HA2 improves the endosomal escape activity of the peptide. It is however worth noting that while each component serves independent functions, they can also influence one another. For instance, Tat appears to modulate the pH-dependent membrane activity of HA2 analogs. Likewise, HA2 likely impacts the interaction of Tat with cell surface components.

The addition of the sequence FFLIPKG, identified as the retro version of the cathepsin cleavable peptide GKPILFF, enhances the cellular penetration ability of the CPP R8 [74]. The relative hydrophobicity of this peptide, named penetrating-accelerating sequence or Pas, is likely a determining factor in this enhancement [74, 75]. In particular, the number of phenylalanine residues correlates with the uptake of peptide, as measured by flow cytometry. Likewise, the addition of hydrophobic aromatic rings containing amino acids (tryptophan or phenylalanine) has the ability to increase the endosomal escape ability of CPPs, with sequences as short as GFWFG or GWWG contributing to substantial enhancements [75–77]. Notably, separation of the hydrophobic residues in the peptide sequence does not result in increased endosomal escape, suggesting that the hydrophobic residues need to be contiguous. Moreover, the presence of three tryptophan residues induces cytotoxicity [77]. These results indicate that a hydrophobicity window is exploitable to improve CPPs but that this window may be relatively narrow. Overall, additional structure-activity relationships are required to fully evaluate the mechanistic steps that hydrophobic residues modulate. In this regard, it is worth noting that histidines residues, fluorophores or photosensitizers, described earlier as other modifications that can enhance CPP penetration, are also relatively hydrophobic. It is therefore possible that their hydrophobicity contributes to some of the endosomal enhancements they provide.

10.1.6.2 Small Molecules

A class of endosomal escape enhancers that can improve CPPs are small molecules. Chloroquine has shown an improvement in the endosomal escape of CPPs/cargo but comes at the expense of increased cytotoxicity [77, 78]. A high-throughput screen identified UNC7938, a compound found to assist in the cytosolic penetration of oligonucleotides in vitro and in vivo [79]. Notably, UNC7938, enhances the endosomal escape of Tat, R8, or dfTat. For instance, the combination of UNC7938 and dfTat achieves the cytosolic delivery of payloads under conditions where dfTat alone fails [80]. Given that these small molecules are generally protonatable at acidic pH, a possible mode of action is through the proton sponge effect. These compounds can also bind to lipid bilayers. They may therefore modulate the biophysical properties of lipid bilayers. In this context, UNC7938 appears to damage the membrane of endosomes, as measured by recruitment of ESCRT complex components and galectin 3 [67]. Hence, this molecule may destabilize endosomes and make the membrane of these organelles more prone to the CPP-induced leakage. In turn, UNC7938 can allow CPPs that are poorly active to gain a disrupting endosomal membrane activity and help already active CPPs exert an effect at a lower endosomal concentration [80].

10.1.7 Drawbacks of Relying on Endosomal Escape

Continuing this idea of the potential cytotoxicity of membrane damage, it is rather surprising that endosomal escape can be tolerated in cells considering how detrimental plasma membrane damage can be. Nonetheless, this potential advantageous approach of endosomal escape has some drawbacks (Figure 10.4). For instance, for an endosomal escape to work, endocytosis has to happen first. However, the level and rate of CPP endocytosis can vary between different cell types. Endocytic uptake can be relatively elevated in cancer cells, relatively low in T-cells, and completely absent in erythrocytes [81]. So, while a CPP may accumulate inside endosomes at a

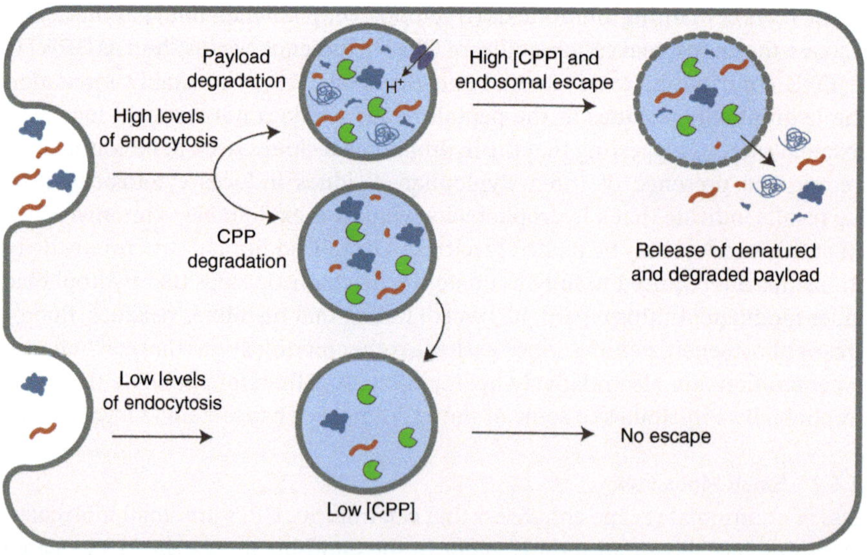

Figure 10.4 Drawbacks of relying on endosomal escape: For a CPP to escape an endosome, it must reach a concentration threshold that allows for an escape to happen. This threshold may not be reached for several reasons. First, not all cells endocytose at the same rate, so if a low amount of CPP is initially endocytosed, this threshold concentration will never be met. Secondly, upon endocytosis, the CPP is likely to encounter a variety of proteases along the endocytic pathway that can degrade the CPP and decrease the amount of active CPP in the endosome. In addition to the CPP, payload stability must be considered. Acidification of the endosome can lead to the unfolding of the payload and proteases can cause payload degradation. Denaturing/degradation of the payload can also lead to deleterious effects, such as dominant-negative versions of the payload or cause cytotoxicity.

concentration that is high enough to trigger endosomal escape in a cancer cell, it may not reach this threshold concentration in a T-cell and fail to deliver a payload. Likewise, various stresses and stimuli can alter the endocytic level of a cell. While this can be potentially exploited to enhance delivery, hard-to-control stimuli that reduce endocytosis can lead to low reproducibility and high variability in cell penetration.

One of the main functions of the endocytic pathway is to traffic material in the cell to be degraded. The late endosomes and lysosomes contain enzymes that can partially or fully degrade CPPs and/or payloads before they enter cells. The acidic pH of endosomes can also denature proteinaceous payloads. Consequently, degradation of CPPs may reduce their endosomal escape efficacy requiring higher CPP doses to compensate for this loss. CPPs made out of D- instead of L-amino acids have been used to prevent the degradation of the CPP within endosomes. Retro-inverso peptides sequences, which also consist of D-amino acids but in a reverse sequence, have also been described. Naturally occurring proteases in the cell are unable to digest peptide bonds between D-amino acids. Therefore, D or retro-inverso peptides are able to resist degradation. For instance, D-dfTat, a retro-inverso version of dfTat (or L-dfTat), resists endosomal degradation [82]. In contrast, approximately 50% of dfTat is degraded by the time it enters the cell. Consequently, D-dfTat displays an endosomal escape activity superior to its L analog [82]. A limitation with D-CPPs however is that their lower propensity for degradation can become a problem once they reach the cytosol of cells. Indeed, while dfTat has little effect on cell growth, D-dfTat can inhibit proliferation. This effect is likely related to the fact that dfTat is degraded in the cytosol within three hours, while D-dfTat is stable for more than three days.

The degradation of payloads that transit through the endocytic pathway prior to endosomal escape can reduce the efficiency with which desired intracellular biological outcomes are achieved. Additionally, it is possible that partial degradation of the payload delivered may generate dominant-negative truncated fragment that could have the opposite effect of what is intended. In this context, it is interesting to speculate that tools that encapsulate payloads while within endosomes but release them once endosomal escape is achieved could prove advantageous.

10.1.8 Conclusion

CPPs are powerful tools for the delivery of cargoes inside cells, and endosomal escape provides a clever mechanism to efficiently penetrate the cytosol of cells while keeping cytotoxicity to a minimum. However, the endocytic pathway is extremely complex and the mechanisms at which CPPs use to escape endosomes remain somewhat of a mystery. This is a difficult task to research due to the low activity of current CPPs and cytotoxicity of endosomolytic agents currently on the market. More work needs to be done in order to elucidate exactly how CPPs can disrupt endosomal membranes in a nondestructive manner. The pursuit of novel endosomolytic sequences must continue to further improve CPP technology. There is still a plethora of knowledge to be discovered on CPPs and endosomal escape, yet the future remains bright for the continued use of CPPs as delivery tools.

Acknowledgment

The authors thank the US National Institute of General Medical Sciences for funding (grant R01GM110137).

References

1 Madani, F., Lindberg, S., Langel, U. et al. (2011). Mechanisms of cellular uptake of cell-penetrating peptides. *Journal of Biophysics* 2011: 414729. https://doi.org/10.1155/2011/414729.

2 Derossi, D., Calvet, S., Trembleau, A. et al. (1996). Cell internalization of the third helix of the Antennapedia homeodomain is receptor-independent. *The Journal of Biolgical Chemistry* 271 (30): 18188–18193. https://doi.org/10.1074/jbc.271.30.18188.

3 Matsuzaki, K., Yoneyama, S., Murase, O. et al. (1996). Transbilayer transport of ions and lipids coupled with mastoparan X translocation. *Biochemistry* 35 (25): 8450–8456. https://doi.org/10.1021/bi960342a.

4 Pouny, Y., Rapaport, D., Mor, A. et al. (1992). Interaction of antimicrobial dermaseptin and its fluorescently labeled analogues with phospholipid membranes. *Biochemistry* 31 (49): 12416–12423. https://doi.org/10.1021/bi00164a017.

5 Wang, T.Y., Libardo, M.D.J., Angeles-Boza, A.M. et al. (2017). Membrane oxidation in cell delivery and cell killing applications. *ACS Chemical Biology* 12 (5): 1170–1182. https://doi.org/10.1021/acschembio.7b00237.

6 Wang, T.Y., Sun, Y., Muthukrishnan, N. et al. (2016). Membrane oxidation enables the cytosolic entry of polyarginine cell-penetrating peptides. *The Journal of Biological Chemistry* 291 (15): 7902–7914. https://doi.org/10.1074/jbc.M115.711564.

7 Hirose, H., Takeuchi, T., Osakada, H. et al. (2012). Transient focal membrane deformation induced by arginine-rich peptides leads to their direct penetration into cells. *Molecular Therapy* 20 (5): 984–993. https://doi.org/10.1038/mt.2011.313.

8 Duchardt, F., Fotin-Mleczek, M., Schwarz, H. et al. (2007). A comprehensive model for the cellular uptake of cationic cell-penetrating peptides. *Traffic* 8 (7): 848–866. https://doi.org/10.1111/j.1600-0854.2007.00572.x.

9 Fischer, R., Fotin-Mleczek, M., Hufnagel, H. et al. (2005). Break on through to the other side-biophysics and cell biology shed light on cell-penetrating peptides. *ChemBioChem* 6 (12): 2126–2142. https://doi.org/10.1002/cbic.200500044.

10 Richard, J.P., Melikov, K., Vives, E. et al. (2003). Cell-penetrating peptides. A reevaluation of the mechanism of cellular uptake. *The Journal of Biological Chemistry* 278 (1): 585–590. https://doi.org/10.1074/jbc.M209548200.

11 Lee, Y.J., Datta, S., and Pellois, J.P. (2008). Real-time fluorescence detection of protein transduction into live cells. *Journal of the American Chemical Society* 130 (8): 2398–2399. https://doi.org/10.1021/ja7102026.

12 Kaksonen, M. and Roux, A. (2018). Mechanisms of clathrin-mediated endocytosis. *Nature Review Molecular Cell Biology* 19 (5): 313–326. https://doi.org/10.1038/nrm.2017.132.

13 Pelkmans, L. and Helenius, A. (2002). Endocytosis via caveolae. *Traffic* 3 (5): 311–320. https://doi.org/10.1034/j.1600-0854.2002.30501.x.

14 Ridley, A.J. (1994). Membrane ruffling and signal transduction. *Bioessays* 16 (5): 321–327. https://doi.org/10.1002/bies.950160506.

15 Mayor, S. and Pagano, R.E. (2007). Pathways of clathrin-independent endocytosis. *Nature Review Molecular Cell Biology* 8 (8): 603–612. https://doi.org/10.1038/nrm2216.

16 Gruenberg, J. (2001). The endocytic pathway: a mosaic of domains. *Nature Review Molecular Cell Biology* 2 (10): 721–730. https://doi.org/10.1038/35096054.

17 Stenmark, H. (2009). Rab GTPases as coordinators of vesicle traffic. *Nature Review Molecular Cell Biology* 10 (8): 513–525. https://doi.org/10.1038/nrm2728.

18 Griffiths, G., Hoflack, B., Simons, K. et al. (1988). The mannose 6-phosphate receptor and the biogenesis of lysosomes. *Cell* 52 (3): 329–341. https://doi.org/10.1016/s0092-8674(88)80026-6.

19 Kobayashi, T., Beuchat, M.H., Lindsay, M. et al. (1999). Late endosomal membranes rich in lysobisphosphatidic acid regulate cholesterol transport. *Nature Cell Biology* 1 (2): 113–118. https://doi.org/10.1038/10084.

20 Kobayashi, T., Stang, E., Fang, K.S. et al. (1998). A lipid associated with the antiphospholipid syndrome regulates endosome structure and function. *Nature* 392 (6672): 193–197. https://doi.org/10.1038/32440.

21 Kobayashi, T., Startchev, K., Whitney, A.J. et al. (2001). Localization of lysobisphosphatidic acid-rich membrane domains in late endosomes. *Biological Chemistry* 382 (3): 483–485. https://doi.org/10.1515/BC.2001.059.

22 Prevette, L.E., Benish, N.C., Schoenecker, A.R. et al. (2015). Cell-penetrating compounds preferentially bind glycosaminoglycans over plasma membrane lipids in a charge density- and stereochemistry-dependent manner. *Biophysical Chemistry* 207: 40–50. https://doi.org/10.1016/j.bpc.2015.08.003.

23 Richard, J.P., Melikov, K., Brooks, H. et al. (2005). Cellular uptake of unconjugated TAT peptide involves clathrin-dependent endocytosis and heparan sulfate receptors. *The Journal of Biological Chemistry* 280 (15): 15300–15306. https://doi.org/10.1074/jbc.M401604200.

24 Granger, E., McNee, G., Allan, V. et al. (2014). The role of the cytoskeleton and molecular motors in endosomal dynamics. *Seminars in Cell and Developmental Biology* 31: 20–29. https://doi.org/10.1016/j.semcdb.2014.04.011.

25 Sato, K., Nagai, J., Mitsui, N. et al. (2009). Effects of endocytosis inhibitors on internalization of human IgG by Caco-2 human intestinal epithelial cells. *Life Science* 85 (23–26): 800–807. https://doi.org/10.1016/j.lfs.2009.10.012.

26 Cantiello, H.F., Stow, J.L., Prat, A.G. et al. (1991). Actin filaments regulate epithelial Na+ channel activity. *American Journal of Physiolpgy* 261 (5 Pt 1): C882–C888. https://doi.org/10.1152/ajpcell.1991.261.5.C882.

27 Kaplan, I.M., Wadia, J.S., and Dowdy, S.F. (2005). Cationic TAT peptide transduction domain enters cells by macropinocytosis. *Journal of Controlled Release* 102 (1): 247–253. https://doi.org/10.1016/j.jconrel.2004.10.018.

28 Wadia, J.S., Stan, R.V., and Dowdy, S.F. (2004). Transducible TAT-HA fusogenic peptide enhances escape of TAT-fusion proteins after lipid raft macropinocytosis. *Nature Medicine* 10 (3): 310–305. https://doi.org/10.1038/nm996.

29 Wang, L.H., Rothberg, K.G., and Anderson, R.G. (1993). Mis-assembly of clathrin lattices on endosomes reveals a regulatory switch for coated pit formation. *Journal of Cell Biology* 123 (5): 1107–1117. https://doi.org/10.1083/jcb.123.5.1107.

30 Dreja, K., Voldstedlund, M., Vinten, J. et al. (2002). Cholesterol depletion disrupts caveolae and differentially impairs agonist-induced arterial contraction. *Arteriosclerosis, Thrombosis, and Vascular Biology* 22 (8): 1267–1272. https://doi.org/10.1161/01.atv.0000023438.32585.a1.

31 Yang, S.T., Zaitseva, E., Chernomordik, L.V. et al. (2010). Cell-penetrating peptide induces leaky fusion of liposomes containing late endosome-specific anionic lipid. *Biophysical Journal* 99 (8): 2525–2533. https://doi.org/10.1016/j.bpj.2010.08.029.

32 Appelbaum, J.S., LaRochelle, J.R., Smith, B.A. et al. (2012). Arginine topology controls escape of minimally cationic proteins from early endosomes to the cytoplasm. *Chemistry & Biology* 19 (7): 819–830.

33 Erazo-Oliveras, A., Najjar, K., Dayani, L. et al. (2014). Protein delivery into live cells by incubation with an endosomolytic agent. *Nature Methods* 11 (8): 861–867. https://doi.org/10.1038/nmeth.2998.

34 Brock, D.J., Kustigian, L., Jiang, M. et al. (2018). Efficient cell delivery mediated by lipid-specific endosomal escape of supercharged branched peptides. *Traffic* 19 (6): 421–435. https://doi.org/10.1111/tra.12566.

35 Jeong, C., Yoo, J., Lee, D. et al. (2016). A branched TAT cell-penetrating peptide as a novel delivery carrier for the efficient gene transfection. *Biomaterials Research* 20 (1): 28. https://doi.org/10.1186/s40824-016-0076-0.

36 Liu, Z., Li, M., Cui, D. et al. (2005). Macro-branched cell-penetrating peptide design for gene delivery. *Journal of Controlled Release* 102 (3): 699–710. https://doi.org/10.1016/j.jconrel.2004.10.013.

37 Erazo-Oliveras, A., Najjar, K., Truong, D. et al. (2016). The late endosome and its lipid BMP act as gateways for efficient cytosolic access of the delivery agent dfTAT and its macromolecular cargos. *Cell Chemical Biology* 23 (5): 598–607.

38 Brock, D.J., Kondow-McConaghy, H., Allen, J. et al. (2020). Mechanism of cell penetration by permeabilization of late endosomes: Interplay between a multivalent TAT peptide and bis(monoacylglycero)phosphate. *Cell Chemical Biology* 27 (10): 1296–1307. e5.

39 Hafez, I.M., Maurer, N., and Cullis, P.R. (2001). On the mechanism whereby cationic lipids promote intracellular delivery of polynucleic acids. *Gene Therapy* 8 (15): 1188–1196. https://doi.org/10.1038/sj.gt.3301506.

40 Wharton, S.A., Martin, S.R., Ruigrok, R.W. et al. (1988). Membrane fusion by peptide analogues of influenza virus haemagglutinin. *Journal of General Virology* 69 (Pt 8): 1847–1857. https://doi.org/10.1099/0022-1317-69-8-1847.

41 Koshman, Y.E., Waters, S.B., Walker, L.A. et al. (2008). Delivery and visualization of proteins conjugated to quantum dots in cardiac myocytes. *Journal of Molecular and Cellular Cardiology* 45 (6): 853–856. https://doi.org/10.1016/j.yjmcc.2008.08.006.

42 Korte, T., Ludwig, K., and Herrmann, A. (1992). pH-dependent hydrophobicity profile of hemagglutinin of influenza virus and its possible relevance in virus fusion. *Bioscience Repports* 12 (5): 397–406. https://doi.org/10.1007/BF01121503.

43 Lee, Y.J., Johnson, G., and Pellois, J.P. (2010). Modeling of the endosomolytic activity of HA2-TAT peptides with red blood cells and ghosts. *Biochemistry* 49 (36): 7854–7866. https://doi.org/10.1021/bi1008408.

44 Lundberg, P., El-Andaloussi, S., Sutlu, T. et al. (2007). Delivery of short interfering RNA using endosomolytic cell-penetrating peptides. *The FASEB Journal* 21 (11): 2664–2671. https://doi.org/10.1096/fj.06-6502com.

45 Ogris, M., Carlisle, R.C., Bettinger, T. et al. (2001). Melittin enables efficient vesicular escape and enhanced nuclear access of nonviral gene delivery vectors. *The Journal of Biological Chemistry* 276 (50): 47550–47555. https://doi.org/10.1074/jbc.M108331200.

46 Boeckle, S., Wagner, E., and Ogris, M. (2005). C- versus N-terminally linked melittin-polyethylenimine conjugates: the site of linkage strongly influences activity of DNA polyplexes. *Journal of Gene Medicine* 7 (10): 1335–1347. https://doi.org/10.1002/jgm.783.

47 Zhang, Q., Tang, J., Ran, R. et al. (2016). Development of an anti-microbial peptide-mediated liposomal delivery system: a novel approach towards pH-responsive anti-microbial peptides. *Drug Delivery* 23 (4): 1163–1170. https://doi.org/10.3109/10717544.2014.1003665.

48 Qian, Z., Martyna, A., Hard, R.L. et al. (2016). Discovery and mechanism of highly efficient cyclic cell-penetrating peptides. *Biochemistry* 55 (18): 2601–2612. https://doi.org/10.1021/acs.biochem.6b00226.

49 Vermeulen, L.M.P., De Smedt, S.C., Remaut, K. et al. (2018). The proton sponge hypothesis: fable or fact? *European Journal of Pharmaceutics and Biopharmaceutics* 129: 184–190. https://doi.org/10.1016/j.ejpb.2018.05.034.

50 Sonawane, N.D., Thiagarajah, J.R., and Verkman, A.S. (2002). Chloride concentration in endosomes measured using a ratioable fluorescent Cl^- indicator: evidence for chloride accumulation during acidification. *The Journal of Biological Chemistry* 277 (7): 5506–5513. https://doi.org/10.1074/jbc.M110818200.

51 Sonawane, N.D., Szoka, F.C. Jr., and Verkman, A.S. (2003). Chloride accumulation and swelling in endosomes enhances DNA transfer by polyamine-DNA polyplexes. *The Journal of Biological Chemistry* 278 (45): 44826–44831. https://doi.org/10.1074/jbc.M308643200.

52 Won, Y.Y., Sharma, R., and Konieczny, S.F. (2009). Missing pieces in understanding the intracellular trafficking of polycation/DNA complexes. *Journal of Controlles Release* 139 (2): 88–93. https://doi.org/10.1016/j.jconrel.2009.06.031.

53 Yamano, S., Dai, J., Hanatani, S. et al. (2014). Long-term efficient gene delivery using polyethylenimine with modified Tat peptide. *Biomaterials* 35 (5): 1705–1715. https://doi.org/10.1016/j.biomaterials.2013.11.012.

54 Chu, D., Xu, W., Pan, R. et al. (2015). Rational modification of oligoarginine for highly efficient siRNA delivery: structure-activity relationship and mechanism of intracellular trafficking of siRNA. *Nanomedicine* 11 (2): 435–446. https://doi.org/10.1016/j.nano.2014.08.007.

55 Midoux, P., Kichler, A., Boutin, V. et al. (1998). Membrane permeabilization and efficient gene transfer by a peptide containing several histidines. *Bioconjugate Chemistry* 9 (2): 260–267. https://doi.org/10.1021/bc9701611.

56 Midoux, P. and Monsigny, M. (1999). Efficient gene transfer by histidylated polylysine/pDNA complexes. *Bioconjugate Chemistry* 10 (3): 406–411. https://doi.org/10.1021/bc9801070.

57 Farber, J.L. (1990). The role of calcium ions in toxic cell injury. *Environmental Health Perspectives* 84: 107–111. https://doi.org/10.1289/ehp.9084107.

58 Eriksson, I., Waster, P., and Ollinger, K. (2020). Restoration of lysosomal function after damage is accompanied by recycling of lysosomal membrane proteins. *Cell Death & Disease* 11 (5): 370. https://doi.org/10.1038/s41419-020-2527-8.

59 Kavcic, N., Butinar, M., Sobotic, B. et al. (2020). Intracellular cathepsin C levels determine sensitivity of cells to leucyl-leucine methyl ester-triggered apoptosis. *The FEBS Journal* 287 (23): 5148–5166. https://doi.org/10.1111/febs.15326.

60 Brojatsch, J., Lima, H. Jr., Palliser, D. et al. (2015). Distinct cathepsins control necrotic cell death mediated by pyroptosis inducers and lysosome-destabilizing agents. *Cell Cycle* 14 (7): 964–972. https://doi.org/10.4161/15384101.2014.991194.

61 Li, J., Cao, F., Yin, H.L. et al. (2020). Ferroptosis: past, present and future. *Cell Death & Disease* 11 (2): 88. https://doi.org/10.1038/s41419-020-2298-2.

62 Berg, K., Selbo, P.K., Prasmickaite, L. et al. (1999). Photochemical internalization: a novel technology for delivery of macromolecules into cytosol. *Cancer Research* 59 (6): 1180–1183.

63 Maiolo, J.R. 3rd, Ottinger, E.A., and Ferrer, M. (2004). Specific redistribution of cell-penetrating peptides from endosomes to the cytoplasm and nucleus upon laser illumination. *Journal of the American Chemical Society* 126 (47): 15376–15377. https://doi.org/10.1021/ja044867z.

64 Choi, Y., McCarthy, J.R., Weissleder, R. et al. (2006). Conjugation of a photosensitizer to an oligoarginine-based cell-penetrating peptide increases the efficacy of photodynamic therapy. *ChemMedChem* 1 (4): 458–463. https://doi.org/10.1002/cmdc.200500036.

65 Srinivasan, D., Muthukrishnan, N., Johnson, G.A. et al. (2011). Conjugation to the cell-penetrating peptide TAT potentiates the photodynamic effect of carboxytetramethylrhodamine. *PLoS One* 6 (3): e17732. https://doi.org/10.1371/journal.pone.0017732.

66 Muthukrishnan, N., Johnson, G.A., Lim, J. et al. (2012). TAT-mediated photochemical internalization results in cell killing by causing the release of calcium into the cytosol of cells. *Biochimica et Biophysica Acta* 1820 (11): 1734–1743. https://doi.org/10.1016/j.bbagen.2012.06.020.

67 Kondow-McConaghy, H.M., Muthukrishnan, N., Erazo-Oliveras, A. et al. (2020). Impact of the endosomal escape activity of cell-penetrating peptides on the endocytic pathway. *ACS Chemical Biology* 15 (9): 2355–2363. https://doi.org/10.1021/acschembio.0c00319.

68 Yong, C.Q.Y., Valiyaveettil, S., and Tang, B.L. (2020). Toxicity of microplastics and nanoplastics in mammalian systems. *International Journal of Environmental Research and Public Health* 17 (5): 1509. https://doi.org/10.3390/ijerph17051509.

69 Skowyra, M.L., Schlesinger, P.H., Naismith, T.V. et al. (2018). Triggered recruitment of ESCRT machinery promotes endolysosomal repair. *Science* 360 (6384): eaar5078. https://doi.org/10.1126/science.aar5078.

70 Radulovic, M., Schink, K.O., Wenzel, E.M. et al. (2018). ESCRT-mediated lysosome repair precedes lysophagy and promotes cell survival. *The EMBO Journal* 37 (21): e99753. https://doi.org/10.15252/embj.201899753.

71 Paz, I., Sachse, M., Dupont, N. et al. (2010). Galectin-3, a marker for vacuole lysis by invasive pathogens. *Cellular Microbiology* 12 (4): 530–544. https://doi.org/10.1111/j.1462-5822.2009.01415.x.

72 Sandvig, K., Bergan, J., Dyve, A.B. et al. (2010). Endocytosis and retrograde transport of Shiga toxin. *Toxicon* 56 (7): 1181–1185. https://doi.org/10.1016/j.toxicon.2009.11.021.

73 Sun, X., Zhang, A., Baker, B. et al. (2011). Development of SNAP-tag fluorogenic probes for wash-free fluorescence imaging. *ChemBioChem* 12 (14): 2217–2226. https://doi.org/10.1002/cbic.201100173.

74 Takayama, K., Nakase, I., Michiue, H. et al. (2009). Enhanced intracellular delivery using arginine-rich peptides by the addition of penetration accelerating sequences (Pas). *Journal of Controlled Release* 138 (2): 128–133. https://doi.org/10.1016/j.jconrel.2009.05.019.

75 Takayama, K., Hirose, H., Tanaka, G. et al. (2012). Effect of the attachment of a penetration accelerating sequence and the influence of hydrophobicity on octaarginine-mediated intracellular delivery. *Molecular Pharmaceutics* 9 (5): 1222–1230. https://doi.org/10.1021/mp200518n.

76 Rydberg, H.A., Matson, M., Amand, H.L. et al. (2012). Effects of tryptophan content and backbone spacing on the uptake efficiency of cell-penetrating peptides. *Biochemistry* 51 (27): 5531–5539. https://doi.org/10.1021/bi300454k.

77 Lonn, P., Kacsinta, A.D., Cui, X.S. et al. (2016). Enhancing endosomal escape for intracellular delivery of macromolecular biologic therapeutics. *Scientific Reports* 6: 32301. https://doi.org/10.1038/srep32301.

78 Yang, S., Coles, D.J., Esposito, A. et al. (2009). Cellular uptake of self-assembled cationic peptide-DNA complexes: multifunctional role of the enhancer chloroquine. *Journal of Controlled Release* 135 (2): 159–165. https://doi.org/10.1016/j.jconrel.2008.12.015.

79 Yang, B., Ming, X., Cao, C. et al. (2015). High-throughput screening identifies small molecules that enhance the pharmacological effects of oligonucleotides. *Nucleic Acids Research* 43 (4): 1987–1996. https://doi.org/10.1093/nar/gkv060.

80 Allen, J., Najjar, K., Erazo-Oliveras, A. et al. (2019). Cytosolic delivery of macromolecules in live human cells using the combined endosomal escape activities of a small molecule and cell penetrating peptides. *ACS Chemical Biology* 14 (12): 2641–2651. https://doi.org/10.1021/acschembio.9b00585.

81 Song, S., Zhang, Y., Ding, T. et al. (2020). The dual role of macropinocytosis in cancers: promoting growth and inducing methuosis to participate in anticancer therapies as targets. *Frontiers in Oncology* 10: 570108. https://doi.org/10.3389/fonc.2020.570108.

82 Najjar, K., Erazo-Oliveras, A., Brock, D.J. et al. (2017). An L- to D-amino acid conversion in an endosomolytic analog of the cell-penetrating peptide TAT influences proteolytic stability, endocytic uptake, and endosomal escape. *The Journal of Biological Chemistry* 292 (3): 847–861. https://doi.org/10.1074/jbc.M116.759837.

11

Pharmacokinetics of Therapeutic and Diagnostic Agents Conjugated with Cell-Penetrating Peptides

Takeshi Fuchigami[1], *Masayuki Munekane*[1], *and Kazuma Ogawa*[1,2]

[1] *Kanazawa University, Graduate School of Medical Sciences, Laboratory of Clinical Analytical Sciences, Kanazawa, Ishikawa, Japan*
[2] *Kanazawa University, Institute for Frontier Science Initiative, Kanazawa, Ishikawa, Japan*

11.1 Introduction

Recently, cell-penetrating peptides (CPPs) have received considerable attention for delivering various molecules, including small hydrophilic molecules, peptides, antibodies, and nucleic acids [1]. CPPs are short peptides with less than 30 amino acids. Drugs that use intracellular molecules as the site of pharmacological effects have difficulty accessing the target and must penetrate the cell membrane to achieve therapeutic efficacy. By conjugating CPPs to drugs with low cell-membrane permeability, they improve the efficiency of intracellular uptake and enhance their biological effects as described elsewhere [2–5]. Therefore, they are being developed as therapeutic agents for various diseases such as cancer, inflammation, central nervous system (CNS) disorders, otolaryngology, eye disorders, and diabetes. However, many issues need to be addressed before proceeding with CPP to the clinic, including evidence of therapeutic effects, metabolic stability, and toxicities [6]. Pharmacokinetics studies addressing bioimaging, biodistribution, and metabolic stability may provide a basis for solving these issues. This chapter discusses pharmacokinetic research focusing on CPPs for treating and diagnosing several diseases. We summarize reports on in vivo biodistribution studies. Molecular imaging can be a powerful tool for tracking the movement of CPPs. Nuclear medicine techniques such as positron emission tomography (PET) and single-photon emission computed tomography (SPECT) can visualize the distribution in any part of the body. However, the resolution of clinical PET and SPECT is at best 1–2 mm. Although PET is capable of highly quantitative imaging, it is still difficult to guarantee the quantitativeness of SPECT. However, the accurate amount of an agent accumulation can be quantitated by measuring the radioactivity of dissected organs. On the other hand,

Cell-Penetrating Peptides: Design, Development and Applications, First Edition. Makoto Oba and Yosuke Demizu.

since only the signals of radionuclides can be captured, metabolites and parent compounds cannot be distinguished from images and measured radioactivity. Optical imaging can be performed with a resolution of 1 μm or higher using a confocal microscope for tissue sections. However, when performing bioimaging in vivo, the resolution is at best 0.5–1.0 mm, depending on the detection depth. Further, the depth of only a few centimeters is achievable, and quantitative measurements are not accurate. The amount of drug in each tissue was measured using the fluorescence intensity of the dissected organs. However, the presence of autofluorescence in organs makes accurate quantification difficult. Ultrasound imaging provides a resolution of 50 μm, and a detection depth of a dozen centimeters is possible. CT and magnetic resonance imaging (MRI) can image deep body tissues at high resolution (CT; 50 μm, MRI; 10–100 μm), but high concentrations of contrast agents are required. [7, 8]. Therefore, methods for evaluating the pharmacokinetics of CPP-conjugated agents need to be selected carefully. In addition, it needs to be considered that labeling may change the pharmacokinetics of molecules.

Herein, we first discuss the in vivo distribution of CPPs. Then, we summarize the changes in the pharmacokinetics of therapeutic agents when conjugated with CPPs.

11.2 Pharmacokinetics of CPPs

Sarko et al. performed an excellent systematic in vitro and in vivo evaluation of the stability and pharmacokinetics of ten radiolabeled CPPs {penetratin, Tat, PreS2-TLM, model amphipathic peptide (MAP), R9, a peptide derived from the murine VE-cadherin (pVEC), MTS, SynB1, transportan 10 (TP10), SV40T-nuclear localization signal (SV40T-NLS)} as shown in Table 11.1 [9]. These peptides showed significantly different stabilities in the human plasma. For example, the blood half-life of MAP, TP10, and SV40T-NLS was over 72 hours, whereas penetratin and R9 had much shorter half-lives of 1.2 and 1.9 hours, respectively. In vivo studies in PC-3 tumor-bearing mice demonstrated that all these CPPs showed fast blood clearance and declined to 1% injected dose per gram of tissue (% ID/g) after four hours. The accumulation of these CPPs in tumor tissue was lower than in other tissues except for the brain and muscle. Each peptide showed a characteristic organ distribution; for example, R9 and TP10 exhibited high accumulation in the liver, and Tat and NSL demonstrated high accumulation in the kidneys [9]. In another reports, Tat peptides simultaneously labeled with ^{99m}Tc and fluorescein showed relatively higher accumulation in the liver than ^{111}In-DOTA-Tat, indicating labeling moieties could influence the pharmacokinetics of CPPs [10, 11]. Penetratin showed relatively high brain permeability. A high accumulation of penetratin and R9 was also observed in the spleen. PET images of ^{68}Ga-labeled CPPs in rats are shown in Figure 11.1. The images indicate that ^{68}Ga-labeled penetratin, Tat, and TP10 were distributed throughout the body, centering on the liver and kidneys. Such plasma stability and biodistribution data are very important for predicting the in vivo properties of CPPs [9].

Table 11.1 Serum stability and biodistribution characteristics of ^{111}In-DOTA-conjugated CPPs.

CPPs (Sequence)	Serum stability ($t_{1/2}$)	Biodistribution characteristics
Penetratin (DOTA-RQIKIWFQNRRMKWKK)	1.2 h	High accumulation in the liver, kidney, and spleen High brain permeability
Tat (DOTA-GRKKRRQRRRPPQ)	8.8 h	High accumulation in the kidney
$PreS_2$-TLM (DOTA-PLSSIFSRIGDP)	4.4 h	Rapidly excreted
MAP (DOTA-KLALKLALKALKAALKLA)	>72 h	High accumulation in the liver and kidney
R9 (DOTA-RRRRRRRRR)	1.9 h	High accumulation in the liver and spleen
pVEC (DOTA-LLIILRRRIRKQAHAHSK)	3.9 h	High accumulation in the liver and kidney
MTS (DOTA-AAVALLPAVLLALLAP)	48.1 h	High accumulation in the liver
$SynB_1$ (DOTA-RGGRLSYSRRRFSTSTGR)	5.2 h	High accumulation in the liver and kidney
TP10 {AGYLLGKε(DOTA) INLKALAALAKKIL}	>72 h	High accumulation in the liver
SV40T-NLS {PKKKRKVKε(DOTA)}	>72 h	High accumulation in the kidney

MAP, Model amphipathic peptide; pVEC, A peptide derived from the murine VE-cadherin; MTS, Membrane translocating sequence of Kaposi fibroblast growth factor (k-FGF); TP10, Transportan 10; NLS, Nuclear localization signal. Source: Data from [9].

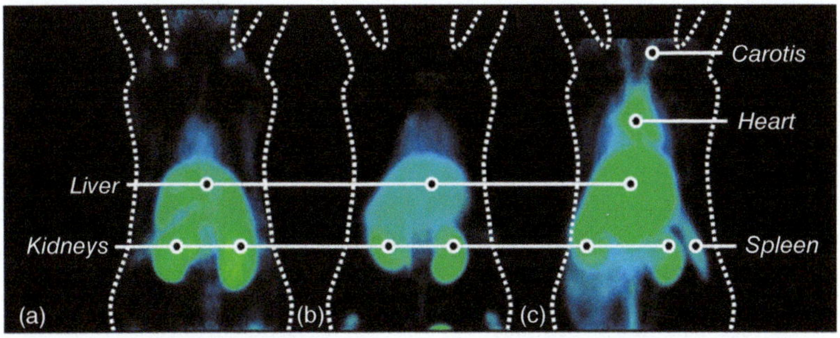

Figure 11.1 In vivo micro-PET imaging of the biodistribution of three labeled CPPs after intravenous injection. Wistar rats were injected with ^{68}Ga-labeled penetratin (a), Tat (b), and TP10 (c) via the tail vein and scanned one hour after injection. Source: [9]/with permission of American Chemical Society.

11.3 Pharmacokinetics of CPP-Conjugates for CNS Disorders

CNS disorders include stroke, Parkinson's disease, and Alzheimer's disease. The blood-brain barrier (BBB) limits the access of drugs to the brain. CPP can translate drugs that rarely pass through the BBB to the brain via the adsorption-mediated transcytosis pathway. Such CPP-conjugates exhibited efficient BBB permeability at concentrations below the micromolar without producing a cytolytic effect [12]. There are several reports about biodistribution studies of CPP-conjugates for CNS disorders as listed in Table 11.2.

DJ-1 is one of the protective effects of oxidative stress associated with PD disease [16]. Lev et al. developed ND-13, a fusion peptide consisting of the 13 amino acid segments of DJ-1-protein and the 7 amino acid segments of Tat. The in vivo brain distribution of fluorescein isothiocyanate (FITC)-labeled ND-13 showed fluorescence signals in the brain tissue. On the other hand, the FITC-ND-13 derivative without the Tat peptide had low BBB permeability. These results indicate that conjugation of the drug with Tat can be a strategy for developing new CNS drugs [13]. K16ApoE is a cationic BBB-penetrating peptide. K16ApoE has been shown to increase cisplatin uptake into the brain 34-fold [14]. Ahischwede et al. developed

Table 11.2 Representative reports of CPP-conjugates for CNS disorders which performed in vivo pharmacokinetics experiments.

CPPs (Sequence)	Targeting molecules	Payloads	Biodistribution characteristics	Refs
Tat (YGRKKRR)	DJ-1 peptide fragment (KGAEEMETVIPVD)	FITC	Time-dependent accumulation of the fluorescent signal in the brain parenchyma around the blood vessels of C57/bl6 mice	[13]
K16ApoE (KKKKKKKK KKKKKKKK LRVRLASH LRKLRKRL LRDA)	Cisplatin, Cetuximab	^{125}I	Around 0.4% dose brain uptake (Both K16ApoE and a mixture of K16ApoE-Cetuximab)	[14]
R9 (RRRRRRRRR)	Generation five poly(amido)amine dendrimers conjugated with gelatinase-activatable cell-penetrating peptides ($SucE_8$-PLGC(Me)AG-R9)	Cy5, gadolinium	Over 40 fmol/mg of uptake of the dendrimers in the brain 24 h post-ischemia mice, representing a 7.6-fold increase compared to the contralateral region	[15]

novel K16ApoE-fused nanoparticles bound to amyloid-oriented curcumin. In vivo evaluation using a mouse model of amyloid β (Aβ) deposition in the brain showed that the intracerebral accumulation of ^{125}I-labeled K16ApoE-introduced nanoparticles was significantly higher than that of nanoparticles without K16ApoE. In addition, Gd-labeled K16ApoE-bound nanoparticles were observed to accumulate specifically in amyloid plaques of brain tissue compared to the control nanoparticles without K16ApoE [17]. Chen et al. developed a gelatinase-activated cell-permeable peptide dendrimer (ACPPD) bound to multiple Cy5 and gadolinium moieties for application in stroke diagnosis. The activating linker contained an enzyme-cleaved sequence between the polyanion (Suc-E8) and polycation (R9). Fluorescence and MRI signals of ACPPD were significantly increased in the ischemic tissue of the stroke mouse model, indicating that activating CPP-tagged dendrimers is useful for stroke monitoring [15]. Waggoner et al. fabricated engineered fluorescent polystyrene nanoparticles with various peptides and compared their biodistribution after direct intracerebral and tail vein administrations. When peptide-modified nanoparticles were administered directly into the brain, the nanoparticles modified with positively charged peptides (R9, K9) exhibited a more restricted distribution from the injection site than the other zwitterionic, neutral, or neutral negatively charged peptide-modified nanoparticles. After intravenous administration in a mouse model of traumatic brain injury (TBI), positively charged peptide-conjugated nanoparticles accumulated more in nontarget tissues such as the heart, lungs, and kidneys than in other peptide-modified nanoparticles. Its accumulation in TBI sites was lower than that of nanoparticles with peptides other than charged peptides. Thus, the pharmacokinetics of CPP-modified nanoparticles vary greatly depending on the administration routes [18]. CPPs are useful tools for delivering drugs into the brain, and their biodistribution may change significantly depending on the administration route. It is of concern that subtle changes in drug levels in the brain can lead to neurotoxicity. Further studies on the pharmacokinetics of CNS target molecules fused with CPP using techniques such as bioimaging are necessary.

11.4 Pharmacokinetics of CPP-Conjugates for Infectious Diseases

Classical therapeutic agents for infectious diseases are often inadequate because many drugs targeting pathogens exhibit hydrophilic properties that limit their therapeutic effects against intracellular pathogens. Therefore, various conjugates of CPPs with drugs, including antibiotics, antimicrobial peptides, antiviral agents, and antibodies, have been developed to enhance therapeutic effects in such diseases [19]. Antibiotics conjugated with CPPs have been extensively developed to treat drug-resistant pathogens effectively, as reviewed elsewhere [20]. Their enhanced therapeutic effects seem to be caused by the increased intracellular internalization and should be quantitatively elucidated. Therefore, several groups have reported the detailed pharmacokinetics of CPP-drug conjugates. Vancomycin (Van) is a glycopeptide antibiotic used to treat infections caused by gram-positive pathogens, including multidrug resistance. However, Van showed poor accessibility to intracellular bacteria and the brain leading to inadequate effectivity in such regions. There are several reports about biodistribution studies of CPP-conjugates for the treatment of infectious diseases as listed in Table 11.3.

Ruczyński et al. fabricated transportan 10 (TP10)-Van conjugates via polyethylene glycol (PEG) linkers. Combined with the antibacterial activity of TP10 itself, these conjugates had increased antibacterial activity against methicillin-resistant Staphylococcus aureus (MRSA) strains. In addition, an in vivo study demonstrated that these conjugates increased penetration into brain tissues 200-fold compared to Van [21]. In another report, hexaarginine-conjugated Van derivatives have been developed as new antibacterial agents. FU002 has a robust antibacterial activity and does not show excessive toxicity, such as hemolytic action. PET imaging studies in rats with ^{124}I-labeled Van and FU002 were performed to elucidate the pharmacokinetics of Van derivatives. While ^{124}I-FU002 exhibited high hepatic retention, ^{124}I-Van showed rapid renal clearance (Figure 11.2) [24]. ^{125}I-labeled triarginine-containing Van lipopeptide conjugates with high antibacterial activity against Van-resistant enterococci showed high uptake in the liver and much lower uptake in other organs, including the kidneys [25]. Such biodistribution and excretion pathways changes compared to Van are expected to be a

Table 11.3 Representative reports of CPP-conjugates for infectious diseases which performed in vivo pharmacokinetics experiments.

CPPs (Sequence)	Target-oriented molecules	Payloads	Biodistribution characteristics	Refs
TP10 (AGYLLGKINLKALAALAKKIL)	Vancomycin	Intact	About 200-fold higher brain uptake of [Lys7(PEG4-Van)] TP10 (2611 ± 120 nM) in mice compared with that of Van (11 ± 2 nM) by using LC/MS method	[21]
Tat (YGRKKRRQRRRGSG)	Self-assembled nanoparticles of fusion inhibitory peptides composed CPP and influenza peptide fragments of hemagglutinin ectodomain	Intact	The highest accumulation in lung (over 4 μM/g of tissue) after intranasal treatment of nanoparticles in cotton rats by ELISA	[22]
p41 (SWLRRIWRWICKVLSRFK)	Galactosylated antiviral peptide nano complexes	^{177}Lu	High accumulation in the liver (6% injected dose/g) after 4 h i.v. injection to BALB/c mice	[23]

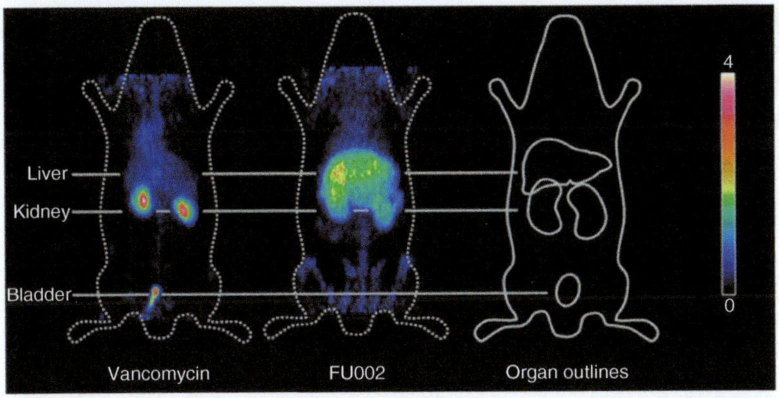

Figure 11.2 Molecular imaging to visualize the pharmacokinetics of vancomycin (Van) and FU002. PET imaging of ^{124}I-labeled Van and its ^{124}I-labeled derivative ^{8}Tyr-FU002 at 10 minutes after injection is shown. While Van shows renal excretion, ^{124}I-^{8}Tyr-FU002 has a broader distribution profile. Source: Umstätter et al. [24], Reproduced with permission from John Wiley & Sons.

new therapeutic option for patients with renal disease who have difficulty using Van and for gram-positive bacterial infections in the liver. Figueira et al. developed influenza fusion inhibitors that conjugated peptide fragments of hemagglutinin with Tat and lipids. The complexes were administered to rats (intranasal, intra-lung, or subcutaneous), and agents accumulated in each organ were quantified by ELISA. The intranasal and intra-lung treatments exhibited good bioavailability. The highest uptake was observed in the lungs, expected to be a highly efficient treatment for antivirus effects [22]. Zhang et al. developed galactosylated antiviral peptide nano complexes (Gal-APN) containing p41, a cationic peptide, as an active ingredient for the antiviral treatment of HCV. Biodistribution studies of ^{177}Lu-labeled Gal-APN in mice (i.v. injection) confirmed high accumulation in the liver, indicating that therapeutic effects can be obtained depending on the amount accumulated in the target. On the other hand, improvement in liver accumulation and therapeutic effects of p41 could be due to Gal rather than CPP [23]. Vaccines also have active ingredients that are often difficult to pass through biological membranes; thus, CPPs are expected to serve as a carrier tool. Numerous CPP-conjugated vaccines for viral infections are under development (hepatitis B, hepatitis C, influenza, HIV, etc.). However, few detailed studies on pharmacokinetics have been reported, and further research is awaited [19].

11.5 Pharmacokinetics of CPP-Conjugates for Inflammatory Diseases

Percutaneous administration is an effective means of topical delivery of anti-inflammatory drugs, but the stratum corneum and mucosa are significant obstacles to delivery. CPP-based therapeutic agents for inflammation have been extensively investigated to improve skin penetration [6]. Henry et al. developed a ^{64}Cu-labeled acid-sensitive cyclic peptide pHLIC c[E4W5C] ^{64}Cu-c[E4W5C] showed significantly higher uptake in tissues such as blood, lungs, liver, and small intestine in lipopolysaccharide (LPS)-treated mice than in non-treated control mice. Furthermore, increased uptake of ^{64}Cu-c[E4W5C] was also observed in the brains of animals treated with LPS ^{64}Cu-pHLIC is expected to be a prospective imaging probe for the noninvasive evaluation of acidosis associated with central and peripheral innate immune activation [26]. Kim et al. evaluated the in vitro and in vivo bioactivity and therapeutic effects of a STAT3-specific high-affinity peptide (APTstat3) after intradermal and transdermal delivery for the treatment of psoriasis. In a lipid formulation, they fabricated discoidal lipid nanoparticles (DLNP) complex of R9-conjugated APTstat3 cell-permeable peptide (APTstat3-9R). The nanoparticles achieved efficient skin penetration of the STAT3 inhibitory peptide after transdermal administration to psoriasis model mice and were consistently able to inhibit psoriatic skin inflammation [27]. In addition, the development of CPP-drug complexes for the treatment of inflammation is being vigorously pursued [28, 29]. There are no reports of biodistribution in these reports. Hence, further research is required.

11.6 Pharmacokinetics of CPP-Conjugates for Cancer Treatment

Specific protein-protein interactions are involved in intracellular signal transduction pathways in tumor cell survival and progression. Therapeutic agents, diagnostic agents, and theranostic agents (enable both therapeutic and diagnosis) intervening in such pathways could improve cancer management. CPPs are expected to be intracellular transporters of these agents [5]. Since small interfering RNA (siRNA) can specifically inhibit the expression of target proteins, they are expected to be cancer-specific therapeutic agents. However, their low membrane permeability and metabolic stability severely limit their clinical translation. Therefore, CPPs have been considered to deliver siRNA into cells in vivo [30]. It has been reported that conjugation of CPPs with various nanoparticles increases accessibility to tumor tissue, often changing to favorable properties that reduce peripheral retention [31]. Table 11.4 listed representative examples of biodistribution studies of CPP-conjugates for the treatment and diagnosis of cancers.

There are many reports on the use of CPPs to deliver intracellular modulators' heat shock protein 70 (Hsp70) [34], protein kinase B [41], cellular Src (c-Src) [42], transducers and activators of transcription 3 (STAT3) [32], survivin [43], and p53 [44]. Tan et al. developed a Tat-conjugated antihuman epidermal growth factor receptor 2 (HER2)/neu peptide mimetic (AHNP)-STAT3-binding peptide (P3-AHNP-STAT3BP). in vivo evaluation using tumor-bearing mice showed that the peptide complexes were selectively incorporated into HER2-expressing breast cancer cells via Tat. The anticancer effects mediated by the effects on STAT3 were

Table 11.4 Representative reports of CPP-conjugates for cancers which performed in vivo pharmacokinetics experiments.

CPPs (Sequence)	Target-oriented molecules	Payloads	Biodistribution characteristics	Refs
Tat-derived peptide (YGRKKRRQR)	Conjugate of anti-HER2/neu peptide mimetic and STAT3–inhibiting peptide (P3-AHNP-STAT3BP)	FITC (in vitro) Intact (in vivo)	Significant accumulation of P3-AHNP-STAT3BP in HER2 overexpressing tumor tissues compared to peptides without CPP (immunohistochemistry)	[32]
PVBLG-8 {Poly(γ-(4-vinylbenzyl)-l-glutamate)}	siRNA against EGFR encapsulated nanoparticles	Cy5	Significant fluorescence signals of nanoparticles in the tumor tissue in mice 4 h after i.v. injection	[33]
C-terminal domain of Hsp70 (TKDNNLLGRFELSG; TPP)	Pegylated TPP (TPP-PEG_{24})	^{89}Zr	High accumulation in the Hsp70-positive tumor tissue (over 6% injected dose/g) in mice 24 h after i.v. injection	[34]
Tetrameric cell-permeable peptides (tCPPs) consisting of penetratin or Tat	Matuzumab (mAb targeting EGFR)	^{124}I	Significant reduction in blood radioactivity without decreased tumor uptake when compared with no tCPP-conjugated matuzumab	[35]
r8 (octa-D-arginine)	r8 and a penta- D-glutamate (e5) masking sequence connected with a PLGC(Me)AG MMP substrate	Cy5/Cy7	About 10-fold higher fluorescence signals in cancer-positive lymph node compared with cancer-negative lymph node in 4T1 metastatic lymph node mouse model	[36]
TP-LyP-1 peptide {GWTLNSAGYLLGKINLKALAALAKKIL-G_4-c(CGNKRTRGC)}	Tumor-penetrating nanocomplexes composed of TP-LyP-1 and siRNA targeting for ID4	FITC	Over three-fold increase in the tumor fluorescence of the TP-LyP nanocomplexes in MDA-MB-435 xenografts compared to untargeted control nanocomplexes 4 h after i.v. injection	[37]

(Continued)

Table 11.4 (Continued)

CPPs (Sequence)	Target-oriented molecules	Payloads	Biodistribution characteristics	Refs
aTAT (Tat derivative that lysine residues' amines were amidized to succinyl amides)	aTAT-functionalized micelles which loaded DOX	DiR or intact	About 8 μg/g tumor of DOX accumulation in the tumor tissue 18 days after i.v. injection of aTAT-functionalized micelles in tumor-bearing mice in which 8-fold higher than Tat-functionalized micelles	[38]
GALA (amphipathic α-helix peptide composed of 30 amino acid residues)	PEG-SS-PEI nanoparticles loaded siRNA for EGFR and BRD4	Cy5	About 2.4-fold higher tumor accumulation 24 h after i.v. injection of functional nanoparticles relative to nanoparticles alone	[39]
Neuropilin-1 receptor-targeted tumor-penetrating peptide (RGERPPR)	Peptide-modified Fe_3O_4@SiO_2@mSiO_2 nanoparticle load DOX	DiR or Gd	Over two-fold higher tumor accumulation (fluorescence signals) of peptide-functionalized nanoparticles during 6–24 h after i.v. injection of tumor-bearing mice than nanoparticle alone	[40]

HER2, Human epidermal growth factor receptor 2; STAT3, Transducers and activators of transcription 3; EGFR, Epidermal growth factor receptor; MMPs, Matrix metalloproteinases; DOX, Doxorubicin; ID4, Inhibitor of DNA binding 4; BRD4, Bromodomain-containing protein 4.

enhanced in correlation with the amount accumulated in the tumor tissue [32]. Brain tumors are refractory; the BBB blocks the transport of drugs. Liu et al. encapsulated siRNA into epidermal growth factor receptor (EGFR) in Cy5-labeled micelles modified with the membrane-permeable peptide PVBLG-8. These nanoparticles inhibited EGFR expression and antitumor effects in cancer-bearing mice transplanted with U87-MG, improved compared to existing micelles using Lipofectamine. Consistent with these results, in vivo imaging demonstrated the accumulation of fluorescently labeled PVBLG-8 micelles in tumor tissue [33]. Zhang et al. developed liposomes (CB5005-LS) fused with CPP and NF-κB peptides. Fluorescent imaging of intracranial glioblastoma-bearing mice showed that the fluorescently labeled CB5005-LS showed higher liver accumulation than CPP-unmodified liposomes but showed adequate accumulation in the tumor regions formed in the brain [45]. Stangl et al. developed an ^{89}Zr-labeled PET probe containing a C-terminal sequence domain targeting Hsp70. This ligand forms a dimer with membrane Hsp70 and is taken up by cells. Biodistribution in mice demonstrated that this tracer showed accumulation in tumors according to the expression of Hsp70 and is expected to be a scaffold for new cancer-imaging agents and cargo of intracellular molecules [34]. CD44 is a transmembrane glycoprotein that is overexpressed in several cell types, including liver cancer, and is thought to be involved in the development and progression of cancer [46]. Hyaluronic acid (HA) binds to CD44 on the surface of membranes such as liver cancer tissues. Zhao et al. demonstrated by ultrasonic imaging that the drug can reach deep tumor tissue by modifying the surface of liposomes conjugated with HA and Tat containing anticancer drugs. The nanoparticles released an anticancer drug in the cancer tissue by applying ultrasound and showed an effect of suppressing the growth of liver cancer tissue [47]. Recently, several CPP-conjugated MRI nanoparticles have been developed for application in tumor imaging, demonstrating higher tumor accumulation and clearer MRI signals than those without CPPs [48].

The monoclonal antibody matuzumab (Mat) targeting EGFR has been conjugated with tetrameric cell-permeable peptides (tCPPs) to improve the intracellular delivery of matuzumab. Pharmacokinetic analysis (96 hours after administration) using ^{131}I-labeled tracers in tumor-bearing mice revealed the radioactivity in the blood pool of penetratin-bound Mat (Mat-tPen) and Tat-bound Mat (Mat-tTAT) was less than half of that of the nonbinding antibody (Mat). On the other hand, the accumulation in tumors was similar between Mat-tTAT and Mat but lower than that of Mat-tPen. As a result, the tumor-to-blood ratio increased to 2.7 for Mat-tTAT and 2.0 for Mat-tPen, but less than 1.3 for unmodified Mat (Figure 11.3, left panel). PET imaging using a ^{124}I-labeled CPP-conjugated matuzumab (48 hours after administration) showed the antibody conjugate in the tumor and other areas (Figure 11.3, right panel) [35]. ^{111}In-labeled nanobodies bound to CPP (r9 or penetratin) have been developed for tumor imaging of EGFR. In vitro studies demonstrated that both CPP-nanobodies showed higher retention in the spheroid core than non-CPP-bound nanobodies. However, SPECT studies showed the uptake of ^{111}In-CPP nanobodies into tumors being masked because of a very high accumulation in the liver [49].

Jang et al. developed matrix metalloproteinases (MMP)-activatable CPP (ACPPs) in which a fluorescently labeled polyarginine-based CPP and a negatively charged

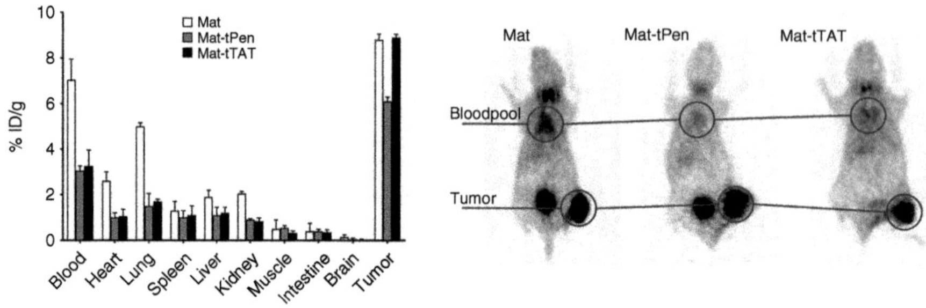

Figure 11.3 Left panel: Tissue distribution of ^{131}I-labeled Mat, Mat-tPen, and Mat-tTAT 96 hours after injection into mice bearing A431 xenografts. Right panel: PET images (maximum intensity projection) of A431 tumor-bearing nude mice 48 hours after injection of ^{124}I-labeled Mat, Mat-tPen, and Mat-tTAT. Substantially higher activity is visible in the blood pool (heart) for Mat, while all substances reveal high tumor accumulation. Source: Sauter et al. [35]/ with permission of ELSEVIER.

polyanion-based peptide were ligated with an enzyme-cleavable linker. When the linker was cleaved by MMPs, each ACPP released a fluorescence-labeled CPP moiety whose membrane permeability was inhibited by binding to a polyanion and entered the cell by the action of CPP. ACPPs with far-red fluorescent cargo showed a 2–3 in vivo contrast ratio and a 3.1-fold increase in tumor standard uptake compared to contralateral normal tissue or control peptides with scrambled linkers. This study demonstrated the usefulness of ACPP as a new enzyme activity-responsive imaging [50]. Inspired by the concept of ACPPs, various imaging probes have been developed for the noninvasive assessment of proteases, including MMPs [51], membrane type 1 MMPs (MT1-MMPs) [52], and legumain [53, 54]. However, the degradation of molecular probes in peripheral systems can make their integration with specific targets difficult. For example, it has been suggested that intracellular uptake via CPP after cleavage of a radioactive probe targeting MMP has a large effect not only on tumor-specific activation but also on activation of the vascular compartment [51]. AVB-620 is an improved fluorescence resonance energy transfer (FRET)-based ACPP with Cy5 and Cy7 introduced into anionic and cationic peptides, respectively. This molecular probe accumulates in malignant tissues with high MMP activity in high contrast. In vivo evaluation of mice showed that MMP hydrolyzes AVB-620 and that D-octaarginine (r8) with cleaved Cy7 accumulates in the primary tumor and lymph node metastasis site with high retention for several tens of hours [36]. Clinical studies have also shown that AVB-620 enables real-time tumor detection by intraoperative imaging of surgical specimens after intravenous infusion of AVB-620 [55]. Li et al. developed a prospective fluorescence probe ACPPplX, which is a fusion of MMP-2-sensitive activated CPP (R9GPLGLAGE8) and protoporphyrin (PpIX) for tumor-targeted photodynamic therapy. Fluorescence imaging of mouse tumor tissues with activated MMP2, ACPPplX exhibited clear fluorescence signals, whereas CPP-PpIX without MMP2-cleavble sequence showed no significant signals [56].

Tumor-homing peptides are CPPs selected by tumor tissue discovered by techniques such as phage display [57]. Although p32 is a mitochondrial protein, it is exposed to the cell membrane of cancer cells. When LyP-1 binds to p32, it is cleaved by an endogenous enzyme and flows into the cell [58]. Lipid nano complexes with LyP-1 bound to the surface have achieved efficient delivery of siRNA into mouse tumors to the inhibitor of DNA binding 4 (ID4), demonstrating the associated high antitumor activity [37]. iRGD is enzymatically cleaved by binding to integrins. The digest is recognized by the neuropilin-1 (NRP-1) receptor and is taken up into cells by increasing the permeability of blood vessels and tissues [59]. Rhodamine B-labeled conjugates of iRGD and the cationic anticancer peptide domain (HPRP-A1) were distributed at high levels in tumor tissues with minimal peripheral organs [60]. Justin et al. have developed a CPP-nanoparticle complex in which transportan is modified from an iRGD-binding nanocomposite targeting integrins that are highly expressed in tumor blood vessels. As a result, the accumulation of anti-Kras siRNA in tumor tissue increased in a mouse model of pancreatic cancer, suggesting a possible treatment improvement [61]. Cationic residues of traditional CPPs exhibit nonspecific interactions with biomolecules, which make them difficult to use in vivo. For example, the β-galactosidase-Tat fusion protein injected intraperitoneally into mice was distributed in virtually all tissues [62]. When administered intravenously (i.v.), CPP is recognized by the reticular endothelial system, transferred from the blood to tissues, and easily penetrates most organs [62]. Jin et al. fabricated micelles modified with a succinyl amide derivative of Tat (aTAT) to prevent nonspecific interactions with the bloodstream of Tat. Upon entering the acidic tumor stroma or cell endo/lysosome, the succinyl amide of aTAT was rapidly hydrolyzed, and Tat function was fully restored. The blood clearance of aTAT-conjugated micelles was much slower than that of Tat-micelles, indicating the high stability of aTAT in blood flow. In addition, the aTAT-micelles exhibited greatly enhanced accumulation in tumor tissues and therapeutic efficacy of DOX [38]. Wan et al. developed PEG disulfide bond-linked polyethylenimine (PEG-SS-PEI) nanoparticles to deliver siRNA co-targeting EGFR bromodomain-containing protein 4 (BRD4) for the therapy of triple negative breast cancer. They used GALA as a CPP and CREKA as the tumor extracellular matrix (ECM). In the in vivo biodistribution studies in tumor-bearing mice, the GALA-and CREKA-conjugated PEG-SS-PEI showed significantly higher tumor uptake and lower liver uptake than nonconjugated PEG-SS-PEI. Its favorable pharmacokinetics profile correlates well with its effective cancer treatment and minimal side effects [39]. SVS-1 is a cancer-targeting peptide that folds preferentially on the surface of cancer cells, changes to a β-hairpin structure, and is taken up into the cells [63, 64]. We have developed ^{67}Ga-labeled SVS-1 derivatives with arginine or histidine substitutions, showing high tumor accumulation, tumor/blood ratio (3.8–8.0), and tumor/muscle ratio (3.3–5.0) 120 minutes after administration. The accessibility of SVS-1 derivatives is much higher than that of the commonly used CPPs mentioned above [9]. These radiolabeled SVS-1 analogs can target the surface membranes of cancer cells and provide scaffolding for in vivo imaging agents [65]. Lymph node metastasis contributes significantly to the distant metastasis of cancer cells in the primary lesion. Hu et al. developed novel nanoparticles in which D-nonaarginine (r9) was conjugated with a lymph node-directed cabazitaxel nano vehicle (r9-CN). Pharmacokinetic studies using fluorescent dyes demonstrated that r9-CN showed long retention and deep penetration into tumor masses in metastatic lymph nodes compared to nanovesicles lacking the r9 peptide (CN). Consistent with the biodistribution results, r9-CN showed excellent inhibitory effects on primary tumor growth, lymph node metastasis, and distant lung metastasis [66]. Fe_3O_4@SiO_2@mSiO_2 nanoparticles conjugated with CPP (RGERPPR) with high tumor tissue selectivity demonstrated significant improvement in tumor accumulation in mice compared to particles without CPP by near-infrared fluorescence in vivo imaging. Small animal MRI studies have shown that CPP-NPs have a T1-T2 dual-mode contrast imaging effect that leads to a more accurate diagnosis than conventional methods [40]. CPPs can be used as next-generation in vivo imaging tools. Semiconductor polymer dots (Pdots) are promising probes for bioimaging. CPP-conjugated Pdots enable bright visualization of cancer cells with low cytotoxicity. With deep-red excitation and bright NIR emission, these Pdots are promising for in vivo systemic fluorescence imaging [67].

Many reports suggest that the introduction of CPPs into intracellular target molecules enhances interference with intracellular molecule binding and interaction. In the future, further contributions to the therapy and diagnosis of cancer are expected by improving the CPPs, optimizing the combination with nanoparticles, and optimizing the administration methods.

11.7 Future Perspective

Many CPP-binding molecules have been developed that have proven to be useful in vitro. In vivo evaluations focused on the therapeutic effects are underway. However, the correlation between the in vivo pharmacokinetics and pharmacological effects of such agents is not well understood. Therefore, in vivo pharmacokinetic studies will be even more important in future research to develop CPP derivatives useful in clinical practice.

Problems with current CPPs include in vivo stability, immunogenicity, intracellular toxicity, and low specificity for target cell membranes [6]. Depending on their conformation, CPPs can be divided into linear and circular CPPs. Generally, cyclic peptides have better metabolic stability than linear peptides [68]. In recent years, cyclic CPPs with improved membrane permeability and intracellular permeability via receptors have been reported [69]. They may become important in the future clinical application of CPP-drug complexes. Activatable CPP derivatives taken up into cells and exert their effects in response to the disease-site environment will become increasingly important, as the development of drugs that are specifically taken up according to the tumor's microenvironment, such as enzyme activity and pH change, is warranted. Tumor-homing peptides have been developed in oncology [57] and are expected to be further applied to the future treatment and diagnosis of cancer. However, further information on the pharmacokinetics and metabolic stability is required.

It should be noted that pharmacokinetics may differ depending on the methods used to determine it. For example, evaluations using radioactive compounds are highly quantitative, whereas autofluorescence may limit the utility of fluorescently labeled compounds. In general, the amount of radiolabeled drugs used as tracers will be much smaller than the actual drug dose unless the two compounds (radiolabeled and unlabeled) are mixed. Therefore, the pharmacokinetics may change depending on the quantity of the compounds administered. In vivo stability tests of each compound are also required to validate the quantified values and/or imaging signals of in vivo studies.

In conclusion, CPPs still have many barriers to their clinical application as therapeutic and diagnostic agents. However, given the recent encouraging research advances, CPP's great potential is expected in this area. To this end, it will be essential to develop CPPs that are stable in the bloodstream and specifically taken up by target-disease cells. Further, optimize the molecular parts to be combined, and select appropriate evaluation methods, including pharmacokinetics.

References

1 Derakhshankhah, H. and Jafari, S. (2018). Cell penetrating peptides: A concise review with emphasis on biomedical applications. *Biomedicine and Pharmacotherapy* 108: 1090–1096.

2 Boisguérin, P., Konate, K., Josse, E. et al. (2021). Peptide-based nanoparticles for therapeutic nucleic acid delivery. *Biomedicines* 9: 583.

3 Collado Camps, E. and Brock, R. (2018). An opportunistic route to success: towards a change of paradigm to fully exploit the potential of cell-penetrating peptides. *Bioorganic and Medicinal Chemistry* 26: 2780–2787.

4 Yokoo, H., Oba, M., and Uchida, S. (2021). Cell-penetrating peptides: emerging tools for mRNA delivery. *Pharmaceutics* 14: 78.

5 Zorko, M., Jones, S., and Langel, Ü. (2022). Cell-penetrating peptides in protein mimicry and cancer therapeutics. *Advanced Drug Delivery Reviews* 180: 114044.

6 Xie, J., Bi, Y., Zhang, H. et al. (2020). Cell-penetrating peptides in diagnosis and treatment of human diseases: from preclinical research to clinical application. *Frontiers in Pharmacology* 11: 697.

7 Rudin, M. and Weissleder, R. (2003). Molecular imaging in drug discovery and development. *Nature Reviews Drug Discovery* 2: 123–131.

8 Willmann, J.K., van Bruggen, N., Dinkelborg, L.M. et al. (2008). Molecular imaging in drug development. *Nature Reviews Drug Discovery* 7: 591–607.

9 Sarko, D., Beijer, B., Garcia Boy, R. et al. (2010). The pharmacokinetics of cell-penetrating peptides. *Molecular Pharmaceutics* 7: 2224–2231.

10 Bullok, K.E., Dyszlewski, M., Prior, J.L. et al. (2002). Characterization of novel histidine-tagged Tat-peptide complexes dual-labeled with ^{99m}Tc-tricarbonyl and fluorescein for scintigraphy and fluorescence microscopy. *Bioconjugate Chemistry* 13: 1226–1237.

11 Polyakov, V., Sharma, V., Dahlheimer, J.L. et al. (2000). Novel Tat-peptide chelates for direct transduction of technetium-99m and rhenium into human cells for imaging and radiotherapy. *Bioconjugate Chemistry* 11: 762–771.

12 Sharma, G., Lakkadwala, S., Modgil, A. et al. (2016). The role of cell-penetrating peptide and transferrin on enhanced delivery of drug to brain. *International Journal of Molecular Sciences* 17: 806.

13 Lev, N., Barhum, Y., Ben-Zur, T. et al. (2015). A DJ-1 based peptide attenuates dopaminergic degeneration in mice models of Parkinson's disease via enhancing Nrf2. *PLoS One* 10: e0127549.

14 Sarkar, G., Curran, G.L., Sarkaria, J.N. et al. (2014). Peptide carrier-mediated non-covalent delivery of unmodified cisplatin, methotrexate and other agents via intravenous route to the brain. *PLoS One* 9: e97655.

15 Chen, S., Cui, J., Jiang, T. et al. (2017). Gelatinase activity imaged by activatable cell-penetrating peptides in cell-based and in vivo models of stroke. *Journal of Cerebral Blood Flow and Metabolism* 37: 188–200.

16 Aleyasin, H., Rousseaux, M.W., Marcogliese, P.C. et al. (2010). DJ-1 protects the nigrostriatal axis from the neurotoxin MPTP by modulation of the AKT pathway. *Proceedings of the National Academy of Sciences of the United States of America USA* 107: 3186–3191.

17 Ahlschwede, K.M., Curran, G.L., Rosenberg, J.T. et al. (2019). Cationic carrier peptide enhances cerebrovascular targeting of nanoparticles in Alzheimer's disease brain. *Nanomedicine* 16: 258–266.

18 Waggoner, L.E., Madias, M.I., Hurtado, A.A. et al. (2021). Pharmacokinetic analysis of peptide-modified nanoparticles with engineered physicochemical properties in a mouse model of traumatic brain injury. *AAPS Journal* 23: 100.

19 Sadeghian, I., Heidari, R., Sadeghian, S. et al. (2022). Potential of cell-penetrating peptides (CPPs) in delivery of antiviral therapeutics and vaccines. *European Journal of Pharmaceutical Sciences* 169: 106094.

20 Zeiders, S.M. and Chmielewski, J. (2021). Antibiotic-cell-penetrating peptide conjugates targeting challenging drug-resistant and intracellular pathogenic bacteria. *Chemical Biology and Drug Design* 98: 762–778.

21 Ruczyński, J., Rusiecka, I., Turecka, K. et al. (2019). Transportan 10 improves the pharmacokinetics and pharmacodynamics of vancomycin. *Scientific Reports* 9: 3247.

22 Figueira, T.N., Augusto, M.T., Rybkina, K. et al. (2018). Effective in vivo targeting of influenza virus through a cell-penetrating/fusion inhibitor tandem peptide anchored to the plasma membrane. *Bioconjugate Chemistry* 29: 3362–3376.

23 Zhang, J., Garrison, J.C., Poluektova, L.Y. et al. (2015). Liver-targeted antiviral peptide nanocomplexes as potential anti-HCV therapeutics. *Biomaterials* 70: 37–47.

24 Umstätter, F., Domhan, C., Hertlein, T. et al. (2020). Vancomycin resistance is overcome by conjugation of polycationic peptides. *Angewandte Chemie - International Edition* 59: 8823–8827.

25 Mühlberg, E., Umstätter, F., Domhan, C. et al. (2020). Vancomycin-lipopeptide conjugates with high antimicrobial activity on vancomycin-resistant enterococci. *Pharmaceuticals* 13: 110.

26 Henry, K.E., Chaney, A.M., Nagle, V.L. et al. (2020). Demarcation of sepsis-induced peripheral and central acidosis with pH (low) insertion cycle peptide. *Journal of Nuclear Medicine* 61: 1361–1368.
27 Kim, J.Y., Ahn, J., Kim, J. et al. (2018). Nanoparticle-assisted transcutaneous delivery of a signal transducer and activator of transcription 3-inhibiting peptide ameliorates psoriasis-like skin inflammation. *ACS Nano* 12: 6904–6916.
28 Gao, S., Tian, B., Han, J. et al. (2019). Enhanced transdermal delivery of lornoxicam by nanostructured lipid carrier gels modified with polyarginine peptide for treatment of carrageenan-induced rat paw edema. *International Journal of Nanomedicine* 14: 6135–6150.
29 Lee, Y.Y., Lee, S.Y., Park, S.Y. et al. (2018). Therapeutic potential of a phospholipase D1 inhibitory peptide fused with a cell-penetrating peptide as a novel anti-asthmatic drug in a Der f 2-induced airway inflammation model. *Experimental and Molecular Medicine* 50: 1–11.
30 Falato, L., Gestin, M., and Langel, Ü. (2021). Cell-penetrating peptides delivering siRNAs: an overview. *Methods in Molecular Biology* 2282: 329–352.
31 Mehrotra, N., Kharbanda, S., and Singh, H. (2020). Peptide-based combination nanoformulations for cancer therapy. *Nanomedicine* 15: 2201–2217.
32 Tan, M., Lan, K.H., Yao, J. et al. (2006). Selective inhibition of ErbB2-overexpressing breast cancer in vivo by a novel TAT-based ErbB2-targeting signal transducers and activators of transcription 3-blocking peptide. *Cancer Research* 66: 3764–3772.
33 Liu, Y., Song, Z., Zheng, N. et al. (2018). Systemic siRNA delivery to tumors by cell-penetrating α-helical polypeptide-based metastable nanoparticles. *Nanoscale* 10: 15339–15349.
34 Stangl, S., Tei, L., De Rose, F. et al. (2018). Preclinical evaluation of the Hsp70 peptide tracer TPP-PEG$_{24}$-DFO[^{89}Zr] for tumor-specific PET/CT imaging. *Cancer Research* 78: 6268–6281.
35 Sauter, M., Strieker, M., Kleist, C. et al. (2020). Improving antibody-based therapies by chemical engineering of antibodies with multimeric cell penetrating peptides for elevated intracellular delivery. *Journal of Controlled Release* 322: 200–208.
36 Miampamba, M., Liu, J., Harootunian, A. et al. (2017). Sensitive in vivo visualization of breast cancer using ratiometric protease-activatable fluorescent imaging agent, AVB-620. *Theranostics* 7: 3369–3386.
37 Ren, Y., Cheung, H.W., von Maltzhan, G. et al. (2012). Targeted tumor-penetrating siRNA nanocomplexes for credentialing the ovarian cancer oncogene ID4. *Science Translational Medicine* 4: 147ra112.
38 Jin, E., Zhang, B., Sun, X. et al. (2013). Acid-active cell-penetrating peptides for in vivo tumor-targeted drug delivery. *Journal of the American Chemical Society* 135: 933–940.
39 Wan, X., Sun, R., Bao, Y. et al. (2021). in vivo delivery of siRNAs targeting EGFR and BRD4 expression by peptide-modified redox responsive PEG-PEI nanoparticles for the treatment of triple-negative breast cancer. *Molecular Pharmaceutics* 18: 3990–3998.
40 Gao, L., Yu, J., Liu, Y. et al. (2018). Tumor-penetrating peptide conjugated and doxorubicin loaded T(1)-T(2) dual mode MRI contrast agents nanoparticles for tumor theranostics. *Theranostics* 8: 92–108.

41 Park, S., Hong, Y., Lee, S. et al. (2020). FCHO1560-571 peptide, a PKB kinase motif, inhibits tumor progression. *Biochemical and Biophysical Research Communications* 528: 478–484.

42 Jaraíz-Rodríguez, M., Talaverón, R., García-Vicente, L. et al. (2021). Connexin43 peptide, TAT-Cx43266-283, selectively targets glioma cells, impairs malignant growth, and enhances survival in mouse models in vivo. *Neuro-Oncology* 22: 493–504.

43 Fuchigami, T., Ishikawa, N., Nozaki, I. et al. (2020). Discovery of inner centromere protein-derived small peptides for cancer imaging and treatment targeting survivin. *Cancer Scence* 111: 1357–1366.

44 Fujiwara, D., Kitada, H., Oguri, M. et al. (2016). A cyclized helix-loop-helix peptide as a molecular scaffold for the design of inhibitors of intracellular protein-protein interactions by epitope and arginine grafting. *Angewandte Chemie - International Edition* 55: 10612–10615.

45 Zhang, Y., Zhang, L., Hu, Y. et al. (2018). Cell-permeable NF-κB inhibitor-conjugated liposomes for treatment of glioma. *Journal of Controlled Release* 289: 102–113.

46 Chen, C., Zhao, S., Karnad, A. et al. (2018). The biology and role of CD44 in cancer progression: therapeutic implications. *Journal of Hematology and Oncology* 11: 64.

47 Zhao, H., Wu, M., Zhu, L. et al. (2018). Cell-penetrating peptide-modified targeted drug-loaded phase-transformation lipid nanoparticles combined with low-intensity focused ultrasound for precision theranostics against hepatocellular carcinoma. *Theranostics* 8: 1892–1910.

48 Gao, P., Mei, C., He, L. et al. (2018). Designing multifunctional cancer-targeted nanosystem for magnetic resonance molecular imaging-guided theranostics of lung cancer. *Drug Delivery* 25: 1811–1825.

49 Collado Camps, E., van Lith, S.A.M., Frielink, C. et al. (2021). CPPs to the test: effects on binding, uptake and biodistribution of a tumor targeting nanobody. *Pharmaceuticals* 14: 602.

50 Jiang, T., Olson, E.S., Nguyen, Q.T. et al. (2004). Tumor imaging by means of proteolytic activation of cell-penetrating peptides. *Proceedings of the National Academy of Sciences of the United States of America* 101: 17867–17872.

51 van Duijnhoven, S.M., Robillard, M.S., Nicolay, K. et al. (2011). Tumor targeting of MMP-2/9 activatable cell-penetrating imaging probes is caused by tumor-independent activation. *Journal of Nuclear Medicine* 52: 279–286.

52 van Duijnhoven, S.M., Robillard, M.S., Nicolay, K. et al. (2015). Development of radiolabeled membrane type-1 matrix metalloproteinase activatable cell penetrating peptide imaging probes. *Molecules* 20: 12076–12092.

53 Fuchigami, T., Itagaki, K., Ishikawa, N. et al. (2019). Synthesis and evaluation of radioactive/fluorescent peptide probes for imaging of legumain activity. *Bioorganic and Medicinal Chemistry Letters* 29: 126629.

54 Jiang, Y., Lu, J., Wang, Y. et al. (2014). Molecular-dynamics-simulation-driven design of a protease-responsive probe for in-vivo tumor imaging. *Advanced Materials* 26: 8174–8178.

55 Unkart, J.T., Chen, S.L., Wapnir, I.L. et al. (2017). Intraoperative tumor detection using a ratiometric activatable fluorescent peptide: a first-in-human phase 1 study. *Annals of Surgical Oncology* 24: 3167–3173.

56 Li, S.Y., Cheng, H., Qiu, W.X. et al. (2015). Protease-activable cell-penetrating peptide-protoporphyrin conjugate for targeted photodynamic therapy in vivo. *ACS Applied Materials and Interfaces* 7: 28319–28329.

57 Kondo, E., Iioka, H., and Saito, K. (2021). Tumor-homing peptide and its utility for advanced cancer medicine. *Cancer Science* 112: 2118–2125.

58 Fogal, V., Zhang, L., Krajewski, S. et al. (2008). Mitochondrial/cell-surface protein p32/gC1qR as a molecular target in tumor cells and tumor stroma. *Cancer Research* 68: 7210–7218.

59 Sugahara, K.N., Teesalu, T., Karmali, P.P. et al. (2010). Coadministration of a tumor-penetrating peptide enhances the efficacy of cancer drugs. *Science* 328: 1031–1035.

60 Hu, C., Huang, Y., and Chen, Y. (2019). Targeted modification of the cationic anticancer peptide HPRP-A1 with iRGD to improve specificity, penetration, and tumor-tissue accumulation. *Molecular Pharmaceutics* 16: 561–572.

61 Lo, J.H., Hao, L., Muzumdar, M.D. et al. (2018). iRGD-guided tumor-penetrating nanocomplexes for therapeutic siRNA delivery to pancreatic cancer. *Molecular Cancer Therapeutics* 17: 2377–2388.

62 Schwarze, S.R., Ho, A., Vocero-Akbani, A. et al. (1999). in vivo protein transduction: delivery of a biologically active protein into the mouse. *Science* 285: 1569–1572.

63 Ishikawa, K., Medina, S.H., Schneider, J.P. et al. (2017). Glycan alteration imparts cellular resistance to a membrane-lytic anticancer peptide. *Cell Chemical Biology* 24: 149–158.

64 Sinthuvanich, C., Veiga, A.S., Gupta, K. et al. (2012). Anticancer β-hairpin peptides: membrane-induced folding triggers activity. *Journal of the American Chemical Society* 134: 6210–6217.

65 Fuchigami, T., Chiga, T., Yoshida, S. et al. (2021). Synthesis and characterization of radiogallium-labeled cationic amphiphilic peptides as tumor imaging agents. *Cancers* 13: 2388.

66 Hu, H., Wang, J., Wang, H. et al. (2018). Cell-penetrating peptide-based nanovehicles potentiate lymph metastasis targeting and deep penetration for anti-metastasis therapy. *Theranostics* 8: 3597–3610.

67 Zhang, Z., Yuan, Y., Liu, Z. et al. (2018). Brightness enhancement of near-infrared semiconducting polymer dots for in vivo whole-body cell tracking in deep organs. *ACS Applied Materials and Interfaces* 10: 26928–26935.

68 Qian, Z., Martyna, A., Hard, R.L. et al. (2016). Discovery and mechanism of highly efficient cyclic cell-penetrating peptides. *Biochemistry* 55: 2601–2612.

69 Dougherty, P.G., Sahni, A., and Pei, D. (2019). Understanding cell penetration of cyclic peptides. *Chemical Reviews* 119: 10241–10287.

Part III

Delivery Tool

12

Drug Delivery

Takashi Misawa

Department of Organic Chemistry, National Institute of Health Sciences, Japan

12.1 Introduction

Drug discovery research has been focused on small molecules due to their specificity, wide range of target molecules, and prominent in vivo half-lives. The water solubility of drugs is an important factor in determining the pharmacokinetics involved in drug discovery, although high hydrophilicity is disadvantageous for intracellular uptake through the cell membrane [1]. Therefore, even compounds with excellent biological activity may not be able to exert their activity due to a lack of cell membrane permeability. In this case, the development of methods to deliver hydrophilic molecules into cells may facilitate biological activity, improve their pharmacokinetics, and lead us to reinvent the therapeutic reagent based on these compounds, which were neglected due to their poor cell-penetrating ability. In the past few decades, cell-penetrating peptides (CPPs) have attracted attention as delivery tools for hydrophilic drugs [2–4]. CPPs are efficiently internalized into living cells with cargo molecules without causing damage to the cell membranes. It has been reported that several cargo molecules, such as fluorescein and bioactive compounds, are delivered into living cells using CPPs (Figure 12.1). To apply this mechanism to drug delivery, it is necessary for the drug and CPPs to form conjugates or complexes. Currently, many studies on drug-CPP conjugates for intracellular delivery have been reported, as shown in Table 12.1. The combination of CPPs and cargo molecules is important for efficient intracellular delivery, and the linker between CPPs and cargo molecules plays a pivotal role in exerting their biological functions. The application of CPPs in the intracellular delivery of hydrophilic molecules is presented in this chapter.

Cell-Penetrating Peptides: Design, Development and Applications, First Edition. Makoto Oba and Yosuke Demizu.

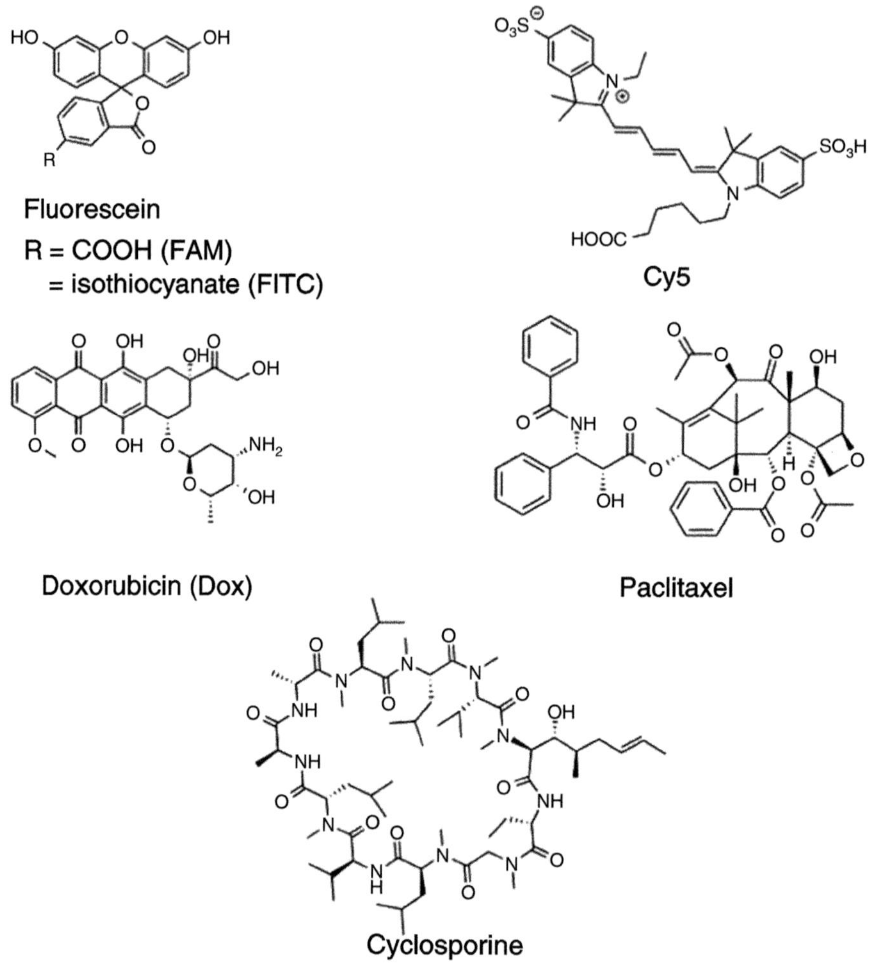

Figure 12.1 The chemical structure of representative cargo molecules.

12.2 Delivery by Arg-Rich Peptides

Arg-rich peptides such as oligoarginine (R_n) are representative CPPs that have been widely utilized for drug delivery. Arginine residues of oligoarginine interact with acidic functional groups present on the cell membrane and induce endocytosis of CPPs and cargo molecules [27, 28]. Fluorescent molecules such as 5- or 6-carboxyfluorescein (FAM) and Cy5 are the most commonly used drugs for drug delivery. Fluorescent molecules are hydrophilic and cannot penetrate the cell membrane. Therefore, fluorescent molecules are conjugated to CPPs and used to ascertain cell membrane permeability and the mechanism of action of intracellular uptake [29]. Futaki et al. reported that the Alexa660-labeled D-form of octaarginine (r8) accumulated in tumor-xenografts. Hence, the delivery of antitumor reagents using oligoarginine is a promising strategy for cancer therapy [27]. Doxorubicin

Table 12.1 List of cell-penetrating peptides for intracellular delivery of drugs.

Classification	Peptides	Cargos	Cell	Ref.
Arg-rich	Oligoarginine	Alexa660	CHOK1	[5]
		Dox	HeLa	[5, 6]
		Cyclosporin A	Jurkat	[7]
Cationic	Tat	FAM	KB-V1, MCF-7	[8]
		Dox	KB-V1, MCF-7	[8]
		Deferoxamine	HeLa, BRE4	[9]
		Fallypride	hCMEC/D3	[10]
	SynB	Dox	BBB penetration	[11, 12]
		Benzylpenicillin	BBB penetration	[23, 24]
Amphipathic	Penetratin	Paclitaxel	MCF7, A549	[13]
	M1	Paclitaxel	Tissue section	[14]
	pVEC	Dox	HeLa U87	[15]
	cW(RW)$_4$K	Dox	HCT116	[16]
Hydrophobic	AA3H	FITC	HeLa	[17]
	Transportant	Vancomycin	Erythrocyte	[18]
		Ciprofloxacin	HepG2, HEK293	[19]
		Levofloxacin	HepG2, HEK293	[19]
Foldamer	dAAs type	FAM	CHO-K1, HeLa, Huh-7, MCF-7	[20–23]
			HeLa	[24–26]
	β-AAs type	FAM		

(Dox) is an anthracycline antitumor antibiotic that suppresses the biosynthesis of DNA and RNA by inserting into the DNA base pairs of tumor cells, inhibiting DNA polymerase, RNA polymerase, and topoisomerase II reactions. Although Dox is broadly used as a therapeutic reagent for breast cancer, the lack of selectivity against target tissues leads to toxicity against normal tissues, such as cardiotoxicity [30]. Additionally, low extracellular pH, a characteristic of several tumors, reduces the intracellular uptake of basic compounds such as Dox, thereby promoting resistance of cancer cells to Dox [31]. Hence, Futaki et al. investigated the effects of the conjugation of r8 to Dox on antitumor activity [5]. A cross-linker was formed between the thiol residue of the C-terminal cysteine and the 13-keto position of Dox using maleimidopropionic acid hydrazone. Hydrazone can be cleaved under acidic conditions, resulting in the release of Dox (Figure 12.2). A study on antitumor activity revealed that r8-conjugated Dox showed higher cytotoxicity against HeLa cells compared to that of co-treatment with r8 and Dox, and exhibited significant antitumor activity in vivo. Alternatively, Peneva et al. reported that R8-conjugated Dox via a disulfide linker could be cleaved under reductive conditions [6]. Their compound was internalized into living cells through the endocytosis pathway of CPPs, and the disulfide bond was cleaved with glutathione within a short period ($t < 5$ hours), releasing Dox. These studies demonstrated the advantages of the conjugation of CPPs to cargo molecules, such as enhancement of

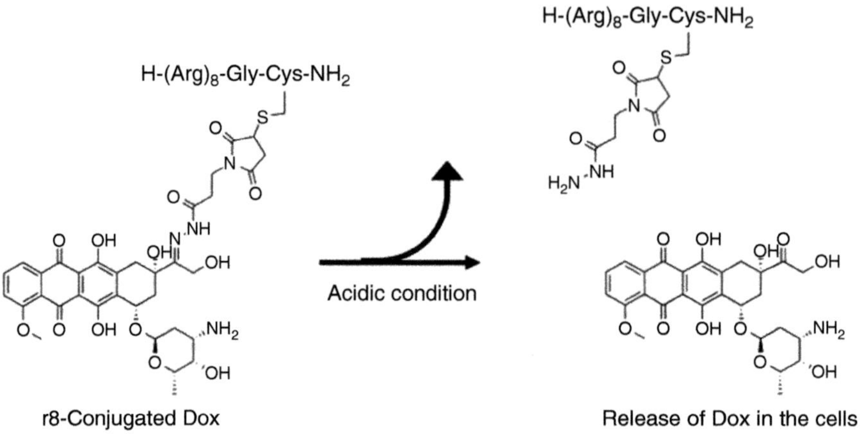

Figure 12.2 The cleavable linker between octa-D-arginine (r8) and doxorubicin and the mechanism of action.

antitumor activity, and can be applicable as the tumor-targeting vector in vivo experiment.

Cyclosporin A (CsA), a calcineurin inhibitor, is widely used as an immunosuppressant. Orally administrated CsA can be absorbed and is effective for a broad range of skin diseases. However, systematic administration is restricted due to its side effects, such as nephrotoxicity [32]. Therefore, the topical application of CsA against several skin disorders was investigated, although the trial failed due to the poor penetration of CsA into the skin. Hence, Khavari et al. reported that conjugation of the heptamer of L- or D-arginine (R7 or r7, respectively) to CsA enhanced penetration into the skin and exerted immunosuppressive activity in Jurkat cells and in vivo [7]. Modification of the secondary alcohol of CsA could decrease its biological activity; therefore, the pH-sensitive cleavable linker was adopted as a crosslink between r7 and CsA (Figure 12.3). The cleavable mechanism is shown in Figure 12.3, whereby the secondary amine of the cleavable linker attacks the ester at

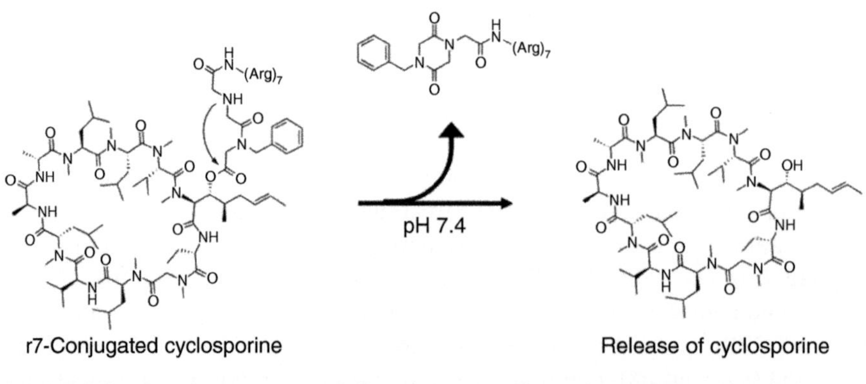

Figure 12.3 Chemical structure of r7-conjugated cyclosporine and cleavage mechanism.

neutral pH, resulting in the release of CsA. The half-life of the cleavable linker was 90 minutes in a PBS solution at pH 7.4. The biological activity of r7-conjugated CsA was investigated using an interleukin 2 (IL-2) secretion assay in Jurkat cells. Treatment with r7-conjugated CsA inhibited IL-2 secretion in a dose-dependent manner, whereas the non-releasable analog did not show inhibitory activity, indicating that the releasing linker plays an important role in exerting inhibitory activity. Treatment with releasable r7-conjugated CsA exhibited anti-inflammatory activity in a mouse model, indicating that the action was local and not systematic.

12.3 Delivery by Cationic Peptides

The most widely used CPP for drug delivery is the Tat peptide, which is derived from HIV-1 Tat proteins [33]. Tat peptide is composed of arginine residues and internalized into cells via endocytosis, similar to oligoarginine. To investigate its intracellular uptake, fluorescence-labeled Tat peptide, Tat-conjugated Dox, and oligoarginine were synthesized [8]. Additionally, it has been reported that small molecules such as deferoxamine [9] and fallypride [10] could be delivered by conjugation of Tat peptides (Figure 12.4). These studies showed that the conjugation of Tat to fluorescein or Dox contributed to efficient intracellular delivery of water-soluble molecules and facilitated antitumor activity, respectively.

The delivery of hydrophilic molecules to the central nervous system (CNS) remains challenging in the field of medicinal chemistry [34]. Hydrophilic molecules show low permeability through the blood-brain barrier (BBB), making it difficult to develop therapeutics for CNS diseases such as Alzheimer's disease. CPPs have been proposed as carriers of hydrophilic molecules in the CNS. In 1999, Schwarze and Dowdy showed that conjugation of the Tat peptide to proteins allows for the transportation of proteins to the brain via the BBB, where Tat-conjugated proteins could exert their biological functions [35]. Based on the BBB penetrating ability, CPP conjugation with drugs is considered a promising approach to delivering cargo to the CNS through the BBB. In contrast, SynB [36], which is derived from the natural antimicrobial peptide (AMP) protegrin 1, is the most widely studied CPP carrier in the CNS. Conjugation of SynB to several antitumor drugs, such as Dox [11, 12], and benzylpenicillin [37, 38], enhances uptake of antitumor drugs in the brain (Figure 12.5). These studies demonstrated that the conjugation of Tat peptide to hydrophilic molecules is a promising strategy for the delivery of hydrophilic cargo molecules to the brain.

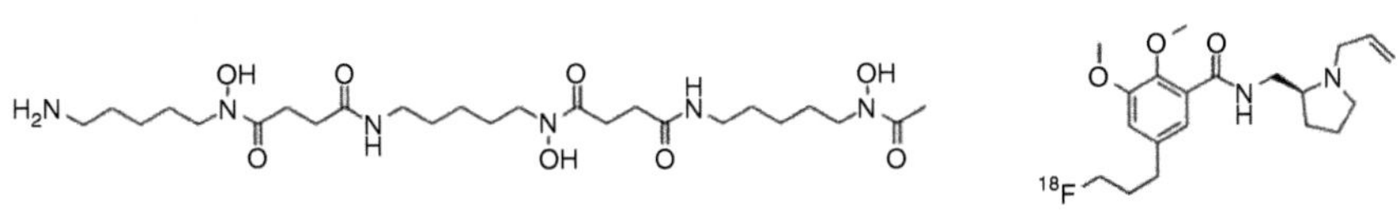

Deforoxamine Fallypride

Figure 12.4 Chemical structure of deferoxamine and fallypride.

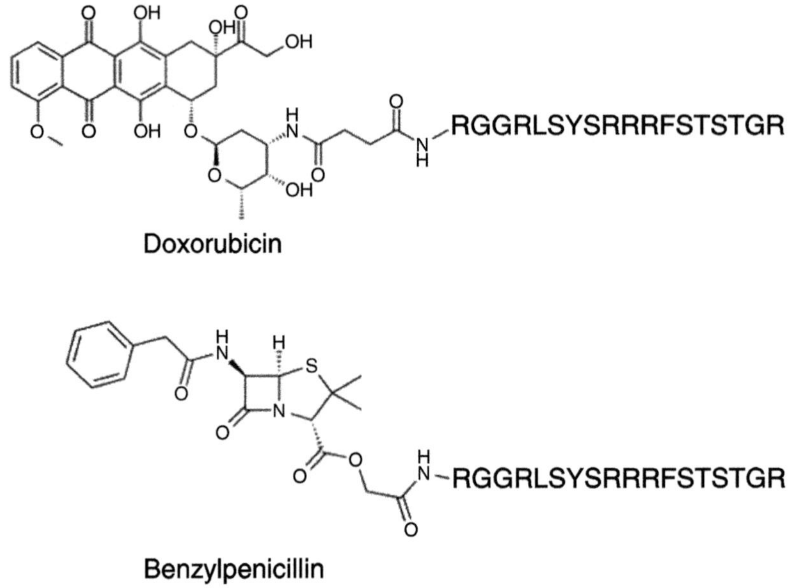

Figure 12.5 The chemical structure of SynB-conjugated doxorubicin and benzylpenicillin.

12.4 Delivery by Amphipathic Peptides

Many amphipathic CPPs were originally identified as AMPs. These molecules interact with the cell membrane via electrostatic interactions between cationic residues of amphipathic peptides and acidic groups. These peptides can interact with the cell membrane based on their hydrophobic residues. Matsuzaki et al. reported that buforin II, which is a representative AMP, showed cell-penetrating activity and is a potential agent for drug delivery [39]. These efforts led us to develop CPPs based on AMPs. Up to date, several amphipathic CPPs such as penetratin, M1 peptide, and pVEC, have been developed and their application for intracellular delivery of paclitaxel and doxorubicin [13–15].

Parang et al. developed a tool to deliver Dox based on cyclic CPPs, which are composed of Arg and Trp residues [W(RW)$_4$] [16]. To investigate the effect of the secondary structure of CPPs on cell-penetrating ability, they synthesized cyclic [W(RW)$_4$] and linear (RW)$_4$ Dox-conjugates, and evaluated their intracellular uptake and antiproliferative activity against several cancer cell lines (Figure 12.6). Cyclic [W(RW)$_4$]-Dox conjugate showed higher antiproliferative activity against HCT-116 cells compared to that of Dox and linear (RW)$_4$-Dox conjugate. Additionally, cyclic [W(RW)$_4$]-Dox conjugate showed resistance against digestive enzymes and longer retention in the cells than Dox alone, indicating that the conjugation of cyclic [W(RW)$_4$] Dox enhanced its accumulation in the cells and promoted its long-lasting effects.

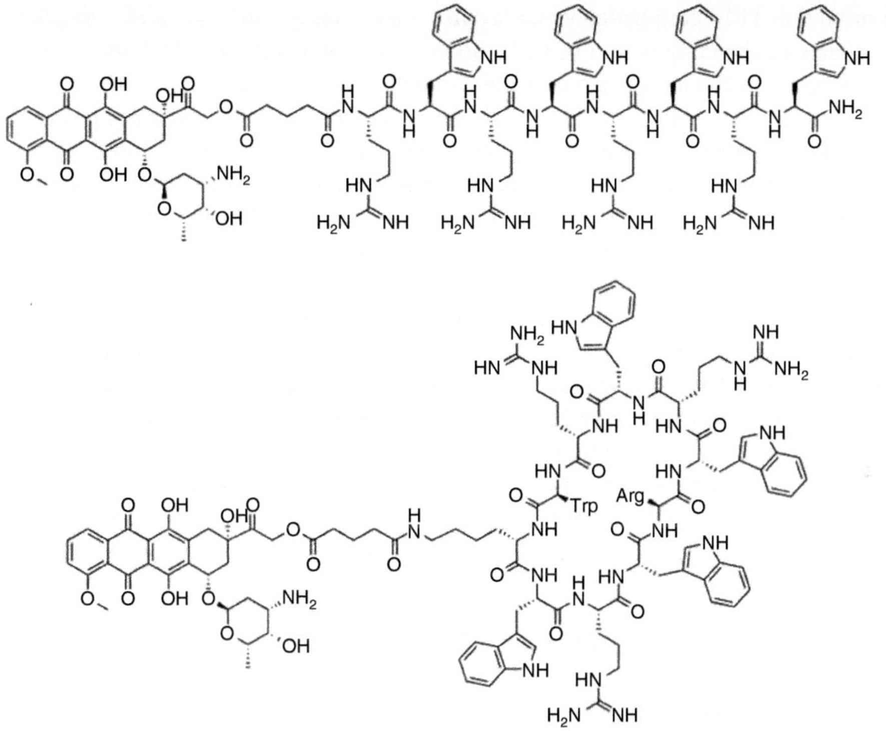

Figure 12.6 Chemical structure of linear $(RW)_4$ and cyclic $[W(RW)_4]$ peptide conjugates with doxorubicin.

12.5 Delivery by Hydrophobic Peptides

Generally, representative CPPs are composed of cationic amino acids such as Lys and Arg, which are important for their cell-penetrating ability. However, these positively charged peptides are prone to interact with the biological milieu and are easily degraded [40]. Thus, a non-cationic CPP (AA3H) derived from annexin, a membrane-interacting human protein, has been reported. AA3H peptides show high hydrophobicity and possess one cationic residue [17]. AA3H-conjugated fluorescein isothiocyanate (FITC) exhibited highly potent cell-penetrating ability via endocytosis and resistance against digestive enzymes. Furthermore, AA3H can deliver cargo molecules even in vivo, suggesting that AA3H could be a candidate carrier for the delivery of macromolecular cargo.

Transportan 10 (TP10) [41], which is derived from the *N*-terminus of the neuropeptide galanin and mastoparan in the *C*-terminus, has potent cell membrane permeability. Mechanistic analysis revealed that TP10 can internalize into living cell independently, and is not restricted by cell type [42]. Therefore, TP10 is utilized as a delivery tool for cargo molecules in a wide range of applications. Resiecka et al.

reported that TP10-conjugated vancomycin via a tetraethylene linker and evaluated the effects of the conjugation of CPPs on antimicrobial activity [18]. Vancomycin was attached to the amino group at the ε-position of the Lys7 residue. The TP10-conjugated vancomycin showed higher antimicrobial activity against gram-positive and gram-negative microbes than that of vancomycin alone, and intracellular methicillin-resistant *Staphylococcus aureus* (MRSA). Furthermore, TP10-conjugated vancomycin possessed BBB penetrating activity and accumulated in the mouse brain (Figure 12.7). In contrast, Lica et al. investigated the effects of TP10-conjugation to ciprofloxacin and levofloxacin, which are fluoroquinolone antibiotics used for the treatment of a number of bacterial infections, on their antibacterial and antifungal activities (Figure 12.8) [19]. Their study revealed that the conjugation of TP10 with ciprofloxacin and levofloxacin decreased the antibacterial activity, whereas TP10-conjugates showed strong antifungal activity against *C. albicans* and could be internalized into mammalian cells such as HEK293, HepG2, and LLC-PK1. The conjugation of CPPs with ciprofloxacin and levofloxacin substantially changed the biological properties of these antibiotics from antimicrobial to antifungal and cytotoxic activities.

12.6 Delivery by Foldamers

Foldamers are oligomers with organized secondary structures that are broadly utilized as organocatalysts [43–45], materials [46, 47], and in medicinal chemistry [48, 49]. Foldamers exploit the stability of bioactive compounds against digestive enzymes and enhance their biological activities. Based on these properties, the development of novel CPP foldamers using several building blocks, such as α,α-disubstituted α-amino acid (dAAs), β-amino acids (β-AAs), and other non-natural amino acid residues, have been reported [50, 51]. Oba et al. reported that the CPP foldamers, composed of L-Arg and several dAA residues such as 2-aminoisobutyric acid (Aib) or 1-aminocyclopentanecarboxyric acid (Ac_5c), showed the efficient intracellular uptake in CHO-K1 cells [20, 21]. Furthermore, Demizu et al. reported the CPP containing Api^{C2Gu} residues, which have a guanidyl group on the piperidine ring, showed potent cell-membrane permeability (Figure 12.9) [22, 23]. On the other hand, Gellman et al. demonstrated that the introduction of cyclic β-AAs residues 2-aminocyclohexylcarboxylic acid (ACHC) into oligoarginine enhanced the intracellular uptake into HeLa cells (Figure 12.10) [24]. These foldamers showed high cell-penetrating ability and resistance to digestive enzymes, resulting in the efficient delivery of cargo molecules. Fluorescein is used for the detection of intracellular uptake and is the main target molecule for delivery into cells using CPP foldamers. The high cell membrane permeability and in vivo stability of foldamers can be utilized in the development of drug delivery systems.

12.6.1 Outlook and Future Direction

CPP-drug conjugates are a promising strategy for target-specific and nontoxic therapy against a variety of diseases. Early research efforts on drug delivery using

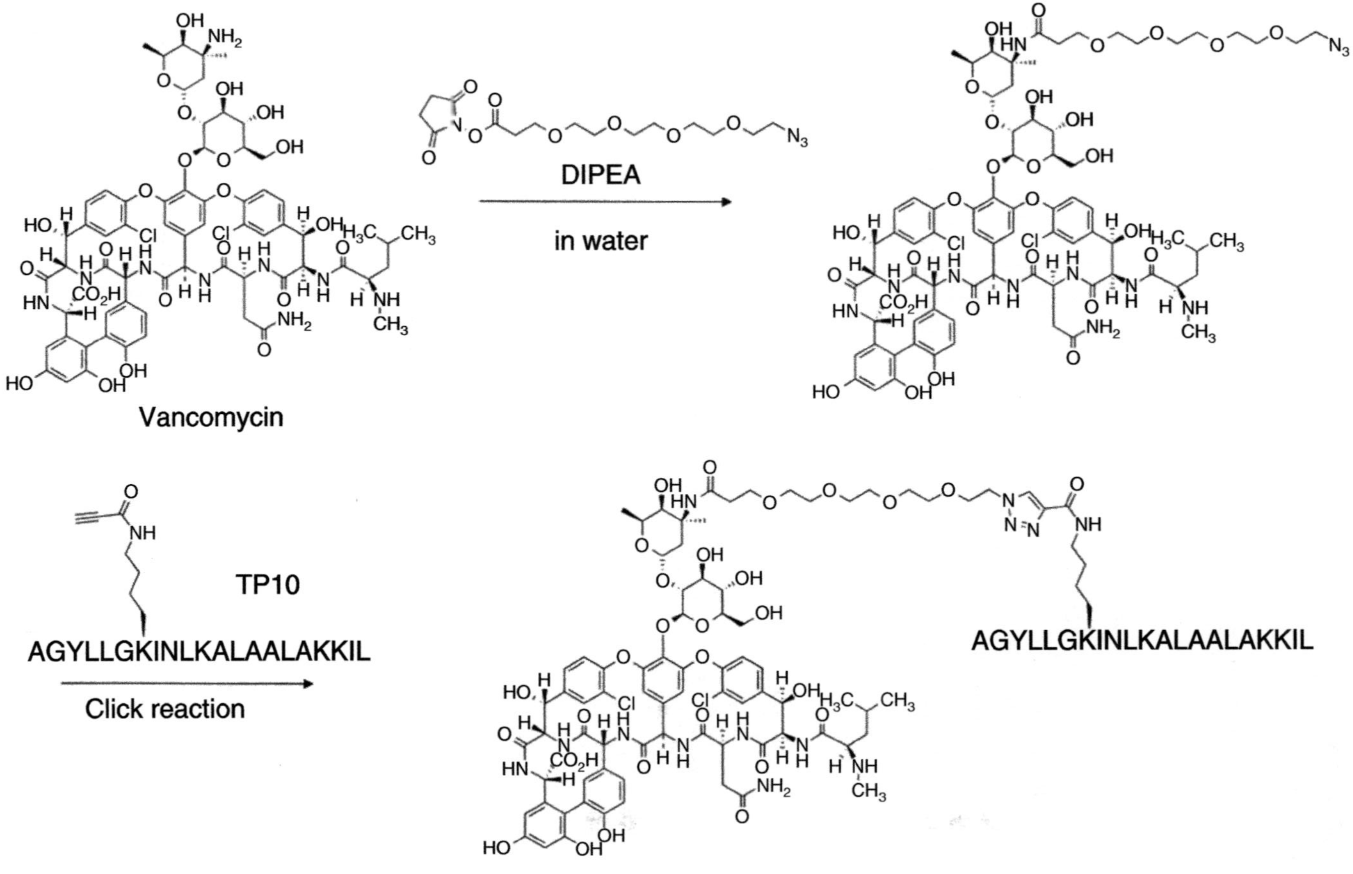

Figure 12.7 Synthesis of TP10-conjugated va ıcomycin.

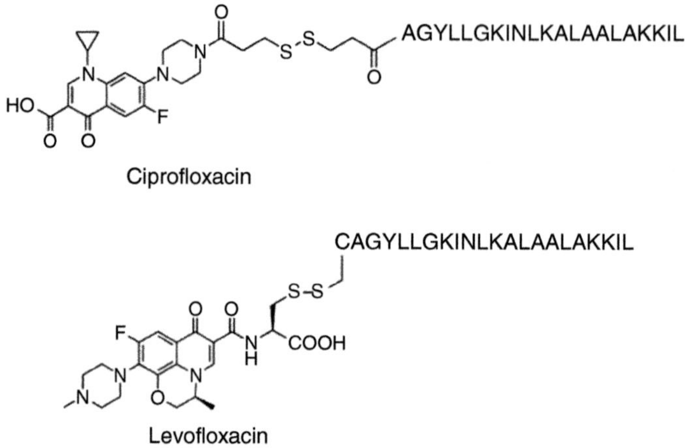

Figure 12.8 Chemical structure of TP10 conjugated with ciprofloxacin and levofloxacin via the disulfide bond.

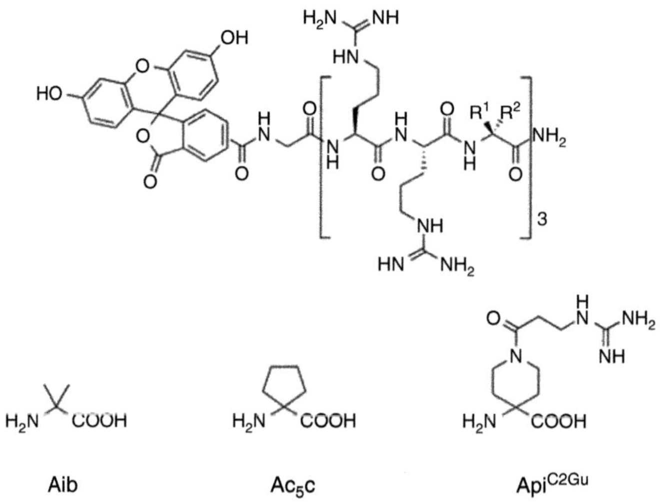

Figure 12.9 The chemical structure of CPP foldamers based on α,α-disubstituted amino acid such as Aib, Ac5c, and Api^{C2Gu}.

CPPs have predominantly focused on some motifs of natural proteins and the delivery of fluorescein to detect intracellular uptake. Recently, several hydrophilic drugs, such as Dox and paclitaxel, have been efficiently delivered into living cells and the CNS via conjugation with CPPs. CPP-drug conjugates are capable of reinventing drugs using compounds which are otherwise ineffective due to their poor cell-penetrating ability. The discovery of various CPPs and the development of foldamers will enable the delivery of a variety of drugs to target tissues for improved treatment efficiency in the future.

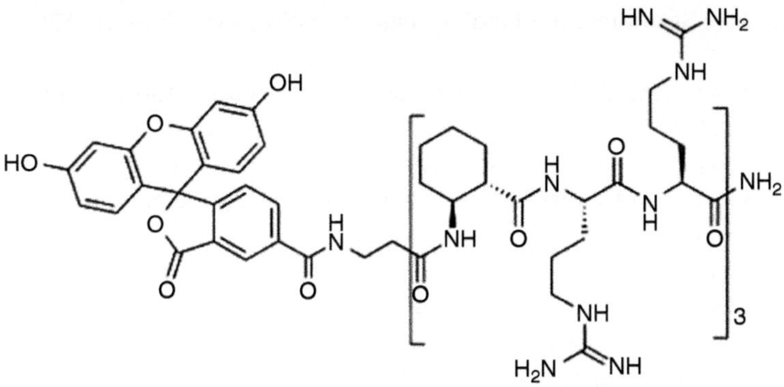

Figure 12.10 The chemical structure of β-peptide foldamers containing ACHC residues.

References

1 Bennion, B.J., Be, N.A., McNerney, M.W. et al. (2017). Predicting a drug's membrane permeability: a computational model validated with *in vitro* permeability assay data. *The Journal of Physical Chemistry B* 121: 5228–5237.

2 Svensen, N., Walton, J.G.A., and Bradly, M. (2012). Peptides for cell-selective drug delivery. *Trends in Pharmacological Sciences* 33: 186–192.

3 Korivi, M., Huang, Y.W., and Liu, B.R. (2021). Cell-penetrating peptides as a potential drug delivery system for effective treatment of diabetes. *Current Pharmaceutical Design* 27: 816–825.

4 Kurrlkoff, K., Vunk, B., and Langel, Ü. (2021). Status update in the use of cell-penetrating peptides for the delivery of macromolecular therapeutics. *Expert Opinion on Biological Therapy* 21: 361–370.

5 Nakase, I., Konishi, Y., Ueda, M. et al. (2012). Accumulation of arginine-rich penetrating peptides in tumors and the potential for anticancer drug delivery in vivo. *Journal of Controlled Release* 159: 181–188.

6 Lelle, M., Frick, S.U., Steinbrink, K. et al. (2014). Novel cleavable cell-penetrating peptide-drug conjugates: synthesis and characterization. *Journal of Peptide Science* 20: 323–333.

7 Rothbard, J.B., Garlington, S., Lin, Q. et al. (2000). Conjugation of arginine oligomers to cyclosporin A facilitates topical delivery and inhibition of inflammation. *Nature Medicine* 6: 1253–1257.

8 Zhang, P., Cheetham, A.G., Lock, L.L. et al. (2013). Cellular uptake and cytotoxicity of drug-peptide conjugates regulated by conjugation site. *Bioconjugate Chemistry* 24: 604–613.

9 Goswami, D., Machini, M.T., Silvestre, D.M. et al. (2014). Cell penetrating peptide (CPP)-conjugated desferrioxamine for enhanced neuroprotection: synthesis and in vitro evaluation. *Bioconjugate Chemistry* 25: 2067–2080.

10 Wängler, C., Chowdhury, S., Höfner, G. et al. (2014). Shuttle-cargo fusion molecules of transport peptides and the hD 2/3 receptor antagonist fallypride: a feasible approach to preserve ligand-receptor binding? *Journal of Medicinal Chemistry* 57: 4368–4381.

11 Rousselle, C., Smirnova, M., Clair, P. et al. Enhanced delivery of doxorubicin into the brain via a peptide-vector-mediated strategy: saturation kinetics and specificity. *Journal of Pharmacology and Experimental Therapeutics* 296: 124–131.

12 Adenot, M., Merida, P., and Lahana, R. (2007). Application of a blood-brain barrier technology platform to predict CNS penetration of various chemotherapeutic agents. *Chemotherapy* 53: 73–76.

13 Wang, S., Susan Duff, N.Z., and Fischer, P.M. (2006). Synthesis and biological activity of conjugates between paclitaxel and the cell delivery vector penetratin. *Bioorganic & Medicinal Chemistry Letters* 16: 2628–2631.

14 Zhang, Y., Guo, P., Ma, Z. et al. (2021). Combination of cell-penetrating peptides with nanomaterials for the potential therapeutics of central nervous system disorders: a review. *Journal of Nanobiotechnology* 19: 255.

15 Eriste, E., Kurrikoff, K., Suhorutšenko, J. et al. (2013). Peptide-based glioma-targeted drug delivery vector gHoPe2. *Bioconjugate Chemistry* 24: 305–313.

16 Shirazi, A.N., Tiwari, R., Chhikara, B.S. et al. (2013). Design and biological evaluation of cell-penetrating peptide-doxorubicin conjugates as prodrugs. *Molecular Pharmaceutics* 10: 488–499.

17 Kim, H.Y., Yum, S.Y., Jang, G. et al. (2015). Discovery of a non-cationic cell penetrating peptide derived from membrane-interacting human proteins and its potential as a protein delivery carrier. *Scientific Reports* 5: 11719.

18 Ruczyński, J., Rusiecka, I., Turecka, K. et al. (2019). Transportan 10 improves the pharmacokinetics and pharmacodynamics of vancomycin. *Scientific Reports* 9: 3247.

19 Ptasyńska, N., Gucwa, K., Olkiewicz, K. et al. (2020). Conjugates of ciprofloxacin and levofloxacin with cell-Penetrating peptide exhibit antifungal activity and mammalian cytotoxicity. *International Journal of Molecular Sciences* 21: 4696.

20 Oba, M., Kunitake, M., Kato, T. et al. (2017). Enhanced and prolonged cell-penetrating abilities of arginine-rich peptides by introducing cyclic α,α-disubstituted α-amino acids with stapling. *Bioconjugate Chemistry* 28: 1801–1806.

21 Oba, M., Nagano, Y., Kato, T. et al. (2019). Secondary structures and cell-penetrating abilities of arginine-rich peptide foldamers. *Scientific Reports* 9: 1349.

22 Yamashita, H., Oba, M., Misawa, T. et al. (2016). A helix-stabilized cell-penetrating peptide as an intracelular delivery tool. *Chembiochem* 17: 137–140.

23 Misawa, T., Ohoka, N., Oba, M. et al. (2019). Development of 2-aminoisobutyric acid (Aib)-rich cell-penetrating foldamers for efficient siRNA delivery. *Chemical Communications Journal* 55: 7792–7795.

24 Potocky, T.B., Menon, A.K., and Gellman, S.H. (2005). Effects of conformational stability and geometry of guanidinium display on cell entry by β-peptides. *Journal of the American Chemical Society* 127: 3686–3687.

25 Potocky, T.B., Menon, A.K., and Gellman, S.H. (2003). Cytoplasmic and nuclear delivery of a TAT-derived peptide and a β-peptide after endocytic uptake into HeLa cells. *Journal of Biological Chemistry* 278: 50188–50194.

26 Demizu, Y., Oba, M., Okitsu, K. et al. (2015). A preorganized β-amino acid bearing a guanidinium side chain and its use in cell-penetrating peptides. *Organic and Biomolecular Chemistry* 13: 5617–5620.

27 Suzuki, T., Futaki, S., Niwa, M. et al. (2002). Possible existance of common internalization mechanisms among arginine-rich peptides. *Journal of Biological Chemistry* 277: 2437–2443.

28 Rothbard, J.B., Jessop, T.C., Lewis, R.S. et al. (2004). Role of membrane potential and hydrogen bonding in the mechanism of translocation of guanidinium-rich peptides into cells. *Journal of the American Chemical Society* 126: 9506–9507.

29 Morishita, M., Kamei, N., Ehara, J. et al. (2007). A novel approach using functional peptides for efficient intestinal absorption of insulin. *Journal of Controlled Release* 118: 177–184.

30 Zhao, L. and Zhang, B. (2017). Doxorubicin induces cardiotoxicity through upregulation of death receptors mediated apoptosis in cardiomyocytes. *Scientific Reports* 7: 44735.

31 Swietach, P., Hulikova, A., and Patiar, S. (2012). Importance of intracellular pH in determining the uptake and efficacy of the weakly basic chemotherapeutic drug, Doxorubicin. *PLoS One* 7: e35949.

32 Busauschina, A., Schnuelle, P., and van der Woude, F.J. (2004). Cyclosporin nephrotoxicity. *Transplantation Proceedings* 36: 229S–233S.

33 Koren, E., Apte, A., Sawant, R.R. et al. (2011). Cell-penetrating TAT peptide in drug delivery systems: proteolytic stability requirements. *Drug Delivery* 18: 377–384.

34 Kimura, S. and Harashima, H. (2020). Current status and challenges associated with CNS-targeted gene delivery across the BBB. *Pharmaceutics* 12: 1216.

35 Schwarze, S.R., Ho, A., Vocero-Akbani, A. et al. (1999). in vivo protein transduction: delivery of a biological active protein into the mouse. *Science* 285: 1569–1572.

36 Drin, G., Cottin, S., Blanc, E. et al. (2003). Studies on the internalization mechanism of cationic cell-penetrating peptides. *Journal of Biological Chemistry* 278: 31192–31201.

37 Razzak, R.A., Florence, G.J., and Gunn-Moore, F.J. (2019). Approaches to CNS drug delivery with a focus on transporter-mediated transcytosis. *International Journal of Molecular Sciences* 20: 3108.

38 Rousselle, C., Clair, P., Temsamani, J. et al. (2002). Improved brain delivery of benzylpenicillin with a peptide-vector-mediated strategy. *Journal of Drug Targeting* 10: 309–315.

39 Park, C.B., Yi, K.S., Matsuzaki, K. et al. (2000). Structure-activity analysis of bufforin II, a histone H2A-derived antimicrobial peptide: the proline hinge is responsible for the cell-penetrating ability of buforin II. *Proceedings of the National Academy of Sciences of the United States of America* 97: 8245–8250.

40 Palm, C., Jayamanne, M., Kjellander, M. et al. (2007). Peptide degradation is a critical determinant for cell-penetrating peptide uptake. *Biochimica et Biophysica Acta* 1768: 1769–1776.

41 Soomets, U., Lindgren, M., Gallet, X. et al. (2000). Deletion analogues of transportan. *Biochimica et Biophysica Acta* 1467: 165–176.

42 Ruseska, I. and Zimmer, A. (2020). Internalization mechanisms of cell-penetrating peptides. *Beilstein Journal of Nanotechnology* 11: 101–123.

43 Girvin, Z.C., Andrewsa, M.K., Liua, X. et al. (2019). Foldamer-templated catalysis of macrocycle formation. *Science* 66: 1528–1531.

44 Bécart, D., Diemer, V., Salaün, A. et al. (2017). Helical oligourea foldamers as powerful hydrogen bonding catalysts for enantioselective C–C bond-forming reactions. *Journal of the American Chemical Society* 139: 12524–12532.

45 Demizu, Y., Yamagata, N., Nagoya, S. et al. (2011). Enantioselective epoxidation of α,β-unsaturated ketones catalyzed by stapled helical l-Leu-based peptides. *Tetrahedron* 67: 6155–6165.

46 Sxefczyk, M. (2021). Peptide foldamer-based self-assembled nanostructures containing cyclic beta-amino acids. *Nanoscale* 13: 11325–11333.

47 Kulkarni, K., Habila, N., Del Borgo, M.P. et al. (2019). Novel materials from the supramolecular self-assembly of short helical β^3-peptide foldamers. *Frontiers in Chemistry* 7: 70.

48 Gopalakrishnan, R., Frolov, A.I., Knerr, L. et al. (2016). Therapeutic potential of foldmaers: from chemical biology tools to drug candidates? *Journal of Medicinal Chemistry* 59: 9599–9621.

49 Pasco, M., Dolain, C., and Guichard, G. (2017). *Comprehensive Supramolecular Chemistry II*, 89–125. Elsevier.

50 Oba, M. (2019). Cell-penetrating peptide foldamers: drug-delivery tools. *Chembiochem* 20: 2041–2045.

51 Yokoo, H., Misawa, T., and Demizu, Y. (2020). De novo design of cell-penetrating foldamers. *The Chemical Record* 20: 912–921.

13

Peptide and Protein Delivery

Hidetomo Yokoo

Kyoto Prefectural University of Medicine, Graduate School of Medical Science, Medical Chemistry, Kyoto, Japan

13.1 Introduction

Bioactive macromolecules such as proteins, peptides, and genetic material have the potential for therapeutic applications [1]. However, the cell membrane does not allow the internalization of large molecules, which makes them unusable as drugs by themselves [2]. To deliver such membrane-impermeable molecules into cells, cell-penetrating peptides (CPPs) have been utilized as delivery systems for various biological cargos. Proteins were considered attractive cargos for therapeutics [3, 4] due to their crucial roles in most biological processes, such as signal transduction, gene transcription and translation, cell proliferation, and apoptosis [5, 6]. The trans-activator of transcription (Tat) protein was discovered as a shuttling protein from the human immunodeficiency 1 (HIV-1) virus to allow entry into cells and translocation into the nucleus [7]. Its 48–57 amino acid peptide is rich in arginine (Arg) and lysine (Lys) residues and is generally considered the minimal sequence for the transduction of Tat protein and is known as the Tat peptide [7, 8]. Since the first application of the Tat peptide for the delivery of proteins of interest in the 1990s, conjugation of the Tat peptide to a protein has often been used for protein delivery into cultured cells and in vivo [9].

With the evolution of CPP-based delivery, numerous studies into de novo CPP design have been carried out to improve their cell penetration and the efficiency of the delivery of bioactive proteins, and some CPPs were designed by mimicking the functional peptide moiety of proteins to control localization, translocation, and transmembrane penetration. To date, conjugates of various CPPs to proteins have been reported to enter cells via several intracellular mechanisms depending on the CPP through endosomal processes and others [10]. Even in last three years, various proteins have been delivered by CPPs, as shown in Table 13.1, for example, green fluorescent protein (GFP), I-SceI restriction enzyme, and Cre recombinases [11], synthetic ubiquitin (Ub), and small Ub-like modifier-2 (SUMO-2) [12], clustered

Cell-Penetrating Peptides: Design, Development and Applications, First Edition. Makoto Oba and Yosuke Demizu.

Table 13.1 Summary of representative cell-penetrating peptides for protein delivery.

Name	Sequence (peptide type)	Cargo	Conjugating method	Delivery site	Ref.
Tat48–57 Tat49–57	GRKKRRQRRR RKKRRQRRR (cationic peptide)	GFP, I-SceI restriction enzyme and Cre recombinase	Covalent (expression)	Cytoplasm	[11]
cR10D	Cyclo(RRRRRRRRRR)-Z[a] (cyclic cationic peptide)	Synthetic ubiquitin and small ubiquitin-like modifier-2 (SUMO-2)	Covalent (S-S)	Cytoplasm, nucleus, and nucleoli	[12]
CPP-DEC	FITC-RRRRRRRRRRK(DEC)[b] (cationic peptide)	Cas9	Covalent (S-S, linker: dithiol-ethyl carbonate combined with a benzyl carbamate)	Striatum distribution in mice brains after injection	[13]
Tat	RKKRRQRRR (cationic peptide)	Fas-associated death domain (FADD)	Covalent (SMCC[c])	Cytoplasm	[14]
cys(Npys)-(D-R)$_9$ cys(Npys)-(BP100)2K$_8$	RRRRRRRRRC(nitropyridyl) (cationic peptide) KKLFKKILKYLKKLFKKILKYLKKKKKKKKC(nitropyridyl) (cationic peptide)	Zinc finger nucleases	Covalent (S-S)	Inside of wheat microspores or embryos	[15]
cR10	PEG2-PEG2- cyclo(KRrRrRrRrRrE)[a] (cationic peptide)	Anti-GFP nanobody GBP1 and mCherry	Covalent (S-S, Halo tag fusion or maleimide functionali-zation)	Nuclear/ nucleolar	[16]
R10	PEG2-PEG2-RRRRRRRRRR (cationic peptide)	Nanobody	Covalent (S-S)	Inside of cells	[17]
K10H16	KKKKKKKKKKGHHHHHHHH-HHHHHHHH (cationic peptide)	Acidic protein MBP-GFP α-Galactosidase A	Noncovalent	Lysosome	[18]

(Continued)

Table 13.1 (Continued)

Name	Sequence (peptide type)	Cargo	Conjugating method	Delivery site	Ref.
Lz(E)-CPP Fam-Lz(K)	EYQALEKEVAQLEAENQALEK-EVAQLEHERRRRRRRRRRRX[e] (cationic peptide) FAM-EYQALKKKVAQLKAKNQ-ALKKKVAQLKHKX[f] (cationic peptide)	Nuclear protein Nanog	Noncovalent (Lz(E) interaction with Lz(K))	Inside of cells	[19]
cLK	Ac-cyclo(CKKLLKLLKKLLKLGGLKKL-LKLLKKLLKLLK) (amphipathic peptides)	Fragment of transcription factor Runx2 (fRunx2)	Noncovalent	Cytoplasm	[20]
HAad	BGGGGIWLTALKFLGKAAAKA-XAKQXLSKL[g] (amphipathic peptides)	IgG	Noncovalent	Cytoplasm	[21]
LK15-oxP4A	KLLKLLLKLLLKLLKGGYR-cyclo(CAC)RFF (amphipathic peptide (LK15) – micropinocytosis-inducing peptide (oxP4A))	Cre recombinase and EGFP derivative	Noncovalent	Cytoplasm	[22]
Tat poly	$(YGRKKRRQRRR)_n$ (various CPP-based polymers)	Cre recombinase	Noncovalent	Cytoplasm	[23]
$BP100\text{-}(KH)_9$ $BP100CH_7$	KKLFKKILKYLKHKHKHKHK-HKHKHKHKH (cationic peptide) KKLFKKILKYLHHcyclo(CRGHTVH-SHHHC)IR (cationic peptide)	Citrine protein	Noncovalent	Cytoplasm of rice callus cells	[24]
Vpr55–91 Vpr55–82	TGVEALIRILQQLLFIHFRIGCR-HSRIGIIQQRRTRN TGVEALIRILQQLLFIHFRIGCR-HSRIG (amphipathic peptide)	Avidin-β-Gal, DTAF-MART-1 epitope and $gp100_{209M}$ epitope	Noncovalent (Avidin-β-Gal, DTAF-MART-1)	Cytoplasm	[25]

a) Z: 4-(dimethylaminoazo)benzene-4-carboxylic acid.
b) DEC: dithiol-ethyl carbonate linker combined with a benzyl carbamate.
c) SMCC: succinimidyl-trans-4-(*N*-maleimidylmethyl)cyclohexane-1-carboxylate).
d) r: D-arginine.
e) X: β-alanine.
f) FAM: carboxyfluorescein.
g) B: Lys(pyrenebutyryl)

regularly interspaced short palindromic repeats (CRISPR)-associated protein 9 (Cas9) [13], Fas-associated death domain (FADD) protein [14], zinc finger nucleases [15], anti-GFP nanobody GBP1 and mCherry [16], nanobody [17], α-galactosidase A [18], transcription factors such as Nanog [19], runt-related transcription factor 2 (Runx2) [20], and acidic proteins such as maltose-binding protein-fused GFP (MBP-GFP) and immunoglobulin G (IgG) [21], Cre recombinase and enhanced GFP (EGFP) derivative [22], Cre recombinase [23], citrine protein [24], and β-galactosidase (β-Gal) and DTAF-MART-1 epitope [25].

When CPPs are utilized as delivery tools for proteins, two methods with covalent conjugation or noncovalent conjugation of the CPP to the cargo have been reported. Covalent conjugation is performed on the N-terminus, C-terminus, or amino acid side chains of the protein cargo. Covalent conjugations are further classified by genetic or chemical methods. The genetic method, which provides CPPs-conjugated protein by gene expression, is utilized to obtain active conjugates of CPPs and proteins [26–37]. A genetic combinatorial library of recombinant conjugates of various CPPs composed of natural amino acids to protein cargos was performed to screen for efficient CPPs [11]. CPPs are introduced genetically at the N-terminus or C-terminus of a protein with consideration of the active conformation of the protein because there are risks from alternating the biological activity of the protein by conjugation of the CPP depending on the sequence and site of conjugation. Co-expression with a chaperone was reported in the synthesis of CPP-conjugated proteins to decrease the structural effect of CPPs [11]. Even though gene engineering is useful for the simplification of purification processes and constructing randomized libraries, the conjugation system of CPPs and proteins in gene engineering is chemically limited due to their composition, which is mainly limited to natural amino acids.

Chemical methods can introduce various CPPs suited for each application (Figure 13.1d). To maintain protein activity following the synthesis of the covalent conjugate, variations in the length and types of linkers and selection of the conjugating site have to be considered. CPPs are also efficient for the alteration of therapeutic proteins into a prodrug. Disulfide bonds are often used for chemical conjugation of CPPs to cargos due to the usefulness of its reductive cleavage in cells, resulting in the release of nonmodified cargo proteins [12]. Such conjugating methods that enable the release of native proteins expands the range of target proteins. A modified chemical linker composed of thiol-ethyl carbonate and benzyl carbamate is degraded by domino-type reactions triggered by disulfide cleavage and release of the aliphatic amine of native proteins [13]. As a conspicuous chemical method, full chemical synthesis can construct conjugate of CPPs and protein [12].

As a noncovalent conjugation method, the ligand of a protein can be used (Figure 13.1c). For example, biotin was introduced onto CPPs to deliver its binding protein, avidin [38]. Another noncovalent conjugation method is based on electrostatic interactions and hydrophobic interactions (Figure 13.1a) [39]. These methods can be performed easily by just mixing CPPs and proteins to form CPP/protein complexes with varied properties of the complexes such as size, stability, and dissociation constant. Modification of the physicochemical properties of complexes is required to

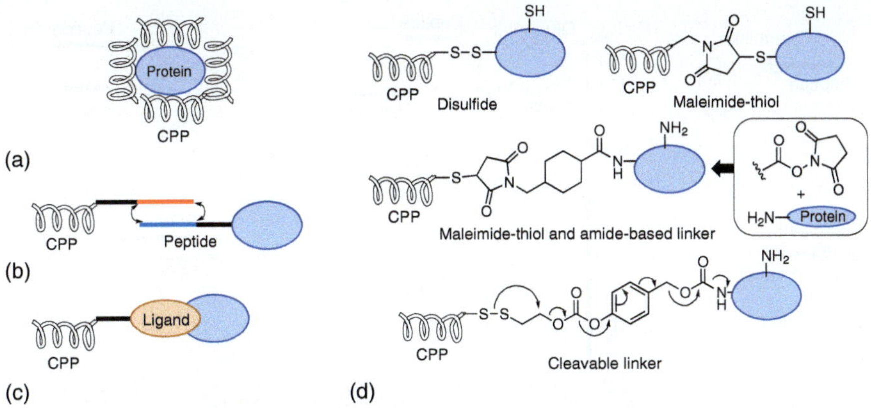

Figure 13.1 Attachment methods of CPPs to protein cargo. Noncovalent methods are shown as (a) electrostatic and hydrophobic interactions, (b) hybridized peptide tag, and (c) protein ligand. Covalent methods are shown as from the viewpoints of (d) chemical linkers.

release cargos in controlled conditions and function in a particular organism. CPPs are also used as targeting moieties in cells and organelles, for example, lysosomes [18] and the nucleus/nucleolus [12, 16, 40]. In both covalent and noncovalent methods for delivery of protein, CPPs are a potent modality with flexibility in the functions enabled through a wide range of potential chemical modifications.

In drug development, the use of peptides as therapeutic agents continues to evolve rapidly. Peptides that mimic natural peptide hormones have offered therapeutic opportunities. Peptides isolated from natural sources were first used in the 1920s, with insulin used as a therapeutic for patients with diabetes who had inadequate hormone production, with this peptide hormone being provided as a drug [41]. In addition to peptide hormones, peptide-based inhibitors against protein-protein interactions (PPIs) have emerged as drug candidates [42, 43]. The interactions between proteins are related to various health and disease mechanisms, and control of these PPIs may result in the development of a therapeutic strategy [44]. If the information on PPIs such as sequences and conformations of proteins related to the interaction is obtained, a mimic of the protein may be useful for the inhibitor design. However, mimicking all aspects of a protein is difficult because PPIs occur through multiple sites and surfaces by means of weak interactions [45]. In addition, PPIs sometimes induce dynamic changes to the structure of interacting proteins such as induced fit PPIs. Peptides can be easily designed as protein mimics from the crucial sequence involved in the PPI with the potential to interact with the surface of target proteins, which may result in a similar function to the original protein and also inhibit the signal pathway [46, 47]. Furthermore, chemical modifications of peptides can be performed to increase the binding affinity and stability [48]. However, such peptides are not generally able to cross cell membranes, so peptides can target only extracellular PPIs by themselves, even though the majority of PPIs occur intracellularly. Thus, to increase their cellular internalization, CPPs have been utilized in the delivery of bioactive peptides via mainly covalent bonds, and noncovalent bonds are also used (Figure 13.2, Table 13.2).

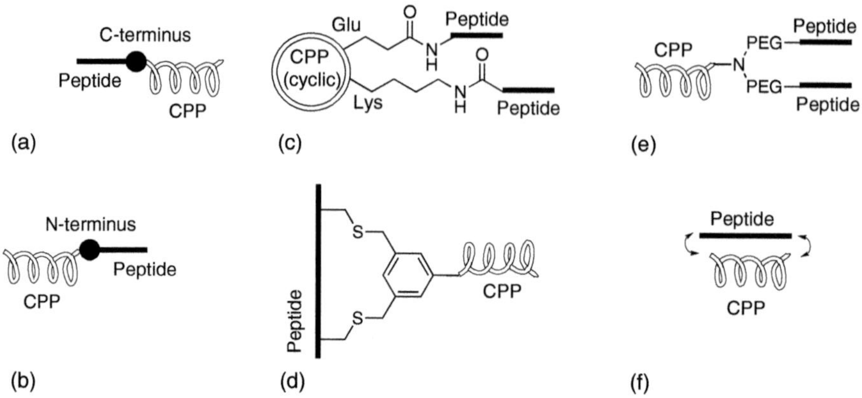

Figure 13.2 Attachment methods of CPPs to peptide cargos. Covalent methods are shown as (a) attachment at the C-terminus, (b) attachment at the N-terminus, (c) attachment on a side chain of amino acid, (d) attachment at a stapling structure, and (e) multiple attachments. Noncovalent methods are shown as (f) electrostatic and hydrophobic interactions.

As mentioned here, CPPs can be utilized to drug delivery systems of bioactive macromolecular cargos such as proteins and peptides, which potentially revolute the therapeutic treatment and biomedical field. Here, I mainly focus on the recent studies in the last three years and try to provide protein and peptide delivery by CPPs from the point of view of the methods of forming conjugates and the classes of peptides such as a cationic or amphipathic peptide.

13.2 The Delivery of Protein by CPPs

13.2.1 Covalent Method

13.2.1.1 Genetic Method

Na et al. utilized a genetic method to construct a library of CPPs for screening efficient CPPs for the delivery of protein cargos into bacteria via electroporation [11]. Covalent conjugates of the Tat peptide with proteins such as GFP, I-SceI, and Cre protein were internalized into *Escherichia coli* (*E. coli*) with no significant cytotoxicity. CPP-conjugates of I-SceI showed plasmid removal in *E. coli* and CPP-conjugated Cre recombinase showed marker gene excision from the genome of *Methylomonas sp.* DH-1.

13.2.1.2 Chemical Method

The chemical method was performed by Brik et al. To enhance cellular uptake of a protein cargo, a cyclic decaarginine (cR10)-modified 4-{[4-(dimethylamino)phenyl]azo}benzoic acid (DABCYL) was synthesized and cR10-DABCYL covalently attached to fully synthetic Ub or SUMO-2 scaffolds via disulfide bonds [12]. The conjugation of cR10-DABCYL resulted in efficient uptake of Ub or SUMO-2 at low

Table 13.2 Summary of representative cell-penetrating peptides for peptide delivery.

Name	Sequence (peptide type)	Cargo	Conjugate method	Delivery site	Ref.
Penetratin	RQIKIWFQNRRMKWKK (cationic peptide)	Insulin or part of parathyroid hormone (PTH(1–34))	Noncovalent	Inside of cells	[49]
Tat	Ac-YGRKKRRQRRR (cationic peptide)	JYX-3 bound to YEATS domains	Covalent (amide bond)	Inside of cells	[50]
Tat	YGRKKRRQRR (cationic peptide)	KSL-128114 bound to PDZ1 domain of syntenin	Covalent (amide bond)	Inside of cells	[51]
Tat	Ac-GRKKRRQRRRPQ (cationic peptide)	GVITAA inhibitor DHHHC21/LCK	Covalent (amide bond)	Not shown	[52]
Tat	YGRKKRRQRRR (cationic peptide)	Partial structure of EV-A71 2C and EV-B 2C	Covalent (amide bond)	Inside of cells	[53]
Tat	GRKKRRQRRR (cationic peptide)	Peptide to induce the degradation of CTDP-43 protein	Covalent (amide bond)	Inside of cells	[54]
Tat	YGRKKRRQRRR (cationic peptide)	NR2B9c corresponding to the C-terminus of the NMDA GluN2B subunit	Covalent (amide bond)	Cytoplasm	[55]
Tat	GRKKRRQRRPPQ (cationic peptide)	Substrate of cathepsin D conjugated with examethasone	Covalent (amide bond)	Inside of cells	[56]
K-FGF	AAVLLPVLLAAP (hydrophobic peptide)				
P7	FNLPLPSRPLLR (not categorized)				
R7	RRRRRRR (cationic peptide)	Protein degradation inducer SNIPER	Covalent (amide bond)	Not shown	[57]

(Continued)

Table 13.2 (Continued)

Name	Sequence (peptide type)	Cargo	Conjugate method	Delivery site	Ref.
CPP9	Cyclo(fΦRrRrQ)[a, b] (cyclic amphipathic peptide)	MDM2 ligand and β-catenin	Covalent (amide bond)	Cytoplasm	[58]
CPP9	Cyclo(fΦRrRrQ)[a, b] (cyclic amphipathic peptide) CPP9	Peptidyl Keap1 inhibitors	Covalent (amide bond)	Cytoplasm	[59]
CPP9	Cyclo(FfΦRrRrQ)[a, b] (cyclic amphipathic peptide)	PNA targeting miR-155	Covalent (amide bond)	Lysosomes and nuclei	[60]
CPP12	Cyclo(FΦR_4)[a, b] (cyclic amphipathic peptide)	Pmp-ct bound to STAT3	Covalent (amide bond)	Cytoplasm	[61]
[R_6W_3]	Cyclo(CRRWWRRWRR) (amphipathic peptide)	Protein kinase C inhibitor with an isotope labelling and affinity tag	Covalent (S-S)	Inside of cells	[62]
[C_{12}-R_4]	Cyclo(CZRRRR)[c] (amphipathic peptide)				
β-hairpin sequence	RWVRVpGRWIRQ[d]	Substrate of DUBs	Covalent (amide bond)	Cytoplasm	[63]

a) Φ: L-2-naphthylalanine.
b) f: D-phenylalanine.
c) Z: (*R*,*S*)-2-amino tetradecanoic acid.
d) p: D-proline, .

micromolar concentrations. Additionally, conjugates of CPPs to Ub or a Ub variant were applied to functional studies of deubiquitinases in live cells.

Cleavable linkers in reducing conditions have the potential to undergo selective cleavage in intracellular environments where the concentration of the reducing agent glutathione is higher than in the extracellular compartment. Murthy et al. developed a method to conjugate CPPs and proteins via a self-immolative linker composed of a dithiol-ethyl carbonate and a benzyl carbamate (DEC) moiety and applied it to the CRISPR–Cas9 system [13]. DEC can be introduced onto aliphatic amines of proteins, and the cleavage of DEC in intracellular reductive environments releases the native protein rapidly. The group introduced decaarginine (R10) into the Lys residues of Cas9 via DEC, and the CPP–Cas9 conjugate was able to diffuse through brain tissue better than unmodified Cas9 after intracranial injection. Furthermore, a conjugate of CPP and Cas9 containing a guide RNA, namely a Cas9 ribonucleoprotein, was able to internalize into cells and edit cells without transfection reagents.

The expression of FADD in cancer cells is critical to the regulation of their proliferation because suppression of FADD expression in cancer cells lets the cells escape apoptosis signaling. Chandramani Pathak et al. conjugated purified FADD protein with Tat peptide covalently via succinimidyl-*trans*-4-(*N*-maleimidylmethyl) cyclohexane-1-carboxylate (SMCC) and iodoacetamide [14]. Tat-conjugated FADD protein internalized through the caveolar endocytic pathway and was retained in the cytosol. Inside cancer cells, Tat-conjugated FADD rapidly constituted a death-inducing signaling complex known as DISC, which induced apoptosis signaling and cell death.

Bilichak et al. developed a conjugate composed of a zinc finger nuclease (ZFN) and CPPs such as cys(Npys)-conjugated to D-R9 or $(BP100)_2K_8$ via disulfide bonds [15]. The conjugate was able to internalize into either wheat microspores and embryos, and the reductive intracellular conditions induced the release of ZFN, resulting in the editing of the inositol pentakisphosphate kinase 1 gene.

Hackenberger et al. demonstrated that thiol reactive or HaloTag reactive CPP (R10 or cyclic R10: cR10) additives can enhance the cellular uptake protein-CPP conjugates at low micromolar concentrations [16]. It should be noted that cR10 contained 10 Arg residues with altering L and D configurations. These additives were anchored covalently on the cell surface, resulting in effectively co-delivery of protein cargos. In particular, the co-addition of a slight excess of reactive CPP, thionitro-benzoic-acid-activated R10 peptide, may increase the delivery of cysteine (Cys)-containing proteins such as functional enzymes, nanobodies, and full-length immunoglobulin-G antibodies into the cytosol.

Hackenberger et al. applied a conjugate of fluorescent nanobodies and R10 containing a Cys via a covalent disulfide bond for live-cell super-resolution techniques, stimulated emission–depletion (STED) imaging targeting intracellular proteins [17]. Photostabilizer Trolox was also introduced into the CPP-nanobody using expressed protein ligation to increase intramolecular photostabilization. These semisynthetic nanobodies enable the imaging of GFP and mCherry, as well as two endogenous targets, nuclear lamins and a DNA replication and repair protein by confocal and STED microscopy.

13.2.2 Noncovalent Method

Iwasaki et al. developed a CPP based on Lys and histidine (His) residues named K10H16, which can form complexes with proteins noncovalently by electrostatic interactions [18]. The K10H16 peptide was able to bind to an acidic protein, MBP-GFP, and transported the protein into intracellular lysosomes. As another application, a complex of K10H16 and α-galactosidase A, which is one of the lysosomal enzymes associated with lysosomal storage disease (LSD), was delivered to intracellular lysosomes and suppressed the growth of LSD cells more slowly than normal cells.

Hakata et al. applied a conjugate of CPP, oligoarginine, and leucine zipper (Lz) peptide analogue containing glutamic acids (Glu, E) residues, designated as Lz(E) peptide, to the delivery of a protein modified with another Lz peptide analogue, termed Lz(K) [19]. Lz(E) hybridized noncovalently with Lz(K), in which Glu residues in the sequence of Lz(E) were replaced with Lys (K) residues. The conjugated Lz(K) group to the transcription factor Nanog, forming a complex of Nanog-Lz(K)/Lz(E)-CPP, and delivered the complex into cells. Subsequently, after transport to the nucleus, Nanog-Lz(K) showed a transcriptional regulation.

Park et al. investigated the intracellular delivery of a transcription factor, fragmented Runx2 (fRunx2), using a cyclic CPP (cLK) composed of Leu and Lys residues [20]. The noncovalently bound complex of fRunx2/cLK increased the gene expression levels of alkaline phosphatase and osteocalcin, which are associated with osteoblast differentiation and induced osteoblast differentiation in rat mesenchymal stem cells.

Futaki et al. reported attenuated cationic amphiphilic lytic peptide, HAad, which showed significant endosomolytic activity at a very early stage of endocytosis for the delivery of human immunoglobulin G into cells [21]. The introduction of pyrenebutyric acid as a membrane-anchoring unit to HAad, resulting in pBu-HAad, increased the delivery of a noncovalently interacting protein at only 1/20 the concentration of HAad.

Futaki et al. also utilized oxP4A peptide, a potent intramolecularly oxidized form of P4A peptide, to induce macropinocytosis [22]. A CPP consisting of oxP4A and the membrane lytic peptide LK15 interacted with GFP fused with a negatively charged protein tag and delivered GFP into cells at submicromolar concentrations.

Blum et al. reported polymerized peptides of various CPPs such as Tat for the efficient delivery of a DNA-modifying enzyme, Cre recombinase [23]. This noncovalent method based on the noncovalent interaction between the CPP and Cre protein was applied to pluripotent stem cells or HEK293T cells, which resulted in recombination of the loxP sites in the cell.

Protein delivery by CPPs via noncovalent interaction was also performed in the agricultural field. Numata et al. used CPPs such as $BP100(KH)_9$ and $BPCH_7$ for the delivery of Citrine, yellow fluorescent protein [24]. Whereas $BP100(KH)_9$ contained a linear cationic moiety to interact with Citrine, $BPCH_7$ contained a cyclic Citrine-binding moiety that could be altered to a linear form in reducing intracellular conditions [64, 65]. Using pressure and vacuum treatment, $BP100(KH)_9$ and $BP100CH_7$

promoted cellular uptake of Citrine into rice callus cells through the cell wall by the interaction between the CPP and the lipid bilayer.

Kichler et al. demonstrated the delivery of proteins and epitopes into cell lines as well as into human primary dendritic cells by using a CPP based on the C-terminal domain of Viral protein R (Vpr) such as the peptides Vpr55–91 and Vpr55–82 [25]. A noncovalent method was utilized to deliver avidin-β-Gal and prototypic melanoma antigen (DTAF-MART-1) epitope. The complex of DTAF-MART-1 epitope and Vpr55–91 showed immunogenicity in HLA-A2 transgenic mice, which induced antigen-specific cytotoxic T lymphocytes against multiple tumor epitopes.

13.3 The Delivery of Peptides by CPPs

13.3.1 Cationic Peptide-Based Method

Nielsen et al. reported micron-sized complexes composed of the CPP penetratin and insulin or a part of parathyroid hormone (PTH(1–34)) via electrostatic interactions [49]. The CPP increased enzymatic stability and transepithelial permeation of membranes in vitro and in vivo. Furthermore, the optimization of sample pH improved those properties.

YEATS domains were identified as epigenetic readers of histone lysine acetylation and crotonylation and considered promising anticancer drug targets. Li et al. developed a cyclopeptide that bound to the YEATS domains by conjugation of Tat to increase cellular uptake [50].

Syntenin, comprising two postsynaptic density protein-95 (PSD-95)/discs-large/zona occludens-1 (PDZ1) domains, has been considered as a novel therapeutic target in malignant phenotypes including glioblastoma. Strømgaard et al. developed KSL-128114 by conjugation of Tat to the peptide inhibitor of syntenin, KSL-128018, which is bound to the PDZ1 domain of the intracellular scaffold protein syntenin [51]. Treatment with KSL-128114 inhibited the viability of primary glioblastoma cells, resulting in prolonged survival time in a murine model with a patient-derived xenograft.

Moorthy et al. covalently introduced Tat peptide to the GVITAA peptide, which inhibited the interaction between aspartate (Asp)-His-His-Cys (DHHC21) and lymphocyte-specific protein kinase (LCK) [52]. LCK is expressed in T cells and plays a crucial role in T cell activation. In T cells from mice, this CPP-conjugated peptide successfully inhibited LCK-mediated T cell receptor signaling, cytokine secretion, and T cell proliferation.

Enteroviral protein 2C is an attractive target for developing anti-enterovirus drugs due to its conservation among all enteroviruses and RNA helicase activity, which is important during enteroviral life cycles. Zhou et al. developed two peptides containing Tat peptide [53]. 2CL peptide was based on the structure of human enterovirus A71 (EV-A71). 2CL inhibited the RNA helicase activities of 2C proteins encoded by EV-A71 and CVA16. Another peptide, B-2CL peptide, was based on the group B enteroviruses (EV-Bs). B-2CL showed antiviral activity against two important EV-Bs, coxsackievirus B3 and echovirus 11.

Peptide-based degrader of TAR DNA binding protein 43 (TDP-43), named D4 peptide, was developed by Li et al. [54]. TDP-43 is a crucial target in amyotrophic lateral sclerosis treatment. In their study, the covalent conjugation of CPPs to hydrophobic tags and the peptide ligand for TDP-43 was performed. The reduction in the expression levels of TDP-43 by D4 peptide led to cytotoxicity.

Inhibition of the interaction between the PDZ1 domains of PSD-95 and the *N*-methyl-D-aspartate (NMDA) receptor has been suggested as a therapeutic approach for relieving neuronal damage. NR2B9c and N-dimer were reported as peptide inhibitors of the interaction [66]. To investigate their therapeutic effect in detail, Kristensen et al. compared blood–brain barrier uptake of NR2B9c and N-dimer derivatives covalently conjugated with the Tat peptide [55]. The examination using co-culture of primary bovine brain capillary endothelial cells and astrocytes from rats and two-photon microscopy imaging in live mice was performed with fluorescently labelled peptides, which suggested that conjugation with Tat peptide affected the chemical stability of the inhibitor, permeation of blood–brain barrier, and neuronal uptake.

There are tissue barriers in the eye that limit drug delivery (e.g. retinal diseases are treated with intravitreal injections). Bhattacharya et al. developed a peptide-based delivery system to the retinal pigment epithelial cells [56]. A cleavable sequence sensitive to the enzymatic action of cathepsin D was conjugated covalently to cationic and hydrophobic CPPs such as Tat, K-FGF, and P7 peptide [67]. These peptides were also conjugated with dexamethasone via hydrazone bonds and released dexamethasone after internalization into the retinal pigment epithelial cells. in vivo studies in rabbits demonstrated that the CPPs increased vitreal retention of dexamethasone.

Yokoo et al. introduced R7 as a CPP to a protein degradation inducers called SNIPER (Specific and Non-genetic IAP-dependent Protein Erasers) targeting the estrogen receptor alpha (ERα) [57]. Designed LCL-stPERML-R7 composed of three moieties comprising heptaarginine (R7), a small molecule LCL161 (LCL) that recruits the E3 ligase IAPs, and α-helical peptide stPERML binding to ERα, induced proteasomal degradation via ubiquitylation and inhibited ERα-mediated transcription.

13.3.2 Amphipathic Peptide-Based Method

The inhibition of PPIs has emerged as a therapeutic approach, for example, in the inhibition of MDM2/p53 (tumor suppressor protein) for the treatment of $p53^{WT}$ cancers via recovery of the growth-suppressive function by p53, and β-catenin/TCF (transcription factor) for the treatment of colorectal cancers by controlling gene expression. Pei et al. developed peptide inhibitors for the interaction of MDM2/p53 and β-catenin/TFC by conjugation of a cyclic CPP, CPP9, to peptidyl ligands for MDM2 and β-catenin at its N- or C-terminus [58]. In addition, the conjugation of CPPs to the stapling moiety of the ligand peptide also increased cytosolic entry and inhibition activity in vitro (Figure 13.2d).

Pei et al. also covalently conjugated CPP9 to a cyclic peptide-based inhibitor against the Kelch-like ECH-associated protein 1 (Keap1)–nuclear factor erythroid-2 (Nrf2) interaction, which is considered as a regulator of inflammation in the treatment of inflammatory diseases [59]. When mammalian cells were treated with the bicyclic peptide, its uptake and activation of transcription of Nrf2 were observed.

Yavin et al. conjugated CPP9 to peptide nucleic acids (PNAs) targeting an oncogenic miRNA, miR-155, via an effective synthetic route [60]. CPP9–PNA was internalized into glioblastoma cells (U87MG) and showed anti-miR activity, which resulted in a reduction in cell viability and proliferation in U87MG cells.

Kritzer et al. conjugated a cyclic cationic CPP, CPP12, to Pmp-ct to increase cytosolic delivery of the Pmp-ct peptide, which binds to signal transducer and activator of transcription 3 (STAT3) protein [61].

Burlina et al. synthesized various cyclic CPPs based on oligoarginine with aliphatic carbon chain that improved CPP interaction with the cell membrane [62]. As a cargo of the CPP, a peptide inhibitor of protein kinase C containing an affinity tag (biotin(sulfone)GGGG) and isotope labelling to quantify the amount of inhibitor inside cells were examined. The introduction of this CPP covalently promoted cellular uptake of the peptide inhibitor via efficient glycosaminoglycan-dependent endocytosis and direct translocation.

Melvin et al. used a β-hairpin sequence motif (RWVRVpGRWIRQ) as a CPP for the development of a peptide-based biosensor of deubiquitinating enzyme (DUB) activity [63, 68]. The intracellular stability and uptake into intact cells of the substrate sequence (LRGG) of DUBs was increased by conjugation to the β-hairpin sequence. Peptides labelled with a fluorophore (AFC) were able to quantify DUB activity in cells by specific cleavage by DUB and subsequent recovery of AFC fluorescent.

References

1 Dietz, G.P. and Bahr, M. (2004). Delivery of bioactive molecules into the cell: the Trojan horse approach. *Mol. Cell. Neurosci.* 27: 85–131. https://doi.org/10.1016/j.mcn.2004.03.005.

2 Kristensen, M., Birch, D., and Morck Nielsen, H. (2016). Applications and challenges for use of cell-penetrating peptides as delivery vectors for peptide and protein cargos. *Int. J. Mol. Sci.* 17: https://doi.org/10.3390/ijms17020185.

3 Pakulska, M.M., Miersch, S., and Shoichet, M.S. (2016). Designer protein delivery: from natural to engineered affinity-controlled release systems. *Science* 351: aac4750. https://doi.org/10.1126/science.aac4750.

4 Timin, A.S., Litvak, M.M., Gorin, D.A. et al. (2018). Cell-based drug delivery and use of nano-and microcarriers for cell functionalization. *Adv. Healthc. Mater.* 7: https://doi.org/10.1002/adhm.201700818.

5 Lee, K., Rafi, M., Wang, X. et al. (2015). *in vivo* delivery of transcription factors with multifunctional oligonucleotides. *Nat. Mater.* 14: 701–706. https://doi.org/10.1038/nmat4269.

6 Mitragotri, S., Burke, P.A., and Langer, R. (2014). Overcoming the challenges in administering biopharmaceuticals: formulation and delivery strategies. *Nat. Rev. Drug Discov.* 13: 655–672. https://doi.org/10.1038/nrd4363.

7 Brooks, H., Lebleu, B., and Vives, E. (2005). Tat peptide-mediated cellular delivery: back to basics. *Adv. Drug Deliv. Rev.* 57: 559–577. https://doi.org/10.1016/j.addr.2004.12.001.

8 Nam, H.Y., Kim, J., Kim, S. et al. (2011). Cell penetrating peptide conjugated bioreducible polymer for siRNA delivery. *Biomaterials* 32: 5213–5222. https://doi.org/10.1016/j.biomaterials.2011.03.058.

9 Anderson, D.C., Nichols, E., Manger, R. et al. (1993). Tumor cell retention of antibody Fab fragments is enhanced by an attached HIV TAT protein-derived peptide. *Biochem. Biophys. Res. Commun.* 194: 876–884. https://doi.org/10.1006/bbrc.1993.1903.

10 Amand, H.L., Fant, K., Norden, B. et al. (2008). Stimulated endocytosis in penetratin uptake: effect of arginine and lysine. *Biochem. Biophys. Res. Commun.* 371: 621–625. https://doi.org/10.1016/j.bbrc.2008.04.039.

11 Lee, H.M., Ren, J., Tran, K.M. et al. (2021). Identification of efficient prokaryotic cell-penetrating peptides with applications in bacterial biotechnology. *Commun Biol* 4: 205. https://doi.org/10.1038/s42003-021-01726-w.

12 Mandal, S., Mann, G., Satish, G. et al. (2021). Enhanced live-cell delivery of synthetic proteins assisted by cell-penetrating peptides fused to DABCYL. *Angew. Chem. Int. Ed. Eng.* 60: 7333–7343. https://doi.org/10.1002/anie.202016208.

13 He, M., Li, J., Han, H. et al. (2020). A traceless linker for aliphatic amines that rapidly and quantitatively fragments after reduction. *Chem. Sci.* 11: 8973–8980. https://doi.org/10.1039/d0sc00929f.

14 Ranjan, K., Waghela, B.N., Vaidya, F.U. et al. (2020). Cell-penetrable peptide-conjugated FADD induces apoptosis and regulates inflammatory signaling in cancer cells. *Int. J. Mol. Sci.* 21: https://doi.org/10.3390/ijms21186890.

15 Bilichak, A., Sastry-Dent, L., Sriram, S. et al. (2020). Genome editing in wheat microspores and haploid embryos mediated by delivery of ZFN proteins and cell-penetrating peptide complexes. *Plant Biotechnol. J.* 18: 1307–1316. https://doi.org/10.1111/pbi.13296.

16 Schneider, A.F.L., Kithil, M., Cardoso, M.C. et al. (2021). Cellular uptake of large biomolecules enabled by cell-surface-reactive cell-penetrating peptide additives. *Nat. Chem.* 13: 530–539. https://doi.org/10.1038/s41557-021-00661-x.

17 Schneider, A.F.L., Benz, L.S., Lehmann, M. et al. (2021). Cell-permeable nanobodies allow dual-color super-resolution microscopy in untransfected living cells. *Angew. Chem. Int. Ed. Eng.* 60: 22075–22080. https://doi.org/10.1002/anie.202103068.

18 Iwasaki, T., Murakami, N., and Kawano, T. (2020). A polylysine-polyhistidine fusion peptide for lysosome-targeted protein delivery. *Biochem. Biophys. Res. Commun.* 533: 905–912. https://doi.org/10.1016/j.bbrc.2020.09.087.

19 Hakata, Y., Michiue, H., Ohtsuki, T. et al. (2019). A leucine zipper-based peptide hybrid delivers functional Nanog protein inside the cell nucleus. *Bioorg. Med. Chem. Lett.* 29: 878–881. https://doi.org/10.1016/j.bmcl.2019.02.004.

20 Nam, S.H., Lee, Y., Ahn, J.H. et al. (2020). Augmented osteogenesis of mesenchymal stem cells using a fragmented Runx2 mixed with cell-penetrating, dimeric a-helical peptide. *Eur. J. Pharm. Sci.* 144: 105210. https://doi.org/10.1016/j.ejps.2019.105210.

21 Sakamoto, K., Michibata, J., Hirai, Y. et al. (2021). Potentiating the membrane interaction of an attenuated cationic amphiphilic lytic peptide for intracellular protein delivery by anchoring with pyrene moiety. *Bioconjug. Chem.* 32: 950–957. https://doi.org/10.1021/acs.bioconjchem.1c00101.

22 Arafiles, J.V.V., Hirose, H., Hirai, Y. et al. (2021). Discovery of a macropinocytosis-inducing peptide potentiated by medium-mediated intramolecular disulfide formation. *Angew. Chem. Int. Ed. Eng.* 60: 11928–11936. https://doi.org/10.1002/anie.202016754.

23 Blum, A.P., Nelles, D.A., Hidalgo, F.J. et al. (2019). Peptide brush polymers for efficient delivery of a gene editing protein to stem cells. *Angew. Chem. Int. Ed. Eng.* 58: 15646–15649. https://doi.org/10.1002/anie.201904894.

24 Guo, B., Itami, J., Oikawa, K. et al. (2019). Native protein delivery into rice callus using ionic complexes of protein and cell-penetrating peptides. *PLoS One* 14: e0214033. https://doi.org/10.1371/journal.pone.0214033.

25 Gross, D.A., Leborgne, C., Chappert, P. et al. (2019). Induction of tumor-specific CTL responses using the C-terminal fragment of Viral protein R as cell penetrating peptide. *Sci. Rep.* 9: 3937. https://doi.org/10.1038/s41598-019-40594-7.

26 Behzadipour, Y., Sadeghian, I., Ghaffarian Bahraman, A. et al. (2021). Introducing a delivery system for melanogenesis inhibition in melanoma B16F10 cells mediated by the conjugation of tyrosine ammonia-lyase and a TAT-penetrating peptide. *Biotechnol. Prog.* 37: e3071. https://doi.org/10.1002/btpr.3071.

27 Davoodi, S., Bolhassani, A., and Namazi, F. (2021). *in vivo* delivery of a multiepitope peptide and Nef protein using novel cell-penetrating peptides for development of HIV-1 vaccine candidate. *Biotechnol. Lett.* 43: 547–559. https://doi.org/10.1007/s10529-020-03060-3.

28 Koo, J.H., Kim, S.H., Jeon, S.H. et al. (2021). Macrophage-preferable delivery of the leucine-rich repeat domain of NLRX1 ameliorates lethal sepsis by regulating NF-kappaB and inflammasome signaling activation. *Biomaterials* 274: 120845. https://doi.org/10.1016/j.biomaterials.2021.120845.

29 Lee, H., Kim, S.H.L., Yoon, H. et al. (2020). Intracellular delivery of recombinant RUNX2 facilitated by cell-penetrating protein for the osteogenic differentiation of hMSCs. *ACS Biomater Sci Eng* 6: 5202–5214. https://doi.org/10.1021/acsbiomaterials.0c00827.

30 Merritt, H.I., Sawyer, N., and Arora, P.S. (2020). Bent into shape: folded peptides to mimic protein structure and modulate protein function. *Pept. Sci. (Hoboken)* 112: https://doi.org/10.1002/pep2.24145.

31 Miliotou, A.N., Papagiannopoulou, D., Vlachaki, E. et al. (2021). PTD-mediated delivery of alpha-globin chain into Kappa-562 erythroleukemia cells and alpha-thalassemic (HBH) patients' RBCs ex vivo in the frame of protein replacement therapy. *J. Biol. Res. (Thessalon)* 28: 16. https://doi.org/10.1186/s40709-021-00148-3.

32 Miwa, T., Tachii, K., Wei, F.-Y. et al. (2019). Intranasal drug delivery into mouse nasal mucosa and brain utilizing arginine-rich cell-penetrating peptide-mediated protein transduction. *Int. J. Pept. Res. Ther.* 26: 1643–1650. https://doi.org/10.1007/s10989-019-09971-8.

33 Schmit, N.E., Neopane, K., and Hantschel, O. (2019). Targeted protein degradation through cytosolic delivery of monobody binders using bacterial toxins. *ACS Chem. Biol.* 14: 916–924. https://doi.org/10.1021/acschembio.9b00113.

34 Suzuki, M., Iwaki, K., Kikuchi, M. et al. (2021). Characterization of the membrane penetration-enhancing peptide S19 derived from human syncytin-1 for the intracellular delivery of TAT-fused proteins. *Biochem. Biophys. Res. Commun.* 586: 63–67. https://doi.org/10.1016/j.bbrc.2021.11.065.

35 Teo, S.L.Y., Rennick, J.J., Yuen, D. et al. (2021). Unravelling cytosolic delivery of cell penetrating peptides with a quantitative endosomal escape assay. *Nat. Commun.* 12: 3721. https://doi.org/10.1038/s41467-021-23997-x.

36 Yin, J., Hou, S., Wang, Q. et al. (2019). Microenvironment-responsive delivery of the Cas9 RNA-guided endonuclease for efficient genome editing. *Bioconjug. Chem.* 30: 898–906. https://doi.org/10.1021/acs.bioconjchem.9b00022.

37 Yu, S.L., Koo, H., Lee, H.Y. et al. (2019). Recombinant cell-permeable HOXA9 protein inhibits NSCLC cell migration and invasion. *Cell Oncol. (Dordr)* 42: 275–285. https://doi.org/10.1007/s13402-019-00424-4.

38 Jain, A. and Cheng, K. (2017). The principles and applications of avidin-based nanoparticles in drug delivery and diagnosis. *J. Control. Release* 245: 27–40. https://doi.org/10.1016/j.jconrel.2016.11.016.

39 Heitz, F., Morris, M.C., and Divita, G. (2009). Twenty years of cell-penetrating peptides: from molecular mechanisms to therapeutics. *Br. J. Pharmacol.* 157: 195–206. https://doi.org/10.1111/j.1476-5381.2009.00057.x.

40 Dissanayake, S., Denny, W.A., Gamage, S. et al. (2017). Recent developments in anticancer drug delivery using cell penetrating and tumor targeting peptides. *J. Control. Release* 250: 62–76. https://doi.org/10.1016/j.jconrel.2017.02.006.

41 Banting, F.G., Best, C.H., Collip, J.B. et al. (1922). Pancreatic extracts in the treatment of diabetes mellitus. *Can. Med. Assoc. J.* 12: 141–146.

42 Birk, A.V., Liu, S., Soong, Y. et al. (2013). The mitochondrial-targeted compound SS-31 re-energizes ischemic mitochondria by interacting with cardiolipin. *J. Am. Soc. Nephrol.* 24: 1250–1261. https://doi.org/10.1681/asn.2012121216.

43 Chang, Y.S., Graves, B., Guerlavais, V. et al. (2013). Stapled α-helical peptide drug development: a potent dual inhibitor of MDM2 and MDMX for p53-dependent cancer therapy. *Proc. Natl. Acad. Sci. U. S. A.* 110: E3445–E3454. https://doi.org/10.1073/pnas.1303002110.

44 Groß, A., Hashimoto, C., Sticht, H. et al. (2015). Synthetic peptides as protein mimics. *Front. Bioeng. Biotechnol.* 3: 211. https://doi.org/10.3389/fbioe.2015.00211.

45 Neduva, V. and Russell, R.B. (2006). Peptides mediating interaction networks: new leads at last. *Curr. Opin. Biotechnol.* 17: 465–471. https://doi.org/10.1016/j.copbio.2006.08.002.

46 Acar, H., Ting, J.M., Srivastava, S. et al. (2017). Molecular engineering solutions for therapeutic peptide delivery. *Chem. Soc. Rev.* 46: 6553–6569. https://doi.org/10.1039/c7cs00536a.

47 Qvit, N., Rubin, S.J.S., Urban, T.J. et al. (2017). Peptidomimetic therapeutics: scientific approaches and opportunities. *Drug Discov. Today* 22: 454–462. https://doi.org/10.1016/j.drudis.2016.11.003.

48 Bruzzoni-Giovanelli, H., Alezra, V., Wolff, N. et al. (2018). Interfering peptides targeting protein-protein interactions: the next generation of drugs? *Drug Discov. Today* 23: 272–285. https://doi.org/10.1016/j.drudis.2017.10.016.

49 Kristensen, M., Guldsmed Diedrichsen, R., Vetri, V. et al. (2020). Increased carrier peptide stability through pH adjustment improves insulin and PTH(1–34) delivery *in vitro* and *in vivo* rather than by enforced carrier peptide-cargo complexation. *Pharmaceutics* 12: https://doi.org/10.3390/pharmaceutics12100993.

50 Jiang, Y., Liu, S., Tian, G. et al. (2021). Concise solid-phase synthesis enables derivatisation of YEATS domain cyclopeptide inhibitors for improved cellular uptake. *Bioorg. Med. Chem.* 45: 116342. https://doi.org/10.1016/j.bmc.2021.116342.

51 Haugaard-Kedstrom, L.M., Clemmensen, L.S., Sereikaite, V. et al. (2021). A high-affinity peptide ligand targeting syntenin inhibits glioblastoma. *J. Med. Chem.* 64: 1423–1434. https://doi.org/10.1021/acs.jmedchem.0c00382.

52 Gauthaman, A., Jacob, R., Pasupati, S. et al. (2021). Novel peptide-based inhibitor for targeted inhibition of T cell function. *J. Cell Commun. Signal.* https://doi.org/10.1007/s12079-021-00660-0.

53 Fang, Y., Wang, C., Wang, C. et al. (2021). Antiviral peptides targeting the helicase activity of enterovirus nonstructural protein 2C. *J. Virol.* 95: https://doi.org/10.1128/JVI.02324-20.

54 Gao, N., Huang, Y.P., Chu, T.T. et al. (2019). TDP-43 specific reduction induced by Di-hydrophobic tags conjugated peptides. *Bioorg. Chem.* 84: 254–259. https://doi.org/10.1016/j.bioorg.2018.11.042.

55 Kristensen, M., Kucharz, K., Fernandes, F.A. et al. (2020). Conjugation of therapeutic PSD-95 inhibitors to the cell-penetrating peptide Tat affects blood-brain barrier adherence, uptake, and permeation. *Pharmaceutics* 12: https://doi.org/10.3390/pharmaceutics12070661.

56 Bhattacharya, M., Sadeghi, A., Sarkhel, S. et al. (2020). Release of functional dexamethasone by intracellular enzymes: a modular peptide-based strategy for ocular drug delivery. *J. Control. Release* 327: 584–594. https://doi.org/10.1016/j.jconrel.2020.09.005.

57 Yokoo, H., Ohoka, N., Takyo, M. et al. (2021). Peptide stapling improves the sustainability of a peptide-based chimeric molecule that induces targeted protein degradation. *Int. J. Mol. Sci.* 22: https://doi.org/10.3390/ijms22168772.

58 Dougherty, P.G., Wen, J., Pan, X. et al. (2019). Enhancing the cell permeability of stapled peptides with a cyclic cell-penetrating peptide. *J. Med. Chem.* 62: 10098–10107. https://doi.org/10.1021/acs.jmedchem.9b00456.

59 Salim, H., Song, J., Sahni, A. et al. (2020). Development of a cell-permeable cyclic peptidyl inhibitor against the Keap1-Nrf2 interaction. *J. Organomet. Chem.* 85: 1416–1424. https://doi.org/10.1021/acs.joc.9b02367.

60 Soudah, T., Khawaled, S., Aqeilan, R.I. et al. (2019). AntimiR-155 cyclic peptide-PNA conjugate: synthesis, cellular uptake, and biological activity. *ACS Omega* 4: 13954–13961. https://doi.org/10.1021/acsomega.9b01697.

61 Cerulli, R.A., Shehaj, L., Tosic, I. et al. (2020). Cytosolic delivery of peptidic STAT3 SH2 domain inhibitors. *Bioorg. Med. Chem.* 28: 115542. https://doi.org/10.1016/j.bmc.2020.115542.

62 Amoura, M., Illien, F., Joliot, A. et al. (2019). Head to tail cyclisation of cell-penetrating peptides: impact on GAG-dependent internalisation and direct translocation. *Chem. Commun. (Camb.)* 55: 4566–4569. https://doi.org/10.1039/c9cc01265f.

63 Safa, N., Pettigrew, J.H., Gauthier, T.J. et al. (2019). Direct measurement of deubiquitinating enzyme activity in intact cells using a protease-resistant, cell-permeable, peptide-based reporter. *Biochem. Eng. J.* 151: https://doi.org/10.1016/j.bej.2019.107320.

64 Chuah, J.A. and Numata, K. (2018). Stimulus-responsive peptide for effective delivery and release of DNA in plants. *Biomacromolecules* 19: 1154–1163. https://doi.org/10.1021/acs.biomac.8b00016.

65 Ng, K.K., Motoda, Y., Watanabe, S. et al. (2016). Intracellular delivery of proteins via fusion peptides in intact plants. *PLoS One* 11: e0154081. https://doi.org/10.1371/journal.pone.0154081.

66 Bach, A., Clausen, B.H., Møller, M. et al. (2012). A high-affinity, dimeric inhibitor of PSD-95 bivalently interacts with PDZ1-2 and protects against ischemic brain damage. *Proc. Natl. Acad. Sci. U. S. A.* 109: 3317–3322. https://doi.org/10.1073/pnas.1113761109.

67 Park, H.Y., Lee, K.J., Lee, S.J. et al. (2012). Screening of peptides bound to breast cancer stem cell specific surface marker CD44 by phage display. *Mol. Biotechnol.* 51: 212–220. https://doi.org/10.1007/s12033-011-9458-7.

68 Safa, N., Anderson, J.C., Vaithiyanathan, M. et al. (2019). CPProtectides: rapid uptake of well-folded beta-hairpin peptides with enhanced resistance to intracellular degradation. *Pept. Sci. (Hoboken)* 111: https://doi.org/10.1002/pep2.24092.

14

Cell-Penetrating Peptides for Nucleic Acid Delivery – An Update

Eric Vivès, Emilie Josse, Karidia Konate, Sébastien Deshayes, and Prisca Boisguérin

PhyMedExp, University of Montpellier, INSERM U1046, CNRS UMR 9214, Montpellier, 34295, France

14.1 Introduction

Therapeutic nucleic acids (NAs), such as antisense oligonucleotides (ASO), small-interfering RNAs (siRNAs), plasmid DNAs (pDNAs) or messenger RNAs (mRNAs), are more and more considered as a new generation of pharmacological drugs beside small molecules, proteins, or antibodies in the biomedical research [1]. RNA or DNA molecules with different functions acting specifically on protein expression could be fully considered therapeutic NAs. Recent optimization of NA synthesis, chemical modifications, and cellular delivery has accelerated their therapeutic development. Up to now, 17 NA-based therapeutics have been approved by the Food and Drug Administration (FDA) or the European Medicines Agency (EMA).

Short NAs are classified as oligonucleotides (ONs) which could be natural or chemically modified. The use of therapeutic ONs to treat a large number of diseases has expanded the range of possible targets beyond what is generally accessible by conventional pharmaceutics, such as gene silencing, splice modulation, or gene activation. Several examples could be mentioned, such as single-stranded antisense oligonucleotides (ASOs) of 15–20 nucleotides, acting by selectively cleaving pre-mRNAs having complementary sites *via* an RNase H-dependent mechanism [2]. To date, four splice-switch ASOs have been approved by the FDA: Eteplirsen [phosphorodiamidate morpholino oligonucleotide (PMO); Sarepta Therapeutics; 2016], Golodirsen [PMO; Sarepta Therapeutics; 2019], Viltolarsen [PMO; Nippon Shinyaku Pharma; 2020] and Nusinersen [2′-O-methyl phosphorothioate (2’OMe PS); Spinraza; 2016] [3]. Additionally, other clinical trials (Phases I to III) are very encouraging for other ASOs in different pathologies such as Huntington’s disease, amyotrophic lateral sclerosis, Dravet syndrome, and Duchene muscular dystrophy (DMD).

Cell-Penetrating Peptides: Design, Development and Applications, First Edition. Makoto Oba and Yosuke Demizu.

Subsequently, siRNAs (20–25 double-stranded nucleotides) were developed as a major therapeutic tool for silencing gene expression *via* degradation of the target mRNA [4, 5]. Today, two therapeutics based on RNA interference (RNAi) are approved: ONPATTRO® (Partisiran) for the treatment of polyneuropathy in hereditary transthyretin-mediated (hATTR) amyloidosis in 2018 [6], and GIVLAARI™ (Givosiran) for treating acute hepatic porphyria (AHP) in 2019 [7].

Single-stranded micro RNAs (miRs, 20–24 nucleotides) are endogenous, highly conserved, small noncoding RNAs that have been implicated as key regulators of target gene expression [8]. At the post-transcriptional level, miRs bind to the 3′-untranslated regions of the corresponding target mRNAs of protein-coding genes, thereby resulting in target mRNA degradation and the lack of mRNA translation.

Longer NAs such as therapeutic DNAs are mainly used in the form of plasmids (pDNA), which encode specific genes or regulatory sequences for endogenous proteins [9]. The suppressor gene p53 is, to our knowledge, the most widely transferred gene in clinical trials to replace the mutated p53 expressed in different types of cancer [10]. In 2003, the State Food and Drug Administration of China approved the recombinant adenovirus-based p53 gene therapy developed by the Chinese company Shenzhen SiBiono GenTech against head and neck squamous cell carcinoma [11]. Another example is the most recently FDA-authorized gene therapy called Zolgensma (Novartis) inducing the expression of the survival motor neuron 1 protein (SMN1) in motor neurons for the treatment of spinal muscular atrophy (SMA) [12]. pDNA can also be used to edit genes *via* the CRISPR mechanism through the internalization of plasmids coding the Cas9 protein and the RNA guide strand corresponding to the targeted gene sequence.

More recently, also the mRNA-based technology was pushed forward with two mRNA vaccines marketed by the companies Pfizer/BioNTech [13] and Moderna [14] to fight against the worldwide COVID-19 pandemic. In addition, several other mRNA candidates are currently in clinical trials for vaccine-based cancer immunotherapy [15].

Even if these NAs are considered "drugs of the future," they require the use of carriers to overcome their weak cell penetration as well as the rapid endo/lysosomal degradation once inside the cell. Several delivery technologies based on viral or nonviral strategies (cationic lipids, polymers, dendrimers or peptides) have been developed [3, 16]. Among these nonviral carriers, cell-penetrating peptides (CPPs) have gained increasing popularity because of their short length and efficient cell translocation ability [17–20]. Usually defined as short (up to 30 amino acids) peptides that originate from different sources (e.g. human, mouse, virus, or purely synthetic), CPPs were developed as one of the most promising nonviral strategies for improving the intracellular routing of NAs. There are two main strategies to vectorize oligonucleotides (ONs) using CPPs: by directly conjugating ONs to CPPs (covalent conjugation approach) or by forming noncovalent complexes between CPPs and ONs (nanoparticle-based approach) (see Section 14.5).

In this chapter, we present the characteristics of CPPs, their use as conjugates or as peptide-based nanoparticles (PBNs) as well as their internalization mechanism. Further on, we have also highlighted the possible modifications/functionalizations of CPPs to increase their specificity and efficacity as well as the clinical perspectives. CPPs used to decorate other nanoparticles such as polymers, nanotubes, or even viruses have been reported elsewhere [21, 22].

14.2 Characteristics of CPPs

CPPs were developed after structure-function analyses of proteins known to efficiently internalize in cells in vitro [23–25] resulting in a short sequence of 16-amino-acid peptide named $pAntp_{43-58}$ or penetratin [26] and another 13 amino acids peptide including a highly cationic region named Tat_{48-60} [27]. A few years later, a multitude of peptides with translocation properties were identified as CPPs [17].

Most CPPs are small, cationic, with a typical length of 5–30 residues and a median charge of +5 [19]. They can pass through cell membranes *via* energy-dependent or independent mechanisms without any interactions with specific receptors [28]. Although several different criteria have been proposed for the classification of CPPs based on their origin, sequence, function, or mechanism of uptake, no unified taxonomy of these peptides presently exists [19]. According to their physical–chemical properties, they can be simply categorized into three main classes: cationic, amphipathic, and hydrophobic peptides [29].

14.3 Mechanism of CPP Internalization

The mechanisms of intracellular internalization of CPPs are still not fully understood mainly due to results dependent on a large heterogeneity of physicochemical properties, sizes, and applied concentrations of CPPs and their corresponding cargoes [30]. However, it is nowadays well accepted that CPPs likely use different pathways which could be considered energy-independent (direct translocation) and energy-dependent (endocytosis) mechanisms (Figure 14.1). Independently of the subsequent internalization mechanism, the first contact of CPPs with the cell membrane appears to be with the endothelial glycocalyx (network of membrane-bound proteoglycans and glycoproteins) covering the endothelium lumen. Glycosaminoglycans (GAGs), generally exposing polyanionic charges, are involved in the cellular binding and uptake of nearly all cationic CPPs or polycationic nanoparticles [31–33].

14.3.1 Endocytosis-Dependent Mechanism

After solving the problem of cell fixation leading to the artefactual observation of cell internalization, the endocytosis-dependent pathway was suggested for several CPPs in the early 2000s [34]. Since now, several CPPs (e.g. Tat and stearylated-R8), CPP-conjugates (e.g. Tat-DRBD and Pip6a-PMO) or CPP-complexes (e.g. PepFect:siRNA and RALA:siRNA) have been shown to use energy-dependent

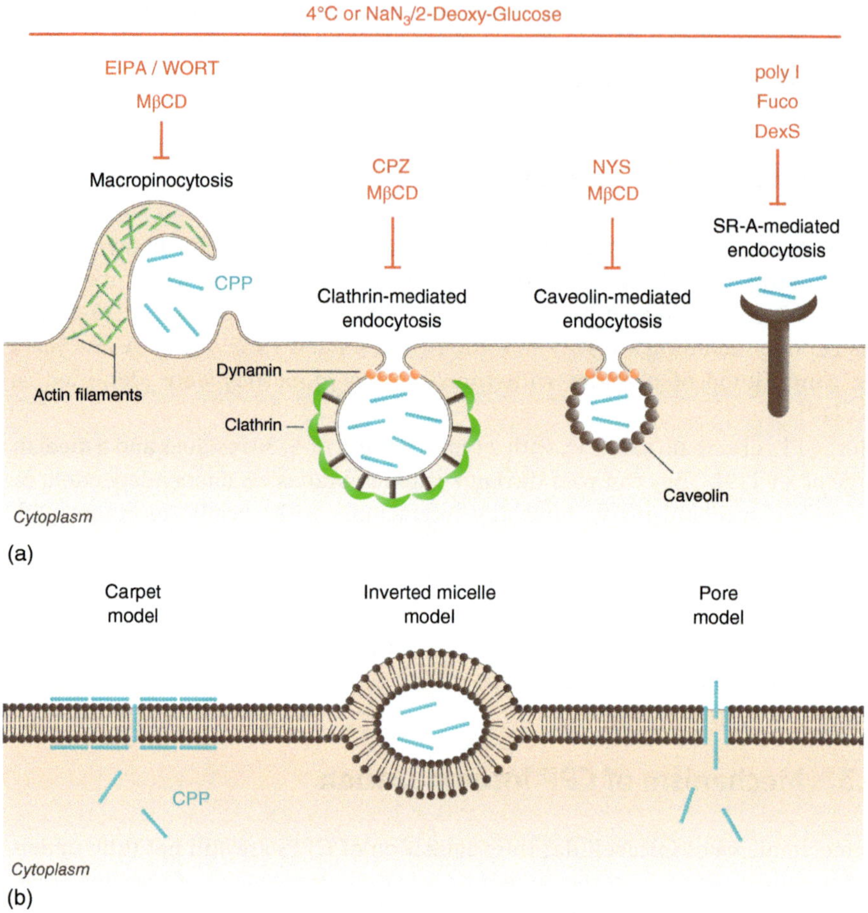

Figure 14.1 Schema of different cellular internalization mechanisms used by CPPs. (**a**) Vectorized therapeutic molecules can be taken up by macropinocytosis or by clathrin- and caveolin-mediated endocytosis. In some cases, cellular internalization can be receptor-mediated as shown for the PepFect family depending on the Scavenger A receptors (SR-A). All these energy-dependent pathways can be inhibited using incubation at 4 °C or with NaN_3/2-deoxyglucose (DG) for ATP depletion (-ATP). More specifically, clathrin-dependent endocytosis will be inhibited by chlorpromazine (CPZ), caveolin-dependent endocytosis by nystatin (NYS), and macropinocytosis by 5-(N-ethyl-N-isopropyl)-amiloride (EIPA) and wortmannin (WORT). Methyl-β-cyclodextrin(MβCD), a compound that sequesters the cholesterolin the cell membrane, is often used as an inhibitor for lipid-raft inducedinternalization (macropinocytosis and endocytosis). (**b**) For direct translocation, there are several proposed models such as the transient pore formation (toroidal pore or barrel stave pore model), the inverted micelle model, or the carpet model leading to membrane rearrangements or transient membrane disruptions. All chemical inhibitors used to inhibit endocytosis-dependent internalization have no effects on direct cellular translocation.

endocytosis pathways (reviewed in [21, 31, 35]). Especially, clathrin- or caveolae-mediated endocytosis, as well as macropinocytosis have been suggested to explain the uptake of different CPPs and their cargoes (Figure 14.1a). These pathways were often characterized using biophysical methods (e.g. infrared spectroscopy, electron microscopy, liposomal leakage assay, etc.), chemical inhibitors (such as nystatin, chlorpromazine, etc.) or co-localization studies with fluorescent-labeled markers (e.g. Transferrin, Cholera toxin subunit B, etc.) [36–39].

Consequently, several approaches have been developed to improve their endosomal escape to limit as much as possible their lysosomal degradation with the general goal to improve intracellular delivery and efficiency. Strategies in the peptide development aimed to facilitate endosomal escape *via* osmotic swelling of the endosome through the "proton sponge effect" (e.g. C6M3 peptide [40]), by induction of conformational changes of the peptide (e.g. RALA peptide [41]), or due to the conjugation of fusogenic peptides (e.g. HA peptide) or fatty acids (stearylation) (Reviewed in [42]).

14.3.2 Direct Translocation

Besides endocytosis-dependent internalization, there is accumulating evidence that arginine-rich CPPs were also able to be internalized by direct translocation through plasma membranes. As a process, which requires no energy, direct translocation was defined as a mechanism involving the formation of inverted micelles, pores, and accumulation of peptides at the membrane surface in the "carpet" model [28]. In all cases, it required temporary destabilization of the membrane in order to enable cellular internalization (Figure 14.1b).

For CPP-conjugates, a correlation existed between the administration of concentrated arginine-rich CPPs and their internalization modes. Brock and co-workers demonstrated that cytosolic translocation of the 3 CPPs Tat, Penetratin, and nona-arginine occurred for concentrations higher than 10 μM [37].

Nevertheless, several CPPs were shown to use also direct translocation in well-defined studies. As described for the MPG peptide, the mechanism of internalization of a CPP, and also of a non-covalent CPP:ON-complex, was associated with transient permeabilization events enabling a direct translocation process [43, 44]. Moreover, CPP-like peptides issued from perforin and granzyme B proteins were able to enter the cells by locally distorting the plasma membrane (transient calcium influx). The resulting membrane damage caused by the uptake was masked by membrane repair response (LAMP-2 detection) as demonstrated by Palm and co-workers [45]. Brock and co-workers also showed that the direct translocation of cationic CPPs depended on the CPP-induced translocation of acid sphingomyelinase to the outer leaflet of the plasma membrane and ceramide formation [46].

Additionally, for siRNA transfection, PBNs formulated with CADY [47], RICK [48], or WRAP [39] were shown to internalize in different cell lines without co-localization with endosomal markers and remained active in the presence of endocytosis inhibitors or at low temperature.

14.4 CPPs Used as Conjugates for Nucleic Acid Delivery

CPPs were first covalently conjugated to different NAs in order to enhance their cellular delivery. However, due to electrostatic interactions between positively charged CPPs and negatively charged NAs, the internalization properties of the CPPs were masked. Therefore, the covalent linkage of CPPs to NAs was essentially restricted to neutral NAs such as peptide NAs (PNAs) or PMO as detailed later.

14.4.1 CPP-PNA

Peptide nucleic acid (PNA) is an artificial DNA mimic with a peptide backbone instead of the original phosphoribosyl backbone, thus combining the characteristics of both peptides and NAs. Unmodified PNAs are also not taken up in cell culture in vitro, and have a poor bioavailability in vivo [49]. To address these problems, PNAs were chemically modified in their backbones and nucleobases (reviewed in [50]), or they were directly conjugated to CPPs [51, 52].

The first demonstration of PNA conjugated to Penetratin was performed by Corey and co-workers [53] followed by Langel and co-workers comparing PNA efficiency when coupled to Penetratin or Transportan [54]. Afterward, several other groups have started to compare PNA delivery upon their modification with Penetratin, nuclear localization sequences (NLS), Tat or lysine residues [55–58]. The first successful report of an in vivo application was demonstrated by Cao and co-workers in 2016. Using an acute Hepatitis B virus (HBV) mouse model, they showed an antiviral effect after an intravenous injection of Tat-PNA targeting the direct repeat sequences of HBV [59].

Based on the work of Engelmann and co-workers [60, 61], Glazer and co-workers took advantage of the pH-low insertion peptides (pHLIP) to target the acidity of the tumor microenvironment and deliver more selectively cargoes at the tumor site. They conjugated pHLIP to a PNA-targeting DNA double-strand break repair factor, KU80, in order to radio-sensitize tumor cells [62].

Another example of PNA transfection was shown by Pentelute and co-workers who performed a sortase-mediated ligation to couple the two nontoxic components of the anthrax toxin, the protective antigen (PA) and the N-terminal domain of lethal factor (LFN, 263-mer) to a PNA suppressing a CYCLOPS (Copy-number alterations Yielding Cancer Liabilities Owing to Partial losS) gene [63]. Even if the transfection was performed by a protein rather than a peptide, the advantage of LFN-/PA-mediated delivery was demonstrated by the from 100- to 1000-fold higher antisense activity in cancer cell lines compared to PNA alone or Tat-PNA conjugates.

Besides natural CPPs, an extended work was first performed on synthetic CPPs such as mono-, di-, or polylysine for the PNA delivery by Kole and co-workers [64] and afterward by Corey and co-workers [65]. Wright et al. further demonstrated that PNA$(Lys)_8$ transfection for the correction of the IVS2-654 β-globin splice defect could be improved by co-administration of a PA (a nontoxic component of anthrax toxin) in cultured erythroid precursor cells from β-thalassemia patients [66]. Octa-Lys conjugated PNAs (K8-PNAs) were first evaluated in vivo (mouse model) by

Wancewicz et al. showing dose-dependent and sequence-specific modulation of splicing [67]. K8-PNA allowed PNA accumulation in liver, kidney, and adipose tissue. However, convincing activity was only observed in adipose tissue.

Several works were performed by Gait, Lebleu, and co-workers on PNA delivery using R6-Penetratin showing in vitro activity for luciferase splicing [68] and inhibition of Tat-dependent trans-activation [58], as well as for miRNA inhibition in vitro and in vivo [69, 70]. These groups also performed structure–activity relationship studies on Pip and RXR series in order to understand the importance of CPP composition for inducing the ideal PNA transfection [36, 71].

Other synthetic CPP-PNA conjugates were reported by Nielson and co-workers using $(KFF)_3K$-PNA to target bacterial genes [72], by Gambari and co-workers using R8-PNA to inhibit onco-miRNA miR-221 [73] and by Pei and co-workers showing that small, amphipathic cyclic arginine-rich peptides were more efficient than the linear Tat for PNA delivery [74].

14.4.2 CPP-PMO

The other promising application of CPPs is the possibility of a covalent chemical conjugation to largely used charge-neutral PMOs. Wood and co-workers have pioneered the use of peptide–PMO (PPMO) conjugates for the treatment of various diseases, most notably for dystrophin splice switching in the context of DMD [3, 31, 75]. Early studies demonstrated the efficacy of dystrophin exon skipping using (RXR)4-PMO and B-peptide [(RXRRBR)2XB] (with X = 6-aminohexanoic acid and B = β-alanine) [36, 71]. This PPMO approach was further optimized by generating a chimeric peptide consisting of B-peptide fused with a muscle-specific peptide (MSP) [76, 77] or by using an analog (M12) identified by phage display [78].

As mentioned earlier for PNA delivery, the "Pip"-series developed by Gait and coworkers were also used to generate Pip-PMO conjugates for in vitro and in vivo DMD exon skipping [71, 79–81]. However, PPMO technology revealed some toxic effects such as renal damage in different animal models (rat and cynomolgus monkey) [82, 83]. Nephrotoxicity was directly correlated to the presence of a high amount of arginine residues within the CPP sequences and therefore, research efforts were focused on peptide sequence optimization in order to avoid renal toxicity without compromising splice correction efficacy.

PPMOs are also promising antibacterial and antiviral agents as demonstrated for (RXR)4–PMO conjugates targeting an essential bacterial Acyl carrier (acpP) gene in murine infection models [84] or arginine-rich peptide–PMO conjugates exerting protective effects in murine viral infection models [85–87].

14.5 CPPs Used as Nanoparticles for Nucleic Acid Delivery

Based on their amphipathic properties or modifications (e.g. fatty acid chain grafting), a subset of CPPs was used to encapsulate NAs by forming nanoparticles for triggering their cellular delivery (Figure 14.2). For example, three CPP families will

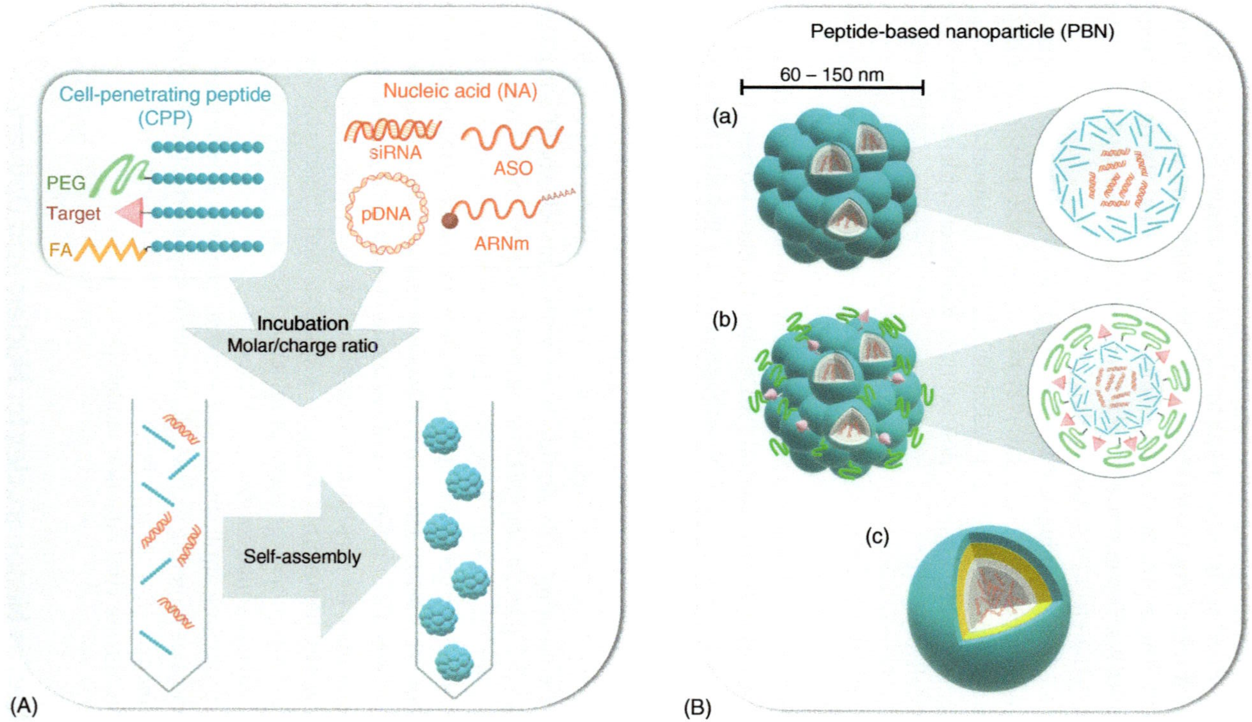

Figure 14.2 **Formulation of peptide-based nanoparticles in the presence of different nucleic acids.** (**a**) Peptide-based nanoparticles (PBNs) are formulated by mixing a CPP or a grafted CPP [PEGylated, targeting sequence or fatty acid (FA)] with an NA, such as pDNA, mRNA, siRNA, or ASO at a given molar or charge ratio. (**b**) By mixing these two compounds, the nanoparticle is formed by self-assembling into naked PBNs (a), multigrafted PBNs (b) or prospective micelle-like PBNs (no model available) (c). In all cases, the PBNs of mean size between 60 and 150 nm encapsulated several NAs for cellular transfection. Source: Boisguérin et al. [88] / MDPI / CC BY-4.0.

be detailed in the next sections, whereas other nanoparticle-forming CPPs could be found in different recent reviews [21, 88, 89].

14.5.1 The PepFect Family

Langel and co-workers developed a subset of different CPPs for NA delivery based on stearylation of the short transportan-derived peptide TP10 to improve non-covalent ON-complex formation and enhance peptide-membrane interactions [90]. Among them, the peptide called PepFect 3 (PF3) was further modified by the incorporation of an endosomolytic trifluoromethylquinoline moiety to increase the endosomal escape of the peptide [91]. Further on, the PepFect 14 (PF14) peptide was designed with non-encoded ornithine residues for increased stability and improved uptake [92]. Evaluation of the internalization mechanism revealed that all PepFect:NA complexes were being taken up via receptor-mediated endocytosis involving class-A scavenger receptors (SCARAs) [93, 94] (**see Figure 14.1b**). More recently, Kurrikoff and colleagues showed that PepFect in vivo gene delivery can be enhanced using an optimal combination of overall charge and hydrophobicity in the peptide backbone as well as through fatty acid alteration [95].

Brock and co-workers published recently the use of PepFect14 to formulate CPP:mRNA nanoparticles, showing efficient reporter protein expression in two- and three-dimensional cancer cell cultures [96]. More importantly, they could reveal an important mCherry protein expression within the tumors of the treated mice after injection of PBNs encapsulating mCherry coding mRNA.

In parallel to the PepFect family development, Langel and co-workers developed the NickFect family by modifying the PF3 sequence [97]. In detail, by replacing Lys7 with ornithine and continuing the synthesis by coupling Gly6 to the δ-NH2 group of ornithine, the authors obtained first the NickFect 51 (NF51) for PBN-delivery of pDNA in different cell types [98]. Afterward, they designed a novel amphipathic α-helical peptide, NF55, with modified net charge and helicity for efficient in vivo pDNA delivery [97, 99]. More recently, they presented a new formulation approach called cryo-concentration for obtaining stable and homogeneous nanoparticles showing significantly higher bioactivity in vivo [100].

14.5.2 The WRAP Family

Based on the knowledge that most amphipathic peptides have distinct hydrophobic and hydrophilic domains for cargo interactions and nanoparticle formation, Deshayes and co-workers have designed the W- and R-rich amphipathic peptides (WRAP) for NA delivery. These CPPs (15-/16-mers) composed of only three amino acids (leucine, arginine, and tryptophan) were able to form stable PBNs in the presence of siRNA and to trigger specific protein silencing in different cell lines [101]. This knock-down efficiency resulted from a rapid PBN internalization following a 5–15 minutes incubation and was associated with a cellular internalization mechanism mainly driven by direct translocation of PBNs through the plasma membrane [39]. Only a minor part of the WRAP-PBNs entered cells through endocytosis-dependent internalization as revealed by a set of experimental approaches combining biophysical (leakage assay) and biological (dynamin triple-KO cells) methods as well as confocal (endocytosis and vesicle markers), and electron microscopy experiments. More interestingly, even if some WRAP-PBNs were internalized by endocytosis-dependent mechanisms, they could rapidly escape from endosomes as suggested by leakage assays using large unilamellar vesicles (LUVs) reflecting the endosomal membrane composition.

Based on a structure-activity relationship (SAR) study correlating the amino acid composition with nanoparticle formation and cellular siRNA-internalization using 13 new WRAP-analogues, Konate and co-workers further determined the sequential features for obtaining the most efficient CPPs for siRNA delivery [102]. More importantly, they confirmed that these WRAP peptides induced an equivalent level of targeted gene silencing compared to other non-peptide transfection agents but in most cases with lower cell toxicity. Finally, Couillaud and co-workers established the proof of concept for specific gene silencing in solid tumors through in vivo molecular imaging [103].

Costa and co-workers proved that WRAP peptide was also suitable as a gene delivery system in the context of cancer gene therapy, as shown by the WRAP-mediated delivery of a p53 encoding pDNA [104]. Through the design of an experimental tool, the optimal ratio of nitrogen to phosphate groups (N/P) was determined for WRAP:pDNA in comparison with the complex formed by the RALA peptide as discussed later. Moreover, the same group demonstrated that WRAP-PBNs could be used as a therapeutic approach to treat mitochondrial genome (mtDNA) based diseases [105]. WRAP-PBNs grafted with a mitochondrial targeting sequence (MTS) revealed a clear mitochondrial targeting as shown by confocal microscopy and *Western blot* evaluation on mitochondrial cell fractions.

14.5.3 The RALA Family

The fusogenic peptide GALA (30-mer) was designed as a simple model of a viral fusion peptide but due to its anionic nature, it was not able to condense NAs [106]. Consequently, the GALA peptide was modified by replacing glutamate residues with lysine residues, resulting in the KALA peptide, which was then able to condense NAs to nanoparticles [107]. KALA has also been used in combination with polylysine [108], polyethylenimine (PEI) [109], and (poly (DMAEMA-NVP))-block-PEG-galactose [110] for DNA gene delivery. More recently, Katayama and co-workers found that a liposome modified with the KALA peptide was the most effective drug delivery system for mitochondrial targeting in C2C12 cells [111].

McCarthy and co-workers developed the RALA peptide by substituting lysine residues with arginine known to improve DNA condensing ability and cell penetration [41]. This new secondary amphipathic peptide folded in an α-helix revealed enhanced NA-condensation, improved cell penetration, and minimal cytotoxicity compared to the previous generation. RALA PBNs were internalized via the clathrin- and caveolin-mediated endocytosis pathways, resulting in endosomal encapsulation. However, pH-drop (pH <) in the endosomes induced an increasing α-helicity of RALA, triggering the endosomal release of the transfected cargo before its degradation [41]. Modification of the RALA sequence in terms of amino acid composition and sequence length failed to improve the functional characteristics of RALA, confirming its more suitable sequence for nontoxic gene delivery [112]. RALA has become a widely used PBN-delivery system as exemplified for the transfection of different cargoes such as therapeutic plasmids [41, 104, 113], siRNA [114–116], mRNA [117], or for DNA vaccination [118], in addition to the anionic small molecule (such as bisphosphonates) demonstrating its broad utility [119].

14.6 CPP Functionalization for a More Specific Delivery

Peptides in general, and CPPs or PBNs in particular, offer the possibility to be grafted in different ways to a large variety of molecules to modify or improve their overall characteristics (Figure 14.3). These include their stability and availability in biological fluids, their cell-penetrating properties to increase their internalization into the cells, and also the boost of their cellular selectivity to address the transported drugs to a specific cell type. These modifications rely on the grafting or the insertion of aliphatic moieties to increment the hydrophobicity of the CPPs and to later better interfere with cellular membranes, either the plasma membrane itself to promote a direct membrane translocation, or at several levels all along the endocytosis pathway to destabilize endosomal or lysosomal membranes once the drug conjugate is entrapped in these vesicles.

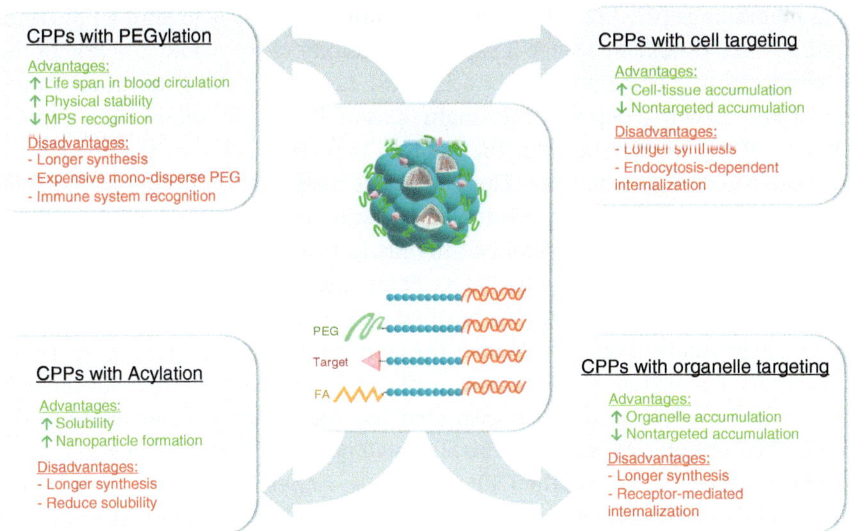

Figure 14.3 Examples of functionalized CPPs and PBNs. NA-conjugates of CPPs or NA-encapsulating PBNs could be functionalized with different molecules such as PEG moiety, acyl derivates, or grafting cell or organelle targeting sequences (as exemplified here for siRNA delivery). For each modification, some advantages or disadvantages are provided. Source: Boisguérin et al. [88] / MDPI / CC BY-4.0.

We can also consider the covalent grafting of targeting peptides (or sugars) to selectively address the drug to a specific receptor expressed by a peculiar targeted cell.

But we can also further project to target intracellular compartments thanks to specific targeting peptides for several applications. For instance, the targeting of the nucleus or, more recently of the mitochondria, upon the grafting of NLS or MTS has been already performed [105, 120]. Modification of CPPs upon the grafting of PEG units has been also developed, mainly not only in order to avoid the opsonization of drug-CPP conjugates onto serum proteins, but also to limit their elimination through the hepatic and the renal systems [121]. Altogether, PEGylation of CPPs is critical to increasing their blood circulation time and, consequently, improving significantly their chance to reach ultimately the targeted cell or tissue. In general, all these modifications have been done independently for each class of these molecules (fatty acids, targeting units, and PEGs). But, ultimately, the synthesis of "smart nanoparticles" built up with an accurate balance of these molecules upon the incorporation of differing amounts of each focuses the attention of the drug delivery community. In this section, we illustrate the most recent modifications of the functionalization of peptide-based delivery vehicles.

14.6.1 Lipid-Grafting

Since, at one stage, CPPs have to interfere with cell lipid bilayer to enter cell cytoplasm, great attention has been considered to increasing their hydrophobic character. Grafting of a lipidic tail has been thus early explored in the CPP history to improve the behavior of different CPPs in order to better fuse with the hydrophobic plasma membranes [92, 122, 123]. Chemically talking, the way to graft a lipid onto a peptide is well established since it exploits the ability of the acidic function of the fatty acid to form a stable amide bond in most cases with the N-terminal amino group of the growing peptide. The main reason is that in solid-phase peptide synthesis (SPPS), the synthesis of the peptide is performed from the C-terminal toward the N-terminal extremity. Therefore, this modification occurs at the final step while the peptide is already assembled and still linked to the resin. The quantitative coupling of the fatty acid can be thus easily and clearly monitored using, for instance, a conventional ninhydrin test. Then, the alkylated peptide has to be cleaved and deprotected, and further purified and characterized. Several investigations have been made upon, for instance, the coupling of acyl chains ranging from 2 to 22 carbons. For instance, in the case of pDNA, it has been shown that the efficiency of the transfection of the transported NA was directly correlated with the increased hydrophobic character of the alkylated peptide [123]. The location of the hydrophobic character within the CPP can be also rather variable. Along this line, peptide chemistry offers several alternative strategies to graft acyl chains elsewhere than at the N-terminal end, like, for instance, onto the amine side chain of lysine or an ornithine residue [100]. Such modifications have been also evaluated to internalize both pDNAs and ONs, such as splicing correcting nucleotides or siRNAs.

Finally, with the prospect of developing in a near-future peptide-based vaccines, a plethora of potential modifications at different levels with one or several acyl chains has been reported in a very recent review [124]. These modifications could indeed be exploited to modify potentially CPPs used for the delivery of NAs to modify/alter/improve the hydrophobic nature of the formed nanoparticles.

On the other hand, the hydrophobic character is also closely in balance with the cationic nature of the CPP. For instance, a lipopeptide was demonstrated as a good transfection peptide for pDNA delivery in the lungs [100]. But an analog with two additional cationic charges showed a higher gene transfection efficiency [95]. In addition, the transfection efficacy is more probably linked to a very accurate ratio between the hydrophobicity and the cationic characters of CPPs to ensure an appropriate nanoparticle formation and, ultimately, efficient transfection of NA molecules.

14.6.2 PEGylation

PEG is one of the most widely used "stealth" polymers in the drug delivery field. The FDA classified it in the eighties as a Generally Regarded as Safe (GRAS) compound. One of the seminal works using PEG grafting described that its covalent attachment to bovine serum albumin and liver catalase proteins increased their systemic circulation time and decreased their immunogenicity without significantly altering their biological function [125]. It was later shown that PEGylation also improved the stability and circulation time of various nonviral drug delivery systems. Finally, in the nineties, several PEGylated drugs were already marketed with the agreement of the FDA (for a review, see the work of Suk and colleagues [126]).

It can be also understood that PEGylation of lipid-based nanoparticles (LNPs) ensures their better solubility in biological environments. This is probably the rationale for using a low percentage of PEGylated lipids in the formulation of both Moderna and Pfizer vaccines developed against the COVID-19 worldwide pandemic [13, 14]. But more importantly, the reduction of the opsonization of the nanoparticles by serum proteins remains the most rational reason to graft PEG chains onto nanoparticles [127]. We have also shown in a previous work that PEGylation of PBNs to vectorize siRNAs significantly improved their residency time in the bloodstream following intravenous injection [128]. But more specifically, in the case of peptide-mediated delivery of NAs, it should be also largely admitted that PEGylation reduced immunogenicity against the peptide itself [129]. For an optimal benefit, it has been calculated that the appropriate PEG density on PEG-PLGA nanoparticles was 20 PEG chains per 100 nm^2 (for a 5 kDa PEG unit) [130].

On the other hand, it is increasingly suspected that PEGylated drug delivery systems can also induce the formation of anti-PEG antibodies that specifically recognize and bind to PEG (for recent reviews, see [131, 132]). Therefore, the question of the effect of PEG-surface density, the content, the molecular weight and the type of functional groups is also a crucial issue since they are critical factors that have to be considered to ensure a minimal PEG-associated immune response. These aspects have been deeply developed and discussed in a section of a recent review by Hoang Thi and co-workers [131] and will not be further developed here. As a consequence, the development of alternative polymers to replace PEG is already engaged.

14.6.3 Homing Peptide Sequences

Several peptides (with a length ranging from 3 to 12 amino acids) are known to bind to specific receptors almost exclusively expressed by some cells, and more especially to tumor cells. They can be used directly coupled to a chelator associated with a radionuclide as a peptide receptor radionuclide therapy (PRRT) (for a review see [133] and references herein). The way to identify these "homing" peptides is almost endless since several techniques of combinatorial libraries allow their selection and their evaluation to trigger a cell-specific binding (for a review [134]). Therefore, these peptides represent extremely interesting binding and targeting units to concentrate to the surface of these cells a radio-element, a drug, a nanoparticle, or an NA in any type of drug-delivery system.

These homing peptide sequences can be synthesized under different forms, such as linear, cyclic, bicyclic, or branched depending on the nature of the interaction between the ligand and the cell receptor. In general, a cyclic form of such a short ligand is expected to show a higher affinity for the corresponding receptor thanks to the higher rigidity of the cyclic structure. A better biostability is also expected upon cyclization of a short peptide. But non-natural amino acids can be integrated to ensure a higher stability of the targeting peptide against proteases. These homing peptides can be also glycosylated if needed. As already mentioned earlier, with the coupling of fatty acids, the peptide chemistry indeed offers a large diversity for directly grafting these targeting peptides onto PBNs. The coupling between the targeting peptide and the peptide interacting with the NA can be covalent or can be reducible, for instance, to allow the release of the PBNs once the particles reach the reducing environment within the cytosol [135].

In some cases, the subcellular delivery of an NA toward a well-defined compartment can be required. For instance, this could be indeed the nucleus thanks to the insertion of an NLS [136]. Alternatively, the mitochondria could be also targeted upon the coupling of an MTS to a PBN as recently demonstrated for the delivery of a pDNA [105].

Another approach to increasing the specific delivery of therapeutic molecules is the identification of the so-called cell-specific CPPs through different screening methods (different phage display techniques or large peptide libraries). The advantage of these cell-specific CPPs is to reduce nonspecific cellular uptake leading to potent side effects and to reduce administered doses of the therapeutic compound [137]. Also, the enzyme-controlled delivery is used to increase specific drug delivery by adding to the CPP a unique peptide sequence for a specific disease as exemplified by the cleavage-sides for urokinase-type plasminogen activator [138] or for matrix metalloprotease-2 [139], both overexpressed in tumors.

14.7 Conclusions

Considerable developments have occurred in the field of NA-delivery since its establishment in the 1970s. Clinically validated platform technologies, including viral vectors, LNPs, and chemical conjugates, have made it possible to exploit the delivery of different therapeutic NAs. We believe that the recent surge in the LNP development due to the COVID pandemic will help pave the way for future CPP-based advances in medicine. Since the presentation of CPPs is a valuable delivery system for a variety of therapeutic molecules, extensive research has been directed toward their use in gene therapy. The clinical translation of CPP-mediated therapeutics is still ongoing and huge efforts are being made to achieve success. The highly efficient and selective CPPs have a bright future in the treatment of life-threatening diseases, especially brain tumors or rare genetic diseases.

Finally, the functionalization of CPPs or PBNs for improving the delivery of NAs will probably be multifaceted, to conciliate together the transfection efficiency of the NA (its accessibility or release inside the cells), the residence time and the pharmacologic kinetics of CPPs/PBNs in biological fluids (PEG grafting or alternative molecules), and the targeting of CPPs/PBNs to cells/tissues to be treated upon the high-affinity interaction between a targeting unit and the specific cell-type receptor. There will probably not only be a plethora of combinations based on the nature and the ratio of the different components but it will also be based on the number of these to obtain the highest equilibrium between composition and efficiency of these types of "smart delivery vectors."

Acknowledgments

This work was financially supported by the foundation "La Ligue contre le Cancer," the "Fondation ARC pour la Recherche sur le Cancer," the "Accélérateur d'Innovation, Montpellier Université d'Excellence 2021," and the "Centre National de la Recherche Scientifique" (CNRS).

References

1 Talap, J., Zhao, J., Shen, M. et al. (2021). Recent advances in therapeutic nucleic acids and their analytical methods. *Journal of Pharmaceutical and Biomedical Analysis 206*: 114368. https://doi.org/10.1016/j.jpba.2021.114368.

2 Crooke, S.T. (2004). Antisense strategies. *Current Molecular Medicine 4* (5): 465–487. https://doi.org/10.2174/1566524043360375.

3 Roberts, T.C., Langer, R., and Wood, M.J.A. (2020). Advances in oligonucleotide drug delivery. *Nature Reviews. Drug Discovery 19* (10): 673–694. https://doi.org/10.1038/s41573-020-0075-7.

4 Tolia, N.H. and Joshua-Tor, L. (2007). Slicer and the argonautes. *Nature Chemical Biology 3* (1): 36–43. https://doi.org/10.1038/nchembio848.

5 Whitehead, K.A., Langer, R., and Anderson, D.G. (2009). Knocking down barriers: advances in siRNA delivery. *Nature Reviews. Drug Discovery 8* (2): 129–138. https://doi.org/10.1038/nrd2742.

6 Hoy, S.M. (2018). Patisiran: first global approval. *Drugs 78* (15): 1625–1631. https://doi.org/10.1007/s40265-018-0983-6.

7 Scott, L.J. (2020). Givosiran: First approval. *Drugs 80* (3): 335–339. https://doi.org/10.1007/s40265-020-01269-0.

8 Reschke, C.R. and Henshall, D.C. (2015). MicroRNA and epilepsy. *Advances in Experimental Medicine and Biology 888*: 41–70. https://doi.org/10.1007/978-3-319-22671-2_4.

9 Gaspar, R., Coelho, F., and Silva, B.F.B. (2020). Lipid-nucleic acid complexes: physicochemical aspects and prospects for cancer treatment. *Molecules 25* (21): https://doi.org/10.3390/molecules25215006.

10 Ginn, S.L., Amaya, A.K., Alexander, I.E. et al. (2018). Gene therapy clinical trials worldwide to 2017: an update. *The Journal of Gene Medicine 20* (5): e3015. https://doi.org/10.1002/jgm.3015.

11 Pearson, S., Jia, H., and Kandachi, K. (2004). China approves first gene therapy. *Nature Biotechnology 22* (1): 3–4. https://doi.org/10.1038/nbt0104-3.

12 Waldrop, M.A. and Kolb, S.J. (2019). Current treatment options in neurology – SMA therapeutics. *Current Treatment Options in Neurology 21* (6): 25. https://doi.org/10.1007/s11940-019-0568-z.

13 Polack, F.P., Thomas, S.J., Kitchin, N. et al. (2020). Safety and efficacy of the BNT162b2 mRNA Covid-19 vaccine. *The New England Journal of Medicine* https://doi.org/10.1056/NEJMoa2034577.

14 Ledford, H. (2020). Moderna COVID vaccine becomes second to get US authorization. *Nature* https://doi.org/10.1038/d41586-020-03593-7.

15 Di Trani, C.A., Fernandez-Sendin, M., Cirella, A. et al. (2021). Advances in mRNA-based drug discovery in cancer immunotherapy. *Expert Opinion on Drug Discovery 17*: 1–13. https://doi.org/10.1080/17460441.2021.1978972.

16 Senapati, S., Mahanta, A.K., Kumar, S. et al. (2018). Controlled drug delivery vehicles for cancer treatment and their performance. *Signal Transduction and Targeted Therapy 3*: 7. https://doi.org/10.1038/s41392-017-0004-3.

17 Langel, Ü. (ed.) (2015). *Cell-Ppenetrating Peptides*, vol. *1324*. New York: Springer http://link.springer.com/10.1007/978-1-4939-2806-4.

18 Ramsey, J.D. and Flynn, N.H. (2015). Cell-penetrating peptides transport therapeutics into cells. *Pharmacology & Therapeutics 154*: 78–86. https://doi.org/10.1016/j.pharmthera.2015.07.003.

19 Kauffman, W.B., Fuselier, T., He, J. et al. (2015). Mechanism matters: a taxonomy of cell penetrating peptides. *Trends in Biochemical Sciences 40* (12): 749–764. https://doi.org/10.1016/j.tibs.2015.10.004.

20 Stanzl, E.G., Trantow, B.M., Vargas, J.R. et al. (2013). Fifteen years of cell-penetrating, guanidinium-rich molecular transporters: basic science, research tools, and clinical applications. *Accounts of Chemical Research 46* (12): 2944–2954. https://doi.org/10.1021/ar4000554.

21 Tarvirdipour, S., Huang, X., Mihali, V. et al. (2020). Peptide-based nanoassemblies in gene therapy and diagnosis: paving the way for clinical application. *Molecules (Basel, Switzerland) 25* (15): https://doi.org/10.3390/molecules25153482.

22 Desale, K., Kuche, K., and Jain, S. (2021). Cell-penetrating peptides (CPPs): an overview of applications for improving the potential of nanotherapeutics. *Biomaterials Science 9* (4): 1153–1188. https://doi.org/10.1039/d0bm01755h.
23 Frankel, A.D. and Pabo, C.O. (1988). Cellular uptake of the tat protein from human immunodeficiency virus. *Cell 55* (6): 1189–1193.
24 Green, M. and Loewenstein, P.M. (1988). Autonomous functional domains of chemically synthesized human immunodeficiency virus tat trans-activator protein. *Cell 55* (6): 1179–1188.
25 Joliot, A., Pernelle, C., Deagostini-Bazin, H. et al. (1991). Antennapedia homeobox peptide regulates neural morphogenesis. *Proceedings of the National Academy of Sciences 88* (5): 1864–1868. https://doi.org/10.1073/pnas.88.5.1864.
26 Derossi, D., Calvet, S., Trembleau, A. et al. (1996). Cell internalization of the third helix of the Antennapedia homeodomain is receptor-independent. *The Journal of Biological Chemistry 271* (30): 18188–18193. https://doi.org/10.1074/jbc.271.30.18188.
27 Vivès, E., Brodin, P., and Lebleu, B. (1997). A truncated HIV-1 Tat protein basic domain rapidly translocates through the plasma membrane and accumulates in the cell nucleus. *The Journal of Biological Chemistry 272* (25): 16010–16017. https://doi.org/10.1074/jbc.272.25.16010.
28 Ruseska, I. and Zimmer, A. (2020). Internalization mechanisms of cell-penetrating peptides. *Beilstein Journal of Nanotechnology 11*: 101–123. https://doi.org/10.3762/bjnano.11.10.
29 Milletti, F. (2012). Cell-penetrating peptides: classes, origin, and current landscape. *Drug Discovery Today 17* (15–16): 850–860. https://doi.org/10.1016/j.drudis.2012.03.002.
30 Mueller, J., Kretzschmar, I., Volkmer, R. et al. (2008). Comparison of cellular uptake using 22 CPPs in 4 different cell lines. *Bioconjugate Chemistry 19* (12): 2363–2374. https://doi.org/10.1021/bc800194e.
31 Boisguérin, P., Deshayes, S., Gait, M.J. et al. (2015). Delivery of therapeutic oligonucleotides with cell penetrating peptides. *Advanced Drug Delivery Reviews 87*: 52–67. https://doi.org/10.1016/j.addr.2015.02.008.
32 Lehto, T., Ezzat, K., Wood, M.J.A. et al. (2016). Peptides for nucleic acid delivery. *Advanced Drug Delivery Reviews 106* (Pt A): 172–182. https://doi.org/10.1016/j.addr.2016.06.008.
33 Amoura, M., Illien, F., Joliot, A. et al. (2019). Head to tail cyclisation of cell-penetrating peptides: impact on GAG-dependent internalisation and direct translocation. *Chemical Communications (Cambridge, England) 55* (31): 4566–4569. https://doi.org/10.1039/c9cc01265f.
34 Richard, J.P., Melikov, K., Vives, E. et al. (2003). Cell-penetrating peptides. A reevaluation of the mechanism of cellular uptake. *The Journal of Biological Chemistry 278* (1): 585–590. https://doi.org/10.1074/jbc.M209548200.
35 Klabenkova, K., Fokina, A., and Stetsenko, D. (2021). Chemistry of peptide-oligonucleotide conjugates: a review. *Molecules (Basel, Switzerland) 26* (17): 5420. https://doi.org/10.3390/molecules26175420.
36 Abes, S., Moulton, H.M., Clair, P. et al. (2006). Vectorization of morpholino oligomers by the (R-Ahx-R)4 peptide allows efficient splicing correction in the

absence of endosomolytic agents. *Journal of Controlled Release 116* (3): 304–313. https://doi.org/10.1016/j.jconrel.2006.09.011.

37 Duchardt, F., Fotin-Mleczek, M., Schwarz, H. et al. (2007). A comprehensive model for the cellular uptake of cationic cell-penetrating peptides. *Traffic (Copenhagen, Denmark) 8* (7): 848–866. https://doi.org/10.1111/j.1600-0854.2007.00572.x.

38 Vercauteren, D., Vandenbroucke, R.E., Jones, A.T. et al. (2010). The use of inhibitors to study endocytic pathways of gene carriers: optimization and pitfalls. *Molecular Therapy 18* (3): 561–569. https://doi.org/10.1038/mt.2009.281.

39 Deshayes, S., Konate, K., Dussot, M. et al. (2020). Deciphering the internalization mechanism of WRAP: siRNA nanoparticles. *Biochimica Et Biophysica Acta. Biomembranes 1862* (6): 183252. https://doi.org/10.1016/j.bbamem.2020.183252.

40 Xu, W., Pan, R., Zhao, D. et al. (2015). Design and evaluation of endosomolytic biocompatible peptides as carriers for siRNA delivery. *Molecular Pharmaceutics 12* (1): 56–65. https://doi.org/10.1021/mp500429u.

41 McCarthy, H.O., McCaffrey, J., McCrudden, C.M. et al. (2014). Development and characterization of self-assembling nanoparticles using a bio-inspired amphipathic peptide for gene delivery. *Journal of Controlled Release 189*: 141–149. https://doi.org/10.1016/j.jconrel.2014.06.048.

42 Erazo-Oliveras, A., Muthukrishnan, N., Baker, R. et al. (2012). Improving the endosomal escape of cell-penetrating peptides and their cargos: strategies and challenges. *Pharmaceuticals 5* (11): 1177–1209. https://doi.org/10.3390/ph5111177.

43 Deshayes, S., Gerbal-Chaloin, S., Morris, M.C. et al. (2004). On the mechanism of non-endosomial peptide-mediated cellular delivery of nucleic acids. *Biochimica Et Biophysica Acta 1667* (2): 141–147. https://doi.org/10.1016/j.bbamem.2004.09.010.

44 Deshayes, S., Plénat, T., Charnet, P. et al. (2006). Formation of transmembrane ionic channels of primary amphipathic cell-penetrating peptides. Consequences on the mechanism of cell penetration. *Biochimica Et Biophysica Acta 1758* (11): 1846–1851. https://doi.org/10.1016/j.bbamem.2006.08.010.

45 Palm-Apergi, C., Lorents, A., Padari, K. et al. (2009). The membrane repair response masks membrane disturbances caused by cell-penetrating peptide uptake. *FASEB Journal 23* (1): 214–223. https://doi.org/10.1096/fj.08-110254.

46 Verdurmen, W.P.R., Thanos, M., Ruttekolk, I.R. et al. (2010). Cationic cell-penetrating peptides induce ceramide formation via acid sphingomyelinase: Implications for uptake. *Journal of Controlled Release 147* (2): 171–179. https://doi.org/10.1016/j.jconrel.2010.06.030.

47 Rydström, A., Deshayes, S., Konate, K. et al. (2011). Direct translocation as major cellular uptake for CADY self-assembling peptide-based nanoparticles. *PLOS ONE 6* (10): e25924. https://doi.org/10.1371/journal.pone.0025924.

48 Vaissière, A., Aldrian, G., Konate, K. et al. (2017). A retro-inverso cell-penetrating peptide for siRNA delivery. *Journal of Nanobiotechnology 15* (1): 34. https://doi.org/10.1186/s12951-017-0269-2.

49 Nielsen, P.E. (2005). Addressing the challenges of cellular delivery and bioavailability of peptide nucleic acids (PNA). *Quarterly Reviews of Biophysics 38* (4): 345–350. https://doi.org/10.1017/S0033583506004148.

50 Brodyagin, N., Katkevics, M., Kotikam, V. et al. (2021). Chemical approaches to discover the full potential of peptide nucleic acids in biomedical applications. *Beilstein Journal of Organic Chemistry 17*: 1641–1688. https://doi.org/10.3762/bjoc.17.116.

51 Saarbach, J., Sabale, P.M., and Winssinger, N. (2019). Peptide nucleic acid (PNA) and its applications in chemical biology, diagnostics, and therapeutics. *Current Opinion in Chemical Biology 52*: 112–124. https://doi.org/10.1016/j.cbpa.2019.06.006.

52 Gait, M.J., Arzumanov, A.A., McClorey, G. et al. (2019). Cell-penetrating peptide conjugates of steric blocking oligonucleotides as therapeutics for neuromuscular diseases from a historical perspective to current prospects of treatment. *Nucleic Acid Therapeutics 29* (1): 1–12. https://doi.org/10.1089/nat.2018.0747.

53 Simmons, C.G., Pitts, A.E., Mayfield, L.D. et al. (1997). Synthesis and membrane permeability of PNA-peptide conjugates. *Bioorganic & Medicinal Chemistry Letters 7* (23): 3001–3006. https://doi.org/10.1016/S0960-894X(97)10136-6.

54 Pooga, M., Soomets, U., Hällbrink, M. et al. (1998). Cell penetrating PNA constructs regulate galanin receptor levels and modify pain transmission in vivo. *Nature Biotechnology 16* (9): 857–861. https://doi.org/10.1038/nbt0998-857.

55 Cutrona, G., Carpaneto, E.M., Ulivi, M. et al. (2000). Effects in live cells of a c-myc anti-gene PNA linked to a nuclear localization signal. *Nature Biotechnology 18* (3): 300–303. https://doi.org/10.1038/73745.

56 Braun, K., Peschke, P., Pipkorn, R. et al. (2002). A biological transporter for the delivery of peptide nucleic acids (PNAs) to the nuclear compartment of living cells. *Journal of Molecular Biology 318* (2): 237–243. https://doi.org/10.1016/S0022-2836(02)00031-1.

57 Koppelhus, U., Awasthi, S.K., Zachar, V. et al. (2002). Cell-dependent differential cellular uptake of PNA, peptides, and PNA-peptide conjugates. *Antisense and Nucleic Acid Drug Development 12* (2): 51–63. https://doi.org/10.1089/108729002760070795.

58 Turner, J.J., Ivanova, G.D., Verbeure, B. et al. (2005). Cell-penetrating peptide conjugates of peptide nucleic acids (PNA) as inhibitors of HIV-1 Tat-dependent trans -activation in cells. *Nucleic Acids Research 33* (21): 6837–6849. https://doi.org/10.1093/nar/gki991.

59 Zeng, Z., Han, S., Hong, W. et al. (2016). A Tat-conjugated peptide nucleic acid Tat-PNA-DR inhibits hepatitis B virus replication in vitro and in vivo by targeting LTR direct repeats of HBV RNA. *Molecular Therapy – Nucleic Acids 5*: e295. https://doi.org/10.1038/mtna.2016.11.

60 An, M., Wijesinghe, D., Andreev, O.A. et al. (2010). pH-(low)-insertion-peptide (pHLIP) translocation of membrane impermeable phalloidin toxin inhibits cancer cell proliferation. *Proceedings of the National Academy of Sciences 107* (47): 20246–20250. https://doi.org/10.1073/pnas.1014403107.

61 Andreev, O., Engelman, D., and Reshetnyak, Y. (2014). Targeting diseased tissues by pHLIP insertion at low cell surface pH. *Frontiers in Physiology 5*: 97. https://doi.org/10.3389/fphys.2014.00097.

62 Kaplan, A.R., Pham, H., Liu, Y. et al. (2020). Ku80-targeted pH-sensitive peptide–PNA conjugates are tumor selective and sensitize cancer cells to ionizing radiation. *Molecular Cancer Research 18* (6): 873–882. https://doi.org/10.1158/1541-7786.MCR-19-0661.

63 Lu, Z., Paolella, B.R., Truex, N.L. et al. (2020). Targeting cancer gene dependencies with anthrax-mediated delivery of peptide nucleic acids. *ACS Chemical Biology 15* (6): 1358–1369. https://doi.org/10.1021/acschembio.9b01027.

64 Sazani, P., Kang, S.-H., Maier, M.A. et al. (2001). Nuclear antisense effects of neutral, anionic and cationic oligonucleotide analogs. *Nucleic Acids Research 29* (19): 3965–3974. https://doi.org/10.1093/nar/29.19.3965.

65 Hu, J. and Corey, D.R. (2007). Inhibiting gene expression with peptide nucleic acid (PNA)–peptide conjugates that target chromosomal DNA. *Biochemistry 46* (25): 7581–7589. https://doi.org/10.1021/bi700230a.

66 Wright, D.G., Zhang, Y., and Murphy, J.R. (2008). Effective delivery of antisense peptide nucleic acid oligomers into cells by anthrax protective antigen. *Biochemical and Biophysical Research Communications 376* (1): 200–205. https://doi.org/10.1016/j.bbrc.2008.08.124.

67 Wancewicz, E.V., Maier, M.A., Siwkowski, A.M. et al. (2010). Peptide nucleic acids conjugated to short basic peptides show improved pharmacokinetics and antisense activity in adipose tissue. *Journal of Medicinal Chemistry 53* (10): 3919–3926. https://doi.org/10.1021/jm901489k.

68 Abes, S., Turner, J.J., Ivanova, G.D. et al. (2007). Efficient splicing correction by PNA conjugation to an R6-Penetratin delivery peptide. *Nucleic Acids Research 35* (13): 4495–4502. https://doi.org/10.1093/nar/gkm418.

69 Fabani, M.M. and Gait, M.J. (2008). MiR-122 targeting with LNA/2′-O-methyl oligonucleotide mixmers, peptide nucleic acids (PNA), and PNA–peptide conjugates. *RNA 14* (2): 336–346. https://doi.org/10.1261/rna.844108.

70 Fabani, M.M., Abreu-Goodger, C., Williams, D. et al. (2010). Efficient inhibition of miR-155 function in vivo by peptide nucleic acids. *Nucleic Acids Research 38* (13): 4466–4475. https://doi.org/10.1093/nar/gkq160.

71 Ivanova, G.D., Arzumanov, A., Abes, R. et al. (2008). Improved cell-penetrating peptide–PNA conjugates for splicing redirection in HeLa cells and exon skipping in mdx mouse muscle. *Nucleic Acids Research 36* (20): 6418–6428. https://doi.org/10.1093/nar/gkn671.

72 Good, L., Awasthi, S.K., Dryselius, R. et al. (2001). Bactericidal antisense effects of peptide–PNA conjugates. *Nature Biotechnology 19* (4): 360–364. https://doi.org/10.1038/86753.

73 Brognara, E., Fabbri, E., Bazzoli, E. et al. (2014). Uptake by human glioma cell lines and biological effects of a peptide-nucleic acids targeting miR-221. *Journal of Neuro-Oncology 118* (1): 19–28. https://doi.org/10.1007/s11060-014-1405-6.

74 Qian, Z., Martyna, A., Hard, R.L. et al. (2016). Discovery and mechanism of highly efficient cyclic cell-penetrating peptides. *Biochemistry 55* (18): 2601–2612. https://doi.org/10.1021/acs.biochem.6b00226.

75 Aoki, Y., Yokota, T., and Wood, M.J.A. (2013). Development of multiexon skipping antisense oligonucleotide therapy for Duchenne muscular dystrophy. *BioMed Research International 2013*: 402369. https://doi.org/10.1155/2013/402369.

76 Yin, H., Moulton, H.M., Betts, C. et al. (2009). A fusion peptide directs enhanced systemic dystrophin exon skipping and functional restoration in dystrophin-deficient mdx mice. *Human Molecular Genetics 18* (22): 4405–4414. https://doi.org/10.1093/hmg/ddp395.

77 Yin, H., Boisguerin, P., Moulton, H.M. et al. (2013). Context dependent effects of chimeric peptide morpholino conjugates contribute to dystrophin exon-skipping efficiency. *Molecular Therapy – Nucleic Acids 2*: e124. https://doi.org/10.1038/mtna.2013.51.

78 Gao, X., Zhao, J., Han, G. et al. (2014). Effective dystrophin restoration by a novel muscle-homing peptide-morpholino conjugate in dystrophin-deficient mdx mice. *Molecular Therapy: The Journal of the American Society of Gene Therapy 22* (7): 1333–1341. https://doi.org/10.1038/mt.2014.63.

79 Betts, C., Saleh, A.F., Arzumanov, A.A. et al. (2012). Pip6-PMO, A new generation of peptide-oligonucleotide conjugates with improved cardiac exon skipping activity for DMD treatment. *Molecular Therapy – Nucleic Acids 1*: e38. https://doi.org/10.1038/mtna.2012.30.

80 Betts, C.A., Saleh, A.F., Carr, C.A. et al. (2015). Prevention of exercised induced cardiomyopathy following Pip-PMO treatment in dystrophic mdx mice. *Scientific Reports 5*: 8986. https://doi.org/10.1038/srep08986.

81 Lehto, T., Alvarez, A.C., Gauck, S. et al. (2014). Cellular trafficking determines the exon skipping activity of Pip6a-PMO in mdx skeletal and cardiac muscle cells. *Nucleic Acids Research 42* (5): 3207–3217. https://doi.org/10.1093/nar/gkt1220.

82 Amantana, A., Moulton, H.M., Cate, M.L. et al. (2007). Pharmacokinetics, biodistribution, stability and toxicity of a cell-penetrating peptide-morpholino oligomer conjugate. *Bioconjugate Chemistry 18* (4): 1325–1331. https://doi.org/10.1021/bc070060v.

83 Moulton, H.M. and Moulton, J.D. (2010). Morpholinos and their peptide conjugates: therapeutic promise and challenge for Duchenne muscular dystrophy. *Biochimica Et Biophysica Acta 1798* (12): 2296–2303. https://doi.org/10.1016/j.bbamem.2010.02.012.

84 Geller, B.L., Marshall-Batty, K., Schnell, F.J. et al. (2013). Gene-silencing antisense oligomers inhibit acinetobacter growth in vitro and in vivo. *The Journal of Infectious Diseases 208* (10): 1553–1560. https://doi.org/10.1093/infdis/jit460.

85 Burrer, R., Neuman, B.W., Ting, J.P.C. et al. (2007). Antiviral effects of antisense morpholino oligomers in murine coronavirus infection models. *Journal of Virology 81* (11): 5637–5648. https://doi.org/10.1128/JVI.02360-06.

86 Neuman, B.W., Stein, D.A., Kroeker, A.D. et al. (2006). Inhibition and escape of SARS-CoV treated with antisense morpholino oligomers. *Advances in Experimental Medicine and Biology 581*: 567–571. https://doi.org/10.1007/978-0-387-33012-9_103.

87 Enterlein, S., Warfield, K.L., Swenson, D.L. et al. (2006). VP35 knockdown inhibits Ebola virus amplification and protects against lethal infection in mice.

Antimicrobial Agents and Chemotherapy 50 (3): 984–993. https://doi.org/10.1128/AAC.50.3.984-993.2006.

88 Boisguérin, P., Konate, K., Josse, E. et al. (2021). Peptide-based nanoparticles for therapeutic nucleic acid delivery. *Biomedicines 9* (5): https://doi.org/10.3390/biomedicines9050583.

89 Xu, J., Khan, A.R., Fu, M. et al. (2019). Cell-penetrating peptide: a means of breaking through the physiological barriers of different tissues and organs. *Journal of Controlled Release 309*: 106–124. https://doi.org/10.1016/j.jconrel.2019.07.020.

90 Mäe, M., El Andaloussi, S., Lundin, P. et al. (2009). A stearylated CPP for delivery of splice correcting oligonucleotides using a non-covalent co-incubation strategy. *Journal of Controlled Release 134* (3): 221–227. https://doi.org/10.1016/j.jconrel.2008.11.025.

91 Andaloussi, S.E.L., Lehto, T., Mäger, I. et al. (2011). Design of a peptide-based vector, PepFect6, for efficient delivery of siRNA in cell culture and systemically in vivo. *Nucleic Acids Research 39* (9): 3972–3987. https://doi.org/10.1093/nar/gkq1299.

92 Ezzat, K., Andaloussi, S.E.L., Zaghloul, E.M. et al. (2011). PepFect 14, a novel cell-penetrating peptide for oligonucleotide delivery in solution and as solid formulation. *Nucleic Acids Research 39* (12): 5284–5298. https://doi.org/10.1093/nar/gkr072.

93 Ezzat, K., Helmfors, H., Tudoran, O. et al. (2012). Scavenger receptor-mediated uptake of cell-penetrating peptide nanocomplexes with oligonucleotides. *FASEB Journal 26* (3): 1172–1180. https://doi.org/10.1096/fj.11-191536.

94 Lindberg, S., Regberg, J., Eriksson, J. et al. (2015). A convergent uptake route for peptide- and polymer-based nucleotide delivery systems. *Journal of Controlled Release 206*: 58–66. https://doi.org/10.1016/j.jconrel.2015.03.009.

95 Kurrikoff, K., Veiman, K.-L., Künnapuu, K. et al. (2017). Effective in vivo gene delivery with reduced toxicity, achieved by charge and fatty acid -modified cell penetrating peptide. *Scientific Reports 7* (1): 17056. https://doi.org/10.1038/s41598-017-17316-y.

96 van den Brand, D., Gorris, M.A.J., van Asbeck, A.H. et al. (2019). Peptide-mediated delivery of therapeutic mRNA in ovarian cancer. *European Journal of Pharmaceutics and Biopharmaceutics 141*: 180–190. https://doi.org/10.1016/j.ejpb.2019.05.014.

97 Freimann, K., Arukuusk, P., Kurrikoff, K. et al. (2016). Optimization of in vivo DNA delivery with NickFect peptide vectors. *Journal of Controlled Release 241*: 135–143. https://doi.org/10.1016/j.jconrel.2016.09.022.

98 Arukuusk, P., Pärnaste, L., Oskolkov, N. et al. (2013). New generation of efficient peptide-based vectors, NickFects, for the delivery of nucleic acids. *Biochimica et Biophysica Acta (BBA) – Biomembranes 1828* (5): 1365–1373. https://doi.org/10.1016/j.bbamem.2013.01.011.

99 Freimann, K., Arukuusk, P., Kurrikoff, K. et al. (2018). Formulation of stable and homogeneous cell-penetrating peptide NF55 nanoparticles for efficient gene delivery in vivo. *Molecular Therapy – Nucleic Acids 10*: 28–35. https://doi.org/10.1016/j.omtn.2017.10.011.

100 Kurrikoff, K., Freimann, K., Veiman, K.-L. et al. (2019). Effective lung-targeted RNAi in mice with peptide-based delivery of nucleic acid. *Scientific Reports 9* (1): 19926. https://doi.org/10.1038/s41598-019-56455-2.

101 Konate, K., Dussot, M., Aldrian, G. et al. (2019). Peptide-based nanoparticles to rapidly and efficiently "Wrap 'n roll" siRNA into cells. *Bioconjugate Chemistry 30* (3): 592–603. https://doi.org/10.1021/acs.bioconjchem.8b00776.

102 Konate, K., Josse, E., Tasic, M. et al. (2021). WRAP-based nanoparticles for siRNA delivery: a SAR study and a comparison with lipid-based transfection reagents. *Journal of Nanobiotechnology 19* (1): 236. https://doi.org/10.1186/s12951-021-00972-8.

103 Ferreiro, I., Genevois, C., Konate, K. et al. (2021). in vivo follow-up of gene inhibition in solid tumors using peptide-based nanoparticles for siRNA delivery. *Pharmaceutics 13* (5): 749. https://doi.org/10.3390/pharmaceutics13050749.

104 Sousa, Â., Almeida, A.M., Faria, R. et al. (2019). Optimization of peptide-plasmid DNA vectors formulation for gene delivery in cancer therapy exploring design of experiments. *Colloids and Surfaces B: Biointerfaces 183*: 110417. https://doi.org/10.1016/j.colsurfb.2019.110417.

105 Faria, R., Vivés, E., Boisguerin, P. et al. (2021). Development of peptide-based nanoparticles for mitochondrial plasmid DNA delivery. *Polymers 13* (11): 1836. https://doi.org/10.3390/polym13111836.

106 Li, W., Nicol, F., and Szoka, F.C. (2004). GALA: a designed synthetic pH-responsive amphipathic peptide with applications in drug and gene delivery. *Advanced Drug Delivery Reviews 56* (7): 967–985. https://doi.org/10.1016/j.addr.2003.10.041.

107 Wyman, T.B., Nicol, F., Zelphati, O. et al. (1997). Design, synthesis, and characterization of a cationic peptide that binds to nucleic acids and permeabilizes bilayers. *Biochemistry 36* (10): 3008–3017. https://doi.org/10.1021/bi9618474.

108 Lee, H., Jeong, J.H., and Park, T.G. (2002). PEG grafted polylysine with fusogenic peptide for gene delivery: high transfection efficiency with low cytotoxicity. *Journal of Controlled Release 79* (1–3): 283–291. https://doi.org/10.1016/s0168-3659(02)00002-0.

109 Lee, H., Jeong, J.H., and Park, T.G. (2001). A new gene delivery formulation of polyethylenimine/DNA complexes coated with PEG conjugated fusogenic peptide. *Journal of Controlled Release 76* (1–2): 183–192. https://doi.org/10.1016/s0168-3659(01)00426-6110.

110 Lim, D.W., Yeom, Y.I., and Park, T.G. (2000). Poly(DMAEMA-NVP)-b-PEG-galactose as gene delivery vector for hepatocytes. *Bioconjugate Chemistry 11* (5): 688–695. https://doi.org/10.1021/bc000014u.

111 Katayama, T., Kinugawa, S., Takada, S. et al. (2019). A mitochondrial delivery system using liposome-based nanocarriers that target myoblast cells. *Mitochondrion 49*: 66–72. https://doi.org/10.1016/j.mito.2019.07.005.

112 McErlean, E.M., McCrudden, C.M., McBride, J.W. et al. (2021). Rational design and characterisation of an amphipathic cell penetrating peptide for non-viral gene

delivery. *International Journal of Pharmaceutics 596*: 120223. https://doi.org/10.1016/j.ijpharm.2021.120223.
113 McCrudden, C.M., McBride, J.W., McCaffrey, J. et al. (2018). Gene therapy with RALA/iNOS composite nanoparticles significantly enhances survival in a model of metastatic prostate cancer. *Cancer Nanotechnology 9* (1): 5. https://doi.org/10.1186/s12645-018-0040-x.
114 Bennett, R., Yakkundi, A., McKeen, H.D. et al. (2015). RALA-mediated delivery of FKBPL nucleic acid therapeutics. *Nanomedicine (London, England) 10* (19): 2989–3001. https://doi.org/10.2217/nnm.15.115.
115 Mulholland, E.J., Ali, A., Robson, T. et al. (2019). Delivery of RALA/siFKBPL nanoparticles via electrospun bilayer nanofibres: an innovative angiogenic therapy for wound repair. *Journal of Controlled Release 316*: 53–65. https://doi.org/10.1016/j.jconrel.2019.10.050.
116 Yan, L.-P., Castaño, I.M., Sridharan, R. et al. (2020). Collagen/GAG scaffolds activated by RALA-siMMP-9 complexes with potential for improved diabetic foot ulcer healing. *Materials Science & Engineering C: Materials for Biological Applications 114*: 111022. https://doi.org/10.1016/j.msec.2020.111022.
117 Udhayakumar, V.K., De Beuckelaer, A., McCaffrey, J. et al. (2017). Arginine-rich peptide-based mRNA nanocomplexes efficiently instigate cytotoxic T cell immunity dependent on the amphipathic organization of the peptide. *Advanced Healthcare Materials 6* (13): 1601412. https://doi.org/10.1002/adhm.201601412.
118 Cole, G., Ali, A.A., McErlean, E. et al. (2019). DNA vaccination via RALA nanoparticles in a microneedle delivery system induces a potent immune response against the endogenous prostate cancer stem cell antigen. *Acta Biomaterialia 96*: 480–490. https://doi.org/10.1016/j.actbio.2019.07.003.
119 Jena, L., McErlean, E., and McCarthy, H. (2020). Delivery across the blood-brain barrier: nanomedicine for glioblastoma multiforme. *Drug Delivery and Translational Research 10* (2): 304–318. https://doi.org/10.1007/s13346-019-00679-2.
120 Lam, A.P. and Dean, D.A. (2010). Progress and prospects: nuclear import of nonviral vectors. *Gene Therapy 17* (4): 439–447. https://doi.org/10.1038/gt.2010.31.
121 Hatakeyama, H., Akita, H., and Harashima, H. (2013). The polyethyleneglycol dilemma: Advantage and disadvantage of PEGylation of liposomes for systemic genes and nucleic acids delivery to tumors. *Biological and Pharmaceutical Bulletin 36* (6): 892–899. https://doi.org/10.1248/bpb.b13-00059.
122 Katayama, S., Hirose, H., Takayama, K. et al. (2011). Acylation of octaarginine: implication to the use of intracellular delivery vectors. *Journal of Controlled Release 149* (1): 29–35. https://doi.org/10.1016/j.jconrel.2010.02.004.
123 Lehto, T., Vasconcelos, L., Margus, H. et al. (2017). Saturated fatty acid analogues of cell-penetrating peptide PepFect14: role of fatty acid modification in complexation and delivery of splice-correcting oligonucleotides. *Bioconjugate Chemistry 28* (3): 782–792. https://doi.org/10.1021/acs.bioconjchem.6b00680.

124 Hamley, I.W. (2021). Lipopeptides for vaccine development. *Bioconjugate Chemistry 32* (8): 1472–1490. https://doi.org/10.1021/acs.bioconjchem.1c00258.

125 Abuchowski, A., McCoy, J.R., Palczuk, N.C. et al. (1977). Effect of covalent attachment of polyethylene glycol on immunogenicity and circulating life of bovine liver catalase. *The Journal of Biological Chemistry 252* (11): 3582–3586.

126 Suk, J.S., Xu, Q., Kim, N. et al. (2016). PEGylation as a strategy for improving nanoparticle-based drug and gene delivery. *Advanced Drug Delivery Reviews 99* (Pt A): 28–51. https://doi.org/10.1016/j.addr.2015.09.012.

127 Romberg, B., Hennink, W.E., and Storm, G. (2008). Sheddable coatings for long-circulating nanoparticles. *Pharmaceutical Research 25* (1): 55–71. https://doi.org/10.1007/s11095-007-9348-7.

128 Aldrian, G., Vaissière, A., Konate, K. et al. (2017). PEGylation rate influences peptide-based nanoparticles mediated siRNA delivery in vitro and in vivo. *Journal of Controlled Release 256*: 79–91. https://doi.org/10.1016/j.jconrel.2017.04.012.

129 Pisal, D.S., Kosloski, M.P., and Balu-Iyer, S.V. (2010). Delivery of therapeutic proteins. *Journal of Pharmaceutical Sciences 99* (6): 2557–2575. https://doi.org/10.1002/jps.22054.

130 Bertrand, N., Grenier, P., Mahmoudi, M. et al. (2017). Mechanistic understanding of in vivo protein corona formation on polymeric nanoparticles and impact on pharmacokinetics. *Nature Communications 8* (1): 777. https://doi.org/10.1038/s41467-017-00600-w.

131 Thi, T.T.H., Pilkington, E.H., Nguyen, D.H. et al. (2020). The importance of poly(ethylene glycol) alternatives for overcoming PEG immunogenicity in drug delivery and bioconjugation. *Polymers 12* (2): E298. https://doi.org/10.3390/polym12020298.

132 Shi, D., Beasock, D., Fessler, A. et al. (2021). To PEGylate or not to PEGylate: Immunological properties of nanomedicine's most popular component, poly(ethylene) glycol and its alternatives. *Advanced Drug Delivery Reviews 114079*: https://doi.org/10.1016/j.addr.2021.114079.

133 Lozza, C., Teulon, I., Pelegrin, A. et al. (2013). Peptides in receptor-mediated radiotherapy: from design to the clinical application in cancers. *Frontiers in Oncology 3*: 247. https://doi.org/10.3389/fonc.2013.00247.

134 Liu, R., Li, X., Xiao, W. et al. (2017). Tumor-targeting peptides from combinatorial libraries. *Advanced Drug Delivery Reviews 110–111*: 13–37. https://doi.org/10.1016/j.addr.2016.05.009.

135 Son, S., Singha, K., and Kim, W.J. (2010). Bioreducible BPEI-SS-PEG-cNGR polymer as a tumor targeted nonviral gene carrier. *Biomaterials 31* (24): 6344–6354. https://doi.org/10.1016/j.biomaterials.2010.04.047.

136 Xu, Y., Liang, W., Qiu, Y. et al. (2016). Incorporation of a nuclear localization signal in pH responsive LAH4-L1 peptide enhances transfection and nuclear uptake of plasmid DNA. *Molecular Pharmaceutics 13* (9): 3141–3152. https://doi.org/10.1021/acs.molpharmaceut.6b00338.

137 Taylor, R.E. and Zahid, M. (2020). Cell penetrating peptides, novel vectors for gene therapy. *Pharmaceutics 12* (3): E225. https://doi.org/10.3390/pharmaceutics12030225.

138 Braun, G.B., Sugahara, K.N., Yu, O.M. et al. (2016). Urokinase-controlled tumor penetrating peptide. *Journal of Controlled Release 232*: 188–195. https://doi.org/10.1016/j.jconrel.2016.04.027.

139 Veiman, K.-L., Künnapuu, K., Lehto, T. et al. (2015). PEG shielded MMP sensitive CPPs for efficient and tumor specific gene delivery in vivo. *Journal of Controlled Release 209*: 238–247. https://doi.org/10.1016/j.jconrel.2015.04.038.

15

PPMOs: A Case Study for Cell-Penetrating Peptide Application

Scott Bittner[1], *Zoe X. Moulton*[2,3], *and Illstrated by Hong M. Moulton*[1,2,3]

[1] *Oregon State University, Department of Biomedical Sciences, Carlson College of Veterinary Medicine, Corvallis, OR, 97331, USA*
[2] *Oregon State University, Department of Biochemistry and Biophysics, Corvallis, OR, 97331, USA*
[3] *Oregon State University, Department of Music, Corvallis, OR, 97331, USA*

15.1 Introduction

Control over biology is a longstanding aim pursued by humanity. As we have come to a deeper understanding of the mechanisms at play beneath the surface of the natural world, we have also endeavored to put our understanding to use in order to reduce suffering and provide for our needs. While many attempts to control biology have been focused on exerting influence over the function of gene products, such as through the use of small-molecule drugs interacting with proteins, the rapid expansion of biological knowledge in recent times has revealed a panoply of potential rational ways to exert influence outside of this narrow protein-bound view. Targeting the flow of information within the cell rather than the output of that information promises to greatly expand our influence upon biology and allow novel solutions to recalcitrant challenges in medicine, biological research, and a host of other arenas.

Although the concept of control of gene expression through the use of antisense oligonucleotides and their analogs, collectively referred to as antisense oligomers (oligos), holds endless promise, the realization of this promise has been largely incomplete. This is due, in large part, to the difficulty of delivering these large molecules into the cellular compartments in which they may encounter their RNA targets. Cell-penetrating peptides (CPPs) have become established as a powerful tool for enabling intracellular delivery of antisense oligos. This indispensable development has brought antisense technologies closer to the realization of their promise. We shall examine the development and use of CPPs for the intracellular delivery of a class of antisense oligomers named phosphorodiamidate morpholino oligomers (PMOs or Morpholinos) as a case study, demonstrating the successful adaptation of

Cell-Penetrating Peptides: Design, Development and Applications, First Edition. Makoto Oba and Yosuke Demizu.

CPP technology to the unique challenge of oligo delivery. The lessons that have been learned here in adapting the unique capabilities of CPPs to intracellular delivery of oligos may provide useful intelligence to those attempting to harness CPP capabilities for delivery of other cargoes, and provide a fascinating instance of researchers "connecting the dots" over years of dedicated work, adapting existing technology from a distant field to their particular technical aim.

PMOs are uncharged, synthetic nucleic acid analogues that are designed to complement and bind cellular RNA targets through Watson-Crick base pairing [1], and PPMOs are CPP-conjugated PMOs (Figure 15.1). Modulation of native RNA function is achieved by steric blockage of natural interactions between the targeted RNA and cellular components. Figure 15.2 illustrates four major mechanisms of action of PMOs/PPMOs: (i) targeted to a translation start codon or a 5′-untranslated region on a messenger RNA (mRNA), a PMO can be used to downregulate translation by preventing assembly and function of the mature ribosome; (ii) targeted to a splice site on a precursor mRNA, a PMO can be used to interfere with the natural splicing process through steric blockage of the binding of proteins and small nuclear ribonucleoproteins (snRNPs) leading to the assembly of the spliceosome; (iii) targeted to immature or mature miRNAs PMOs have been used to interfere with microRNA processing and activity [2]; and (iv) targeted to poly-A signal sequences PMOs can inhibit mRNA polyadenylation [3]. Thus, the flow of information within the cell may be targeted in multiple ways, allowing control of gene expression through both downregulation and influence upon precursor mRNA processing.

The use of PMO antisense technology has been explored within a broad range of applications, including therapies for genetic disease resulting in four approved drugs for Duchenne muscular dystrophy [4–7], cell-based and in vivo biological

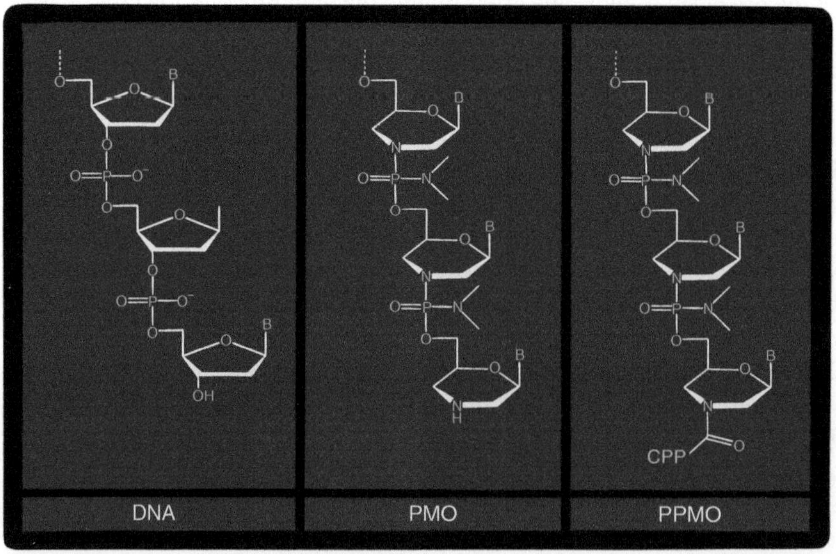

Figure 15.1 Structures of DNA, PMO, and PPMO.

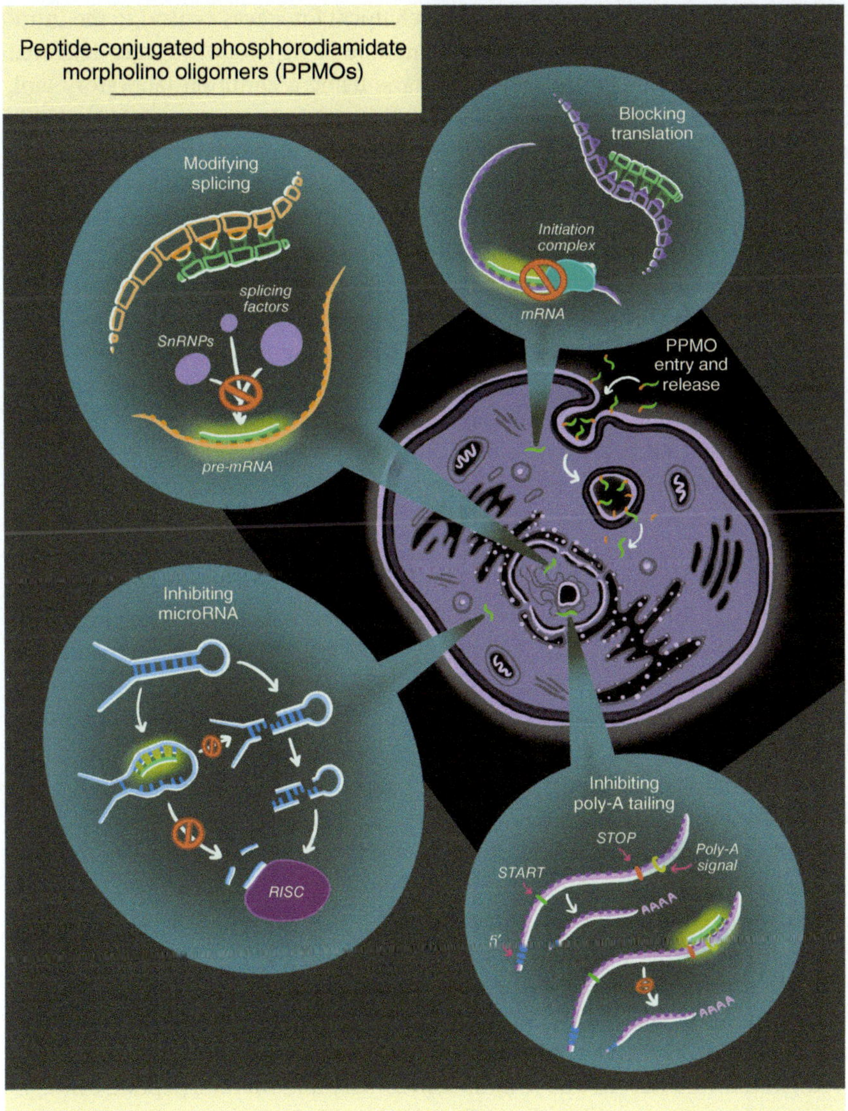

Figure 15.2 Mechanisms of action of PMOs/PPMOs.

research [8–11], and detection of nucleic acids [12]. This charge-neutral antisense oligomer type holds endless promise for manipulation of the biological environment, but many demonstrations of its use heavily depend on delivery methods (e.g. microinjection and electroporation) that are cumbersome to perform and often irrelevant for in vivo applications. Consistent with the difficulty of delivering PMOs to targeted RNA without assistance, limited efficacy has been a prominent subject of discussion in the therapeutic application of the technology [5].

The use of PMOs within many applications is severely limited by an inability to reach the site of intracellular activity without special transfection techniques. In the early days of PMO technology, the use of mechanical force was central to the intracellular delivery of PMOs. Scrape loading, a method in which PMOs were delivered by means of perturbation of the cell membrane using a common laboratory tool, was developed as a simple option for PMO delivery for research work [13]. While the method was appropriate for research work on adherent cell lines, it was minimally relevant to delivery within the organism. The scrape loading method was thought to cause transient openings in the cell membrane, allowing a rapid influx of PMO-containing extracellular fluids. A similar method using mechanical stress imposed on the cell by passage through a syringe was also developed [14]. It is not easy to imagine such methods being effectively employed for therapeutic purposes within the body, and irregularities in PMO delivery using this method additionally make it unattractive for cell-based research. Electroporation has also been used for the delivery of PMOs [15–17], and a method of delivery through the complexation of PMOs with DNA and transfection reagents has been developed [18]. Notably, microinjection of PMOs into embryonic model organisms, particularly zebrafish, has enabled extensive developmental biology research using spatially controlled knockdown of gene expression [9, 19]. However, delivery of inconsistent volumes corresponding to variability in dosage is a concern for use of this method [20].

While these delivery methods have enabled many important research applications of the PMO technology, they fall short of enabling effective delivery for most in vivo uses of PMOs in adult organisms. Despite the excitement engendered by the birth of this revolutionary approach, it was clear that the many potential uses for PMOs within both research and therapeutic contexts demanded further innovation with respect to delivery methods.

The HIV Tat peptide was the first CPP used to deliver PMOs to cells. In 2003, Moulton et al. reported that conjugation of PMOs to the Tat peptide resulted in greatly increased antisense activity within cultured cells [21]. Demonstrated both through downregulation of a reporter gene and through a positive-readout splice modification assay, Tat peptide-conjugated PMOs (PPMOs) displayed dramatic increases in activity over the corresponding unconjugated PMOs, which showed little activity in simple incubation with cells. Specificity of effects was demonstrated through the use of PPMOs comprising irrelevant sequences, establishing that the observed effects were indeed due to antisense activity and that the peptide conjugation did not show nonspecific effects. Co-addition of the Tat peptide and a PMO without covalent conjugation did not increase PMO activity. Evaluated alongside the Tat peptide were several other cationic peptides. Despite their similarity with respect to size and charge, Tat far outperformed these peptides, intimating that the ability of a peptide to cause intracellular delivery of a PMO is a nuanced trait that depends on more esoteric structural features than simple charge or size. The authors suggested that arginine content could be the key variable explaining the difference in effects between peptides. The Tat peptide contains significantly more arginine residues than the other peptides analyzed in their work, and it had been previously suggested that the spacing of charges on the guanidinium group of the arginine side chain could bring about favorable interactions with phosphates within cell membranes via "bidentate" hydrogen bonding, facilitating subsequent uptake [22].

Interestingly, it was observed that the concentrations needed to cause a given degree of activity were higher for the Tat peptide-PMO conjugates than for unconjugated PMO delivered using the previously mentioned mechanical delivery tactics.

Fluorescence microscopy performed during this early assessment of the impact of peptide conjugation to PMOs revealed a punctate fluorescence signal following treatment of cells using a fluorophore-labeled PPMO. This finding was an early indication of a critical challenge faced by antisense oligomer approaches in general: entrapment within the endosome. Research following this finding would confirm that endosomal entrapment presented a major hurdle to optimal intracellular delivery of PPMOs [23], and while progressive refinement of the technology has resulted in an impressive intracellular delivery capacity, this hurdle remains a limiting factor to activity within the cell.

Conjugation to cationic CPPs represented a turning point for the PMO technology, enabling what had largely been theoretical possibilities to become practical realities. The chemical traits needed for optimal performance of PMOs were preserved, including (i) water solubility is favorable for most of the peptides that have been employed; (ii) stability is high in a wide variety of environments; (iii) activity is highly specific; (iv) conjugation is simple to perform; and (v) purification is straightforward due to the uncharged PMO versus the cationic PPMO. Biological compatibility is remarkable, with generally low toxicity and low efficacious concentrations. The discovery that conjugation to CPPs profoundly enhances the intracellular delivery of PMOs was the event that began to propel this technology toward its full therapeutic promise, enabling the realistic pursuit of many innovative applications.

15.2 Peptide Development

While conjugation of PMOs to the HIV Tat peptide represented major progress toward enabling the widespread implementation of the PMO technology, the Tat peptide exhibited some weaknesses. The concentrations necessary to achieve the observed effects were high, and cytotoxicity attended treatment using these concentrations [21]. Despite its weaknesses, Tat peptide conjugation was an important proof of principle that pointed the way toward the development of peptides showing excellent performance as PMO conjugates.

The prescient observation that arginine was a key component to an effective PMO delivery peptide set into motion a path of inquiry that led to the development of delivery peptides with far more effective designs. A 2004 study revealed striking differences in antisense activity upon PMO conjugation to several peptides of distinct structure [24]. An R_9F_2C peptide was found to be several-fold more effective for PMO delivery over the Tat peptide, PTD4 (a Tat peptide derivative), a *Drosophila* antennapedia peptide, R_6F_2C and R_5F_2C. This finding again suggested the importance of arginine as a component of peptides for intracellular PMO delivery.

A 2002 study by Rothbard et al. identified peptide characteristics beyond simple arginine content that enable increased intracellular access [22]. The insertion of non-α-amino acids to create aliphatic spacers between arginine residues was shown to increase cellular uptake of a peptide relative to a peptide without the insertion, with a direct relationship between cellular uptake and the length of spacers observed up to a maximum with the use of 6-aminohexanoic acid (Ahx). 8-Aminooctanoic acid spacers showed relatively low uptake, suggesting that this trend has an upper limit around a six-carbon spacer length. The authors suggested that optimal spacing between charges could facilitate a more flexible, dynamic interaction between the peptide and the exterior of the cell, and that this contributes to better capacity for cellular uptake.

A 2006 comparison of the PMO delivery capacity of the Tat peptide, an oligoarginine peptide (R_9F_2) and a peptide containing arginine residues separated by Ahx spacers ($(R\text{-}Ahx\text{-}R)_4$) concurred with the findings of Rothbard et al., demonstrating that conjugation of a splice-modulating PMO to the $(R\text{-}Ahx\text{-}R)_4$ peptide facilitated greater specific PMO activity than conjugation to either the Tat or oligoarginine peptides [23]. Interestingly, the relative increase in oligonucleotide activity afforded by this peptide was not attended by a relative increase in cell membrane permeability as assessed by propidium iodide uptake. Coupled with evidence showing that conjugate internalization was due to endocytic activity, this finding was consistent with the general transition in understanding at that time toward a more complex picture of the mechanisms of CPP activity. Rather than simple perturbation of the plasma membrane, it was becoming apparent that CPPs could aid intracellular access by PMO cargoes through endocytosis, and that multiple modes of action by peptide-cargo conjugates were involved in their delivery capacity [23, 25].

Further examination of the capability of the $(R\text{-}Ahx\text{-}R)_4$ peptide for intracellular PMO delivery was carried out in a 2008 study focused on elucidating structure-activity relationships for this peptide [25]. The role of affinity for heparan sulfate was highlighted here; it was known that arginine-based cell-penetrating peptide-oligonucleotide conjugates enter cells through an energy-dependent mechanism, and thus potential interactions with cell surface features such as proteoglycans could influence the capacity for cellular uptake. Using heparin chromatography, it was shown that increases in the length of the hydrophobic spacer between arginine residues produced decreases in affinity for heparin, yet increasing the length of the spacer also resulted in increased oligonucleotide activity, with a maximum activity level observed for the six-carbon spacer Ahx. These results were taken to suggest that while some affinity is necessary for initial interaction with the cell surface and subsequent uptake, too strong an interaction could prevent the release of the oligonucleotide into the cytosol. Again, the uniquely strong contribution of the R-Ahx-R motif to delivery capacity was shown.

While the general structural characteristics of multiple arginine residues and relatively hydrophobic spacer regions have been solidly determined to lend efficacy to CPPs for delivery of oligonucleotide cargo, the space containing all potential structures is enormous and cannot be said to have been fully probed or even significantly explored. The structural motifs found to perform well here were only one possible solution, and it is likely that there are many such paths available within the universe of possible CPP structures. The application of machine learning to this challenge has begun to enable the evaluation of a greater portion of this space and has yielded some interesting conclusions [26, 27].

Using a deep learning approach in which training input generated through physical experimentation with large peptides (termed "miniproteins") of modular construction informed the prediction of miniprotein structures with high nuclear delivery capacity, Schissel et al. discovered a series of miniproteins able to deliver PMOs to the nucleus [26]. Sequence optimization included minimizing arginine content in an effort to reduce the cytotoxicity of output sequences. Interestingly, lysine residues were an important component of miniproteins with high delivery activity and, despite the limitation on arginine content imposed by the optimization process, effective delivery miniproteins were cationic. Consistent with foundational work in this area, the contribution of aliphatic spacer regions between charged residues to efficacy in cargo delivery was highlighted, with Ahx emerging as an attribute that strongly influenced predicted activity. This approach was extended to identify a short peptide retaining PMO delivery capacity with minimized arginine content, demonstrating the use of the approach for identifying peptides with high practical value [27].

While machine learning has certainly opened up new terrain, it is valuable to consider the ways in which human thinking informs the use of this tool. For instance, the prima facie assignment of arginine content as a characteristic to be minimized [26] canalizes optimization toward low arginine content, which could obscure potential arginine-containing structures with low toxicity. Data sets employed for training carry the limitations of the traditional empirical process along with them, and the algorithmic process is vulnerable to investigator bias. It is clear that expanding the breadth of the search for optimal CPPs for PMO delivery using computers is an apt tactic, but the perceptiveness of the human empirical process should not be underestimated. It is likely that future research in this area will consist of a give-and-take between the two modes of analysis, and it should be remembered that "there is more than one way to skin a cat." Here, the findings of old-school trial-and-error research were the input that was needed to effectively direct machines, and their output is likely to be taken back into that setting.

It is clear that there are many possible paths to the effective delivery of PMOs through conjugation to CPPs. Evidence supporting the effectiveness of a given structure does not rule out the effectiveness of other possible structures, and small combinatorial changes to a structure may produce large differences in activity. The practical infinitude of the chemical space available makes a complete definition of the structural principles behind effective peptide performance currently impossible. So, efforts such as those highlighted here employing both systematic testing and creativity are likely to succeed.

The challenge of intracellular delivery of oligonucleotides for eukaryotic systems can be reduced to two distinct physical barriers: the first is the cell membrane, which must be passed either through the usage of normal cellular uptake processes such as endocytosis or through direct translocation, and the second is the endosomal membrane, which confronts entities that have gained access to the interior of the cell through the use of endocytic processes. Effective use of endocytic pathways without a means to overcome the endosomal membrane barrier still leaves cargo sequestered from RNA targets, and the task falls to the CPP to facilitate both cellular uptake and passage from the endosome into the cytosol.

The finding that conjugation of PMOs to the $(R\text{-}Ahx\text{-}R)_4$ peptide enables a high level of activity without the help of endosomolytic compounds such as chloroquine was a significant development over the preceding peptide conjugates, but the considerable enhancement of activity seen with chloroquine treatment indicated that entrapment within the endosome following endocytosis remained as an imposing hurdle to the optimal implementation of the PPMO platform [23]. Alongside observations of increased oligomer activity with chloroquine treatment, microscopy studies have consistently indicated that fluorophore-bound PPMOs are largely confined to vesicular structures within the cell, consistent with endosomal entrapment [21, 23, 28, 29]. This hurdle presently remains as a limitation to RNA target access by PPMOs, and while efforts have been aimed toward establishing an effective and nontoxic strategy for enabling escape from the endosome, PPMO activity for a given extracellular dose applied, while relatively high, may still be pressed to a higher level. Increases in target access will likely contribute to increases in the therapeutic indices of PPMOs as the extracellular concentrations required for a given level of oligonucleotide activity are reduced.

15.3 Platform Technology

Adapting the PPMO technology from one task to the next merely requires a change in the sequence of bases. This simple fact implies one of the great strengths of PPMO technology: it is a platform that can be rapidly adapted to a very broad range of applications. The broad utility is advantageous because if the delivery, pharmacokinetics, and safety of the platform itself are optimized, the development of tools such as new therapeutics as a class of drugs may follow a shorter, generalized path. Additionally, a sequence-based targeting mechanism allows for the development of new tools at a relatively blistering pace. To illustrate, PPMOs that showed significant inhibitory activity toward SARS-CoV-2 in cultured cells [30] were successfully designed and synthesized within weeks of receipt of genomic sequence data for SARS-CoV-2. Rapidly adaptable tools such as this are sorely needed for response to threats that emerge and change quickly, such as zoonotic pathogens. Technologies such as the PPMO platform are the leading edge of a new paradigm for human influence over the intracellular domain, poised to take advantage of new biological knowledge that has accompanied technological advances. Here we will discuss three major areas of PPMO applications (Figure 15.3): viruses, bacteria, and genetic diseases.

15.3.1 Applications of the PPMO Platform: Viruses

Viruses represent an attractive target for the PPMO platform. Their economic mode of existence depends heavily on a usually very limited set of RNAs, many of which

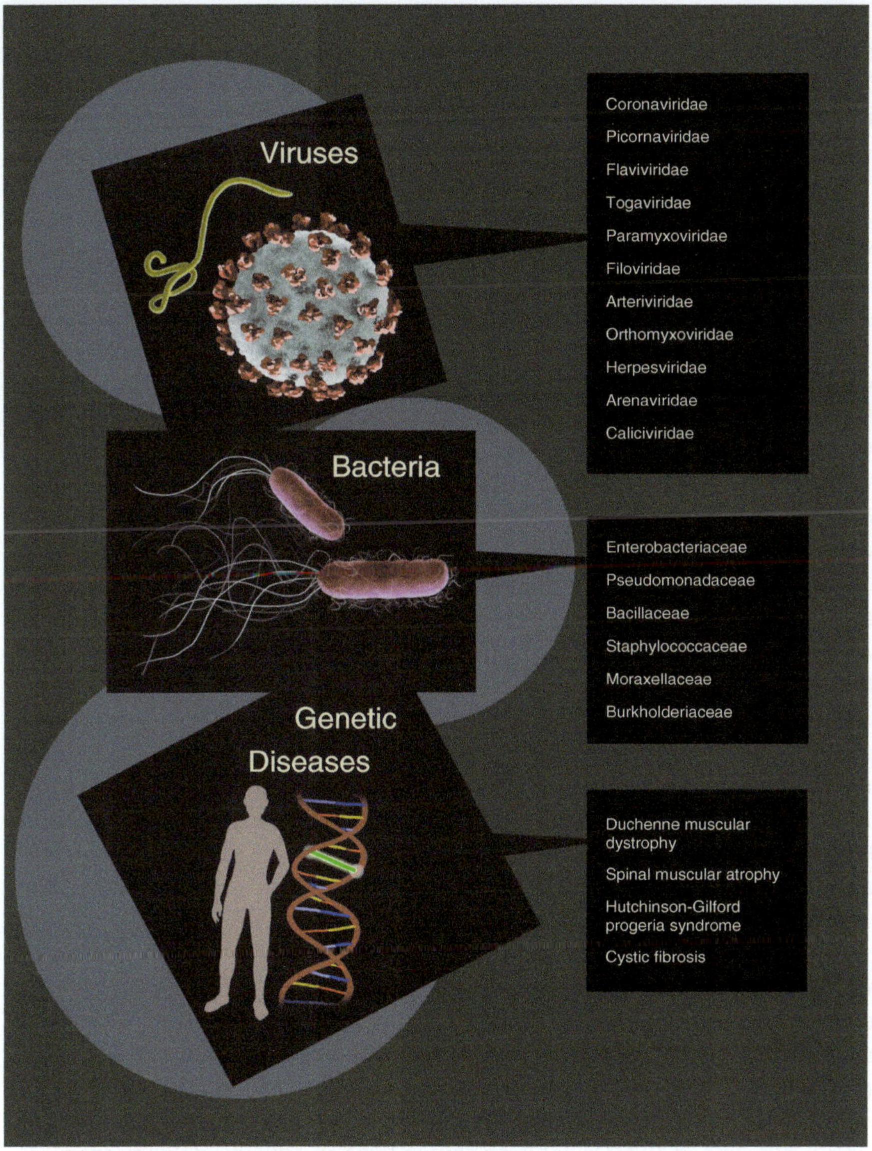

Figure 15.3 Diversity of major PPMO applications.

display a high level of conservation between taxa. Viral RNA targets within the host cell are accessible to PPMOs, and beyond the simple blockage of translation of mRNAs, viruses offer a host of other targets including genomic RNA, complementary RNAs and structural RNA elements such as genome cyclization sequences.

PPMOs have seen effective use against a diverse set of viruses, including influenza viruses [31–34], respiratory syncytial virus [28], dengue virus [35–37], West Nile virus [38], Japanese encephalitis virus [39], herpes simplex virus type 2 [40], Ebola virus [41], various arenaviruses [42], equine arteritis virus [43], porcine reproductive and respiratory syndrome virus [44, 45], assorted picornaviruses [46, 47], foot-and-mouth disease virus [48], mouse hepatitis virus [49], SARS-CoV [50], SARS-CoV-2 [30], measles virus [51], Sindbis virus, Venezuelan equine encephalitis virus [52], and murine norovirus [53]. The success of the PPMO platform against such a varied assortment of entities is a strong illustration of its universality.

The connection between the number of arginine repeats present in the CPP and PPMO activity that had previously been established in cellular assays [24] was recapitulated in a study employing PPMOs to treat mice infected with the Ebola virus [41]. Treatment of mice using a single PMO conjugated to peptides containing an increasing number of arginine-Ahx repeats revealed a clear direct relationship between the number of repeats and survival of mice following challenge with the Ebola virus. Here, peptide characteristics that had been shown to increase in vitro PPMO activity were directly connected to an increase in therapeutic benefit within a mammal. This finding provided validity to the application of knowledge gained through prior structure-activity work to in vivo therapeutic settings.

An exciting possibility was raised in the application of the PPMO technology to arenavirus infection: high levels of sequence conservation within a group of viruses could allow for the development of PPMOs effective across multiple viral species [42]. Here, single PPMO sequences displayed antiviral efficacy against the Junín virus, Tacaribe virus, Pichinde virus and lymphocytic choriomeningitis virus. A similar tactic proved effective in targeting picornaviruses [46]. Such an approach could be valuable for the efficient development of antivirals against viral groups whose constituents do not have the prevalence to attract attention for drug development alone, or for defense against biological warfare. Additionally, the general tactic of targeting conserved sequences to enable the broad application of a single tool could be valuable for targeting other groups of pathogens such as bacteria or eukaryotic parasites.

The work done in this area is extensive, and for further exploration, the interested reader is directed to Stein's expert review on the use of PPMOs against RNA viruses [29]. While PPMOs have been demonstrated to be effective in targeting numerous viral species, especially RNA viruses, a rich untested landscape remains. Application of the technology against DNA viruses has been relatively sparse and the use for targeting papillomaviruses and poxviruses, in particular, is inviting. Topical application of PPMOs has been shown to be an effective delivery approach for PPMO treatment of herpes simplex virus type 2 infection [40], and applying similar methods for papillomavirus infection could be a powerful therapeutic tactic. Poxviruses are an attractive target because many of the methods used for PPMO treatment of RNA virus infections could be directly applied to this neglected group of pathogens.

The PPMO platform is well-suited for use in host-directed antiviral therapies. This indirect tactic for addressing viral infection depends upon influencing the biological environment within or upon the host cell through modification of host features, rather than those of the pathogen. Such an approach, beyond simply broadening the limited arsenal available to address viral infection, is attractive in that it does not overtly introduce direct selective pressure upon the pathogen. Moreover, because a varied set of viruses may depend upon a single feature of the host cell for their infective cycles, there is potential for the development of host-directed antiviral agents that are active against multiple pathogens.

Host-directed PPMOs have been shown to perform impressively against both influenza viruses and SARS-CoV-2. The transmembrane protease TMPRSS2 has been shown to facilitate influenza virus infectivity through cleavage of hemagglutinin, and also enables SARS-CoV-2 cellular entry by cleavage of the spike glycoprotein [33, 54]. Using a single PPMO designed to modify TMPRSS2 precursor mRNA splicing resulting in the expression of a nonfunctional protein, our lab and collaborators have demonstrated that virus titers for both influenza A virus and SARS-CoV-2 were significantly reduced in human lung cell cultures [33, 54]. The work done here represents a valuable approach enabled by the PPMO platform: knockdown of host factor expression allowed for elucidation of pathogenesis alongside evaluation of a potential therapeutic approach.

Similarly, in vivo knockdown of host factors influencing the influenza virus infective cycle has revealed the landscape of the host-pathogen relationship within influenza virus infection [55, 56]. The use of PPMOs allowed interrogation of the role of the host factor UBR4 in the infective cycle through transient knockdown where the generation of knockout mice was lethal, and it was demonstrated that host-directed in vivo PPMO treatment can reduce lung virus titer and improve survival [55]. A similar approach allowed for in vivo demonstration of K-homology splicing regulatory protein (KHSRP) function in the retinoic acid-inducible gene I (RIG-I) response to influenza virus infection [56]. Additionally, PPMO-mediated host factor knockdown facilitated the development of an animal model of respiratory syncytial virus infection [57]. In these studies, the strength of the PPMO platform as both a research tool and a potential therapeutic measure was shown.

15.3.2 Applications of the PPMO Platform: Bacteria and Protists

Bacteria represent a challenging target for oligonucleotide applications. While the development of oligonucleotide and CPP technologies has been largely focused on optimization for applications within mammalian cells, bacteria present additional and unique physical obstacles to intracellular delivery. Beyond the cytoplasmic membrane, additional barriers are present which, depending on the bacterial species, may include thick cell wall elements, outer membranes and capsules. Additional physical obstacles are often present in natural bacterial populations and communities, where complex biofilm elements create separate microenvironments. Beyond the structural differences, the lack of endocytic uptake further complicates effective oligonucleotide delivery. Because most efforts have been focused on developing the technology for delivery to mammalian cells, the story of the application of the PPMO technology to bacterial applications has mostly consisted of trial-and-error repurposing of existing entities. Despite this strategic disadvantage, several groups have successfully demonstrated the use of PPMOs to control gene expression in bacteria for a variety of different aims, and their creativity has introduced some exciting possibilities.

Compared to the PPMOs used within mammalian systems, the most striking difference in the PPMO platform in the bacterial setting is the optimal length of the oligomer. Using unconjugated PMOs, it was found that optimal activity levels were reached with 11-mer oligomers and that oligomers greater than 14 bases in length were inactive [58]. A marked contrast was seen between PMO activity within a cell-free assay system and within *Escherichia coli* culture: longer oligomers showed greater activity in the cell-free system, but shorter oligomers performed much better within the bacterial culture. The authors suggested that this difference could be due to the physical barrier presented by the outer membrane, or to the involvement of transport mechanisms across the plasma membrane. The use of 11-mer oligomers has become common practice for those employing the PPMO platform for bacterial applications [59–63].

It is notable that despite the differences in the physical barriers to PPMO entry between mammalian cells and bacteria, the general CPP framework of alternating arginine residues with hydrophobic spacers (particularly Ahx) emerges as a uniquely effective delivery tool within either setting [59–61, 63]. This consistent theme suggests that the unique effectiveness of this peptide for PMO delivery may be due to its interactions with purely physical barriers because the physiological similarities it may exploit are few.

Various peptide compositions were tested for their capacity to facilitate the delivery of a PMO targeting the essential gene *acpP* in *E. coli* [59]. While preserving the general characteristic of alternating cationic and hydrophobic residues within the peptide, variations in repeating motif construction and amino acid composition were employed in a comparison of minimum inhibitory concentrations within pure bacterial cultures. Repeating motifs of the form cationic-nonpolar-cationic clearly showed stronger inhibitory activity than those of the forms cationic-nonpolar-nonpolar or cationic-nonpolar. When taken to the in vivo setting using *E. coli*-infected mice, a peptide composed of four R-Ahx-R repeats outperformed alternatives. Treatment by intraperitoneal injection of PPMOs resulted in both increased survival and dose-dependent decreases in bacteremia, while treatment using water or an irrelevant control sequence PPMO resulted in much higher bacterial loads and dramatically decreased survival. This work again highlighted the strength of the R-Ahx-R structural motif, with results generated in pure bacterial cultures mirrored within a model therapeutic setting.

A similar trend was found when PPMOs were used to target *Pseudomonas aeruginosa* [60]. Again, a delivery peptide containing R-Ahx-R repeats was found to facilitate greater inhibitory activity than several alternatives when conjugated to PMOs targeting the expression of essential genes. Interestingly, PPMO activity increased significantly when measures were taken to destabilize or permeabilize the bacterial outer membrane. The use of a growth medium containing reduced divalent cation content as well as co-treatment using polymyxin B or colistin resulted in greatly decreased minimum inhibitory concentrations in pure culture. These findings point to the significance of the physical barriers to successful bacterial PPMO delivery. Importantly, the efficacy seen in pure culture experiments carried through into experiments using biofilms, a key feature of *P. aeruginosa* pathogenicity. A PPMO using the R-Ahx-R motif and targeting the expression of *acpP* was effective at both preventing biofilm formation and reducing biofilms present prior to treatment. Here, the PPMO was able to traverse the barriers imposed by both the structure of the bacterium and the structure of the biofilm. The same PPMO showed great efficacy against infection in mice, causing a significant reduction in lung bacterial load which was attended by obvious improvements in lung pathology. Importantly, these effects were seen under postinfection PPMO treatment. The experimental setup used here was valuable for producing a realistic indication of the therapeutic potential for the approach, with PPMO effects observed under conditions that correspond to potential human use. This work was followed by further application of effective PPMOs to multidrug-resistant *P. aeruginosa* strains and further exploration of delayed treatment, in which PPMOs were again found to be highly effective [64].

While the technology was primarily developed using mammalian cells, PPMOs have been found to be effective for controlling bacterial gene expression in a diverse array of bacterial species [61–69], and this has opened many useful lines of inquiry and revealed valuable therapeutic approaches that are beyond the capacity of small-molecule antimicrobials.

Beyond the applications of PPMOs to viruses and bacteria highlighted here, the technology has also been used to target and investigate the biology of eukaryotic parasites. PPMOs were effectively used to inhibit replication of the protist *Toxoplasma gondii* and reduced parasitic load in infected mice [70]. The PPMO platform was also used to successfully inhibit growth of *Plasmodium falciparum* [71]. Again, PPMOs demonstrated utility not only as potential therapeutic agents but also as a tool for investigating pathogen biology.

15.3.3 Applications of the PPMO Platform: Genetic Disease

It is intuitive to apply technologies that target the flow of information within the cell to conditions that involve defects of information in their pathophysiology. PPMOs have seen dramatically effective use within numerous models of genetic disease, and the promise that the technology offers for application within this broad area is staggering.

15.3.3.1 Duchenne Muscular Dystrophy (DMD)

Much of the framing for research employing PPMOs has been derived from work done using the precursor technology, the PMO platform. A prominent focus for the use of PMOs has been the treatment of DMD, a progressive muscle-wasting disease arising from genetic mutations that cause failure in the production of dystrophin. Lack of functional dystrophin results in progressive loss of muscle function, leading to disability and premature death. Unconjugated PMOs used to direct splicing to exclude mutation-containing portions of transcripts have been extensively developed as therapeutics showing modest efficacy in DMD, offering therapeutic measures where there were previously none [4–7]. PPMOs have been demonstrated thoroughly to induce a much greater therapeutic change in model systems with dramatically improved and persistent efficacy, pointing the way toward an effective progression of DMD therapeutics and entering the clinical phase of development [72–77]. Research in this area has clearly shown the transformative effect of CPP conjugation upon PMO efficacy, and ongoing translational efforts will reveal the impact this will have on the therapeutic utility of the antisense approach.

15.3.3.2 Spinal Muscular Atrophy (SMA)

A unique application of the PPMO platform is found in its use within models of SMA. For those affected by this disease, mutations within the *SMN1* gene prevent the expression of the survival of motor neuron protein, which is necessary for the continued viability of motor neurons. The presence of a very similar second gene, *SMN2*, which has normally limited function, offers an opportunity for addressing the cause of the disease through upregulation of expression of the functional survival of motor neuron protein. Only a very small portion of *SMN2* transcripts is translated to functional protein naturally, but blockade of a splice site on precursor mRNAs by a PPMO causes greater expression of the functional protein. The therapeutic benefit that this approach produces is unequivocal: with a single PPMO treatment administered by intravenous injection, the survival of SMA model mice was increased from a median of 12 days to a median of 167 days [78]. Without conjugation to the CPP, the PMO produced no change in survival time. Beyond the drastic effect on survival caused by the PPMO, it is notable that intravenous injection was used to administer the PPMO, which contrasts with the currently approved oligonucleotide therapy administered through intrathecal injection [79]. Not only does the PPMO platform offer a highly efficacious approach to treatment, but it may also make possible a less invasive and safer mode of administration.

There is demonstrated potential for PPMOs to allow effective treatment of SMA during the symptomatic course of the disease. Delayed treatment of model mice using PPMOs employing both an arginine-spacer-arginine peptide and a D-oligoarginine peptide resulted in significantly increased lifespan and produced improvements in tests for motor capacity [80]. Effective use of intraperitoneal injection again demonstrated that PPMOs afford the ability to use systemic modes of administration, an important consideration for SMA therapies. Interestingly, the D-oligoarginine peptide outperformed alternatives within this application. Observable efficacy following treatment during symptomatic disease here indicates that there is potential to address SMA using PPMOs in realistic clinical scenarios.

15.3.3.3 Hutchinson-Gilford Progeria Syndrome (HGPS)

Another powerful application of PPMOs for therapeutic modulation of RNA splicing was demonstrated within the context of HGPS [81]. Within the majority of HGPS cases, a mutation within the *LMNA* gene results in aberrant splicing of transcripts, causing translation of a shortened product called progerin that retains a post-translational prenylation modification. Progerin accumulates on the nuclear lamina and exerts toxic effects that ultimately result in accelerated aging of the affected individual. Preventing aberrant splicing of *LMNA* transcripts using a PPMO blocked the production of progerin significantly within fibroblasts taken from an individual with HGPS, and treatment within a transgenic mouse model resulted in a significant increase in lifespan [81].

15.3.3.4 Cystic Fibrosis

PPMOs also show promise for addressing cystic fibrosis for which existing therapies are not effective. A minority of cystic fibrosis patients have a loss of function of the cystic fibrosis transmembrane conductance regulator (CFTR) due to mutations causing defective splicing, and existing therapies poorly address the disease under this etiology. Dang et al. demonstrated using patient-derived cells that it is possible to correct defective splicing and restore CFTR function using PPMOs in combination with a small molecule that enhances intracellular delivery of PPMOs [82]. Beyond introducing a useful strategy for the treatment of cystic fibrosis within this subgroup of patients, this work is notable for the innovative delivery strategy that was used. High-throughput screening for small molecules that enhance intracellular delivery of PPMOs enabled this group to dramatically increase the activity of PPMOs, and this unique approach is likely to have a beneficial impact on the PPMO therapeutic window [82–85].

While it is impossible to ignore the truly revolutionary therapeutic potential of the PPMO platform, there are some limitations to consider and address. PPMOs employing cationic CPPs show nephrotoxicity at high doses, narrowing the therapeutic window [86]. Additionally, PPMOs, as with other macromolecular technologies, are subject to endosomal entrapment following uptake by the cell [21, 23, 28, 29]. This also carries a deleterious impact on the therapeutic window by blunting the efficacy that would be possible with unhindered cytosolic access. In addition to a careful determination of optimal dosing regimens, efforts to increase the PPMO therapeutic window by enabling endosomal escape or by identification of CPPs with reduced impact upon the kidney will help propel the platform toward more widespread clinical use.

15.3.4 Future Applications of the PPMO Platform

Perhaps the most exciting aspect of the PPMO platform is its infinite adaptability. Targeting the flow of information within the cell rather than the products of information flow (as is typically done using tools such as small-molecule drugs targeting proteins) expands the set of potential ways to influence the biological milieu enormously. The biology that composes the terrain over which these tools may be used to exert influence is currently in a state of rapid discovery, with the function of entities such as noncoding RNAs little described. As understanding of the biology of gene expression grows, so will the set of targets that may be intelligently addressed using this platform technology.

Of note, the potential for use of this platform for labeling RNAs, both intracellularly and within extracellular fluid environments such as the blood, has been so far little explored. The use of PPMOs as probes for nucleic acid detection presents key advantages over more prevalent methods, such as PCR-dependent assays: detection is possible with stoichiometric amounts of the detection target, obviating target amplification, and the penetrative capacity of peptide conjugates could enable the detection of targets within samples of heterogeneous composition, or within intact tissues or organisms.

As pathogens accumulate resistance to agents used for their control, it is increasingly necessary to put into use new entities that have not yet been employed to induce selective pressure. While simply generating new chemical entities with similar modes of action is immediately effective for overcoming resistance mechanisms, the antisense approach represents an entirely different way of approaching this challenge. The introduction of new molecular entities in the small-molecule drug arena has declined as the chemical space available for drug development is increasingly well-trodden. New approaches for targeting pathogens are desperately needed. The PPMO platform has repeatedly demonstrated its capacity to effectively and specifically target pathogens of all ilks, and it stands as a well-validated and realistic tool to combat infectious diseases. Employing this platform will provide a highly selective alternative approach for addressing infectious diseases without generating resistance.

Beyond the context of infectious disease, PPMOs invite numerous therapeutic applications. Impressive advances have been made in addressing genetic disorders such as DMD, SMA, and HGPS in animal models, and PPMOs have now begun to be tested clinically for DMD [77]. The universality of the platform, however, begs its application to many other disorders of genetic origin that are mechanistically well understood. Research investigating the treatment of cancer using PPMOs, for example, is surprisingly scarce [87]. Research efforts have shown repeatedly that, given a well-understood pathological state and a rationale for its mitigation through RNA targeting, it is possible to quickly develop an effective intervention using the PPMO platform.

While therapeutic applications of this powerful technology have experienced the greatest emphasis in research, control of the biological environment through targeted influence on gene expression has the potential for a much broader set of applications. Use in plant systems has barely been explored [69], and control of gene expression without genome modification could be a useful approach where selective breeding or genome modification is not possible, or where transient alterations are desired.

CPPs have been indispensable to the development of morpholino oligomer antisense technology, enabling its application to a far broader set of tasks than would have been possible without the effective intracellular delivery that they promote.

Acknowledgments

The authors express heartfelt gratitude to Jon Moulton for his valuable help with the manuscript.

References

1 Summerton, J. and Weller, D. (1997). Morpholino antisense oligomers: design, preparation, and properties. *Antisense and Nucleic Acid Drug Development* 7: 187–195.

2 Kloosterman, W.P., Lagendijk, A.K., Ketting, R.F. et al. (2007). Targeted inhibition of miRNA maturation with morpholinos reveals a role for miR-375 in pancreatic islet development. *PLoS Biology* 5: e203.

3 Gong, Q., Stump, M.R., and Zhou, Z. (2014). Upregulation of functional Kv11.1 isoform expression by inhibition of intronic polyadenylation with antisense morpholino oligonucleotides. *Journal of Molecular and Cellular Cardiology* 76: 26–32.

4 Heo, Y.A. (2020). Golodirsen: first approval. *Drugs* 80: 329–333.

5 Aartsma-Rus, A. and Krieg, A.M. (2017). FDA approves eteplirsen for Duchenne muscular dystrophy: the next chapter in the eteplirsen saga. *Nucleic Acid Therapeutics* 27: 1–3.

6 Dhillon, S. (2020). Viltolarsen: first approval. *Drugs* 80: 1027–1031.

7 Shirley, M. (2021). Casimersen: first approval. *Drugs* 81: 875–879.

8 Nutt, S.L., Bronchain, O.J., Hartley, K.O. et al. (2001). Comparison of morpholino based translational inhibition during the development of Xenopus laevis and Xenopus tropicalis. *Genesis* 30: 110–113.

9 Bedell, V.M., Westcot, S.E., and Ekker, S.C. (2011). Lessons from morpholino-based screening in zebrafish. *Briefings in Functional Genomics* 10: 181–188.

10 Chen, J.C., King, O.D., Zhang, Y. et al. (2016). Morpholino-mediated knockdown of DUX4 toward facioscapulohumeral muscular dystrophy therapeutics. *Molecular Therapy* 24: 1405–1411.

11 Sekhon, H.S., London, C.A., Sekhon, M. et al. (2008). c-MYC antisense phosphosphorodiamidate morpholino oligomer inhibits lung metastasis in a murine tumor model. *Lung Cancer* 60: 347–354.

12 Xiong, Y., McQuistan, T.J., Stanek, J.W. et al. (2018). Detection of unique Ebola virus oligonucleotides using fluorescently-labeled phosphorodiamidate morpholino oligonucleotide probe pairs. *Analytical Biochemistry* 557: 84–90.

13 Partridge, M., Vincent, A., Matthews, P. et al. (1996). A simple method for delivering morpholino antisense oligos into the cytoplasm of cells. *Antisense and Nucleic Acid Drug Development* 6: 169–175.

14 Ghosh, C. and Iversen, P.L. (2000). Intracellular delivery strategies for antisense phosphorodiamidate morpholino oligomers. *Antisense and Nucleic Acid Drug Development* 10: 263–274.

15 Thummel, R., Bailey, T.J., and Hyde, D.R. (2011). in vivo electroporation of morpholinos into the adult zebrafish retina. *Journal of Visualized Experiments* (58): e3603.

16 Thummel, R., Bai, S., Sarras, M.P. Jr. et al. (2006). Inhibition of zebrafish fin regeneration using in vivo electroporation of morpholinos against fgfr1 and msxb.

Developmental Dynamics: An Official Publication of the American Association of Anatomists 235: 336–346.

17 Mellitzer, G., Hallonet, M., Chen, L. et al. (2002). Spatial and temporal "knock down" of gene expression by electroporation of double-stranded RNA and morpholinos into early postimplantation mouse embryos. *Mechanisms of Development* 118: 57–63.

18 Morcos, P.A. (2001). Achieving efficient delivery of morpholino oligos in cultured cells. *Genesis* 30: 94–102.

19 Bill, B.R., Petzold, A.M., Clark, K.J. et al. (2009). A primer for morpholino use in zebrafish. *Zebrafish* 6: 69–77.

20 Umemoto, N., Nishimura, Y., Shimada, Y. et al. (2013). Fluorescent-based methods for gene knockdown and functional cardiac imaging in zebrafish. *Molecular Biotechnology* 55: 131–142.

21 Moulton, H.M., Hase, M.C., Smith, K.M. et al. (2003). HIV Tat peptide enhances cellular delivery of antisense morpholino oligomers. *Antisense and Nucleic Acid Drug Development* 13: 31–43.

22 Rothbard, J.B., Kreider, E., VanDeusen, C.L. et al. (2002). Arginine-rich molecular transporters for drug delivery: role of backbone spacing in cellular uptake. *Journal of Medicinal Chemistry* 45: 3612–3618.

23 Abes, S., Moulton, H.M., Clair, P. et al. (2006). Vectorization of morpholino oligomers by the (R-Ahx-R)4 peptide allows efficient splicing correction in the absence of endosomolytic agents. *Journal of Controlled Release* 116: 304–313.

24 Moulton, H.M., Nelson, M.H., Hatlevig, S.A. et al. (2004). Cellular uptake of antisense morpholino oligomers conjugated to arginine-rich peptides. *Bioconjugate Chemistry* 15: 290–299.

25 Abes, R., Moulton, H.M., Clair, P. et al. (2008). Delivery of steric block morpholino oligomers by (R-X-R)4 peptides: structure-activity studies. *Nucleic Acids Research* 36: 6343–6354.

26 Schissel, C.K., Mohapatra, S., Wolfe, J.M. et al. (2021). Deep learning to design nuclear-targeting abiotic miniproteins. *Nature Chemistry* 13: 992–1000.

27 Lopez-Vidal, E.M., Schissel, C.K., Mohapatra, S. et al. (2021). Deep learning enables discovery of a short nuclear targeting peptide for efficient delivery of antisense oligomers. *JACS Au* 1: 2009–2020.

28 Lai, S.H., Stein, D.A., Guerrero-Plata, A. et al. (2008). Inhibition of respiratory syncytial virus infections with morpholino oligomers in cell cultures and in mice. *Molecular Therapy* 16: 1120–1128.

29 Stein, D.A. (2008). Inhibition of RNA virus infections with peptide-conjugated morpholino oligomers. *Current Pharmaceutical Design* 14: 2619–2634.

30 Rosenke, K., Leventhal, S., Moulton, H.M. et al. (2021). Inhibition of SARS-CoV-2 in Vero cell cultures by peptide-conjugated morpholino oligomers. *Journal of Antimicrobial Chemotherapy* 76: 413–417.

31 Lupfer, C., Stein, D.A., Mourich, D.V. et al. (2008). Inhibition of influenza A H3N8 virus infections in mice by morpholino oligomers. *Archives of Virology* 153: 929–937.

32 Ge, Q., Pastey, M., Kobasa, D. et al. (2006). Inhibition of multiple subtypes of influenza A virus in cell cultures with morpholino oligomers. *Antimicrobial Agents and Chemotherapy* 50: 3724–3733.

33 Bottcher-Friebertshauser, E., Stein, D.A., Klenk, H.D. et al. (2011). Inhibition of influenza virus infection in human airway cell cultures by an antisense peptide-conjugated morpholino oligomer targeting the hemagglutinin-activating protease TMPRSS2. *Journal of Virology* 85: 1554–1562.

34 Gabriel, G., Nordmann, A., Stein, D.A. et al. (2008). Morpholino oligomers targeting the PB1 and NP genes enhance the survival of mice infected with highly pathogenic influenza A H7N7 virus. *Journal of General Virology* 89: 939–948.

35 Stein, D.A., Huang, C.Y., Silengo, S. et al. (2008). Treatment of AG129 mice with antisense morpholino oligomers increases survival time following challenge with dengue 2 virus. *Journal of Antimicrobial Chemotherapy* 62: 555–565.

36 Holden, K.L., Stein, D.A., Pierson, T.C. et al. (2006). Inhibition of dengue virus translation and RNA synthesis by a morpholino oligomer targeted to the top of the terminal 3′ stem-loop structure. *Virology* 344: 439–452.

37 Kinney, R.M., Huang, C.Y., Rose, B.C. et al. (2005). Inhibition of dengue virus serotypes 1 to 4 in Vero cell cultures with morpholino oligomers. *Journal of Virology* 79: 5116–5128.

38 Deas, T.S., Bennett, C.J., Jones, S.A. et al. (2007). in vitro resistance selection and in vivo efficacy of morpholino oligomers against West Nile virus. *Antimicrobial Agents and Chemotherapy* 51: 2470–2482.

39 Anantpadma, M., Stein, D.A., and Vrati, S. (2010). Inhibition of Japanese encephalitis virus replication in cultured cells and mice by a peptide-conjugated morpholino oligomer. *Journal of Antimicrobial Chemotherapy* 65: 953–961.

40 Eide, K., Moerdyk-Schauwecker, M., Stein, D.A. et al. (2010). Reduction of herpes simplex virus type-2 replication in cell cultures and in rodent models with peptide-conjugated morpholino oligomers. *Antiviral Therapy* 15: 1141–1149.

41 Swenson, D.L., Warfield, K.L., Warren, T.K. et al. (2009). Chemical modifications of antisense morpholino oligomers enhance their efficacy against Ebola virus infection. *Antimicrobial Agents and Chemotherapy* 53: 2089–2099.

42 Neuman, B.W., Bederka, L.H., Stein, D.A. et al. (2011). Development of peptide-conjugated morpholino oligomers as pan-arenavirus inhibitors. *Antimicrobial Agents and Chemotherapy* 55: 4631–4638.

43 Zhang, J., Stein, D.A., Timoney, P.J. et al. (2010). Curing of HeLa cells persistently infected with equine arteritis virus by a peptide-conjugated morpholino oligomer. *Virus Research* 150: 138–142.

44 Opriessnig, T., Patel, D., Wang, R. et al. (2011). Inhibition of porcine reproductive and respiratory syndrome virus infection in piglets by a peptide-conjugated morpholino oligomer. *Antiviral Research* 91: 36–42.

45 Patel, D., Opriessnig, T., Stein, D.A. et al. (2008). Peptide-conjugated morpholino oligomers inhibit porcine reproductive and respiratory syndrome virus replication. *Antiviral Research* 77: 95–107.

46 Stone, J.K., Rijnbrand, R., Stein, D.A. et al. (2008). A morpholino oligomer targeting highly conserved internal ribosome entry site sequence is able to inhibit multiple species of picornavirus. *Antimicrobial Agents and Chemotherapy* 52: 1970–1981.

47 Yuan, J., Stein, D.A., Lim, T. et al. (2006). Inhibition of coxsackievirus B3 in cell cultures and in mice by peptide-conjugated morpholino oligomers targeting the internal ribosome entry site. *Journal of Virology* 80: 11510–11519.

48 Vagnozzi, A., Stein, D.A., Iversen, P.L. et al. (2007). Inhibition of foot-and-mouth disease virus infections in cell cultures with antisense morpholino oligomers. *Journal of Virology* 81: 11669–11680.

49 Neuman, B.W., Stein, D.A., Kroeker, A.D. et al. (2004). Antisense morpholino-oligomers directed against the 5′ end of the genome inhibit coronavirus proliferation and growth. *Journal of Virology* 78: 5891–5899.

50 Neuman, B.W., Stein, D.A., Kroeker, A.D. et al. (2005). Inhibition, escape, and attenuated growth of severe acute respiratory syndrome coronavirus treated with antisense morpholino oligomers. *Journal of Virology* 79: 9665–9676.

51 Sleeman, K., Stein, D.A., Tamin, A. et al. (2009). Inhibition of measles virus infections in cell cultures by peptide-conjugated morpholino oligomers. *Virus Research* 140: 49–56.

52 Paessler, S., Rijnbrand, R., Stein, D.A. et al. (2008). Inhibition of alphavirus infection in cell culture and in mice with antisense morpholino oligomers. *Virology* 376: 357–370.

53 Bok, K., Cavanaugh, V.J., Matson, D.O. et al. (2008). Inhibition of norovirus replication by morpholino oligomers targeting the 5’-end of the genome. *Virology* 380: 328–337.

54 Bestle, D., Heindl, M.R., Limburg, H. et al. (2020). TMPRSS2 and furin are both essential for proteolytic activation of SARS-CoV-2 in human airway cells. *Life Science Alliance* 3.

55 Tripathi, S., Pohl, M.O., Zhou, Y. et al. (2015). Meta- and orthogonal integration of influenza “OMICs” data defines a role for UBR4 in virus budding. *Cell Host & Microbe* 18: 723–735.

56 Soonthornvacharin, S., Rodriguez-Frandsen, A., Zhou, Y. et al. (2017). Systems-based analysis of RIG-I-dependent signalling identifies KHSRP as an inhibitor of RIG-I receptor activation. *Nature Microbiology* 2: 17022.

57 Green, G., Johnson, S.M., Costello, H. et al. (2021). CX3CR1 is a receptor for human respiratory syncytial virus in cotton rats. *Journal of Virology* 95: e0001021.

58 Deere, J., Iversen, P., and Geller, B.L. (2005). Antisense phosphorodiamidate morpholino oligomer length and target position effects on gene-specific inhibition in *Escherichia coli*. *Antimicrobial Agents and Chemotherapy* 49: 249–255.

59 Mellbye, B.L., Puckett, S.E., Tilley, L.D. et al. (2009). Variations in amino acid composition of antisense peptide-phosphorodiamidate morpholino oligomer affect potency against *Escherichia coli* in vitro and in vivo. *Antimicrobial Agents and Chemotherapy* 53: 525–530.

60 Howard, J.J., Sturge, C.R., Moustafa, D.A. et al. (2017). Inhibition of Pseudomonas aeruginosa by peptide-conjugated phosphorodiamidate morpholino oligomers. *Antimicrobial Agents and Chemotherapy* 61: e01938-16.

61 Daly, S.M., Sturge, C.R., Felder-Scott, C.F. et al. (2017). MCR-1 inhibition with peptide-conjugated phosphorodiamidate morpholino oligomers restores sensitivity to polymyxin in Escherichia coli. *mBio* 8: e01315–e01317.

62 Daly, S.M., Sturge, C.R., Marshall-Batty, K.R. et al. (2018). Antisense inhibitors retain activity in pulmonary models of Burkholderia infection. *ACS Infectious Diseases* 4: 806–814.

63 Geller, B.L., Marshall-Batty, K., Schnell, F.J. et al. (2013). Gene-silencing antisense oligomers inhibit Acinetobacter growth in vitro and in vivo. *The Journal of Infectious Diseases* 208: 1553–1560.

64 Moustafa, D.A., Wu, A.W., Zamora, D. et al. (2021). Peptide-conjugated phosphorodiamidate morpholino oligomers retain activity against multidrug-resistant Pseudomonas aeruginosa in vitro and in vivo. *mBio* 12: e02411–e02420.

65 Geller, B.L., Li, L., Martinez, F. et al. (2018). Morpholino oligomers tested in vitro, in biofilm and in vivo against multidrug-resistant Klebsiella pneumoniae. *Journal of Antimicrobial Chemotherapy* 73: 1611–1619.

66 Greenberg, D.E., Marshall-Batty, K.R., Brinster, L.R. et al. (2010). Antisense phosphorodiamidate morpholino oligomers targeted to an essential gene inhibit Burkholderia cepacia complex. *Journal of Infectious Diseases* 201: 1822–1830.

67 Mitev, G.M., Mellbye, B.L., Iversen, P.L. et al. (2009). Inhibition of intracellular growth of Salmonella enterica serovar Typhimurium in tissue culture by antisense peptide-phosphorodiamidate morpholino oligomer. *Antimicrobial Agents and Chemotherapy* 53: 3700–3704.

68 Sturge, C.R., Felder-Scott, C.F., Pifer, R. et al. (2019). AcrAB-TolC inhibition by peptide-conjugated phosphorodiamidate morpholino oligomers restores antibiotic activity in vitro and in vivo. *ACS Infectious Diseases* 5: 1446–1455.

69 Sandoval-Mojica, A.F., Altman, S., Hunter, W.B. et al. (2020). Peptide conjugated morpholinos for management of the huanglongbing pathosystem. *Pest Management Science* 76: 3217–3224.

70 Lai, B.S., Witola, W.H., El Bissati, K. et al. (2012). Molecular target validation, antimicrobial delivery, and potential treatment of Toxoplasma gondii infections. *Proceedings of the National Academy of Sciences of the United States of America* 109: 14182–14187.

71 Garg, A., Wesolowski, D., Alonso, D. et al. (2015). Targeting protein translation, RNA splicing, and degradation by morpholino-based conjugates in Plasmodium falciparum. *Proceedings of the National Academy of Sciences of the United States of America* 112: 11935–11940.

72 Wu, B., Lu, P., Cloer, C. et al. (2012). Long-term rescue of dystrophin expression and improvement in muscle pathology and function in dystrophic mdx mice by peptide-conjugated morpholino. *The American Journal of Pathology* 181: 392–400.

73 Jearawiriyapaisarn, N., Moulton, H.M., Buckley, B. et al. (2008). Sustained dystrophin expression induced by peptide-conjugated morpholino oligomers in the muscles of mdx mice. *Molecular Therapy* 16: 1624–1629.

74 Echigoya, Y., Nakamura, A., Nagata, T. et al. (2017). Effects of systemic multiexon skipping with peptide-conjugated morpholinos in the heart of a dog model of

Duchenne muscular dystrophy. *Proceedings of the National Academy of Sciences of the United States of America* 114: 4213–4218.

75 Wu, B., Moulton, H.M., Iversen, P.L. et al. (2008). Effective rescue of dystrophin improves cardiac function in dystrophin-deficient mice by a modified morpholino oligomer. *Proceedings of the National Academy of Sciences of the United States of America* 105: 14814–14819.

76 Fletcher, S., Honeyman, K., Fall, A.M. et al. (2007). Morpholino oligomer-mediated exon skipping averts the onset of dystrophic pathology in the mdx mouse. *Molecular Therapy* 15: 1587–1592.

77 Sheikh, O. and Yokota, T. (2021). Pharmacology and toxicology of eteplirsen and SRP-5051 for DMD exon 51 skipping: an update. *Archives of Toxicology* 1–9.

78 Hammond, S.M., Hazell, G., Shabanpoor, F. et al. (2016). Systemic peptide-mediated oligonucleotide therapy improves long-term survival in spinal muscular atrophy. *Proceedings of the National Academy of Sciences of the United States of America* 113: 10962–10967.

79 Hoy, S.M. (2017). Nusinersen: first global approval. *Drugs* 77: 473–479.

80 Bersani, M., Rizzuti, M., Pagliari, E. et al. (2021). Cell-penetrating peptide-conjugated morpholino rescues SMA in a symptomatic preclinical model. *Molecular Therapy* 30: 1288–1299.

81 Erdos, M.R., Cabral, W.A., Tavarez, U.L. et al. (2021). A targeted antisense therapeutic approach for Hutchinson-Gilford progeria syndrome. *Nature Medicine* 27: 536–545.

82 Dang, Y., van Heusden, C., Nickerson, V. et al. (2021). Enhanced delivery of peptide-morpholino oligonucleotides with a small molecule to correct splicing defects in the lung. *Nucleic Acids Research* 49: 6100–6113.

83 Yang, B., Ming, X., Cao, C. et al. (2015). High-throughput screening identifies small molecules that enhance the pharmacological effects of oligonucleotides. *Nucleic Acids Research* 43: 1987–1996.

84 Wang, L., Ariyarathna, Y., Ming, X. et al. (2017). A novel family of small molecules that enhance the intracellular delivery and pharmacological effectiveness of antisense and splice switching oligonucleotides. *ACS Chemical Biology* 12: 1999–2007.

85 Juliano, R.L., Wang, L., Tavares, F. et al. (2018). Structure-activity relationships and cellular mechanism of action of small molecules that enhance the delivery of oligonucleotides. *Nucleic Acids Research* 46: 1601–1613.

86 Amantana, A., Moulton, H.M., Cate, M.L. et al. (2007). Pharmacokinetics, biodistribution, stability and toxicity of a cell-penetrating peptide-morpholino oligomer conjugate. *Bioconjugate Chemistry* 18: 1325–1331.

87 Ma, W., Lin, Y., Xuan, W. et al. (2012). Inhibition of p53 expression by peptide-conjugated phosphorodiamidate morpholino oligomers sensitizes human cancer cells to chemotherapeutic drugs. *Oncogene* 31: 1024–1033.

16

Cell-Penetrating Peptide-Conjugated Polymer Micelle for pDNA/siRNA Delivery

Takanori Kanazawa

University of Shizuoka, School of Pharmaceutical Sciences, 52-1 Yada, Suruga-ku, Shizuoka, 422-8526, Japan

16.1 Introduction

16.1.1 Cell-Penetrating Peptide

The intracellular uptake of macromolecules, such as proteins and genes/nucleic acids, can be enhanced by cell-penetrating peptides (CPPs) [1, 2]. The first two CPPs are identified: the 11 amino acid cationic domain of trans-activator of transcription (Tat) protein of the human immunodeficiency virus protein (YGRKKRRQRRR) [3] and the 16 amino acid sequence from the third helix of the homeobox Antennapedia transcription factor of Drosophila melanogaster domain (RQIKIWFQNRRMKWKK) termed Antp or penetratin [4]. A basic domain by 6 arginine and 2 lysine residues of Tat has shown to be the minimal sequence responsible for cellular uptake mediated membrane permeation. The interaction of arginine-rich peptides with membrane-associated proteoglycans results in activation of Rac, leading to F-actin reorganization and macropinocytosis [5, 6]. Specifically, the guanidine group of arginine interacts with carbonate, sulfate, and phosphate groups of proteoglycans on the cellular surface, and thus plays an important role in intracellular delivery carriers of peptides, nucleotides, and nanoparticles such as liposomes and polymer micelles by endocytosis [7–9]

A lot of CPPs with basic amino acids, in particular, oligoarginine are essential for membrane permeability. After internalization into the cell, it is very important that plasmid DNA (pDNA)/small interfering RNA (siRNA) could be escaped from the endosome into the cytoplasm and released from the CPPs in the cytoplasm to achieve efficient gene expression/RNA interfering. The disulfide bond is stable in the extracellular environment. So, the CPPs consisting of disulfide bonds can be used to prepare a stable complex with pDNA/siRNAs. In addition, the disulfide bond can be cleaved rapidly by glutathione (GSH) in the cytoplasm. Thus, pDNA/

Cell-Penetrating Peptides: Design, Development and Applications, First Edition. Makoto Oba and Yosuke Demizu.

siRNA can be rapidly released from the carrier to provide an efficient gene expression or RNA interference. Therefore, we designed a newly multifunctional CPP, CH2R4H2C (CHHRRRRHHC), in which 4 Arg (R) can induce endocytosis with higher cellular uptake efficacy, 4 His (H) can promote the early endosomal escape by the proton sponge effect, and 2 Cys (C) can control the greater intracellular release of pDNA/siRNA than possible using only the Tat peptide (Figure 16.1) [10]. The CH2RH2C peptide can form tighter complexes with pDNA/siRNA through ionic interactions and by an intermolecular disulfide cross-linkage between 2 Cys (C) residues in nonreducing environments such as the extracellular space and blood (GSH: 0.5–10 μM), resulting in effective protection of nucleic acids against nucleases. Subsequently, siRNA can be easily released in the reducing environment inside the cell (GSH: 0.5–10 mM in the cytosol; 20 mM in the nucleus), which promotes cleavage of the disulfide linkages, following higher uptake by the target cells through endocytosis triggered with oligoarginine. Moreover, the efficacy of stearic acid (STR) conjugated with a CH2R4H2C was also elucidated as siRNA [10–12] and pDNA carriers [13, 14]. STR-CH2R4H2C/siRNA complexes showed significantly higher cellular uptake primarily by a macropinocytosis route and stronger silencing effects than non-cross-linked carrier complexes; they exerted higher antitumor effects following direct injection into the tumor in S-180 sarcoma-bearing mice [10] and enhanced transdermal RNAi agent delivery by topical application [11, 12]. This carrier also resulted in increased gene expression of pDNA [13] and promoted an immune response by a needle-free jet injector of the DNA vaccination through the rabbit vaginal membrane [14].

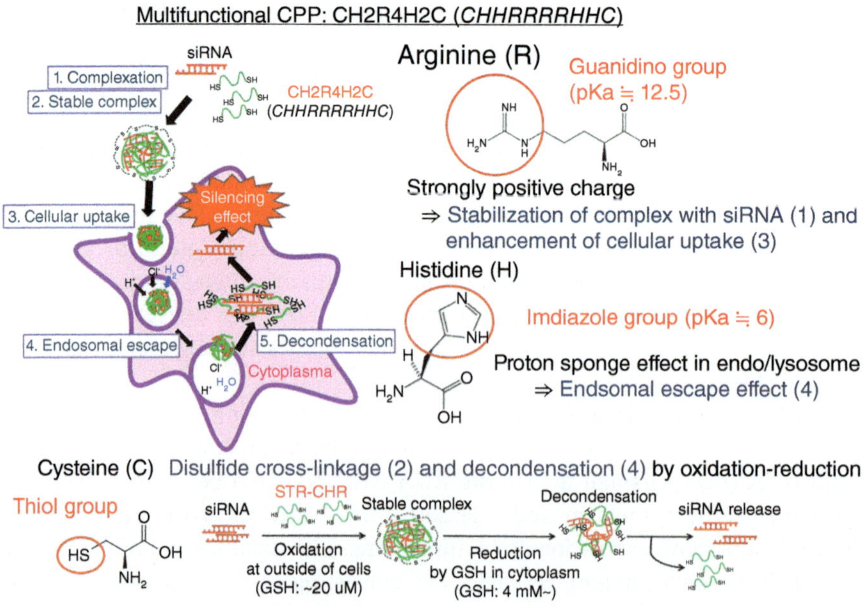

Figure 16.1 Scheme of multifunctional CPP; CH2R4H2C for siRNA delivery [10]. Source: Takanori Kanazawa.

16.1.2 Polymer Micelles

An amphiphilic block copolymer composed of hydrophilic and hydrophobic segments has the tendency to self-assemble into micelles in a selective solvent [15–17]. The polymeric micelles formed through the multimolecular assembly of the block copolymer can then be utilized for core-shell-type colloidal carriers for drugs and genes [18–20]. In particular, poly(ethylene glycol) (PEG) has been used to improve the solubility and steric stability of many pDNA/siRNA delivery systems including liposomes and micelles [17]. It is possible to enhance the circulation time by coating the delivery system with bioinert water-compatible polymers such as PEG, which can give rise to steric stabilization of the delivery vehicle against undesirable aggregation and nonspecific electrostatic interactions with the surroundings [19, 21, 22]. The surface modification of nanoparticles with PEG can extend the circulation of polymer micelle/pDNA complexes. Furthermore, keeping the size of the carriers to less than 200 nm enables them to efficiently accumulate in tumor and inflamed tissue as a result of the enhanced permeability and retention (EPR) effect [23–27]. The PEG-shielded polyplex structures are typically constructed by complexation of DNA/siRNA with PEG-modified polycations, largely due to the strong hydrophilicity of the PEG segments, which causes a significant weakening of the binding of the polycation to DNA. Furthermore, the PEGylation of polyplexes has an inhibitive effect on the cellular uptake of the polyplexes [28]. Recently, many studies have demonstrated the concept of multicomponent polymer-based formulations of pDNA/siRNA delivery complexes. These polymers incorporate one or more additional components into the PEGylated polycation platform to enhance the performance of the PEGylated polyplexes in various aspects, including cellular uptake [29, 30], endosomal escape of polyplexes [31], and timely release of the encapsulated DNA into the cytosol [32].

Kataoka et al. also reported the preparation of core-shell-type poly ion complex micelles (PICs) with a disulfide cross-linked core by the self-assembly of iminothiolane-modified poly (ethylene glycol)-block-poly(L-lysine) [PEG-b-(PLL-IM)] to form an siRNA carrier [33, 34]. These micelles, which respond to a reducing environment, achieved 100-fold higher siRNA transfection efficacy compared with non-cross-linked PICs prepared from PEG-b-poly(L-lysine). Furthermore, the triblock copolymer, PEG, poly(propylene sulfide) (PPS), and a positively charged Tat peptide or oligolysine (PEG-PPS-SS-peptide), could transport siRNA inside the cell and mediate silencing directly proportional to the amount of internalization and gene transfer [35]. The polymer micelles of Methoxy PEG (MPEG)- poly(lactic-co-glycolic acid) (PLGA)-b-poly-L-Lys (PLL), including adriamycin or siRNA, have enhanced targeting delivery and cellular uptake and have improved the antitumor effects in Huh-7 hepatic carcinoma-bearing mice after intravenous injection [36].

In this chapter, two CPP (Tat or CH2R4H2C)-conjugated polymer micelles, which can regulate both pharmacokinetics by polymer micelles and intracellular dynamics by CPP for systemic, nose-to-brain, oral delivery to achieve pDNA/siRNA delivery system to tumor, inflamed, and brain tissue, were introduced.

16.2 Tat-Conjugated Polymer Micelles

16.2.1 Systemic Delivery

16.2.1.1 pDNA Delivery

Gene therapy refers to the treatment of human diseases by transfer of therapeutic genes to specific tissues or cells in a patient [37, 38]. It has the potential to cure both inherited and acquired genetic disorders by supplying functional copies of the defective genes to the diseased cells. An ideal nonviral gene carrier would be capable of performing the multiple functions required for the precise delivery of systemically administered DNA to the nucleus of the targeted cell; these required functions include protection of the therapeutic genes in the extracellular environment, specific cell targeting, early endosomal escape, nuclear entry, and DNA release [39, 40]. The polymer micelles formed through the multimolecular assembly of the block copolymer can then be utilized for core-shell type colloidal carriers for drugs and genes [18–20]. In particular, the PEG-shielded polyplex structures are typically constructed by complexation of DNA with PEG-modified polycations, largely due to the strong hydrophilicity of the PEG segments, which causes a significant weakening of the binding of the polycation to DNA. Furthermore, the PEGylation of polyplexes has an inhibitive effect on the cellular uptake of the polyplexes [28].

A disulfide bond is stable in the extracellular environment but this bond can be cleaved rapidly by GSH in the cytoplasm and nucleus. In order to achieve efficient transcription, it is very important that pDNAs can be released from the vector once inside the cell. The disulfide bond is important for gene delivery because of the decondensation and release of pDNA that is expected to be triggered by the reducing environment of the cytosol and nucleus. Therefore, we designed Tat-conjugated MPEG-poly(caprolactone) (PCL) block copolymers through the disulfide linkage to incorporate pDNA into polymeric nanoparticles by physical entrapment and electrostatic interaction and to improve intracellular uptake and gene expression of pDNA (Figure 16.2) [41].

The particle size of the MPEG-PCL micelles was larger than that of the Tat analog-conjugated MPEG-PCL, and the zeta potential of MPEG-PCL showed a negative charge whereas that of Tat analog-conjugated MPEG-PCL showed a positive charge. This suggests that these synthesized Tat-conjugated polymers can form nanoparticles, possibly like a polymer micelle, and which are different from MPEG-PCL nanoparticles and have a slightly positive charge in water, most likely because the Tat is presented on the surface of the nanoparticles. As shown in Figure 16.3a, MPEG-PCL without Tat did not show a delay of pDNA containing firefly luciferase cDNA driven by CMV promoter (pCMV-Luc) bands, indicating that MPEG-PCL cannot form a complex with pCMV-Luc at any weight ratio due to the negative charge of the MPEG-PCL surface. In contrast, MPEG-PCL-ester-Tat and MPEG-PCL-SS-Tat showed a delay of pCMV-Luc bands at up to a 5/1 weight ratio, indicating that Tat analog-conjugated MPEG-PCL can form a stable complex with pCMV-Luc.The results of the in vitro transfection experiments in COS7 and S-180 cancer cells

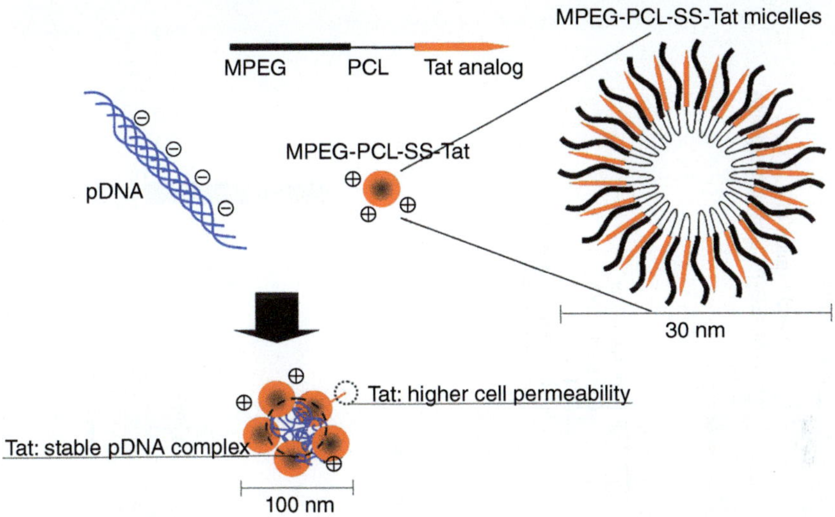

Figure 16.2 Scheme of Tat-conjugated polymer micelles through disulfide linkage (PEG-PCL-SS-Tat) for pDNA delivery. Source: Adapted from Tanaka et al. [41].

indicate that MPEG-PCL-SS-Tat also significantly increases the transfection efficiency at 10/1 and 15/1 weight ratios compared to naked pCMV-Luc. MPEG-PCL-ester-Tat did not increase luciferase activity, compared to naked pCMV-Luc (Figure 16.3c). As shown in Figure 16.3b, bands of released pCMV-Luc were observed at the weight ratios of PEG-PCL-SS-Tat to pCMV-Luc of 5/1 and 10/1 in the gel electrophoresis analysis of MPEG-PCL-Tat and pCMV-Luc complexes after treatment for three hours with the reducing regent, 10 mM GSH. These results suggested that MPEG-PCL-ester-Tat and MPEG-PCL-SS-Tat formed a nano-sized complex with pDNA due to not only the electrostatic interaction by Tat but also constructive and physical complexation by micelles. So, in the reductive environment, the interaction between MPEG-PCL-SS-Tat and pDNA was reduced because the ability of constructive and physical complexation disappeared by cleavage disulfide bonds. As a result, pDNA was easily released from MPEG-PCL-SS-Tat.

In order for a systemic gene carrier to be effective, the degradation that occurs by interaction with serum proteins and enzymes needs to be avoided. Thus, the luciferase activity in COS7 cells transfected with Tat analog, MPEG-PCL-SS-Tat in the presence of serum was determined (Figure 16.4a). MPEG-PCL-SS-Tat after complexation with pCMV-Luc significantly and strongly increased luciferase activity dose-dependently, even in the presence of serum, compared to naked pDNA and Tat analog complex. This is most likely due to nonionic water-soluble PEG chains acting as a hydrophilic shell to stabilize and protect the pDNA from serum proteins. Consequently, a number of MPEG-PCL-SS-Tat molecules might survive to then produce systemic transfection in the presence of serum. The transfection of pCMV-Luc into tumor tissue inoculated with S-180 cells increased 24 hours after intravenous injection of the naked pCMV-Luc or MPEG-PCL-SS-Tat/pCMV-Luc complex

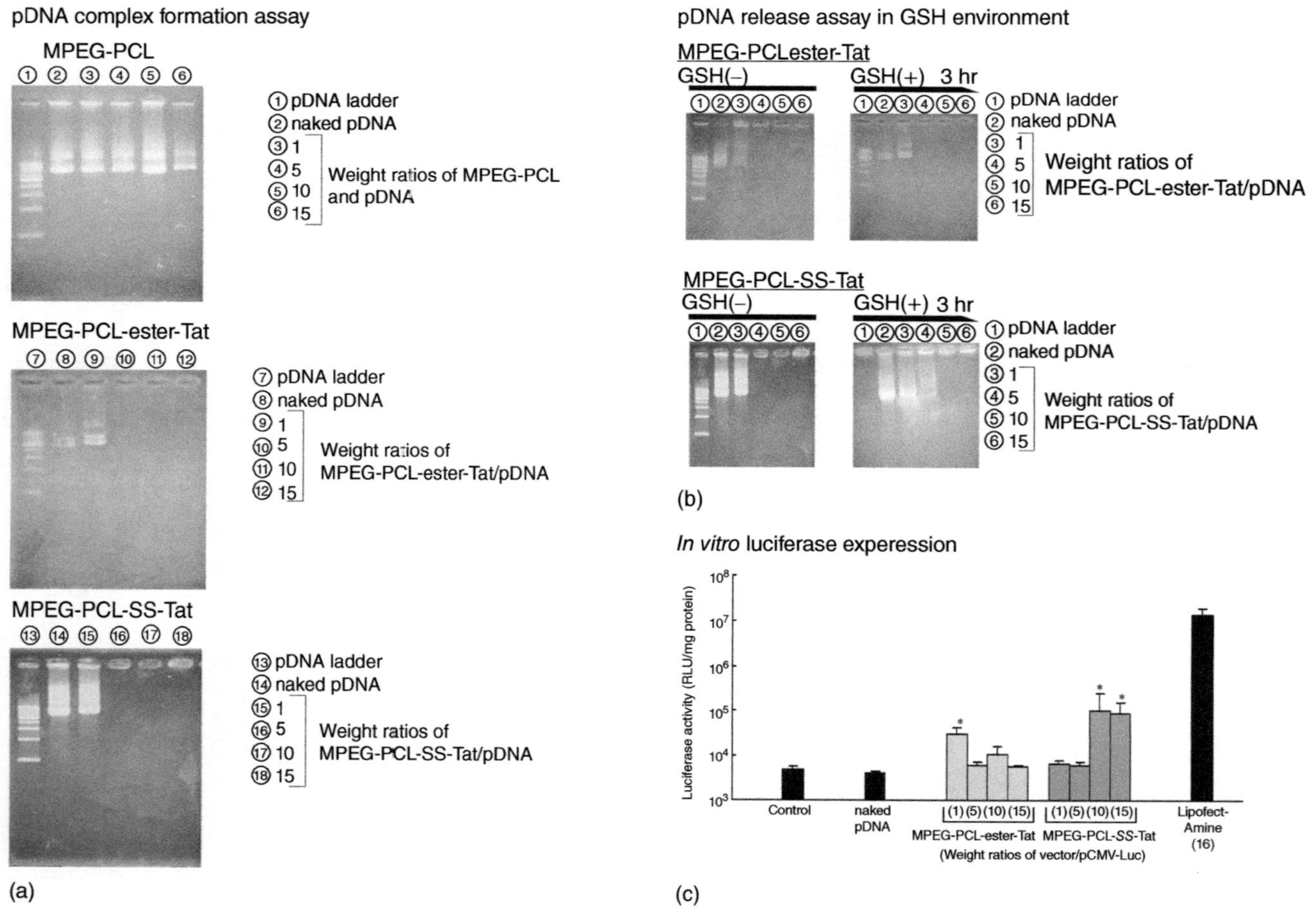

Figure 16.3 Complex formation, pDNA release in reducing environment, and luciferase expression of MPEG-PCL-ester-Tat and MPEG-PCL-SS-Tat with pCMV-Luc. Source: Tanaka et al.[41]/with permission of Elsevier.

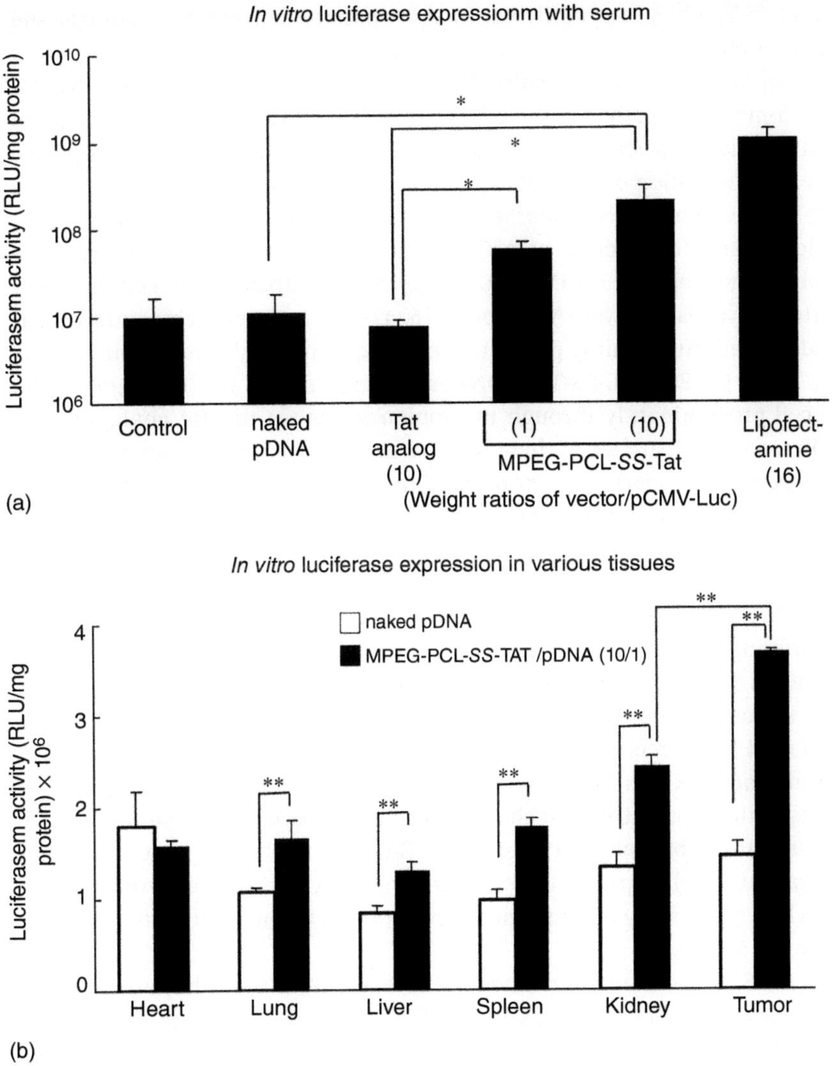

Figure 16.4 Luciferase activities in vitro in presence of serum transfected, and in various tissues of tumor-bearing mice intravenously injected, with MPEG-PCL-SS-Tat/pCMV-Luc complex. Source: Tanaka et al. [41] /with permission of Elsevier.

(Figure 16.4b). In the naked pCMV-Luc-injected mice, the luciferase activity did not advance above the control level in any tissue, possibly due to degradation by nucleases in the serum. In contrast, when the MPEG-PCL-SS-Tat/pCMV-Luc complex was injected into the mice, the luciferase activity in the lung, liver, spleen, kidney and tumor was significantly higher than that produced in the naked pCMV-Luc-injected mice. In particular, the luciferase activity in tumor tissue after intravenous injection of MPEG-PCL-SS-Tat/pCMV-Luc complex was significantly higher than that in the other normal tissues. The MPEG-PCL-SS-Tat/pCMV-Luc (10/1 weight ratio) complex decreased to a particle diameter of about 60 nm and a zeta potential of about +10 mV, which was sufficient to allow passive delivery into tumor tissue using the EPR effect.

Most nonviral vectors are internalized by endocytosis such as macropinocytosis, the clathrin-mediated pathway, the caveola-mediated pathway and fusion. Gene expression efficiency is strongly related to the cellular uptake pathway of genes. Therefore, determination of the cellular uptake pathway of MPEG-PCL-SS-Tat was required. The luciferase activity markedly decreased in the presence of each inhibitor. Amiloride and transfection at 4 °C strongly and significantly inhibited gene expression, suggesting that MPEG-PCL-SS-Tat mainly transfects the gene through the macropinocytosis pathway. As reported [6, 42–44], arginine-rich peptides such as Tat and oligoarginine mainly pass into the cytosol through this macropinocytosis pathway. Thus, MPEG-PCL-SS-Tat/pDNA complexes might also undergo uptake into the cell predominantly through macropinocytosis via the Tat analog, which may be present on the surface of these nanoparticle complexes. Furthermore, these results suggest that a marked proportion of these nanoparticles were transfected through the other endocytosis processes, via the caveola- and clathrin-mediated pathways.

16.2.1.2 siRNA Delivery

Small interfering RNAs (siRNAs) can suppress the expression of a specific gene and are considered potential therapeutic agents for many diseases, including cancer, infection, and other gene-related diseases [45–47]. However, there are lots of barriers that should be overcome for efficient systemic delivery of siRNA to the target cells. For instance, siRNAs are unstable in vivo environment and easily degraded by endosomes in the cells. Further, it shows low uptake to cytoplasm and nucleus. To breakthrough these barriers, MPEG-PCL-SS-Tat were applied to systemic siRNA delivery for cancer therapy by targeting the vascular endothelial growth factor (VEGF), which plays a key role in cancer development through promote angiogenesis (Figure 16.5) [48].

MPEG-PCL-SS-Tat can readily condense with siRNA due to that arginine has strong compaction ability with siRNA; therefore, MPEG-PCL-SS-Tat can condense them at the low N/P ratio. The cellular uptake of carrier/siRNA complexes and the cellular uptake of MPEG-PCL-SS-Tat complexes at N/P ratio 30 was higher than the uptake of Tat complexes (Figure 16.6a). This suggested that MPEG-PCL-SS-Tat complexes had a higher density of arginine residues on the surface of the particle than that of Tat complexes because of forming micelle. Kim et al. reported that the presence of flexible PEG chains on the surface of the polyelectrolyte complex (PEC) micelles could increase the stability of the polyelectrolyte complexes [49]. Thus, the physically stabilized MPEG-PCL-SS-Tat complexes could be readily taken up by cells via endocytosis. Intracellular uptake of MPEG-PCL-SS-Tat/ fluorescein amidites (FAM)-siRNA complexes increased with an increase in the N/P ratio. We hypnotized that the number of free Tat peptides on the surface of MPEG-PCL-SS-Tat complexes increased with an increase in the N/P ratio, and MPEG-PCL-SS-Tat/ FAM-siRNA complexes (N/P ratio: 30) showed the highest cellular uptake. The results of in vitro ELISA were consistent with the cellular uptake assays which

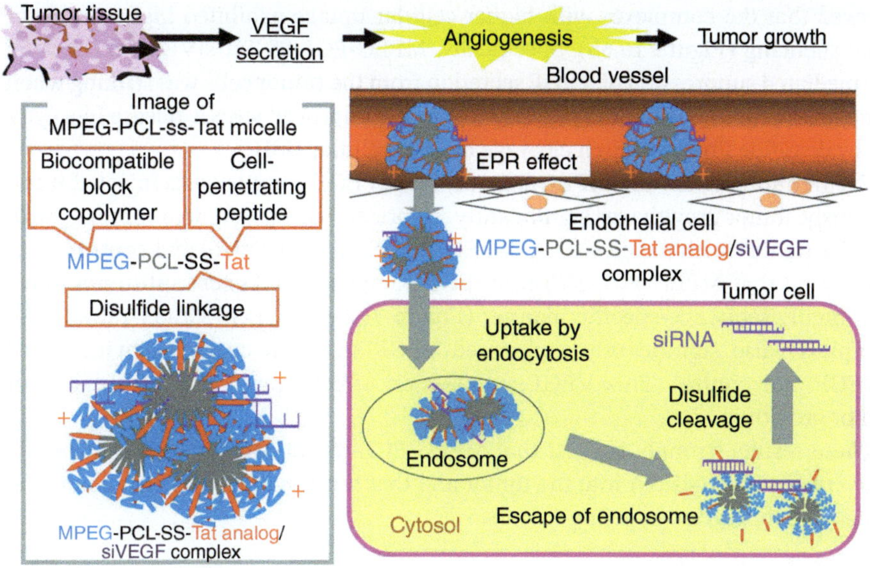

Figure 16.5 Scheme of systemic delivery of PEG-PCL-SS-Tat/anti-VEGF siRNA (siVEGF) for cancer therapy. Source: [48] /with permission of Elsevier.

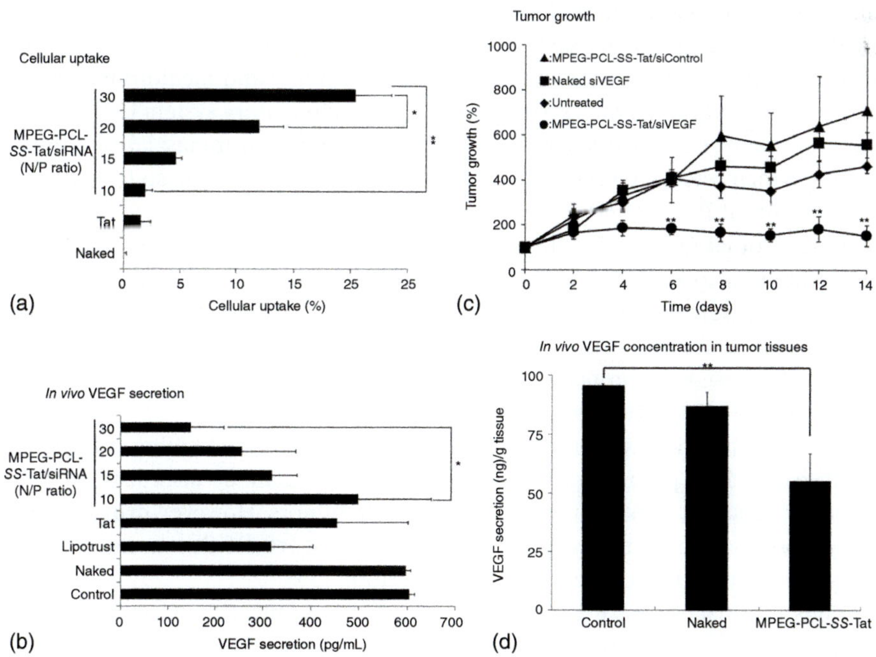

Figure 16.6 in vitro siRNA cellular uptake and inhibition of VEGF secretin, and in vivo systemic antitumor effects in tumor-bearing mice of MPEG-PCL-SS-Tat/siVEGF complex. Source: Kanazawa et al. [48] /with permission of Elsevier.

showed that the complexes with higher cellular uptake exhibited higher levels of gene silencing (Figure 16.6b). Specifically, MPEG-PCL-SS-Tat/siVEGF (N/P ratio: 30) mediated suppression of VEGF secretion from the tumor cells was striking when compared with the untreated possibly due to the improved siRNA release in a cytosol by disulfide linkage which cleavage by intracellular GSH.

Furthermore, when the MPEG-PCL-SS-Tat/siVEGF complex was injected intravenously, tumor growth was significantly suppressed compared with tumor growth in mice treated with naked siRNA, MPEG-PCL-SS-Tat/siControl and control mice, suggesting that MPEG-PCL-SS-Tat/siVEGF complex could exert antitumor effect in a highly sequence-specific manner (Figure 16.6c). MPEG-PCL-SS-Tat/siVEGF complexes also significantly suppressed VEGF secretion compared with naked siVEGF and control. This VEGF suppression effect caused the suppression of tumor growth in vivo.

These results strongly suggest that MPEG-PCL-SS-Tat can promote the specific delivery of pDNA/siRNA into the tumor and that this might be useful as a nonviral systemic nanocarrier targeting tumors.

16.2.2 Nose-to-Brain Delivery

Current drug therapies for neurodegenerative diseases, such as Alzheimer's disease, Parkinson's disease, and amyotrophic lateral sclerosis, are primarily administered orally. After administration, oral drugs are transported across the blood-brain barrier (BBB) via the circulatory system. However, the BBB and blood–cerebrospinal-fluid-barrier (BCSFB) limit the transfer of drugs with low and medium molecular weights to the central nervous system (CNS) via the circulatory system [50].

In recent years, extensive research has been conducted on intranasal drug administration, which offers a direct route of drug delivery to the CNS (nose-to-brain) by bypassing the BBB and BCSFB. Drugs can reportedly reach the brain parenchyma and cerebrospinal fluid (CSF) through the interstitial fluid that fills the perineural and perivascular spaces [51–53]. The intranasal administration of biologically active drugs with medium and high molecular weights, which have difficulty penetrating the BBB, such as inulin, insulin, and antibodies, have been reported to work effectively via the nose-to-brain pathway in rodents [53–55], nonhuman primates [56], and humans [57, 58]. Therefore, nose-to-brain delivery is a highly versatile route, which, in combination with novel drugs being developed for treating intractable CNS diseases, is a promising approach for the treatment of disorders such as Alzheimer's disease, Parkinson's disease, and brain tumors. However, drug delivery via the nose-to-brain pathway has a bioavailability of $<0.1\%$ due to mucociliary clearance and nasal mucosal epithelium permeation [59].

Various approaches have been introduced to overcome this obstacle, one of which is the use of nanocarriers [60–62]. The use of nanocarriers could increase the availability of drugs to the CNS, since their small diameter would potentially allow the nanoparticles to be transported transcellularly along the neurons to the brain through various endocytic pathways of sustentacular or neuronal cells present in the nasal membrane. Additionally, conjugation of the CPP to the surface on the nanocarriers was expected to increase the penetration across the nasal mucosa and target cell membrane into the brain. In our study, the comparison of nose-to-brain delivery between MPEG-PCL and MPEG-PCL-Tat micelles was investigated [63]. At four hours following intranasal administration in rats, MPEG-PCL-Tat was significantly higher than MPEG-PCL polymer micelles. We subsequently determined the biodistribution of coumarin in rats after intranasal administration of coumarin-loaded MPEG-PCL-Tat micelles. The concentrations of coumarin in nontargeted tissues, such as liver, lung, heart, kidney, and spleen, in rats administered coumarin-loaded MPEG-PCL-Tat were lower than those measured following administration of coumarin solution without the polymer micelles. These findings indicate that CPP-conjugated polymer micelles can improve nose-to-brain drug delivery.

Next, we applied MPEG-PCL-Tat to the siRNA nose-to-brain delivery system [64–66]. Significantly higher brain penetration was observed following intranasal administration of MPEG-PCL-Tat to rats than with intravenous injection, suggesting that superior delivery of siRNA to the brain is possible using CPP-conjugated polymer micelles [64]. Additionally, in our study of the mechanisms that promote siRNA transfer to the brain following intranasal administration using MPEG-PCL-Tat, we focused on the olfactory and trigeminal nerves, which are the two reported pathways of drug transport to the brain from the nasal cavity. First, nasal mucosal tissue was observed 15 minutes after intranasal administration of naked siRNA and complexes of siRNA with MPEG-PCT-Tat micelles (Figure 16.7a). Rats administered siRNA complexed with MPEG-PCT-Tat exhibited more intense fluorescence in the mucosal epithelial surface and lamina propria mucosae than the animals administered naked siRNA, indicating that MPEG-PCL-Tat exhibits high mucosal penetration. Next, we assessed the distribution of the siRNA to the olfactory bulb after intranasal administration and found a significantly higher transfer of the siRNA to the olfactory bulb in the MPEG-PCL-Tat-treated animals, as compared to those administered the naked siRNA (Figure 16.7b). This finding suggests that the use of MPEG-PCL-Tat may increase the transfer of nucleic acid to the brain via the olfactory nerve pathway in the nasal mucosal tissue. Additionally, the trigeminal nerve after intranasal administration detected a higher transfer of the fluorescein-labeled model siRNA (anionic dextran, MW: 10 000) to the trigeminal nerve in the MPEG-PCL-Tat than in animals administered fluorescein-labeled model siRNA alone. This suggests that, as with the olfactory nerve, MPEG-PCL-Tat may also enable the nucleic acid to be delivered to the brain via the trigeminal nerve. Figure 16.7c demonstrates the distribution in the whole brain, from the olfactory bulb to the brainstem, over time, measured following intravenous or intranasal administration of fluorecein-labeled model siRNA with and without MPEG-PCL-Tat. While the fluorescence of the Alexa-labeled model siRNA did not distribute in the whole brain following intravenous administration, high accumulation was observed in the olfactory bulb 15 minutes after administration via the nose-to-brain pathway, both with or without MPEG-PCL-Tat. One hour after intranasal administration with MPEG-PCL-Tat, high accumulation was observed in the rostral brain tissue that appeared

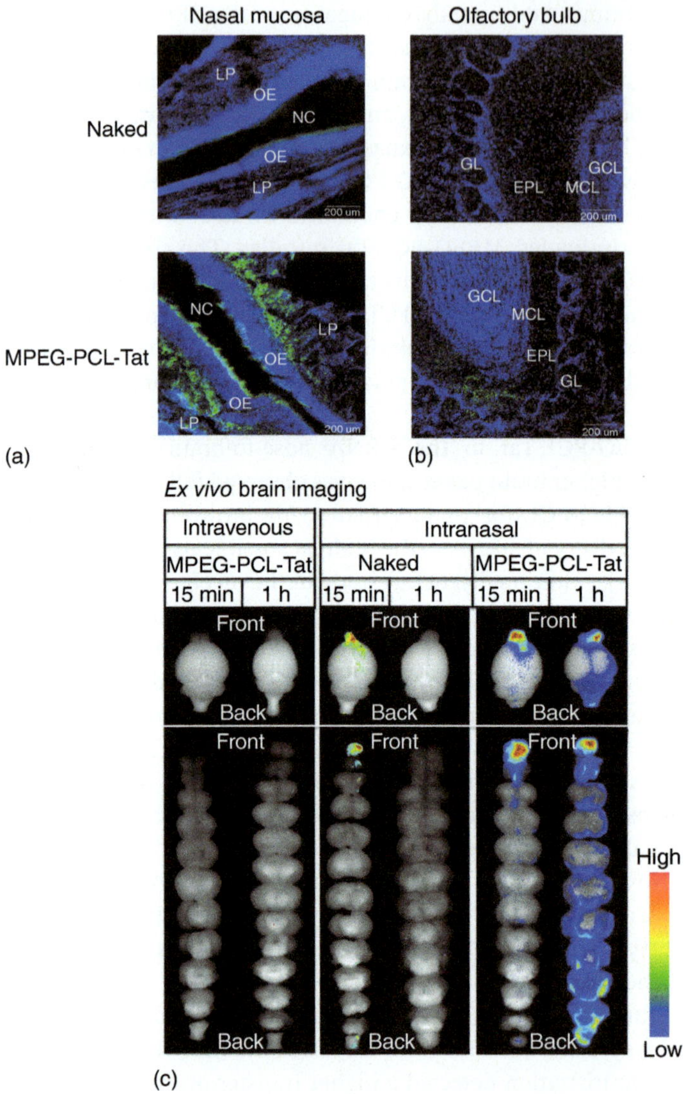

Figure 16.7 Distribution in the nasal mucosa and olfactory bulb tissue, and ex vivo brain imaging of fluorecein-siRNA following intranasal administration with MPEG-PCL-Tat. Source: Kanazawa et. al. [64]/with permission of Elsevier.

to follow the distribution trend into the olfactory bulb, while the caudal brain tissue and brain stem exhibited a high concentration that appeared to follow a distribution trend into the trigeminal nerve. Taken together, these observations suggest that, once transported to the olfactory bulb and brain stem via the olfactory nerve and trigeminal nerve pathways, nucleic acid is subsequently transported to other brain tissues. This shows that the proposed system of delivery of nucleic acid to the brain by intranasal administration using MPEG-PCL-Tat relies primarily on an increase in the transfer of nucleic acid to the brain via the olfactory nerve and trigeminal nerve pathways.

We investigated intranasal delivery of siRNA/drug co-loaded MPEG-PCL-Tat micelles using a rat model of malignant glioma [65]. The cellular uptake of siRNA and evaluated its in vitro cytotoxicity in rat glioma C6 cells. The intracellular siRNA uptake of siRNA/MPEG-PCL-Tat complexes increased with increasing N/P ratio, with the highest intracellular uptake efficacy observed with the complexes at an N/P ratio of 30. To evaluate the in vitro transfection efficiency of the MPEG-PCL-Tat/siRNA complexes at an N/P ratio of 30, we assessed the cytotoxicity induced by Raf-1 gene silencing in C6 cells. These results showed that the cell viability significantly decreased with an increase in the concentration of Raf-1 siRNA (siRaf-1), indicating that MPEG-PCL-Tat/siRaf-1 induces cell death in rat glioma cells due to the high cellular uptake of siRaf-1 by MPEG-PCL-Tat carrier. We finally evaluated the in vivo therapeutic effects in glioma model rats of intranasally administered naked siRaf-1, MPEG-PCL-Tat/control siRNA, MPEG-PCL-Tat/siRaf-1 complex, camptothecin (CPT)-loaded-MPEG-PCL-Tat micelles/control siRNA complex, and CPT-loaded MPEG-PCL-Tat micelles/siRaf-1 complexes (Figure 16.8). MPEG-PCL-Tat/siRaf-1 complex, CPT-loaded-MPEG-PCL-Tat micelles/control siRNA complex, and CPT-loaded MPEG-PCL-Tat micelles/siRaf-1 complexes showed high therapeutic efficacy after seven days of continuous delivery. These results indicate that MPEG-PCL-Tat improved the delivery of siRaf-1 and CPT to the brain, and showed a marked prolongation of the mean survival period (Figure 16.8b). No cytotoxicity was detected in neuronal cells and no signs of macroscopic damage in rat nasal mucosa, olfactory and trigeminal nerves, and brain tissue were observed. Moreover, this approach showed excellent therapeutic efficacy in rat models with cerebral ischemia-reperfusion injury [66].

16.3 Multifunctional Peptide-Conjugated Polymer Micelles

16.3.1 Systemic Delivery

16.3.1.1 Tumor Tissue

To develop a systemic siRNA carrier for tumor tissue, a block-copolymer conjugated with our developed multifunctional CPP (CH2R4H2C), MPEG-PCL-CH2R4H2C was developed (Figure 16.9) [67].

MPEG-PCL-CH2R4H2C can rigidly condense and shield the oligonucleotides in preparations and blood by disulfide bonding and protecting it from nucleases. This delivers the nucleotides to the tumor by the EPR effect, transfers into the cells by endocytosis triggered by the CPP, and triggers early endosomal escape by the proton sponge effect of histidine. The disulfide bonds are cleaved in the reducing environment (about 1000-fold higher concentration of GSH than that in the blood) of the cytoplasm and efficiently release the nucleotides into the cytoplasm to silence the target mRNA [67, 68].

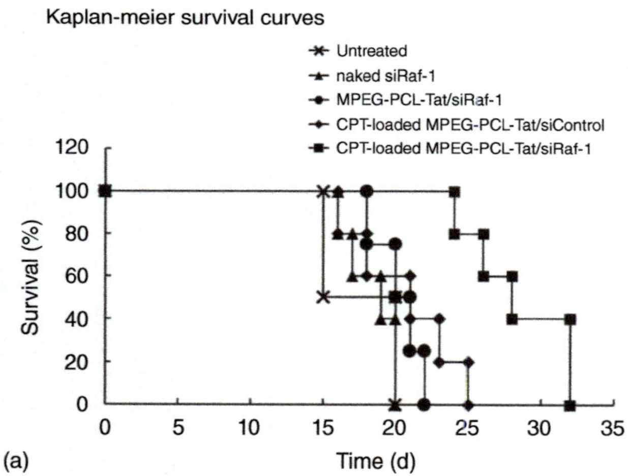

(a)

Median survival period

Treatment	Number of rats	Average of survival period (days)	*p* value
Untreated	5	16.6	
Naked siRaf-1	5	18.4	
MPEG-PCL-Tat/siRaf-1	5	20.4	<0.05 [a,b]
CPT-loaded MPEG-PCL-Tat /siControl	5	20.6	<0.05 [a,b]
CPT-loaded MPEG-PCL-Tat /siRaf-1	5	28.4	<0.01 [a,b], <0.05 [c,d]

[a] Compared to untreated rats, [b] Compared to naked siRaf-1,
[c] Compared to MPEG-PCL-Tat/siRaf-1, [d] Compared to CPT-loaded MPEG-PCL-Tat/siControl.

(b)

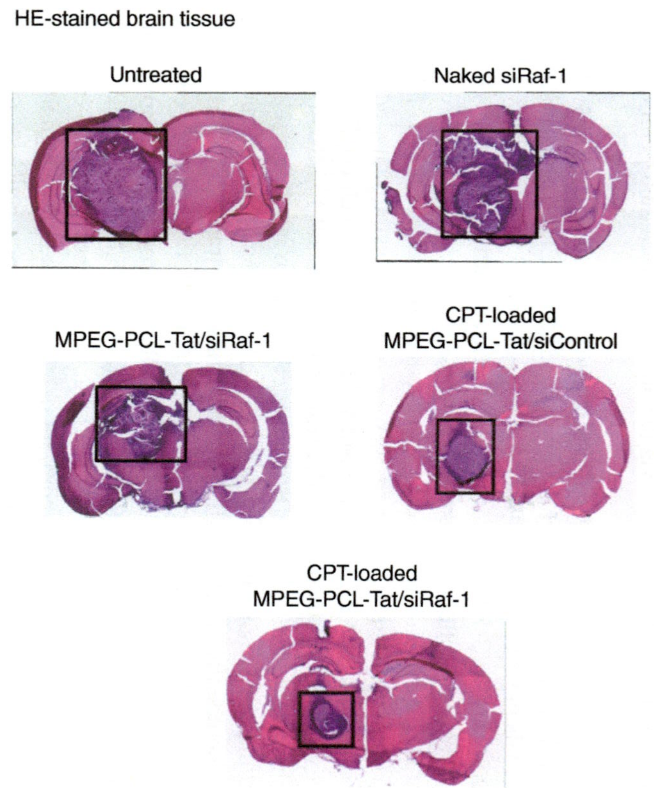

(c)

Figure 16.8 Kaplan-Meier survival curves, median survival period, and HE-stained brain tissue of rats bearing intracranial C6 glioma treated with intranasal camptothecin (CPT)-loaded MPEG-PCL-Tat/Anti-Raf-1 siRNA. Source: Kanazawa et al. [65]/with permission of American Chemical Society.

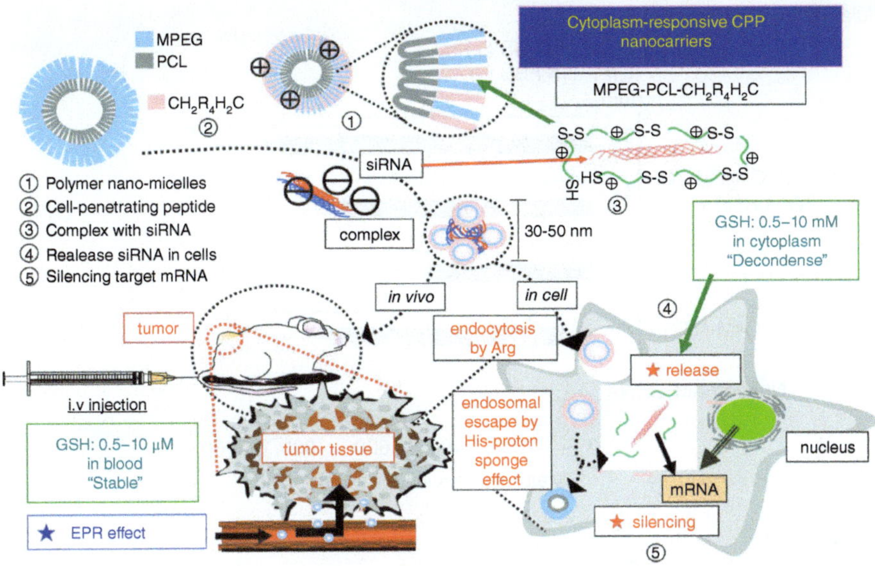

Figure 16.9 Scheme of multifunctional CPP (CH2R4H2C)-conjugated polymer micelles for systemic siRNA delivery. Source: Tanaka et al. [67]/with permission of Elsevier.

The FAM-siRNA cellular uptake by MPEG-PCL-CH2R4H2C with an N/P ratio of 20 was higher than the uptake by CH2R4H2C (Figure 16.10a). Naked FAM-siRNA could barely enter the cells at this condition because of enzymatic degradation and/or no endocytosis. These results suggest that MPEG-PCL-CH2R4H2C/siRNA complexes have a higher density of Arg residues on the surface of the particle than do CH2R4H2C/siRNA complexes because MPEG-PCL-CH2R4H2C/siRNA complexes form rigid micelles owing to their hydrophobic PCL chain. It was also clarified that clusters of Arg side chains in MPEG-PCL-CH2R4H2C complexes, as well as disulfide bonds through Cys aid in peptide penetration into cells, whereas the His residue allows for rapid endosomal escape by the proton sponge effect at the stage of early endosomal formation. Furthermore, although cellular uptake of CH2R4H2C/siRNA complexes significantly decreased in the presence of serum, MPEG-PCL-CH2R4H2C/siRNA complexes maintained high cellular uptake ability even when cultured with serum proteins (Figure 16.10a). These results suggest that a PEG chain provides both resistance toward interactions with serum proteins and enzymatic attack in serum. We hypothesized that our micelles form a "V" shape that exposes CH2R4H2C and the PEG chain on the surface of the micelle, allowing each moiety to exert its respective function as shown in Figure 16.10. Mao et al. also proposed that a hydrophilic-hydrophobic-hydrophilic tri-block polymer micelle also formed "V"-shaped polymers that exposed the cationic polymer and PEG chain on the surface of the micelle, conferring a high silencing effect on the micelles and siRNA complexes both in vitro and in vivo [69]. In addition, the free siRNA could be easily released from the MPEG-PCL-CH2R4H2C/siRNA complexes in the cytoplasm of GSH-rich environment.

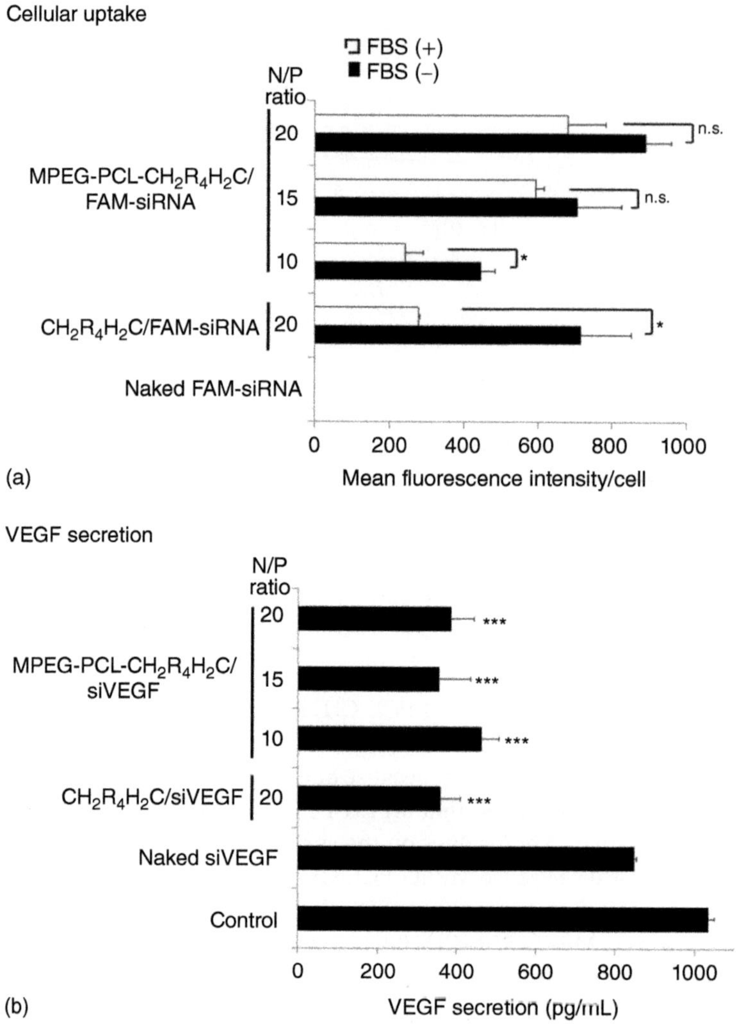

Figure 16.10 In vitro siRNA cellular uptake and inhibition of VEGF secretin transfected with MPEG-PCL-CH2R4H2C/siVEGF complex. Source: Tanaka et al. [67]/with permission of Elsevier.

The results of in vitro ELISA of VEGF secreted in the S-180 cells culture system were consistent with the cellular uptake assays, which showed that the MPEG-PCL-CH2R4H2C/anti-VEGF siRNA (siVEGF) complexes at any N/P ratio between 10 and 20 exhibited significantly higher levels of gene silencing than naked siVEGF (Figure 16.10b). We speculated that there is little difference in the in vitro ELISA of both carrier/siRNA complexes because CH2R4H2C complexes (N/P ratio = 20) and MPEG-PCL-CH2R4H2C complexes (N/P ratio = 10–20) showed greater than 80% cellular uptake. Both CPP carriers were confirmed to be noncytotoxic and safe carriers in cell proliferation and cytotoxicity assays.

The biodistribution by MPEG-PCL-CH2R4H2C was determined using a fluorescent-labeled model siRNA. Figure 16.11a, b indicated that this MPEG-PCL-CH2R4H2C provided relatively sustained blood levels and much higher tumor levels of the anionic marker for two hours, attributed to the effects of CPP (high cellular uptake), PEGylation of self-assembled nanomicelles (avoidance of uptake by the reticuloendothelial system) and the EPR effects. To evaluate the in vivo antitumor effect, we intravenously injected the MPEG-PCL-CH2R4H2C/siVEGF complex via the tail vein of S-180 tumor-bearing mice. The tumor volume strikingly increased in mice injected with naked siVEGF as well as in non-treated control mice. It is possible that naked siVEGF was rapidly degraded in blood by nucleases, and decreased endocytosis in the tumor tissue may have limited cellular uptake of siVEGF. In contrast, when the MPEG-PCL-CH2R4H2C/siVEGF complex was intravenously injected, tumor growth was significantly suppressed compared with mice treated with naked siVEGF and siControl (nonsense sequence oligonucleotide) as well as untreated control mice (Figure 16.11c). We also evaluated VEGF secretion from tumor tissue after intravenous injection and found that the MPEG-PCL-CH2R4H2C/siVEGF complex significantly suppressed VEGF secretion compared with naked siVEGF and untreated control (Figure 16.11d). This suppression effect should sufficiently suppress neovascularization on the tumor surface, causing a persistent regression of tumor growth.

Systemic siRNA delivery with MPEG-PCL-CH2R4H2C also suppressed a cell migration/invasion effect in melanoma B16F10 cells, and in vivo intravenous administration of siRelA/MPEG-PCL-CH2R4H2C micelles in a lung metastasis mouse model increased MPEG blood retention and accumulation at metastatic sites, as well as increased siRNA introduction into cancer cells. Inhibition of RelA affected the expression of cell adhesion factors, cell migration and invasion, and significant suppression of metastasis compared to the control conditions [70].

16.3.1.2 Inflamed Tissue

NF-κB is one of the key therapeutic targets by siRNA in inflammatory diseases such as atopic dermatitis [11, 12, 71–73], rheumatoid arthritis (RA) [74, 75], inflammatory bowel disease (IBD) [76] since this transcription factor controls the production of inflammatory cytokines; these include IL-1, IL-6, and TNF-α, which are produced by macrophages and lymphocytes in many inflammatory diseases. Recently, siRNA for RelA which is a subunit of NF-κB (siRelA)-mediated gene silencing has been investigated as a novel treatment for chronic inflammatory diseases. In our study of a potential systemic treatment system for RA by intravenous administration of siRelA complexed with MPEG-PCL-CH2R4H2C, which can form nano-sized (approximately 50 nm) stable complexes with siRNA were investigated in experimental mice with collagen induced arthritis (CIA) [75].

The MPEG-PCL-CH2R4H2C produced effective cellular uptake of siRNA and reduced the levels of several inflammatory cytokines, including TNF-α, IL-6, and IL-12, in synoviocytes. These are the target cells for RA therapy because they are known to play an important role in controlling the allergic inflammatory responses

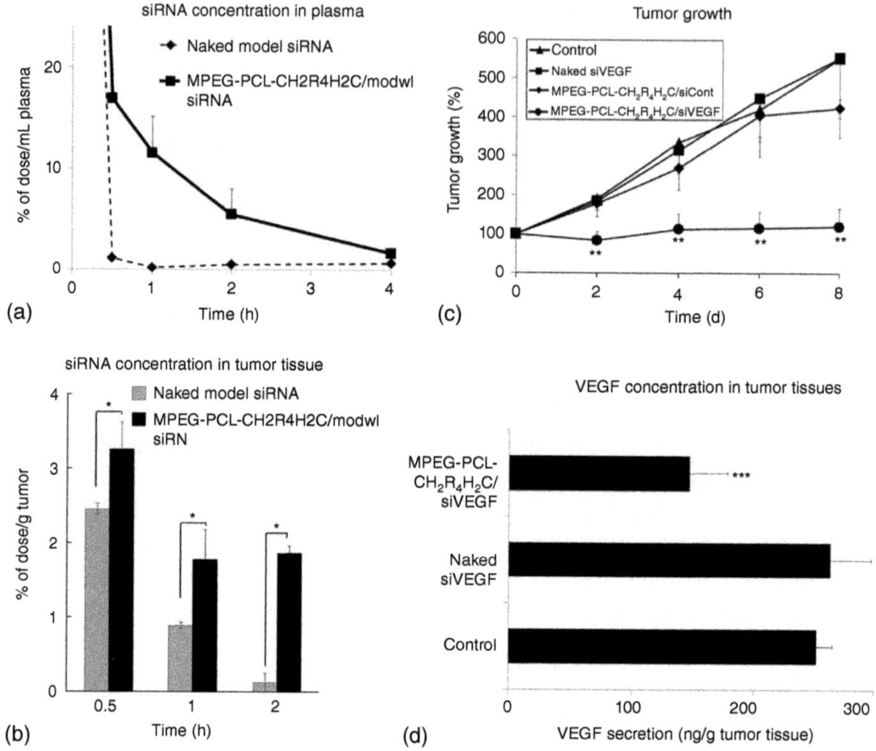

Figure 16.11 In vivo model-siRNA concentration in plasma and tumor tissue, and systemic antitumor effects and silencing effect on VEGF secretion in tumor-bearing mice intravenously injected of MPEG-PCL-SS-Tat/siVEGF complex. Source: Tanaka et al. [67]/with permission of Elsevier.

in RA disease. In addition, the versatile peptide-modified polymer micelles were not cytotoxic at any N/P ratios of less than 20. We also studied the cellular uptake and suppression of RelA mRNA and TNF-α levels in dendritic cells transfected with MPEG-PCL-CH2R4H2C/siRelA complexes. As a result, MPEG-PCL-CH2R4H2C micelles also strikingly increased the intracellular siRNA uptake and suppressed the RelA mRNA and TNF-α levels as the N/P ratio of MPEG-PCL-CH2R4H2C increased.

Next, in vivo imaging of the versatile peptide-modified polymer micelle/model siRNA complexes showed remarkably high accumulation in arthritic paws, as compared with naked model siRNA (Figure 16.12a). This significant difference lasted for up to 168 hours in all four paws. On the other hand, normal healthy mice (without CIA) showed weaker fluorescence intensity in their limbs following the administration of versatile peptide-modified polymer micelle/model siRNA complexes. These results indicated a clear difference between the microenvironment of the inflamed and normal paw tissues. A recent study reported that inflamed tissues are characterized by enhanced vascular permeability, which allows small drug carrier systems to extravasate at these sites via the EPR effect. Subsequently, these are

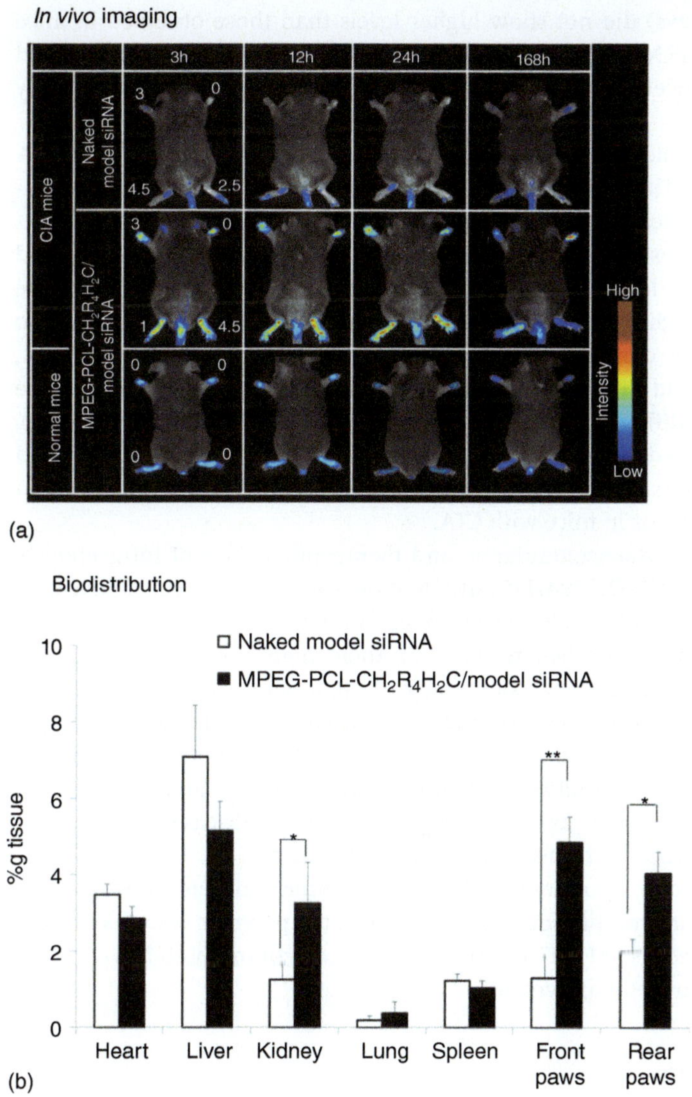

Figure 16.12 In vivo imaging and biodistribution of model siRNA in CIA mice intravenously injected of MPEG-PCL-CH2R4H2C/siRNA. Source: Kanazawa et al. [75]/with permission of Elsevier.

retained in the extravascular space [25–27], and a large portion are taken up by target cells such as synoviocytes, dendritic cells, and macrophages in the synovial layer [26, 27]. These features indicate that the use of PEG-conjugated nano-sized stable micelles with sustained half-lives in circulation could provide an attractive approach to improving the systemic therapeutic index of RA drugs. We also determined the biodistribution of a model siRNA in CIA mice (Figure 16.12b) and found that this was improved in inflamed paw tissue at eight hours post-injection by the use of versatile peptide-modified polymer micelles. In addition, other tissues (except for the kidneys) did not show higher levels than those observed in mice treated with naked siRNA. The reason that the kidney levels were elevated related to the size of the micelles (approximately 100 nm), which rendered them easily excreted.

Finally, we determined the therapeutic effects of these complexes in the CIA mouse model of RA. The versatile peptide-modified polymer micelle/siRelA complexes produced significantly stronger suppression of the increase in paw thickness and the clinical score, as well as a stronger reduction of RelA mRNA expression and inflammatory cytokine levels, than any of the other treatment groups (including the MTX group) (Figure 16.13a–e). Furthermore, although erosion of the synovium to the bone and cartilage damage were observed in mice treated with naked siRelA, siControl complex, and MTX, this was hardly observed in mice treated with the versatile peptide-modified polymer micelle/siRelA complexes (Figure 16.13f–g). These results indicated that MPEG-PCL-CH2R4H2C/siRelA complexes exerted anti-arthritic effects by suppressing multiple inflammatory cytokines, which play an important role in arthritis in mice with CIA.

We also investigated the accumulation and therapeutic effects of intravenously administered MPEG-PCL-CH2R4H2C/siRNA at the site using IBD (ulcerative colitis; UC) model mice [76]. The intravenous administration of MPEG-PCH-CH2R4H2C/siRNA nanomicelles resulted in their high accumulation in the inflamed large intestine, decreased shortening of the inflamed large intestine and clinical score, and suppressed TNF-α production in UC model mice, indicating a positive inhibitory effect of the nanocarriers. Hence, intravenous administration of these nanomicelles may be used as a potential treatment for UC and other inflammation-related diseases, such as allergic and autoimmune diseases. In addition, MPEG-PCL-CH2R4H2C can be used as a novel siRNA carrier.

We have introduced in vivo therapeutic effects of intravenous administration of siRNA complexed with multifunctional CPP-conjugated polymer micelles on cancers, RA, and IBD because MPEG-PCL-CH2R4H2C can control both in vivo pharmacokinctics and intracellular dynamics.

16.3.2 Oral Delivery

Oral administration of siRNAs is desirable because it is the most efficient, patient-friendly, and direct route of drug administration for the treatment of UC [77]. However, the clinical administration of siRNA is currently limited to intravenous or subcutaneous delivery routes. Nucleic acids such as siRNA cannot withstand the harsh environment of the gastrointestinal (GI) tract not only because of the digestion by nucleases but also the excretion by mucous capture [78]. Moreover, regardless of gastric acidity, siRNAs are highly unstable and easily degraded in vivo [79], requiring nanocarriers that protect them from the action of biological enzymes. An ideal delivery vehicle protects the siRNA from degradation in the GI environment and allows the drug to enter the cytoplasm of enterocytes where gene silencing occurs at the mRNA level. For example, in recent years, it has been reported that hyaluronic acid-polymer nanoparticles encapsulated in chitosan/alginate

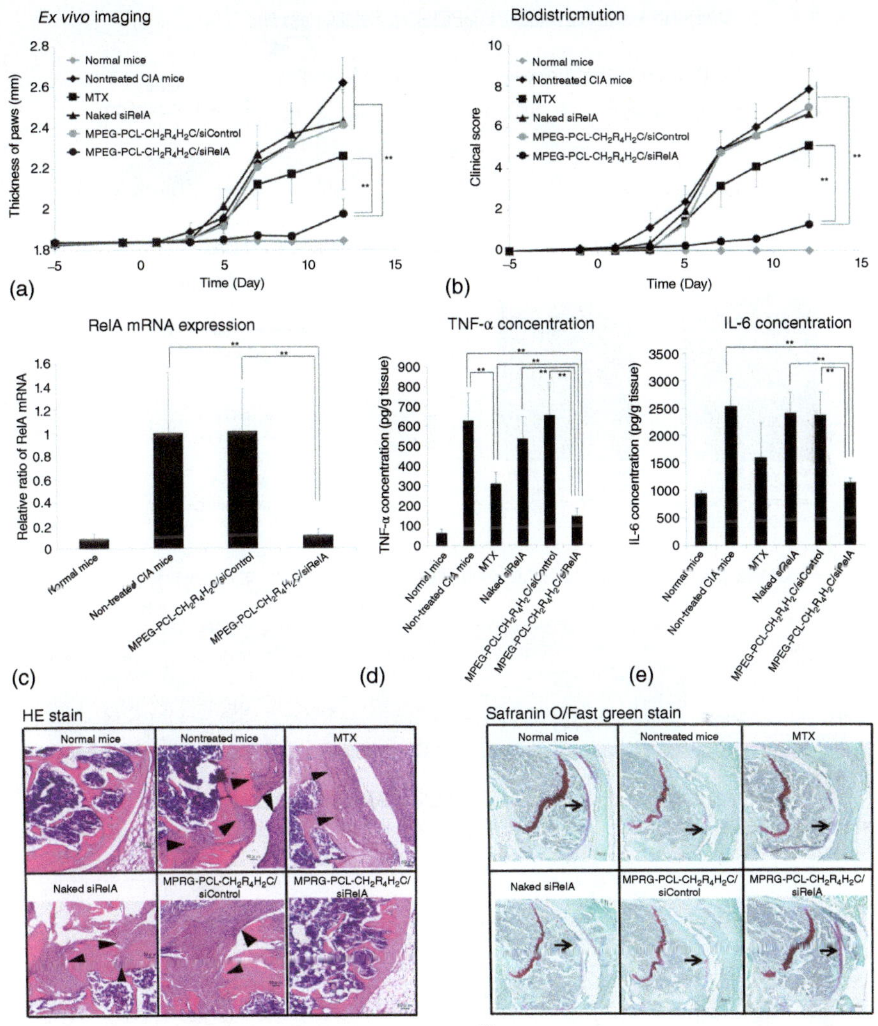

Figure 16.13 Anti-arthritis effects, suppression of RelA and inflammatory cytokines, and histological analysis of knee joints in CIA mice intravenously injected of MPEG-PCL-CH2R4H2C/siRelA. Source: Kanazawa et al. [75]/with permission of Elsevier.

hydrogels, nanoparticle-in-microsphere oral system particles, and polylactide- or poly (lactide-co-glycolide acid)-based nanoparticles have been used as oral siRNA carriers [80–83]. Additionally, osmotically sensitive liposomes have been developed as peptide-drug carriers for oral administration [84]. However, the number of such applications is limited.

MPEG-PCL-CH2R4H2C was applied and evaluated as a carrier for oral siRNA delivery targeted to the colon to treat UC (Figure 16.14). The resistance of the siRNA degradation in the gastric and intestinal pH environment, cellular uptake and viability in macrophage and Caco-2 cells, in vivo distribution of siRNA in the colon

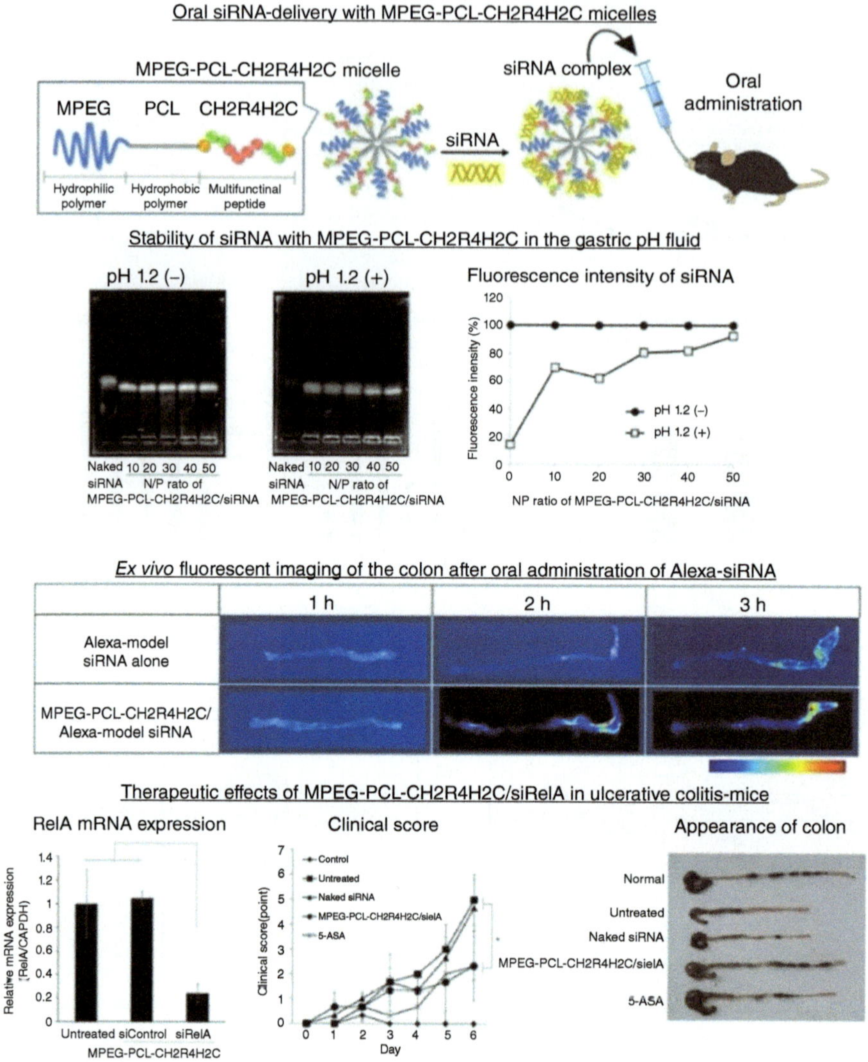

Figure 16.14 Oral delivery of MPEG-PCL-CH2R4H2C/siRNA complex for IBD. Source: Ibaraki et al. [85].

after oral administration, and the therapeutic effects of the RelA-targeted siRNA by oral administration in a UC-induced mouse model were investigated.

While passing through the GI tract, therapeutic agents are exposed to a wide range of pH values (pH 1–2 in the stomach) to approximately 8 in the large intestine and rectum [86]. One of the most pressing challenges is the degradation of siRNA by potent enzymatic and mechanical processes in the GI tract.

The observed degradation of siRNA in the gastric pH environment suppressed when administered as a complex is thought to be related to the N/P ratio of MPEG-PCL-CH2R4H2/siRNA, resulting in a strong complex that protects the siRNA from degradation (Figure 16.14). On the other hand, under the intestinal pH generated by adding a pH 6.8 fluid, no degradation of either naked siRNA or MPEG-PCL-CH2R4H2 was observed even after 12 hours, suggesting that siRNA is relatively stable in the intestinal pH environment. Further, colonic accumulation of MPEG-PCL-CH2R4H2/siRNA after its oral administration was evaluated in healthy mice. MPEG-PCL-CH2R4H2C/model siRNA was accumulated in the large intestine two hours after administration (Figure 16.14). In general, the residence time of orally administered drug carriers in the stomach and small intestine is approximately two and three hours, respectively [87]. These results suggest that the use of MPEG-PCL-CH2R4H2C would make the particles tolerant to the gastric and intestinal pH, and thus resulted in their accumulation in the colon.

Finally, the therapeutic effects of orally administered MPEG-PCL-CH2R4H2/siRelA on UC model mice were evaluated. The target RelA expression and therapeutic effects have been achieved by the oral delivery of MPEG-PCL-CH2R4H2C/siRelA complex. Therefore, we demonstrated that MPEG-PCL-CH2R4H2 is also useful as a carrier for oral siRNA delivery for the treatment of UC; it knocks down the target mRNA and shows therapeutic effects on UC.

16.4 Conclusions and Future Perspective

Through the chapter, we have shown that the CPP-conjugated polymer micelles enable the systemic pDNA/siRNA delivery to tumor and inflamed tissues by regulating both pharmacokinetics by MPEG-PCL and intracellular dynamics by CPPs, and their potential for application to nose-to-brain and oral delivery. We feel that this review offers important insight that can be applied to improve drug delivery based on CPPs.

References

1 Wender, P.A., Mitchell, D.J., Pattabiraman, K. et al. (2000). The design, synthesis, and evaluation of molecules that enable or enhance cellular uptake: peptoid molecular transporters. *Proceedings of the National Academy of Sciences U.S.A.* 97: 13003–13008.

2 Taylor, R.E. and Zahid, M. (2020). Cell penetrating peptides, novel vectors for gene therapy. *Pharmaceutics* 12: 225.

3 Green, M., Ishino, M., and Loewenstein, P.M. (1989). Mutational analysis of HIV-1 Tat minimal domain peptides: identification of trans-dominant mutants that suppress HIV-LTR-driven gene expression. *Cell* 58: 215–223.

4 DeRossi, D., Joliot, A.H., Chassaing, G. et al. (1994). The third helix of the Antennapedia homeodomain translocates through biological membranes. *Journal of Biological Chemistry* 269: 10444–10450.

5 Futaki, S., Ohashi, W., Suzuki, T. et al. (2001). Stearylated arginine-rich peptides: a new class of transfection systems. *Bioconjugate Chemistry* 12: 1005–1011.

6 Futaki, S., Nakase, I., Tadokoro, A. et al. (2007). Arginine-rich peptides and their internalization mechanisms. *Biochemical Society Transactions* 35: 784–787.

7 Nakamura, Y., Kogure, K., Futaki, S. et al. (2007). Octaarginine-modified multifunctional envelope-type nano device for siRNA. *Journal of Controlled Release* 119: 360–367.

8 Furuhata, M., Danev, R., Nagayama, K. et al. (2008). Decaarginine-PEG-artificial lipid/DNA complex for gene delivery: nanostructure and transfection efficiency. *Journal of Nanoscience and Nanotechnology* 8: 2308–2315.

9 Kim, T.I., Ou, M., Lee, M. et al. (2009). Arginine-grafted bioreducible poly(disulfide amine) for gene delivery systems. *Biomaterials* 30: 658–664.

10 Tanaka, K., Kanazawa, T., Ogawa, T. et al. (2010). Disulfide crosslinked stearoyl carrier peptides containing arginine and histidine enhance siRNA uptake and gene silencing. *International Journal of Pharmaceutics* 398: 219–224.

11 Kanazawa, T., Hamasaki, T., Endo, T. et al. (2015). Functional peptide nanocarriers for delivery of novel anti-RelA RNA interference agents as a topical treatment of atopic dermatitis. *International Journal of Pharmaceutics* 489: 261–267.

12 Ibaraki, H., Kanazawa, T., Takashima, Y. et al. (2016). Development of an innovative intradermal siRNA delivery system using a combination of a functional stearylated cytoplasm-responsive peptide and a tight junction-opening peptide. *Molecules* 21: 1279.

13 Tanaka, K., Kanazawa, T., Ogawa, T. et al. (2011). A novel, bio-reducible gene vector containing arginine and histidine enhances gene transfection and expression of plasmid DNA. *Chemical and Pharmaceutical Bulletin* 59: 202–207.

14 Kanazawa, T., Tamura, T., Yamazaki, M. et al. (2013). Needle free intravaginal DNA vaccination using a stearoyl oligopeptide carrier promotes local gene expression and immune responses. *International Journal of Pharmaceutics* 447: 70–74.

15 Allen, C., Yu, Y., Maysinger, D. et al. (1998). Polycaprolactone-b-poly(ethylene oxide) block copolymer micelles as a novel drug delivery vehicle for neurotrophic agents FK506 and L 685,818. *Bioconjugate Chemistry* 9: 564–572.

16 Otsuka, H., Nagasaki, Y., and Kataoka, K. (2003). PEGylated nanoparticles for biological and pharmaceutical applications. *Advanced Drug Delivery Reviews* 55: 403–419.

17 Jang, J.S., Kim, S.Y., Lee, S.B. et al. (2006). Poly(ethylene gricol)/poly(ε-caprolactone) diblock copolymeric nanoparticles for non-viral gene delivery: the role of charge group and molecular weight in particle formation, cytotoxicity and transfection. *Journal of Controlled Release* 113: 173–182.

18 Shin, I.G., Kim, S.Y., Lee, Y.M. et al. (1998). Methoxy poly (ethylene gricol)/ε-caprolactone amphiphilic block copolymeric micelle containing indomethacin: I. preparation and characterization. *Journal of Controlled Release* 51: 1–11.

19 Kakizawa, Y. and Kataoka, K. (2002). Block copolymer micells for delivery of gene and related compounds. *Advanced Drug Delivery Reviews* 54: 203–222.

20 Kim, S.Y., Lee, Y.M., Baik, D.J. et al. (2003). Toxic characteristics of methoxy poly (ethylene gricol)/poly (ε-caprolactone) nanospheres; in vitro and in vivo studies in the normal mice. *Biomaterials* 24: 55–63.

21 Templeton, N.S., Lasic, D.D., Frederik, P.M. et al. (1997). Improved DNA: liposome complexes for increased systemic delivery and gene expression. *Nature Biotechnology* 15: 647–652.

22 Park, M.R., Han, K.O., Han, I.K. et al. (2005). Degradable polyethylenimine-alt-poly (ethylene gricol) copolymers as a novel gene carrier. *Journal of Controlled Release* 105: 367–380.

23 Matsumura, Y. and Maeda, H. (1986). A new concept for macromolecular therapeutics in cancer chemotherapy: mechanism of tumoritropic accumulation of proteins and the antitumor agent smancs. *Cancer Research* 46 (12 Pt 1): 6387–6392.

24 Hatakeyama, H., Akita, H., Kogure, K. et al. (2007). Development of a novel systemic gene delivery system for cancer therapy with a tumor-specific cleavable PEG-lipid. *Gene Therapy* 14: 68–77.

25 van den Hoven, J.M., Van Tomme, S.R., Metselaar, J.M. et al. (2011). Liposomal drug formulations in the treatment of rheumatoid arthritis. *Molecular Pharmaceutics* 8: 1002–1015.

26 Hofkens, W., Storm, G., van den Berg, W.B. et al. (2011). Liposomal targeting of glucocorticoids to the inflamed synovium inhibits cartilage matrix destruction during murine antigen-induced arthritis. *International Journal of Pharmaceutics* 416: 486–492.

27 Metselaar, J.M., van den Berg, W.B., Holthuysen, A.E. et al. (2004). Liposomal targeting of glucocorticoids to synovial lining cells strongly increases therapeutic benefit in collagen type II arthritis. *Annals of the Rheumatic Diseases* 63: 348–353.

28 Sharma, R., Lee, J.S., Bettencourt, R.C. et al. (2008). Effects of the incorporation of hydrophobic middle block into a PEG-polycation diblock copolymer on the physicochemical and cell interaction properties of the polymer-DNA complexes. *Biomacromolecules* 9: 3294–3307.

29 Gebhart, C.L. and Kabanov, A.V. (2001). Evaluation of polyplexes as gene transfer agents. *Journal of Controlled Release* 73: 401–416.

30 Alvarez-Lorenzo, C., Barreiro-Iglesias, R., Concheiro, A. et al. (2005). Biophysical characterization of complexation of DNA with block copolymers of poly (2-dimethylaminoethyl) methacrylate, poly (ethylene oxide), and poly (propylene oxide). *Langmuir* 21: 5142–5148.

31 Oishi, M., Kataoka, K., and Nagasaki, Y. (2006). pH-Responsive three-layered PEGylated polyplex micelle based on a lactosylated ABC triblock copolymer as a targetable and endosome-disruptive nonviral gene vector. *Bioconjugate Chemistry* 17: 677–688.

32 Bikram, M., Ahn, C.H., Chae, S.Y. et al. (2004). Biodegradable poly (ethylene gricol)-co-poly (L-lysine)-g- histidine multiblock copolymers for nonviral gene delivery. *Macromolecules* 37: 1903–1916.

33 Matsumoto, S., Christie, R.J., Nishiyama, N. et al. (2009). Environment-responsive block copolymer micelles with a disulfide cross-linked core for enhanced siRNA delivery. *Biomacromolecules* 10: 119–127.

34 Takemoto, H., Ishii, A., Miyata, K. et al. (2010). Polyion complex stability and gene silencing efficiency with a siRNA-grafted polymer delivery system. *Biomaterials* 31: 8097–8105.

35 Segura, T. and Hubbell, J.A. (2007). Synthesis and in vitro characterization of an ABC triblock copolymer for siRNA delivery. *Bioconjugate Chemistry* 18: 736–745.

36 Liu, P., Yu, H., Sun, Y. et al. (2012). A mPEG-PLGA-b-PLL copolymer carrier for adriamycin and siRNA delivery. *Biomaterials* 33: 4403–4412.

37 Anderson, W.F. (1992). Human gene-theraphy. *Science* 256: 808–813.

38 Luo, D. and Saltzman, W.M. (2000). Synthetic DNA delivery systems. *Nature Biotechnology* 18: 33–37.

39 Wiethoff, C.M. and Middaugh, C.R. (2003). Barriers to nonviral gene delivery. *Journal of Pharmaceutical Sciences* 92: 203–217.

40 Putnam, D. (2006). Polymers for gene delivery across length scales. *Nature Materials* 5: 439–451.

41 Tanaka, K., Kanazawa, T., Shibata, Y. et al. (2010). Development of cell-penetrating peptide-modified MPEG-PCL diblock copolymeric nanoparticles for systemic gene delivery. *International Journal of Pharmaceutics* 396: 229–238.

42 Khalil, I.A., Kogure, K., Futaki, S. et al. (2006). High density of octaarginine stimulates macropinocytosis leading to efficient intracellular trafficking for gene expression. *Journal of Biological Chemistry* 281: 3544–3551.

43 Nakase, I., Tadokoro, A., Kawabata, N. et al. (2007). Interaction of arginine-rich peptides with membrane-associated proteoglycans is crucial for induction of actin organization and macropinocytosis. *Biochemistry* 46: 492–501.

44 Chang, M., Chou, J.C., Chen, C.P. et al. (2007). Noncovalent protein transduction in plant cells by macropinocytosis. *New Phytologist* 174: 46–56.

45 Pappas, T.C., Bader, A.G., Andruss, B.F. et al. (2009). Applying small RNA molecules to the directed treatment of human diseases: realizing the potential. *Expert Opinion on Therapeutic Targets* 12: 115–127.

46 Dykxhoorn, D.M. (2009). RNA interference as an anticancer therapy:patent perspective. *Expert Opinion on Therapeutic Patents* 19: 475–491.

47 Ponnappa, B.C. (2009). siRNA for inflammatory diseases. *Current Opinion in Investigational Drugs* 10: 418–424.

48 Kanazawa, T., Sugawara, K., Tanaka, K. et al. (2012). Suppression of tumor growth by systemic delivery of anti-VEGF siRNA with cell-penetration peptide-modified MPEG-PCL nanomicelles. *European Journal of Pharmaceutics and Biopharmaceutics* 81: 470–477.

49 Kim, S.H., Jeong, J.H., Lee, S.H. et al. (2008). Local and systemic delivery of VEGF siRNA using polyelectrolyte complex micelles for effective treatment of cancer. *Journal of Controlled Release* 129: 107–116.

50 Pardridge, W.M. (2005). The blood–brain barrier: Bottleneck in brain drug development. *NeuroRx* 2: 3–14.

51 Emad, N.A., Ahmed, B., Alhalmi, A. et al. (2021). Recent progress in nanocarriers for direct nose to brain drug delivery. *Journal of Drug Delivery Science and Technology* 64: 102642.

52 Dhuria, S.V., Hanson, L.R., and Frey, W.H. (2010). Intranasal delivery to the central nervous system: mechanisms and experimental considerations. *Journal of Pharmaceutical Sciences* 99: 1654–1673.

53 Thorne, R.G., Pronk, G.J., and Padmanabhan, V. (2004). Delivery of insulin-like growth factor-I to the rat brain and spinal cord along olfactory and trigeminal pathways following intranasal administration. *Neuroscience* 127: 481–496.

54 Kanazawa, T., Fukuda, M., Suzuki, N. et al. (2018). Novel methods for intranasal administration under inhalation anesthesia to evaluate nose-to-brain drug delivery. *Journal of Visualized Experiments* 141: 2–6.

55 Fukuda, M., Kanazawa, T., Iioka, S. et al. (2021). Quantitative analysis of inulin distribution in the brain focused on nose-to-brain route via olfactory epithelium by reverse esophageal cannulation. *Journal of Controlled Release* 332: 493–501.

56 Lee, M.R., Shnitko, T.A., Blue, S.W. et al. (2020). Labeled oxytocin administered via the intranasal route reaches the brain in rhesus macaques. *Nature Communications* 11: 2783.

57 Born, J., Lange, T., Kern, W. et al. (2002). Sniffing neuropeptides: a transnasal approach to the human brain. *Nature Neuroscience* 5: 514–516.

58 Wingrove, J., Swedrowska, M., Scherließ, R. et al. (2019). Characterisation of nasal devices for delivery of insulin to the brain and evaluation in humans using functional magnetic resonance imaging. *Journal of Controlled Release* 302: 140–147.

59 Illum, L. (2000). Transport of drugs from the nasal cavity to the central nervous system. *European Journal of Pharmaceutical Sciences* 11: 1–18.

60 Pires, P.C. and Santos, A.O. (2018). Nanosystems in nose-to-brain drug delivery: a review of non-clinical brain targeting studies. *Journal of Controlled Release* 28: 89–100.

61 Feng, Y., He, H., Li, F. et al. (2018). An update on the role of nanovehicles in nose-to-brain drug delivery. *Drug Discovery Today* 23: 1079–1088.

62 Samaridou, E. and Alonso, M.J. (2018). Nose-to-brain peptide delivery – the potential of nanotechnology. *Bioorganic and Medicinal Chemistry* 26: 2888–2905.

63 Kanazawa, T., Taki, H., Tanaka, K. et al. (2011). Cell-penetrating peptide-modified block copolymer micelles promote direct brain delivery via intranasal administration. *Pharmaceutical Research* 28: 2130–2139.

64 Kanazawa, T., Akiyama, F., Kakizaki, S. et al. (2013). Delivery of siRNA to the brain using a combination of nose-to-brain delivery and cell-penetrating peptide-modified nano-micelles. *Biomaterials* 34: 9220–9226.

65 Kanazawa, T., Morisaki, K., Suzuki, S. et al. (2014). Prolongation of life in rats with malignant glioma by intranasal siRNA/drug codelivery to the brain with cell-penetrating peptide-modified micelles. *Molecular Pharmaceutics* 11: 1471–1478.

66 Kanazawa, T., Kurano, T., Ibaraki, H. et al. (2019). Therapeutic effects in a transient middle cerebral artery occlusion rat model by nose-to-brain delivery of anti-TNF-alpha siRNA with cell-penetrating peptide-modified polymer micelles. *Pharmaceutics* 11: 478.

67 Tanaka, K., Kanazawa, T., Horiuchi, S. et al. (2013). Cytoplasm-responsive nanocarriers conjugated with a functional cell-penetrating peptide for systemic siRNA delivery. *International Journal of Pharmaceutics* 455: 40–47.

68 Okada, H., Ogawa, T., Tanaka, K. et al. (2014). Cytoplasm-responsive delivery systems for siRNA using cell-penetrating peptide nanomicelles. *Journal of Drug Delivery Science and Technolog* 24: 3–11.

69 Mao, C.Q., Du, J.Z., Sun, T.M. et al. (2011). A biodegradable amphiphilic and cationic triblock copolymer for the delivery of siRNA targeting the acid ceramidase gene for cancer therapy. *Biomaterials* 32: 3124–3133.

70 Ibaraki, H., Kanazawa, T., Owada, M. et al. (2020). Anti-metastatic effects on melanoma via intravenous administration of anti-NF-kappaB siRNA complexed with functional peptide-modified nano-micelles. *Pharmaceutics* 12: 64.

71 Uchida, T., Kanazawa, T., Kawai, M. et al. (2011). Therapeutic effects on atopic dermatitis by anti-RelA short interfering RNA combined with functional peptides Tat and AT1002. *Journal of Pharmacology and Experimental Therapeutics* 338: 443–450.

72 Kanazawa, T., Shizawa, Y., Takeuchi, M. et al. (2015). Topical anti-nuclear factor-kappa B small interfering RNA with functional peptides containing sericin-based hydrogel for atopic dermatitis. *Pharmaceutics* 7: 294–304.

73 Ibaraki, H., Kanazawa, T., Kurano, T. et al. (2019). Anti-RelA siRNA-encapsulated flexible liposome with tight junction-opening peptide as a non-invasive topical therapeutic for atopic dermatitis. *Biological and Pharmaceutical Bulletin* 42: 1–11.

74 Kanazawa, T., Tamano, T., Sogabe, K. et al. (2017). Intra-articular retention and anti-arthritic effects in collagen-induced arthritis model mice by injectable small interfering RNA containing hydrogel. *Biological and Pharmaceutical Bulletin* 40: 1929–1933.

75 Kanazawa, T., Endo, T., Arima, N. et al. (2016). Systemic delivery of small interfering RNA targeting nuclear factor κB in mice with collagen-induced arthritis using arginine-histidine-cysteine based oligopeptide-modified polymer nanomicelles. *International Journal of Pharmaceutics* 515: 315–323.

76 Ibaraki, H., Hatakeyama, N., Arima, N. et al. (2022). Systemic delivery of siRNA to the colon using peptide modified PEG-PCL polymer micelles for the treatment of ulcerative colitis. *European Journal of Pharmaceutics and Biopharmaceutics* 170: 170–178.

77 Pridgen, E.M., Alexis, F., and Farokhzad, O.C. (2014). Polymeric nanoparticle technologies for oral drug delivery. *Clinical Gastroenterology and Hepatology* 12: 1605–1610.

78 Han, L., Tang, C., and Yin, C. (2014). Oral delivery of shRNA and siRNA via multifunctional polymeric nanoparticles for synergistic cancer therapy. *Biomaterials* 35: 4589–4600.

79 Higuchi, Y., Kawakami, S., and Hashida, M. (2010). Strategies for in vivo delivery of siRNAs: recent progress. *BioDrugs* 24: 195–205.

80 Naeem, M., Lee, J., Oshi, M.A. et al. (2020). Colitis-targeted hybrid nanoparticles-in-microparticles system for the treatment of ulcerative colitis. *Acta Biomaterialia* 116: 368–382.

81 Laroui, H., Theiss, A.L., Yan, Y. et al. (2011). Functional TNF**α** gene silencing mediated by polyethyleneimine/TNFα siRNA nanocomplexes in inflamed colon. *Biomaterials* 32: 1218–1228.

82 Frede, A., Neuhaus, B., Klopfleisch, R. et al. (2016). Colonic gene silencing using siRNA-loaded calcium phosphate/PLGA nanoparticles ameliorates intestinal inflammation in vivo. *Journal of Controlled Release* 222: 86–96.

83 Ali, H., Weigmann, B., Collnot, E.M. et al. (2016). Budesonide loaded PLGA nanoparticles for targeting the inflamed intestinal mucosa – pharmaceutical characterization and fluorescence imaging. *Pharmaceutical Research* 33: 1085–1092.

84 Minami, K., Takazawa, A., Taniguchi, Y. et al. (2020). Challenge for oral delivery of middle-molecular drugs: use of osmolarity-sensitive liposome as a drug carrier in the GI tract. *Journal of Drug Delivery Science and Technology* 56: 101041.

85 Ibaraki, H., Hatakeyama, N., Takeda, A. et al. (2022). Multifunctional peptide carrier-modified polymer micelle accelerates oral siRNA-delivery to the colon and improves gene silencing-mediated therapeutic effects in ulcerative colitis. *Journal of Drug Delivery Science and Technology* 73: 103481. https://doi.org/10.1016/j.jddst.2022.103481.

86 Van de Graaff, K.M. (1986). Anatomy and physiology of the gastrointestinal tract. *Pediatric Infectious Disease* 5 (1): S11–S16.

87 Huang, Y., Guo, J., and Gui, S. (2018). Orally targeted galactosylated chitosan poly(lactic-co-glycolic acid) nanoparticles loaded with TNF-α siRNA provide a novel strategy for the experimental treatment of ulcerative colitis. *European Journal of Pharmaceutical Sciences* 125: 232–243.

17

Lipid-Based Nanoparticles Using Cell-Penetrating Peptides

Ikramy A. Khalil[1,2] *and Hideyoshi Harashima*[1]

[1] *Hokkaido University, Laboratory for Molecular Design of Pharmaceutics, Faculty of Pharmaceutical Sciences, Kita-12, Nishi-6, Kita-ku, Sapporo, 060-0812, Japan*
[2] *Assiut University, Department of Pharmaceutics, Faculty of Pharmacy, Assiut, 71526, Egypt*

17.1 Introduction

Some drugs need to be internalized into cells to exert their action. Drugs having small molecular weights are easily internalized into cells through the plasma membrane if they are lipophilic or through pores and channels in the membrane if they are hydrophilic and small enough to pass. However, cellular internalization is more difficult when we consider large molecular weight drugs such as proteins and nucleic acids. Although their target sites are mostly located within cells, these macromolecules cannot penetrate the plasma membrane due to their large size and hydrophilic properties [1]. Therefore, different strategies must be developed to enhance the cellular uptake of these macromolecules.

Cell-penetrating peptides (CPPs) are promising biomaterials that can bind and interact with the plasma membrane and can be internalized into cells with their attached cargos [2–5]. They are short peptides that mostly carry a positive charge that facilitates their binding to plasma membranes and their internalization into cells. They can internalize associated molecules and even large drug nanocarriers into cells. Therefore, they have great potential for use in the diagnosis and treatment of diseases. Although their surface charges may limit their use in the systemic administration of drugs due to nonspecific interactions with serum components, they still have the potential for in vivo use if they are incorporated into smart nanocarriers that have the ability to resist degradation and target specific tissues in the body.

Lipid nanoparticles (LNPs) are important drug carriers that need to be internalized into cells and deliver their payload to intracellular targets. LNPs can carry various hydrophilic and lipophilic drugs and can encapsulate even macromolecules such as DNA or RNA [6, 7]. They provide stability to drug molecules and their surface can be further modified to overcome various extracellular barriers for drug delivery to specific cells in the body. Modification of different LNPs with CPPs is an important approach to enhancing their cellular uptake and drug delivery. Since CPPs are nonspecific and can bind to most cell types, additional strategies are needed to develop multifunctional nanoparticles designed to combine beneficial aspects of CPPs with other ligands used for tissue targeting.

In this chapter, we discuss CPPs and their use for enhancing the cellular uptake of different lipid NPs, with an emphasis on the use of these systems for nucleic acid delivery. We also discuss the in vivo use of CPP-modified LNPs and how to design smart systems that can overcome different barriers to optimal drug delivery.

17.2 Cell-Penetrating Peptides

17.2.1 Overview

CPPs are short peptides (5–30 amino acids) that can bind to the plasma membrane, interact with its components, penetrate it in a noninvasive way, and internalize into most cell types efficiently. These peptides share the presence of high content of positive amino acids and most of them carry a net positive charge. They were originally derived from natural proteins; hence they are called protein transduction domains (PTDs). They are also called "Trojan Peptides" owing to early results that reported their ability to translocate through cell membranes and internalize into cells without using the endocytosis pathway [8]. These results were obscure at that time because these early used peptides were hydrophilic and positively charged, meaning that they could not easily penetrate the lipid nature of the plasma membrane. The main interest in these peptides has been in the field of drug delivery to intracellular targets since these peptides were able to internalize attached drugs and cargos as large as a few hundred nanometers vesicles into cells efficiently and safely. Thus, these peptides had great promise for application in the field of diagnosis and therapy of different diseases.

The first known CPP was the transacting activator of transcription (Tat) peptide; an 86 amino acid peptide derived from human immunodeficiency virus-1 (HIV-1), discovered in 1988 [9]. Subsequently, shorter peptide sequences of only 9–12 amino acids (Tat-48–60 or 49–57) were identified to be responsible for the cellular uptake of Tat protein [10]. The next discovered peptide was the homeodomain of Antennapedia (pAntp); a 60 amino acid peptide discovered in 1991 [11]. Similar to Tat, a shorter peptide sequence of 16 amino acid sequence (pAntp-43–58) named penetratin was identified to be responsible for the cellular internalization of the Antennapedia homeodomain [12]. Both Tat and penetratin share the presence of several positive amino acids such as lysine and arginine. Subsequently, it was found that even oligoarginine peptides consisting of only arginine residues (8–16 amino acids) had similar cell-penetrating properties [13]. At least 8 arginine residues were required for cell uptake of oligoarginines [14]. Since that time, many other peptides have either been discovered from natural sources or prepared synthetically. Other commonly used CPPs include model amphipathic peptide (MAP), transportan, Pep-1, and VP22 (Table 17.1). These peptides have been further modified with different chemical entities to facilitate their binding to different types of cargos. For example, modification of the R8 peptide with a stearyl group facilitates the modification of liposomes and other lipid vesicles through mixing stearylated-R8 with the lipids forming the lipid vesicle [21–23].

17.2.2 Types

CPPs can be classified into different ways. Based on their physicochemical properties, CPPs are classified as cationic (~83%), hydrophobic (~15%), or amphipathic (~40%). Cationic CPPs contain several positively charged amino acids (arginine, lysine, and histidine) in their sequences. Tat, penetratin, and oligoarginines are examples of cationic CPPs [2, 5]. The presence of arginine residues appears to be more critical for the cellular internalization process compared to lysine [24]. This can be explained by the presence of the guanidinium head group in arginine which donates two positive charges while binding to negatively charged cell surface molecules, in contrast to lysine, which donates only one positive charge [24]. Endocytosis-independent entry is usually linked to cationic arginine-rich CPPs [5]. Histidine-rich CPPs are cationic only in an acidic environment such as inside the endosomes after endocytosis. Cationic CPPs bind strongly to negatively charged cell surface components and facilitate cellular internalization in vitro. However, they are generally unsuitable for in vivo application due to their high positive charge [25]. Hydrophobic CPPs are less common and contain high content of hydrophobic amino acids while generally possessing a low net charge. Examples of hydrophobic CPPs include Pep-7 and C105Y. The mode of internalization of this class depends largely on the hydrophobic sequence [5]. The third category is the amphipathic CPPs, which contain both polar and nonpolar segments in their sequence. The polar segments are rich in arginine and lysine while the nonpolar segments are rich in leucine, isoleucine, valine, and alanine. Examples of amphipathic CPPs include Pep-1, MAP, transportan, and VP22. Amphipathic peptides can be arranged in primary sequences or arranged in secondary α-helices or β-sheets.

CPPs can be also classified based on their origin. CPPs are either derived from proteins (~42%), synthetic (~55%), or chimeric (~3%). Tat and penetratin are derived from natural proteins while oligoarginines are synthetic. Chimeric peptides such as transportan are formed by the combination of two or more naturally occurring sequences.

17.2.3 Mechanism of Cellular Internalization

When CPPs were first discovered, the common belief was that these peptides had the ability to pass the cell membrane in an energy-independent manner without the involvement of endocytosis (i.e. direct penetration) [26]. Even larger cargos were reported to be internalized through direct penetration [27]. However, since 2003,

Table 17.1 Examples of common CPPs.

Class	Peptide	Sequence	Origin	Ref.
Cationic	Tat (48–60)	GRKKRRQRRRPPQ	Natural	[10]
	Penetratin	RQIKIWFQNRRMKWKK	Natural	[12]
	Octaarginine (R8)	RRRRRRRR	Synthetic	[13]
	DPV3	RKKRRRESRKKRRRES	Natural	[15]
Hydrophobic	Pep-7	SDLWEMMMVSLACQY	Natural	[16]
	C105Y	CSIPPEVKFNKPFVYLI	Natural	[17]
Amphipathic	Pep-1	KETWWETWWTEWSQPKKKRKV	Natural	[5]
	MAP	KLALKLALKALKAALKLA	Synthetic	[18]
	VP22	NAATATRGRSAASRPTQRPRAPARSASRPRRPVQ	Natural	[19]
	Transportan	GWTLNSAGYLLGKINLKALAALAKKIL	Synthetic	[20]

cumulative evidence has emerged showing that internalization is energy-dependent and that different types of endocytosis are involved in cellular entry [28]. This was supported by the fact that the addition of agents that enhance the endosomal escape usually improves the cytosol delivery of peptides or their cargos [29]. The presence of CPP may even have the ability to stimulate different types of endocytosis through interaction with plasma membrane components [23].

It is now clear that the entry process is quite complex and that various factors influence the entry mode. For example, a free peptide may have a higher capacity for direct penetration than peptides bound to other molecules or cargos. Free R8 peptide was efficiently internalized into cells in the presence of endocytosis inhibitors, while R8 bound to plasmid DNA (pDNA) or attached to liposome surface was not internalized under similar conditions [21]. Moreover, many other factors influence the entry mode and no single entry pathway can be emphasized. For example, the entry mode is influenced by the type, size, charge, and hydrophobic region of the peptide. It is also affected by the type of cargo attached, as explained earlier, and by the cell type and cell surface components. The peptide-to-cell ratio also influences the internalization mechanism as higher amounts of the peptides are generally internalized through endocytosis [13].

Endocytosis-independent entry or direct penetration can be explained by several models. The most studied is the formation of inverted micelles in the plasma membrane [8]. The binding of the peptide to the plasma membrane causes a rearrangement of the lipids of the membrane resulting in the formation of inverted micelles whose inner hydrophilic core contains the peptide. These micelles are then destabilized to release the peptide into the cytosol. Other mechanisms have been proposed to explain the endocytosis-independent uptake including pore formation, membrane thinning, and carpet model. The presence of arginine with its guanidine groups is critical for direct penetration.

Endocytosis-dependent internalization is supported by the punctuate appearance of the peptide in the cytoplasm indicating localization in endosomes or intracellular vesicles [21, 28]. Endocytosis is the most widely accepted model of internalization, especially if the peptide is bound to a larger cargo. The peptide electrostatically binds to negative phosphate groups of the membrane and other negative components of the membrane such as heparan sulfate proteoglycans. This binding is essential to stimulate uptake through endocytosis because depletion or blocking of these proteoglycans greatly inhibited the uptake [23]. Binding to proteoglycans is associated with changes in the cytoskeleton and stimulation of endocytosis. Various endocytosis mechanisms have been proposed for different peptides and different cargos including macropinocytosis, clathrin-mediated endocytosis or caveolae-mediated endocytosis [30]. The peptide density appears to influence the endocytic pathway [23]. When using liposomes modified with a high density of R8 peptide, the uptake was found to be largely dependent on macropinocytosis. In the case of liposomes modified with a low density of R8, clathrin-mediated endocytosis was the main entry pathway. Cumulative data indicate that several factors influence the type of endocytosis involved. A single uptake pathway may not be appropriate to describe the entry pathway since several mechanisms may be involved to different degrees.

Regardless of the type of endocytosis involved, the internalized cargos are trapped in endosomes and still need a strategy to escape from endosomes and reach the cytosol [23].

17.3 Lipid-based Nanoparticles

17.3.1 Overview

LNPs are lipid-based spherical vesicles that can be used to deliver different drugs to their target sites. Both hydrophilic and lipophilic drugs, small and high molecular weight drugs can be encapsulated in LNPs. The lipid layer protects the drug from enzymatic degradation and its surface can be modified with various ligands such as CPP to improve cellular internalization, polyethylene glycol (PEG) for prolonged circulation through escape from opsonization in the serum and uptake by phagocytes, or various targeting ligands to target specific cells in the body. In general, the most important component of LNPs is the main lipid, which may be a cationic or pH-sensitive lipid. The latter type has the advantage of being neutral at physiologic pH, which is required to decrease nonspecific interactions and being cationic at acidic pH for interaction with negative nucleic acids and negatively charged endosomal membranes. In addition to the main lipid, LNPs often contain a helper lipid to facilitate fusion with the endosomal membrane for cytosol delivery or cholesterol to increase their stability [6, 7].

17.3.2 Types

Commonly used LNPs include liposomes, niosomes, exosomes, and solid lipid NPs (SLNPs). Liposomes are spontaneously formed spherical lipid vesicles composed of phospholipids that are arranged in a bilayer structure surrounding an aqueous core [31]. Hydrophilic drugs can be encapsulated in the aqueous core while lipophilic drugs can be solubilized in the lipid bilayer. Liposomes may be unilamellar or multilamellar vesicles. Small unilamellar vesicles (SUVs, size ~20–100 nm) or large unilamellar vesicles (LUVs, size ~100–1000 nm) are more suitable for encapsulation of hydrophilic drugs while multilamellar vesicles (MLVs, size 100 nm to several micrometers) are more suitable for lipophilic drugs. Cationic liposomes are commonly used for gene delivery, where they can bind to and condense different nucleic acids to form lipoplexes. As an alternative to liposomes, niosomes are spherical vesicles composed of nonionic surfactants and cholesterol [32]. They are cheaper and more stable than liposomes and can provide a sustained drug release over a prolonged time. Exosomes on the other hand are extracellular vesicles that are released from cells during cell communication [33]. They have a lipid bilayer and provide several advantages as drug carriers including biocompatibility, high encapsulation and low immunogenicity. Similar to liposomes and niosomes, the surface of exosomes can be modified using different ligands for different purposes such as cell-targeting or prolonged circulation. Exosomes are ~200 nm in size and carry a net negative charge, which limits their cellular binding and internalization. SLNP is a newly developed class of LNPs. They are spherical nanocarriers composed of solid lipids at room temperature stabilized by suitable surfactants [34]. The solid core can encapsulate lipophilic drugs between the hydrophobic tails of the lipids. Stable nucleic acid lipid particles (SNALPs) are being developed as an alternative to conventional cationic liposomes for gene delivery [35, 36]. They are composed of ionizable cationic lipids with fusogenic lipids that bind to nucleic acids and form lipid NPs, which are usually stabilized with a PEG-lipid. They have a neutral charge in the biological fluid, which makes them more favorable for systemic administration of nucleic acids [37–39].

17.4 CPP-Modified Lipid Nanoparticles

17.4.1 CPP-LNPs for Cellular Internalization

Tat peptide was the first peptide to be used to enhance the cellular uptake of liposomes. Torchilin et al. developed Tat-liposomes (Tat-Lips) of ~200 nm which have the ability to efficiently internalize into cells [27]. Tat was attached to the surface of the liposome via a PEG spacer. The binding of Tat to the liposome surface without a spacer reduced cellular uptake, indicating that CPP peptides should interact freely with cell surface components. Tat-Lips have been used for the efficient delivery of various drugs to different cell types including tumor cells and airway epithelial cells. R8-modified liposomes (R8-Lips) also showed efficient cellular uptake within a few minutes after incubation with cells [23, 40]. The R8 peptide was attached to a stearyl group to form stearyl R8 (STR-R8) which is mixed with other phospholipids during the preparation of liposomes. During liposome assembly, the R8 peptide is located on the outer surface of the liposome while the hydrophobic group is inserted in the lipid bilayer. R8-Lip showed efficient cellular uptake even though R8 was attached directly to the liposome surface without a spacer. R8-Lip failed to internalize into cells at low temperature and in the presence of endocytosis inhibitors, indicating internalization through endocytosis [22]. Even a low peptide amount on the surface of the liposome (0.8 mol% of total lipid) was sufficient to cause liposome internalization through clathrin-mediated endocytosis. However, a higher peptide amount (5 mol% of total lipid) was required to stimulate cellular uptake through macropinocytosis. It appears that a high peptide concentration is important for the clustering of cell surface proteoglycans resulting in the stimulation of cytoskeleton rearrangement and induction of macropinocytosis. R8-Lip has been the main platform for the development of a lipid-based, multifunctional envelope-type nanodevice (MEND) that encapsulates DNA or RNA [41, 42]. Nucleic acids are first condensed using polycations to form a polymeric core which is coated with a lipid membrane modified with R8. Confocal microscopy of live cells and flow cytometry analysis revealed that R8-MEND was highly internalized into cells and efficiently transfected cells with different nucleic acids [41]. Transfection efficiency is increased by the incorporation of fusogenic lipids such as dioleoyl phosphoethanolamine (DOPE) and fusogenic peptides such as GALA [29, 41]. This confirms that a significant part of internalized NPs are trapped in endosomes and there is a need to enhance endosomal escape to produce efficient transfection. Similarly, different CPPs were added to different lipoplexes to enhance cellular uptake and gene delivery [43].

R8 was also used to target intracellular organelles such as the nucleus and mitochondria. Cationic CPPs can act as a nuclear localization signal (NLS) [5]. R8-Lip was found localized in the perinuclear region in a chase study after the endosomal escape [30]. R8-Lip was also used to deliver drugs to mitochondria [44]. In addition, a dual ligand mitochondria-targeting system was prepared using R8 and RNA aptamers targeting the mitochondria (RNase P aptamer (RP)). The dual ligand system demonstrated both enhanced cellular uptake and mitochondrial targeting in HeLa cells [44].

17.4.2 CPP-LNPs for Tumor Targeting

The presence of cationic amino acids in most CPPs limits their use in tumor targeting after systemic administration. Cationic NPs tend to bind nonspecifically to negatively charged plasma proteins forming aggregates that are either trapped in the lung or taken up by phagocytes for degradation. Following the biodistribution of R8-Lip after intravenous (IV) injection showed that most of the dose accumulated in the liver with almost no tumor delivery [25]. Therefore, a specific design is needed to escape liver uptake and enhance tumor targeting. One strategy relies on coating CPP-LNPs with PEG chains to escape phagocytosis and provide long circulation that allows for accumulation of NPs in the tumor through the enhanced permeability and retention (EPR) effect. However, although coating with PEG significantly increased the delivery to tumor tissue, the cellular uptake into tumor cells was greatly decreased. In addition, the endosomal escape of PEG-NPs was also significantly impaired [45]. To overcome these obstacles, the tips of the PEG chains have been modified with tumor-targeting ligands such as transferrin (Tf). Tf-PEG-NPs produced low transfection activity while the inclusion of R8 in this system increased the gene expression level by ~2.5-fold. Inclusion of the GALA peptide into Tf-PEG-NPs to enhance the endosomal escape improved gene expression by ~2-fold. Meanwhile, incorporation of GALA peptide into Tf-PEG-R8-NPs improved the gene expression by ~460-fold [45]. This indicates that CPP peptides have an important role in tumor gene delivery even if the NPs are coated with PEG. CPPs enhance the interaction between NPs and cell membranes or endosomal membranes, which augments the action of the targeting ligands or the fusogenic peptides. Interestingly, the presence of R8 did not affect the Tf-conferred selectivity of tumor cells because R8 is completely covered with PEG chains and appears to exert most of its activity in the endosome [45].

The concept of using dual ligands for tumor targeting can also be applied with dual-ligand systems of R8 and cyclic RGD (Arginine-Glycine-Aspartic acid) [46]. RGD is used to target integrin $\alpha v\beta 3$ receptors overexpressed in various cancer cells. The presence of RGD or R8 alone improved cellular uptake by ~1.5–2-fold while the presence of both RGD and R8 improved cellular uptake by ~5-fold. As for gene expression, liposomes modified with RDG or R8 alone produced gene expression higher than that of unmodified liposomes by less than 1-order of magnitude. Meanwhile, the dual ligand RGD-R8-Lip produced gene expression 2-order of magnitude higher than unmodified liposomes [46]. The RGD motif binds to its receptor on the surface of tumor cells and the presence of R8 appears to improve the interaction between the liposomal membrane and the plasma membrane. Similarly, dual ligand liposomes were prepared to target tumor endothelial cells (TECs) [47]. The NGR (CYGGRGNG) sequence was used to target CD31 receptors in TECs while the short R4 peptide was used to improve the activity by enhancing cellular interaction with the liposomal membrane. A short peptide was used in this study to confirm the complete shielding of the positive charge by PEG coating. Dual ligand NGR-R4-Lip showed improved tumor accumulation compared to unmodified or single-modified liposomes. Dual ligand liposomes encapsulating doxorubicin (Dox) showed improved antitumor activity compared to liposomes containing a single ligand [47].

R8 can also be attached to the tip of PEG in PEGylated liposomes to enhance the cellular uptake into tumor cells. R8-PEG-Lip encapsulating Dox showed higher tumor accumulation and higher tumor suppression compared to PEG-Lip.

17.4.3 CPP-LNPs for Liver Targeting

Intravenously injected R8-Lip were rapidly cleared from the circulation and accumulated mainly in the liver [25]. However, R8-MEND encapsulating DNA did not show any transfection activities in liver cells. Conventional R8-MEND prepared using a positively charged DNA core coated with a negatively charged lipid envelope did not transfect liver cells even in the presence of GALA peptide, which is used to enhance the endosomal escape [29]. However, a modified R8-GALA-MEND prepared with a negatively charged DNA core coated with a positively charged lipid envelope produced high gene expression in liver cells. DNA appears to be released more easily from negative cores than from positive ones. Neither single ligand R8-MEND nor GALA-MEND produced any activity in the liver, suggesting a possible synergistic role of the GALA peptide and CCPs such as R8. This system has been successfully applied for delivering small interfering RNA (siRNA) to liver cells after IV injection [48]. It was also successful in delivering pDNA encoding growth factor to hepatocytes to protect them from induced liver injury [49].

R8 can also be used to enhance nuclear delivery to hepatocytes. The inclusion of cyclic R8 in the core of Ca-phosphate lipid (LCP) nanoparticles condensing pDNA significantly enhanced gene expression in hepatocytes compared to NPs prepared without R8 [50].

17.4.4 Other Applications of CPP-LNPs

CPPs have the ability to cross the blood–brain barrier (BBB) through transcytosis in vivo [5]. They can also bypass the P-glycoprotein, which increases brain accumulation. Tat-Lips have been used to deliver cytotoxic drugs for brain glioma therapy [51]. Similarly, R8-Lip encapsulating Dox showed improved delivery across the BBB and produced higher cytotoxicity compared to unmodified liposomes.

Tat-Lips were also used for topical delivery to enhance skin permeation. R8-MEND was successfully used for topical gene delivery to hair follicles in the anagen phase for control of hair growth [41].

17.5 Improving CPP-LNPs

One of the most important limitations of CPP-mediated drug delivery is the inefficient endosomal escape process after endocytosis. Additional biomaterials should be combined with CPPs to enhance endosomal escape. One approach is to use fusogenic peptides or lipids to destabilize the endosomal membrane in acidic conditions. The use of GALA peptide with R8 has proven to be an efficient strategy to increase gene expression efficiency [29, 45]. Another strategy is based on the use of recently developed pH-sensitive cationic lipids, which have been shown to be efficient in enhancing the endosomal escape of gene carriers. These ionizable pH-sensitive lipids have gained a lot of attention recently with the approval of Patisiran, the first approved in vivo nonviral gene therapy product. Compared with conventional cationic lipids, the cellular uptake of pH-sensitive lipid vesicles is lower due to the absence of a positive charge at physiologic pH but the gene expression efficiency is higher. Therefore, combining these lipids with CPPs to increase cellular uptake is a promising strategy for improving gene delivery efficiency. However, adding a cationic CPP to neutral NPs prepared from pH-sensitive lipids would increase nonspecific interactions and the particles would not be suitable for administration. Therefore, a smart design is needed to combine the benefits of both pH-sensitive lipids and CPPs in a single system. YSK05 is an example of pH-sensitive lipids which are used in gene delivery [52]. YSK05-NPs modified with R8 produced higher gene expression in vitro compared to YSK05-NPs, but they carry a net positive charge [53]. To mask the positive charge, R8-NPs were first formed using a low amount of R8 to produce negatively charged lipid vesicles encapsulating DNA, which were then coated with YSK05 lipid in an acidic medium. This double-coating strategy significantly improved gene expression by ~70-fold compared to NPs prepared using the same components in a single coating step [53]. Double-coating strategy significantly improved cellular uptake, endosomal escape, and nuclear delivery. It has been hypothesized that the presence of YSK05 in the outer layer, which is not bound to DNA, makes it free to interact more efficiently with the endosomal membrane. After escaping from endosomes, the released DNA core will be modified with R8, which can improve nuclear delivery. This double-coating strategy has been successfully applied to target lung endothelial cells in vivo after IV injection [54].

17.6 Conclusion

CPPs are promising tools for improving cellular uptake and enhancing interactions with cellular membranes. Meanwhile, LNPs are versatile nanocarriers for drug and gene delivery. Modification of LNPs with CPPs is a promising strategy to enhance the cellular uptake of these NPs, especially when used for in vitro drug delivery into cultured cells. However, modifying LNPs with cationic CPPs may not be suitable for in vivo administration. Limitations of the use of CPPs in vivo include rapid clearance, lack of specificity, instability, and low endosomal escape. Strategies should be developed to incorporate CPPs into more advanced smart lipid-based nanocarriers that can avoid these drawbacks and can target specific cells in the body. The ability of CPPs to cross the BBB is a promising approach in brain delivery that needs further research efforts to develop safe and efficient drug delivery systems to the brain. In addition, the ability of CPPs to target specific organelles is an intriguing approach, which needs to be further explored for site-specific drug delivery.

References

1 Jakutaviciute, M., Ruzgys, P., Tamosiunas, M. et al. (2017). Physical methods for drug and gene delivery through the cell plasma membrane. *Advances in Anatomy, Embryology, and Cell Biology* 227: 73–92.

2 Guidotti, G., Brambilla, L., and Rossi, D. (2017). Cell-penetrating peptides: from basic research to clinics. *Trends in Pharmacological Sciences* 38 (4): 406–424.

3 Xie, J., Bi, Y., Zhang, H. et al. (2020). Cell-penetrating peptides in diagnosis and treatment of human diseases: from preclinical research to clinical application. *Frontiers in Pharmacology* 11: 697.

4 Khan, M.M., Filipczak, N., and Torchilin, V.P. Cell penetrating peptides: A versatile vector for co-delivery of drug and genes in cancer. *Journal of Controlled Release* 330: 1220–1228.

5 Desale, K., Kuche, K., and Jain, S. Cell-penetrating peptides (CPPs): an overview of applications for improving the potential of nanotherapeutics. *Biomaterials Science* 9 (4): 1153–1188.

6 Thi, T.T.H., Suys, E.J.A., Lee, J.S. et al. (2021). Lipid-based nanoparticles in the clinic and clinical trials: from cancer nanomedicine to COVID-19 vaccines. *Vaccines (Basel)* 9 (4): 359.

7 Dhiman, N., Awasthi, R., Sharma, B. et al. (2021). Lipid nanoparticles as carriers for bioactive delivery. *Frontiers in Chemistry* 9: 580118.

8 Derossi, D., Chassaing, G., and Prochiantz, A. (1998). Trojan peptides: the penetratin system for intracellular delivery. *Trends in Cell Biology* 8 (2): 84–87.

9 Frankel, A.D. and Pabo, C.O. (1988). Cellular uptake of the tat protein from human immunodeficiency virus. *Cell* 55 (6): 1189–1193.

10 Sugita, T., Yoshikawa, T., Mukai, Y. et al. (2007). Improved cytosolic translocation and tumor-killing activity of Tat-shepherdin conjugates mediated by co-treatment with Tat-fused endosome-disruptive HA2 peptide. *Biochemical and Biophysical Research Communications* 363 (4): 1027–1032.

11 Joliot, A., Pernelle, C., Deagostini-Bazin, H. et al. (1991). Antennapedia homeobox peptide regulates neural morphogenesis. *Proceedings of the National Academy of Sciences of the United States of America* 88 (5): 1864–1868.

12 Derossi, D., Joliot, A.H., Chassaing, G. et al. (1994). The third helix of the Antennapedia homeodomain translocates through biological membranes. *The Journal of Biological Chemistry* 269 (14): 10444–10450.
13 Futaki, S. (2002). Arginine-rich peptides: potential for intracellular delivery of macromolecules and the mystery of the translocation mechanisms. *International Journal of Pharmaceutics* 245 (1–2): 1–7.
14 Futaki, S., Suzuki, T., Ohashi, W. et al. (2001). Arginine-rich peptides. An abundant source of membrane-permeable peptides having potential as carriers for intracellular protein delivery. *The Journal of Biological Chemistry* 276 (8): 5836–5840.
15 De Coupade, C., Fittipaldi, A., Chagnas, V. et al. (2005). Novel human-derived cell-penetrating peptides for specific subcellular delivery of therapeutic biomolecules. *The Biochemical Journal* 390 (Pt 2): 407–418.
16 Morris, M.C., Deshayes, S., Heitz, F. et al. (2008). Cell-penetrating peptides: from molecular mechanisms to therapeutics. *Biology of the Cell* 100 (4): 201–217.
17 Rhee, M. and Davis, P. (2006). Mechanism of uptake of C105Y, a novel cell-penetrating peptide. *The Journal of Biological Chemistry* 281 (2): 1233–1240.
18 Wada, S., Hashimoto, Y., Kawai, Y. et al. (2013). Effect of Ala replacement with Aib in amphipathic cell-penetrating peptide on oligonucleotide delivery into cells. *Bioorganic & Medicinal Chemistry* 21 (24): 7669–7673.
19 Chen, H.C., Chiou, S.T., Zheng, J.Y. et al. (2011). The nuclear localization signal sequence of porcine circovirus type 2 ORF2 enhances intracellular delivery of plasmid DNA. *Archives of Virology* 156 (5): 803–815.
20 Pae, J., Säälik, P., Liivamägi, L. et al. (2014). Translocation of cell-penetrating peptides across the plasma membrane is controlled by cholesterol and microenvironment created by membranous proteins. *Journal of Controlled Release* 192: 103–113.
21 Khalil, I.A., Futaki, S., Niwa, M. et al. (2004). Mechanism of improved gene transfer by the N-terminal stearylation of octaarginine: enhanced cellular association by hydrophobic core formation. *Gene Therapy* 11 (7): 636–644.
22 Iwasa, A., Akita, H., Khalil, I. et al. (2006). Cellular uptake and subsequent intracellular of R8-liposomes introduced at low temperature. *Biochimica et Biophysica Acta (BBA) – Biomembranes* 1758 (6): 713–720.
23 Khalil, I.A., Kogure, K., Futaki, S. et al. (2006). High density of octaarginine stimulates macropinocytosis leading to efficient intracellular trafficking for gene expression. *The Journal of Biological Chemistry* 281 (6): 3544–3551.
24 Sakai, N., Futaki, S., and Matile, S. (2006). Anion hopping of (and on) functional oligoarginines: from chloroform to cells. *Soft Matter* 2 (8): 636–641.
25 Mudhakir, D., Akita, H., Khalil, I.A. et al. (2005). Pharmacokinetic analysis of the tissue distribution of octaarginine modified liposomes in mice. *Drug Metabolism and Pharmacokinetics.* 20 (4): 275–281.
26 Pooga, M., Hällbrink, M., Zorko, M. et al. (1998). Cell penetration by transportan. *The FASEB Journal* 1: 67–77.
27 Torchilin, V.P., Rammohan, R., Weissig, V. et al. (2001). TAT peptide on the surface of liposomes affords their efficient intracellular delivery even at low temperature

and in the presence of metabolic inhibitors. *Proceedings of the National Academy of Sciences of the United States of America* 98 (15): 8786–8791.

28 Vives, E. (2003). Cellular uptake [correction of utake] of the Tat peptide: an endocytosis mechanism following ionic interactions. *Journal of Molecular Recognition* 16 (5): 265–271.

29 Khalil, I.A., Hayashi, Y., Mizuno, R. et al. (2011). Octaarginine- and pH sensitive fusogenic peptide-modified nanoparticles for liver gene delivery. *Journal of Controlled Release* 156 (3): 374–380.

30 Khalil, I.A., Kogure, K., Akita, H. et al. (2006). Uptake pathways and subsequent intracellular trafficking in nonviral gene delivery. *Pharmacological Reviews* 58 (1): 32–45.

31 Torchilin, V.P. (1985). Liposomes as targetable drug carriers. *Critical Reviews in Therapeutic Drug Carrier Systems* 2 (1): 65–115.

32 Gharbavi, M., Amani, J., Kheiri-Manjili, H. et al. (2018). Niosome: a promising nanocarrier for natural drug delivery through blood-brain barrier. *Advances in Pharmacological Sciences* 2018: 6847971.

33 van Dommelen, S.M., Vader, P., Lakhal, S. et al. (2012). Microvesicles and exosomes: opportunities for cell-derived membrane vesicles in drug delivery. *Journal of Controlled Release* 161 (2): 635–644.

34 Scioli, M.S., Muraca, G., and Ruiz, M.E. (2020). Solid lipid nanoparticles for drug delivery: pharmacological and biopharmaceutical aspects. *Frontiers in Molecular Biosciences* 7: 587997.

35 Rossi, J.J. (2006). RNAi therapeutics: SNALPing siRNAs in vivo. *Gene Therapy* 13 (7): 583–584.

36 Kulkarni, J.A., Cullis, P.R., and van der Meel, R. (2018). Lipid nanoparticles enabling gene therapies: from concepts to clinical utility. *Nucleic Acid Therapeutics* 28 (3): 146–157.

37 Rizk, M. and Tüzmen, Ş. (2017). Update on the clinical utility of an RNA interference-based treatment: focus on Patisiran. *Pharmgenomics and Personalized Medicine* 10: 267–278.

38 Adams, D., Gonzalez-Duarte, A., O'Riordan, W.D. et al. (2018). Patisiran, an RNAi therapeutic, for hereditary transthyretin amyloidosis. *The New England Journal of Medicine* 379 (1): 11–21.

39 Kristen, A.V., Ajroud-Driss, S., Conceicao, I. et al. (2019). Patisiran, an RNAi therapeutic for the treatment of hereditary transthyretin-mediated amyloidosis. *Neurodegenerative Disease Management* 9 (1): 5–23.

40 Khalil, I.A., Kogure, K., Futaki, S. et al. (2008). Octaarginine-modified liposomes: enhanced cellular uptake and controlled intracellular trafficking. *International Journal of Pharmaceutics* 354 (1–2): 39–48.

41 Khalil, I.A., Kogure, K., Futaki, S. et al. (2007). Octaarginine-modified multifunctional envelope-type nanoparticles for gene delivery. *Gene Therapy* 14 (8): 682–689.

42 Kogure, K., Moriguchi, R., Sasaki, K. et al. (2004). Development of a non-viral multifunctional envelope-type nano device by a novel lipid film hydration method. *Journal of Controlled Release* 98 (2): 317–323.

43 Nakase, I., Akita, H., Kogure, K. et al. (2012). Efficient intracellular delivery of nucleic acid pharmaceuticals using cell-penetrating peptides. *Accounts of Chemical Research* 45 (7): 1132–1139.

44 Nakamura, T., Yamada, Y., Sato, Y. et al. (2019). Innovative nanotechnologies for enhancing nucleic acids/gene therapy: controlling intracellular trafficking to targeted biodistribution. *Biomaterials* 218: 119329.

45 Khalil, I.A. and Harashima, H. (2018). An efficient PEGylated gene delivery system with improved targeting: synergism between octaarginine and a fusogenic peptide. *International Journal of Pharmaceutics* 538 (1–2): 179–187.

46 Kibria, G., Hatakeyama, H., Ohga, N. et al. (2011). Dual-ligand modification of PEGylated liposomes shows better cell selectivity and efficient gene delivery. *Journal of Controlled Release* 153 (2): 141–148.

47 Takara, K., Hatakeyama, H., Kibria, G. et al. (2012). Size-controlled, dual-ligand modified liposomes that target the tumor vasculature show promise for use in drug-resistant cancer therapy. *Journal of Controlled Release* 162 (1): 225–232.

48 Hayashi, Y., Yamauchi, J., Khalil, I.A. et al. (2011). Cell penetrating peptide-mediated systemic siRNA delivery to the liver. *International Journal of Pharmaceutics* 419 (1–2): 308–313.

49 Hayashi, Y., Mizuno, R., Ikramy, K.A. et al. (2012). Pretreatment of hepatocyte growth factor gene transfer mediated by octaarginine peptide-modified nanoparticles ameliorates LPS/D-galactosamine-induced hepatitis. *Nucleic Acid Therapeutics* 22 (5): 360–363.

50 Hu, Y., Haynes, M.T., Wang, Y. et al. (2013). A highly efficient synthetic vector: nonhydrodynamic delivery of DNA to hepatocyte nuclei in vivo. *ACS Aano* 7 (6): 5376–5384.

51 Torchilin, V.P. and Levchenko, T.S. (2003). TAT-liposomes: a novel intracellular drug carrier. *Current Protein & Peptide Science* 4 (2): 133–140.

52 Sato, Y., Hatakeyama, H., Sakurai, Y. et al. (2012). A pH-sensitive cationic lipid facilitates the delivery of liposomal siRNA and gene silencing activity in vitro and in vivo. *Journal of Controlled Release* 163 (3): 267–276.

53 Khalil, I.A., Kimura, S., Sato, Y. et al. (2018). Synergism between a cell penetrating peptide and a pH-sensitive cationic lipid in efficient gene delivery based on double-coated nanoparticles. *Journal of Controlled Release* 275: 107–116.

54 Hagino, Y., Khalil, I.A., Kimura, S. et al. (2021). GALA-modified lipid nanoparticles for the targeted delivery of plasmid DNA to the lungs. *Molecular Pharmaceutics* 18 (3): 878–888.

Part IV

Application

18

Cell-Penetrating Peptides as Tools to Facilitate Oral Delivery of Biopharmaceuticals

Camilla Jantzen Asgreen[1], *Freja Fredholt*[1], *Hanne Mørck Nielsen*[1], *and Mie Kristensen*[2,*]

[1] *University of Copenhagen, Center for Biopharmaceuticals and Biobarriers in Drug Delivery, Drug Delivery and Biophysics of Biopharmaceuticals, Department of Pharmacy, Universitetsparken 2, Copenhagen, 2100, Denmark*
[2] *University of Copenhagen, CNS Drug Delivery & Barrier Modelling, Department of Pharmacy, Universitetsparken 2, Copenhagen, 2100, Denmark*

18.1 Introduction

The number of biopharmaceuticals (e.g. peptide- and protein-based drugs) in development is increasing due to, among other reasons, their specific mode of action giving rise to fewer side effects when compared to small-molecule drug entities. Biopharmaceuticals are dosed primarily by injections due to poor absorption across the intestinal mucosa, although patients favor the oral route for drug administration. One of the few exceptions is the GLP-1 receptor agonist semaglutide, which, when formulated with the permeation enhancer sodium N-[8-(2-hydroxybenzoyl)amino] caprylate (SNAC), is sufficiently absorbed across the gastric mucosa upon oral intake [1]. In addition to SNAC and other chemical permeation enhancers [2], cell-penetrating peptides (CPPs) have demonstrated potential as carriers for systemic delivery of biopharmaceuticals by promoting absorption from the small intestine [3]. This chapter outlines the state of the art for CPPs as intestinal permeation enhancers for biopharmaceuticals, including descriptions of in vivo methods of choice in preclinical studies. The proposed mechanisms underlying CPP-mediated transepithelial transport of biopharmaceuticals are presented and discussed. Finally, the authors give their perspectives on CPPs as part of future formulations of biopharmaceuticals for oral administration.

18.2 Challenges for Oral Delivery of Biopharmaceuticals

Oral delivery of biopharmaceuticals is a major challenge in drug delivery due to their hydrophilic nature and large molecular size limiting their permeation across

Cell-Penetrating Peptides: Design, Development and Applications, First Edition. Makoto Oba and Yosuke Demizu.

the intestinal mucosa. In addition, biopharmaceuticals are chemically and physically labile. The primary function of the gastrointestinal (GI) tract is the processing of food for subsequent absorption of nutrients and excretion of leftover materials. The introduction of a biopharmaceutical to the GI tract will thus expose the drug to a hostile environment including high enzymatic activity and pH values spanning from 1–2 in the stomach to 4–7.5 in the intestine [4]. Encapsulation in particles or tablets in combination with enteric coating using a pH-sensitive polymer will protect the biopharmaceutical and circumvent degradation during passage through the stomach. The major barriers to successful oral delivery of biopharmaceuticals therefore reside in the intestine including the limited absorption across the intestinal mucosa comprised of an epithelium surface-covered by a protective mucus layer (Figure 18.1). The epithelium contains multiple cell types exerting different functions, not only primarily absorptive enterocytes, but also mucus-secreting goblet cells, enzyme-secreting Paneth cells, hormone-secreting enteroendocrine cells, chemosensory tuft cells, antigen/pathogen absorbing M cells, and regenerating stem cells. The mucus is a viscoelastic gel layer composed mainly of different negatively charged mucin isoforms, in addition to some lipids and proteins, and serves as an interactive as well as a size-exclusion barrier [5]. The mucus layer thickness differs along the GI tract [6], and is the first barrier that the biopharmaceutical encounters upon deposition at the site of absorption in the intestine. The epithelium beneath the mucus layer represents a physical barrier to absorption: The individual cells in the epithelium are tightly interconnected by tight junction proteins allowing passive paracellular diffusion of small hydrophilic compounds and ions but excludes larger molecules, such as biopharmaceuticals. The epithelium moreover represents a chemical barrier due to the excretion of enzymes not only from especially the Paneth cells, but also from the enterocytes. Transmucosal permeation of biopharmaceuticals is limited by, in addition to their large molecular size,

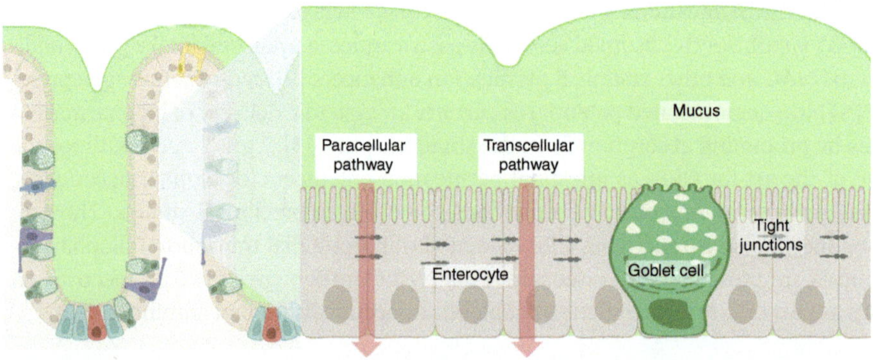

Figure 18.1 Simplified illustration of the intestinal mucosa comprising an epithelium made up of absorptive enterocytes interconnected by tight junctions. Other cell types include mucus-secreting goblet cells (green), enteroendocrine cells (purple), chemosensory tuft cells (blue), antigen/pathogen absorbing M cells (yellow), enzyme-secreting Paneth cells (red), and regenerating stem cells (turquoise). CPPs may facilitate transepithelial cargo permeation via the paracellular pathway or a transcellular pathway.

their high degree of ionization, which potentially facilitates entrapment in the mucus by binding to the charged mucins, and hinders diffusion across the lipophilic cell membranes.

18.3 Cell-Penetrating Peptides as Carriers for Oral Delivery of Biopharmaceuticals

One strategy to enhance the delivery of peptide- and protein-based drugs across the intestinal mucosa includes the use of CPPs as carriers [7, 8]. Though the concept has not yet been proven clinically [9], a number of preclinical studies demonstrate their potential (Table 18.1).

In 2007, a pioneer study by Morishita et al. demonstrated the potential of the CPPs hexa-, octa-, and deca-arginine as carriers for intestinal insulin absorption in rats [26]. That study furthermore demonstrated that the oligoarginines did not enhance intestinal absorption of all hydrophilic macromolecular cargos tested. Subsequent studies exploiting other CPP sequences in combination with various relevant therapeutic or model compound cargos have emerged [10, 12–24, 27]. Taken together, these studies demonstrate promise for CPPs as carriers for future delivery of peptide- and protein drugs across the intestinal mucosa, but also that co-administration with CPPs is not a generic strategy. Several parameters influence the potential of CPPs to facilitate the transepithelial delivery of cargo. These include the primary structure of the CPP and its stability [17, 23], the specific cargo of choice [12, 25, 26], the approach of direct conjugation or co-administration [28], and for the latter also the CPP-to-cargo molar mixing ratio [17] and whether the molecules interact [12, 14, 24, 26].

18.3.1 Parameters Affecting the Potency of Cell-Penetrating Peptides as Carriers for Transepithelial Cargo Absorption

The far majority of CPPs included as carriers for delivery of peptide- and protein-based drugs across the intestinal mucosa belong to the class of cationic CPPs. The cationic residues of the CPP interact with the negatively charged cell-surface constituents via electrostatic interactions prior to CPP and/or cargo translocation across the epithelium. Arginine residues have proven to be important for the initial plasma membrane interaction as their guanidinium head group allows bidentate binding, in contrast to lysine residues that display a monodentate binding site. Tryptophan residues as part of a cationic CPP were demonstrated to further improve the cell-penetrating potential of some CPPs; likely due to its hydrophobic nature facilitating interaction with and/or insertion of the CPP in the plasma membrane [29]. Another important factor is the specific positioning of the amino acid residues in the primary CPP sequence. Khafagy et al. fixed the arginine and lysine residues of the CPP penetratin and randomly repositioned all other residues resulting in the analogue shuffle, which further improved insulin absorption across the nasal epithelium in rats when compared to penetratin [30]. Subsequently, Kamei et al. employed

Table 18.1 Overview of in vivo studies using CPPs as carriers for intestinal absorption of peptide-based drugs or model macromolecules.

CPP	Cargo	Formulation	Animal model	Method	PK	PD	Read-out	Ref.
D-R6	Leuprolide	Conjugate	Rat	In situ loop administration (ileum)	X	X	AUC↓, C_{max}↓, and T_{max}↓ for leuprolide upon conjugation to D-R6. [Testosterone]↓	[10]
L/D-R8, L/D-penetratin, L/D-pVEC, L/D-RRL helix, L-R12, L-HIV-Rev, HIV-Tat, L-Erns, L-PRL4	Insulin	Co-administration	Rat	In situ loop administration (ileum)	X		AUC↑, C_{max}↑, T_{Max}↓, and BA↑ for insulin when co-administered with L-penetratin, D-R8, L-pVEC and L-RRL helix. No CPP-mediated insulin absorption with L-R12, L-HIV-Rev, L-HIV-Tat, L-Erns, L-R8, D-penetratin, D-RRL helix, or D-pVEC.	[11]
D-R8	Gastrin, insulin, exendin-4, calcitonin, GLP-1	Co-administration	Rat	In situ loop administration (ileum)	X		AUC↑ for gastrin, GLP-1, and insulin when co-administered with D-R8. No R8-mediated increase in exendin-4 and calcitonin absorption.	[12]
L/D-R8, L-penetratin	^{68}Ga-DOTA-insulin	Co-administration	Rat	In situ loop administration (ileum)	X	X	AUC↑ for ^{68}Ga-DOTA-insulin and BGL↓ after co-administration with D-R8 and L-penetratin. No L-R8-mediated ^{68}Ga-DOTA-insulin absorption.	[13]

(Continued)

Table 18.1 (Continued)

CPP	Cargo	Formulation	Animal model	Method	PK	PD	Read-out	Ref.
L-Penetratin, L-shuffle, L-penetramax	Insulin	Co-administration	Rat	In situ loop administration (ileum)	X		AUC↑ and C_{max}↑ for insulin when co-administered with L-penetramax > L-shuffle > L-penetratin.	[14]
L-Penetratin, L-penetratin conjugated to β-cyclodextrin	Insulin, FITC-insulin, lipopoly-saccharide	Co-administration (with L-penetratin), nanocomplex (with penetratin conjugated to β-cyclodextrin)	Rat + Mouse	In situ loop administration (ileum), intestinal injection (ileum), oral gavage	X	X	In situ loop administration: Absorption of FITC-insulin in epithelial cells after application with the CPPs. Intestinal injection: AUC↑ and BA↑ for insulin and BGL↓, AOC↑, and BP↑ after application with the CPP as simple physical mixture or nanocomplex. Most notable effect upon nanocomplex administration. Oral gavage: No increase in biomarkers for hepatic damage after 7 consecutive days of nanocomplex administration together with lipopolysaccharide.	[15]

(Continued)

Table 18.1 (Continued)

CPP	Cargo	Formulation	Animal model	Method	PK	PD	Read-out	Ref.
L/D-Penetratin	Insulin	Co-administration	Mouse	Intestinal injection (jejunum), oral gavage	X	X	Intestinal injection: AUC↑ and C_{max}↑ for insulin co-administered with D-penetratin. No L-penetratin-mediated insulin absorption. Oral gavage: AUC↑ for insulin and BGL↓, AOC↑, and BP↑ when co-administered with L- and D-penetratin. Effects more pronounced for D-penetratin.	[16]
L-Penetratin	Insulin	Co-administration	Rat	Intestinal injection (jejunum)		X	BGL↓ after co-administration of insulin with L-penetratin with a sample pH of 5, but 7.4.	[17]
L/D-Penetratin	Lipopoly-saccharide	Co-administration	Mouse	Oral gavage		X	No notable increase in biomarkers for hepatic damage after 7 consecutive days of L- and D-penetratin administration together with lipopolysaccharide.	[18]

(Continued)

Table 18.1 (Continued)

CPP	Cargo	Formulation	Animal model	Method	PK	PD	Read-out	Ref.
L/D-R8, L/D-penetratin L/D-penetramax	Insulin	Co-administration, CPP pretreatment	Rat	In situ loop administration (duodenum, jejunum, ileum, colon)	X		AUC↑, C_{max}↑, T_{max}↓, and BA↑ for insulin when co-administered with L/D-penetramax > L-penetratin > D-R8 > D-penetratin. Region-dependent L/D-penetratin-mediated insulin absorption (AUC); L-penetratin – Ileum > jejunum > duodenum > colon; D-penetratin: Colon > duodenum ≥ jejunum and ileum (AUC). No L-R8-mediated insulin absorption. No CPP-mediated insulin absorption upon L/D-penetramax pretreatment.	[19]
L-Penetratin	Insulin	Co-administration, pHPMA nanoparticle, pHPMA nanocomplex	Rat	Oral gavage	X	X	AUC↑ and BA↑for insulin and BGL↓, AOC↑, and BP↑ when formulated with L-penetratin in pHPMA nanoparticles. No L-penetratin-mediated insulin absorption when formulated pHPMA nanocomplexes or upon co-administration as a simple physical mixture.	[20]

(Continued)

Table 18.1 (Continued)

CPP	Cargo	Formulation	Animal model	Method	PK	PD	Read-out	Ref.
L/D-R8, L/D-Tat, L/D-penetratin, L/D-penetramax	FD-4, FD-10, FD-70	Co-administration	Rat	In situ loop administration (ileum)	X		AUC↑ for FD-4 when co-administered with L/D-R8, L-Tat, L/D-penetratin, and L/D-penetramax, but not D-Tat. AUC↑ for FD-10 when co-administered with L/D-R8, L/D-Tat, L/D-penetratin, and L/D-penetramax. AUC↑ for FD-70 when co-administered with L/D-penetratin. Most pronounced FD-4/FD-10 absorption when co-administered with the D-forms (except for D-Tat)	[21]
L/D-Penetratin, L/D-penetramax	Insulin	Co-administration	Mouse	Oral gavage		X	BGL↓, AOC↑, and BP↑ when insulin was co-administered with L/D-penetratin and L/D-penetramax. Most pronounced effects when co-administered with D-penetratin and D-penetramax, but synergies were not observed upon applying these CPP together.	[22]

(Continued)

Table 18.1 (Continued)

CPP	Cargo	Formulation	Animal model	Method	PK	PD	Read-out	Ref.
L-Penetratin	Insulin, PTH(1-34)	Co-administration	Rat	Intestinal injection (jejunum)	X	X	AUC↑ and C_{max}↑ and for PTH (1-34) when co-administered with L-penetratin. The effect was sample pH-dependent (pH 5 > pH 6.5 > pH 7.4). AUC↑ and C_{max}↑ for insulin and BGL↓ when co-administered with L-penetratin and a sample pH of 5. A slight increase was observed for pH 6.5 samples, but no effect for pH 7.4 samples.	[23]
L-Penetratin, L-shuffle, L-penetramax	Insulin	Co-administration	Rat	Intestinal injection (jejunum)	X	X	AUC↑, C_{max}↑, and BA↑ for insulin and BGL↓, AOC↑, and BP↑ when co-administered with L-penetramax > L-shuffle > L-penetratin, and effects were less pronounce for L-penetratin.	[24]

(Continued)

Table 18.1 (Continued)

CPP	Cargo	Formulation	Animal model	Method	PK	PD	Read-out	Ref.
L/D-Penetratin	GLP-1, exendin-4, IFN-β	Co-administration	Rat	In situ loop administration (ileum)	X		AUC↑ and C_{max}↑ for GLP-1 when co-administered with L-penetratin. C_{max}↑ and T_{max}↑ for exendin-4 when co-administered with L- and D-penetratin. AUC↑ and C_{max}↑ for IFN-β when co-administered with D-penetratin.	[25]
L/D-R6, D-R8, D-R10	Insulin, IFN-β, FD-4	Co-administration	Rat	In situ loop administration	X	X	AUC↑, C_{max}↑, T_{max}↑, and BA↑ for insulin when co-administered with L/D-R6, D-R8, D-R10. No CPP-mediated IFN-β or FD-4 absorption when co-administered with L/D-R6 or D-R10.	[26]

AOC, Area over the curve; AUC, Area under the curve; BA, Bioavailability; BGL, Blood glucose level; BP, Biopotency.

self-organizing map clustering to identify analogues of shuffle with further improved potential as carriers for intestinal absorption of peptide and protein drugs [14]. That led to the discovery of the penetramax analogue. Recently, Diedrichsen et al. confirmed the strong potential of shuffle and penetramax as carriers for intestinal insulin absorption in rats as compared to that of penetratin [24].

The molecular weight of the cargo is another factor to consider. Kamei et al. demonstrated CPP-mediated intestinal absorption of 4- and 10-kDa fluorescein isothiocyanate-labeled (FITC)-dextrans in rats upon co-administration with L- and D-stereoisomers of octaarginine, Tat, penetratin, and penetramax [21]. In contrast, only D-penetratin was demonstrated to increase uptake of 70 kDa FITC-dextran, which was employed as pathogen marker.

Finally, the applied approach, i.e. whether the CPP is co-administered in a physical mixture with the cargo or directly conjugated to the cargo must be considered. Co-administration remains the most widely employed strategy offering ease of sample preparation and the possibility to adjust the molar CPP-to-cargo ratio. For example, the molar mixing ratio was demonstrated to affect the potential of penetratin, and analogues thereof, to facilitate insulin translocation across monolayers of the human intestinal Caco-2 cell culture model [17]. Some studies suggest that the strength of electrostatic interactions between the CPP and its cargo is decisive for the capability of the CPP to facilitate delivery [12]. Then again, a more recent study demonstrated, that limiting the enzymatic degradation of penetratin upon pH lowering, rather than enforcing electrostatic CPP-cargo interactions, improved the epithelial penetratin-mediated insulin and PTH(1-34) translocation in vitro and in vivo [23]. Covalent conjugation can be done post synthesis of the CPP and the cargo or by expression in a suitable host system. Direct conjugation of cargo to the CPP, may, however affect the potency of the cargo as well as the cell-penetrating potential of the CPP [28]. Nevertheless, one study demonstrated that covalent conjugation of Tat to insulin facilitated insulin translocation across Caco-2 cell monolayers [31]. In another study, the concept was tested in vivo but these authors did not observe any difference in rat plasma testosterone concentration upon intestinal administration of hexa-arginine conjugated to leuprolide when compared to administration of leuprolide alone [10].

Due to their peptide nature, CPPs are prone to degradation by enzymes present in the intestinal lumen as well as inside the epithelial cells, and this, at least to some extent, limit their potency as carriers for intestinal absorption of biopharmaceuticals. Various strategies to improve the chemical stability of CPPs are reported. These include changing parts of the primary sequence normally recognized by intestinal enzymes as well as backbone stabilization by including β- and γ-peptoid residues [29]. Another common strategy is the alteration of the amino acid stereochemistry fully or partially from the natural L-form to the D-form. This concept has been demonstrated for penetratin, penetramax, and octaarginine, for which the plasma half-lives were increased upon changing the stereochemical configuration to the all-D form [16, 19]. The efficiency of the CPP as a carrier for intestinal cargo absorption may, however, not be retained upon changing the amino acid stereochemistry from L to D [11, 19].

18.3.2 Formulation of Cell-Penetrating Peptides

So far, the majority of studies investigating CPPs as carriers for intestinal absorption of biopharmaceuticals take a simple approach using a non-modified CPP and cargo as well as not including additional excipients in the test sample. However, some more advanced CPP-cargo formulations are emerging. An example includes penetratin modified with β-cyclodextrin to further improve its carrier potential [15]. That study demonstrated increased intestinal insulin absorption in rats upon co-administration with β-cyclodextrin-penetratin, when compared to co-administration with non-modified penetratin. Another more advanced formulation approach includes the hydrophilic polymer N-(2-hydroxypropyl) methacrylamide (pHPMA) [20]. pHPMA self-assembles into nanostructures in which penetratin and insulin can be loaded for gradual release upon association with the intestinal mucus. That study demonstrated that encapsulation of insulin and penetratin in pHPMA nanoparticles, but not in the simpler pHPMA nanocomplexes, improved the bioavailability of insulin upon oral administration to rats when compared to insulin administered with penetratin in a simple physical mixture.

18.3.3 Region-Dependent CPP-Mediated Intestinal Absorption of Biopharmaceuticals

The anatomy and physiology change throughout the intestine with respect to pH fluctuations, enzymatic activity, mucus thickness, and epithelial tightness [4]. Awareness of these parameters are therefore important during experimental planning and data interpretation. The majority of studies investigating CPP-mediated intestinal absorption of biopharmaceuticals make use of in situ intestinal administration or intestinal injection for early in vivo testing. Test samples are typically injected into the ileum [10–15, 21, 24–26] or jejunum [16, 17, 23], yet the choice hereof likely affects the carrier potential of the CPP as suggested by Khafagy et al. [19]. In that study, the authors demonstrated that L- and D-penetratin-mediated intestinal insulin absorption in rats was region-dependent; and dependent on the penetratin stereochemistry (L-penetratin: Ileum > jejunum > duodenum > colon; D-penetratin: Colon > duodenum > jejunum > ileum). The observed differences in the L-penetratin-mediated insulin delivery may be explained by regional differences in enzymatic activity. In addition, the increased thickness of the mucus and the increasing junction tightness in the colonic epithelium, when compared to the other regions, might explain the observed poor colonic insulin absorption. Finally, studies have demonstrated that sample pH affects the CPP-mediated transepithelial cargo permeation through effects on the CPP-cargo interaction [17, 23, 32] as well as on the chemical stability of the CPP [23].

Thus, detailed planning and reporting of the experimental in vivo setup is highly important to extract the gained knowledge from the specific study. That applies to the e.g. administration method, choice of experimental readouts, and data reporting format including proper statistical analysis.

18.4 In vivo Methods and Data Reporting when Studying Intestinal Absorption of Biopharmaceuticals

Biopharmaceuticals display low bioavailability upon oral administration. Therefore, animal studies reporting on CPPs as carriers for intestinal absorption of biopharmaceuticals must be designed to enable detection of often small changes in, especially pharmacokinetic (PK) and pharmacodynamic (PD) parameters [33]. To the authors' knowledge, only rats and mice have been employed in preclinical testing of CPPs as carriers for intestinal drug absorption. Different methods for administration have been used to evaluate the potential of CPPs as carriers for the delivery of cargo with oral gavage and in situ administration being the most reported (Figure 18.2, Table 18.1).

18.4.1 Oral Gavage

Oral administration to rodents is often performed by oral gavage, during which the animal is briefly restricted and its head is tipped back. A feeding tube is carefully inserted into the esophagus through which the test sample is introduced directly into the stomach [34] (Figure 18.2). The procedure requires a moderate amount of training but is gentle to the animal if done correctly. The dosing should not exceed 5 mg/kg for either mice or rats. Restraining an animal may, however, apply stress to the animal, which again may affect the read-out such as the blood glucose level [35, 36]. Oral gavage is often not the choice when studying simple test samples composed solely of CPP and cargo, as the solution or dispersion must pass the harsh environment of the stomach prior to reaching the intestine. Hence, only a few studies including CPPs as carriers for intestinal absorption of biopharmaceuticals apply

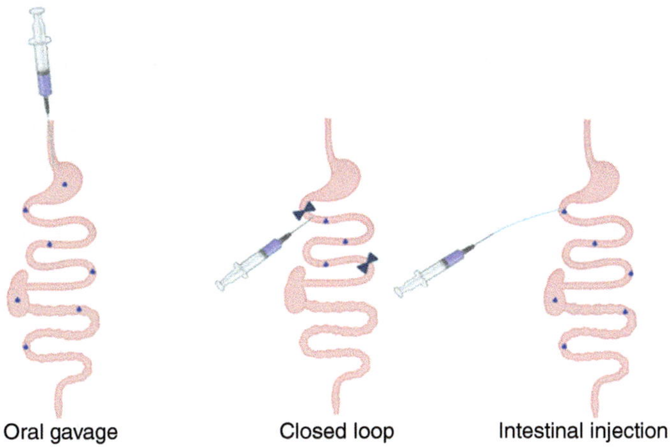

Figure 18.2 Illustration of the most common in vivo methods used to study intestinal CPP-mediated absorption of biopharmaceuticals. These include oral gavage, closed-loop, and intestinal injection.

oral gavage for early in vivo testing. Nielsen et al. did, however, report a decrease in blood glucose upon administration of D-penetratin in solution with insulin to mice via oral gavage [16]. A later stage of in vivo testing may include encapsulation of the CPP and cargo in enteric-coated particles, capsules, or tablets to facilitate the transit of both intact carrier and intact cargo to the intestine.

18.4.2 In situ Administration

Intestinal in situ administration represents experiments where the drug formulation is placed and kept in a defined intestinal segment of interest. A flow of media through the segment may be applied [37]. The method is invasive, when compared to oral gavage, and requires surgical skills and utilization of anesthesia. In situ studies investigating CPPs as intestinal permeation enhancers primarily report the use of intestinal loop administration, where the intestinal segment of interest (i.e. loop) is closed off and filled with the test sample throughout the entire experiment (Figure 18.2) [38, 39]. Another setup is the single-pass perfusion model, where part of the intestine is cannulated and the test sample flushed through the isolated intestinal segment. Hereby, the test sample only passes through the intestinal augment once. [38, 39]. The intestinal loop and the single-pass perfusion models both allow investigation of region-dependent absorption, but with the former being mostly applied when studying CPPs as intestinal absorption enhancers (Table 18.1).

18.4.3 Intestinal Injection

Compared to in situ administration, intestinal injection represents a more simple procedure, where the test sample is injected into the intestine via a catheter (Figure 18.3) [37]. The method is, like the in situ setups, invasive and requires surgical skills and the use of anesthesia. The main difference comparing intestinal injection to in situ administration is that upon intestinal injection, the solution can freely diffuse to adjacent regions and in principle distribute to both more proximal and distal sites from the site of catheter insertion. In contrast, only a defined intestinal segment is exposed to the test sample during in situ administration. In later stages of preclinical studies, intestinal instillation may apply to solid dosage forms placed directly into the intestine.

The terminologies used in in vivo studies are often used interchangeably. Thus, to avoid misinterpretations, it is important that authors thoroughly describe the experimental setup when reporting.

18.4.4 Experimental Readout

Determination of plasma drug concentrations at specific blood-sampling time points provides the option to generate plasma concentration profiles that subsequently are used to extract relevant PK parameters. These include the area under the curve (AUC), the maximal plasma concentration obtained (C_{max}), the time-point at which C_{max} is obtained (T_{max}), and systemic bioavailability. Biopharmaceuticals notoriously show low bioavailability due to poor absorption after oral administration, and in addition, often display short plasma half-lives [40]. Such conditions require sensitive quantification methods for the detection of often minute amounts of the drug compound in plasma. Commonly, compound-specific methods are employed, such as an enzyme-linked immunosorbent assay (ELISA) [37], but these may not be applicable for all drug compounds of interest. In addition, sample handling should avoid the risk of compound loss due to inadequate procedures and/or adhesion to material surfaces, which is a specific risk when handling samples containing biopharmaceuticals.

Due to the low bioavailability of biopharmaceuticals and/or lack of a sufficiently sensitive detection method, PK evaluation can be a true challenge. Biopharmaceuticals are, on the other hand, often highly potent and result in prolonged effects when compared to the time that the drug compound remains in plasma. Therefore, PD evaluation may stand alone in some preclinical studies, though the read-out effect should preferably be validated by PK data to investigate dose-response relations, and to validate that the PD response does not arise from other parameters. In a study by Nielsen et al., no difference in insulin AUC was observed upon oral administration of insulin with L-penetratin when compared to an insulin control [16]. In contrast, a significant effect of administering insulin with L-penetratin was observed upon evaluation of the blood glucose profiles. The data had high variations in the plasma insulin concentrations, and the authors suggested that differences in intestinal enzymatic activity and gastric emptying, combined with the low insulin concentration in the samples, complicated the detection of insulin. The PD data-derived curves can be used for the calculation of the pharmacological availability (biopotency) and insulin, blood glucose profiles used to determine the blood glucose area over the curve (AOC) often provide similar results as the determination of the AUC for the PK curves [24]. The concentration of a biomarker, such as blood glucose, may however be affected by other factors than the test compound. An example is anesthesia. One study investigating different anesthetics demonstrated that the commonly used ketamine/xylazine decreases the blood glucose level [37]. Fentanyl/midazolam did not have the same pronounced effect on the blood glucose but did affect the dynamics of the blood glucose profile when compared to that observed in non-anesthetized animals. Another study showed that intraperitoneal administration of pentobarbital increased the blood glucose level by 33% already three minutes after administration, and then decreased to normal levels after 40 minutes [41]. Stress is another factor to affect blood glucose levels and may originate from cage transportation or handling of the animal [42]. Animal handling (e.g. restraining) and exposure to relevant procedures (e.g. oral gavage) prior to conduction of the actual experiment may to some extent lower stress-induced changes of blood glucose [22].

18.4.5 Safety of Oral Administration of Cell-Penetrating Peptides

A primary safety concern for the use of oral permeation enhancers, including CPPs, is the risk of intestinal mucosal damage, which may cause inflammation and co-absorption of pathogenic bystanders such as toxins [43].

One way to assess in vivo toxicity is to measure lactate dehydrogenase (LDH) leakage into the blood circulation. LDH is released upon plasma membrane damage and thus represents a cell death marker. Previous studies investigated whether various CPPs in situ administered with different peptide cargos to rat ileal segments gave rise to toxicity in terms of LDH release [11, 26]. That was not the case, suggesting that the CPPs did not affect the plasma membrane integrity of the intestinal cells. Histological examination of the intestinal tissue is another way to evaluate the potential adverse effects of CPPs. Morishita et al. did not observe any damaged tissue upon in situ administration of insulin with octaarginine to rat ileal segments [26]. In contrast, Diedrichsen et al. observed adverse tissue effects upon intestinal injection of insulin with penetratin, shuffle, or penetramax [24]. These effects included shedding, surface area reduction and connective tissue retraction; effects that were, though to a minor extent, also observed by histological examination of intestinal tissue subjected to administration of buffer or insulin alone. The CPP-mediated tissue effects were more prominent immediately after administration (five minutes) when compared to after one hour, thus pointing toward reversibility and tissue regeneration. Finally, potential adverse effects have been studied as a result of consecutive oral administration of β-cyclodextrin-penetratin [15] and L- or D-penetratin [18] for seven days. The CPPs were administered by oral gavage together with lipopolysaccharide (LPS). LPS is a component of the bacterial cell wall causing hepatic necrosis upon chronic exposure, which then results in the release of alanine aminotransferase (ALT) and aspartate aminotransferase (ALS). Consecutive oral administration of the CPPs did not increase ALT or ALS levels, thus demonstrating safe use of the CPPs at the selected concentrations in those studies.

18.4.6 Reporting and Statistical Analysis of Data from in vivo Studies

There is an increased focus on the experimental design of animal studies, the statistical analysis, and the reporting of results. This is due to widespread concern about the quality, reproducibility, and translatability of studies involving live animals. The PREPARE and ARRIVE guidelines are useful during experimental planning and reporting of results, respectively [44–46]. Power analyses should be conducted as part of the experimental planning to estimate the sample size needed to detect the smallest effect size of interest. The power of an experiment describes the probability of detecting a prespecified effect at a set significance level. The more powerful a study, the higher the chance of detecting an effect and the lower the risk of making a type II error. A power value between 80% and 90% is optimal [47], but the higher the power value the more animals are needed. Therefore, a dilemma between a statistically strong study and ethical considerations may arise. Unfortunately, many studies are underpowered. That means that there is no realistic chance of finding the effect size of interest with the number of animals used. When investigating oral delivery of biopharmaceuticals, the effect size is often small, thereby making power calculations especially important within this research field.

18.5 Mechanism of Cell-Penetrating Peptide-Mediated Cargo Delivery Across the Intestinal Barrier

In vivo studies are needed for PK/PD evaluations, but do not reveal details on the cellular mechanisms underlying the CPP-mediated intestinal absorption of a co-administered or conjugated cargo. For that, representative in vitro models are essential.

18.5.1 Mucus as Barrier to Cell-Penetrating Peptide-Mediated Cargo Delivery Across the Intestine

Representative cell culture models are key to improving our understanding of mechanistic events at the cellular level. The human intestinal Caco-2 cell culture model is the most commonly employed in vitro model of the intestinal epithelium and is used for screening the potential of CPPs as transepithelial cargo carriers and for mechanistic studies. Caco-2 cells are, however, originally derived from colon, and not the small intestine from where the majority of drugs are absorbed. In addition, the Caco-2 cells do not secrete mucins, the main constituent of mucus besides water, thus rendering them unsuitable for studying the impact of mucus on the CPP and/or cargo transport across the intestinal mucosa. For such studies, the mucin-producing intestinal HT29-MTX cell culture model [5] or Caco-2 cell epithelium supplemented with biosimilar mucus [48, 49] may be implemented. On the other hand, studying the mechanisms underlying CPP interaction with and diffusion through the mucus layer may be more informative using solely ex vivo- or biosimilar mucus to keep the experimental conditions as simple as possible. So far, little is known about the impact of mucus on CPP-mediated transmucosal absorption of biopharmaceuticals. Some studies demonstrating CPP-mediated transepithelial peptide drug permeation across the Caco-2 cell epithelium translate well into expected PK/PD effects in vivo [17, 23, 24]. In contrast, one study including penetratin as a carrier for oral insulin delivery demonstrated improved in vitro mucus penetration of insulin-penetratin complexes upon encapsulation in pHPMA nanocomplexes when compared to nonencapsulated insulin-penetratin complexes [20]. That observation was confirmed in vivo by PK/PD evaluation in diabetic rats, where effects on blood glucose and serum insulin levels were also shown to not differ between insulin administered alone and insulin complexed with penetratin [20]. Thus, for some CPPs, the barrier properties of the mucus layer are questionable, whereas, for others, the mucus may represent a barrier, which needs to be taken into consideration when formulating the CPP and cargo of interest.

18.5.2 Mechanism of Transepithelial Transport of Cell-Penetrating Peptides and Their Cargo

Most mechanistic studies on CPPs report on cellular uptake of the CPP itself or a CPP-conjugated or co-administered cargo molecule. However, studies on the mechanism of CPP and/or cargo permeation across biological barriers, such as the intestinal epithelium, are limited. Two overall routes for translocation across the intestinal epithelium exist: The transcellular route (i.e. transcytosis or direct translocation across the luminal and basolateral membranes) and the paracellular route (i.e. diffusion through the intercellular space) (Figure 18.1). Numerous studies have demonstrated different mechanisms of cellular CPP uptake involving energy-dependent endocytosis or energy-independent direct membrane translocation. These have earlier been excellently reviewed [50], and will therefore not be covered in more detail in this chapter, which will have a strict focus on the mechanism of CPP and/or cargo permeation across the intestinal epithelium. It should, however, be mentioned that the mechanism of cellular CPP uptake depends on various factors such as the specific CPP, its concentration, as well as cell type, cell seeding conditions, and the cellular proliferation/differentiation state. Thus, the mechanism involved for a specific CPP with a specific cargo must be investigated case by case.

As mentioned previously, the far majority of CPPs demonstrating potential as carriers for intestinal delivery of biopharmaceuticals belong to the class of cationic CPPs (Table 18.1). These carry a net positive charge at physiological pH, thereby enabling electrostatic interactions with the negatively charged glycocalyx constituents of the plasma membrane as well as the phospholipid head groups as the initial step prior to cellular uptake and/or barrier permeation.

An early study demonstrated that the CPP octaarginine led to similar levels of insulin uptake in Caco-2 cells seeded at the bottom of a well plate at 37 °C and 4 °C, thereby indicating the involvement of both energy-dependent and independent mechanisms, respectively [51]. However, the mechanism of CPP uptake observed in Caco-2 cells seeded at the bottom of a well plate may differ from the uptake mechanism observed in Caco-2 cell monolayers cultured on permeable filter supports (e.g. Transwells®). Another early study documented that the CPP Tat facilitated insulin transport across Caco-2 cell monolayers as well as showed indications of endocytic uptake into the Caco-2 cells [31]. The authors did not report barrier integrity measurements as part of that study. CPP and/or cargo endocytic uptake into intestinal cell monolayers may, however, not be followed by exocytosis at the abluminal membrane and thus not lead to transcytosis. Instead, the CPP and/or cargo may be segregated to lysosomes for degradation [52], the level of which thus must be investigated in parallel to uptake studies for predicting transcellular permeation. Optimally, the mechanism of CPP and/or cargo uptake into the Caco-2 cells should be investigated in permeable filter support-cultured cell monolayers allowing both quantification of the CPP and/or cargo cellular uptake as well as their transport across the epithelial barrier. In parallel, the barrier integrity should be monitored to reveal whether barrier modulation or breach adds to the net CPP and/or cargo transport. As an alternative to the Caco-2 cell culture model, mucosal explants may be used to study the mechanism of barrier permeation. One study investigated the mechanism underlying enterocytic uptake and barrier permeation of the CPP Tat and the cationic antimicrobial peptide melittin using segments of excised porcine jejunum [53]. That study demonstrated cytosolic uptake of both peptides, but without the involvement of endocytosis. In addition, melittin increased the paracellular permeation of a

3 kDa polar tracer compound, but melittin itself did not permeate the mucosa. Recently, another study demonstrated that the CPPs shuffle and penetramax indeed affected the barrier integrity of Caco-2 cell monolayers [24]. The authors suggested that the observed CPP-mediated transepithelial insulin transport was likely through the paracellular space rather than via a transcellular route. The more recent focus on the involvement of the paracellular route in CPP-mediated transepithelial cargo transport sets new standards for experimental design. Thus, careful monitoring of the barrier integrity, preferably in real-time, and effects on tight junction morphology and expression levels are indeed relevant for future studies [54].

18.6 Perspectives

The application of CPPs as tools to facilitate oral delivery of biopharmaceuticals seems promising based on several in vitro and preclinical in vivo studies, which clearly illustrate that differences in efficacy depends on the specific CPP carrier design, on the investigated cargo, and on the approach by which the combination of the two is applied in the different test methods.

Given the nature of both the CPP and the biopharmaceutical cargo in that both are most often enzymatically labile, prone to self-interactions and to interactions with other molecules, as well as material surfaces, utmost care must be taken in the design of both in vitro and in vivo experiments. The peptide nature of the CPP may in itself be an advantage as this excipient will be degraded in the intestine to biocompatible fragments, and if at all absorbed in an intact form, will be significantly diluted and eliminated from the blood circulation. However, to allow for meaningful interpretation and conclusions on not only the delivery efficacy but also on the mode of action, a series of complementary studies are required taking into account the impact of using often highly different concentrations of the CPP and cargo in in vitro and in vivo studies.

Evaluating the mechanism of action for oral delivery enhancement of biopharmaceuticals by CPPs should be done in tight epithelia resembling the intestine, as interactions with tight junctions may play a larger role in the effect compared to the well-known plasma membrane interaction and subsequent endocytotic uptake. While early studies mainly suggested transcellular transport enhancement, more recent studies point toward the involvement of the paracellular route. Both the transcellular and the paracellular routes may add to the net CPP-mediated transepithelial transport. Further, while the interaction with the intestinal mucus has not been addressed separately in current literature, it needs to be considered prior to the final selection of the lead carrier CPP.

Importantly, both the choice of animal model as well as the mode of test sample administration is crucial for efficacy evaluation. For obvious reasons, there is little to no data on the correlation between oral absorption of biopharmaceuticals in humans and rodents. For small-molecule drugs, such correlations have been made showing that prediction of absorption from mucosa permeability data may be possible in mice [55], while in contrast no correlation was found between oral bioavailability in rats and humans [56]. Several reasons for the latter observation were reported and mostly related to the impact of transporters. If the same correlations are valid for dosing of biopharmaceuticals, it shows a significant disadvantage of oral gavage due to the limited correlation to humans compared to intestinal injection and in situ experimental setups where a good correlation exists. In addition, considering that the stability of biopharmaceuticals is often low, and may be vastly different in animals and humans, this may complicate direct predictions from animal studies. However, comparative studies on the efficacies of CPPs may be done in animals.

Further, it is important to recognize that CPP-cargo mixtures, which for several reasons are preferred over conjugates, are not stand-alone formulations, but must be formulated in applicable dosage forms, which in itself may be challenging. Thus, for direct comparison between different CPP-containing dosage forms, head-to-head studies should be conducted in the same models and same laboratories. No peptide-based excipient, like the CPPs, is yet approved for clinical use for oral delivery of biopharmaceuticals, and a thorough safety evaluation is needed for approval by the regulatory authorities. Evaluation could as a first step include a thorough direct comparison of effects of lipid-based excipients used for similar purposes with a focus on reversibility also considering that although the intestine is rather immune-competent, effects on both the gut microbiome, the mucus, and the epithelium should be evaluated.

References

1 Buckley, S.T., Bækdal, T.A., Vegge, A. et al. (2018). Transcellular stomach absorption of a derivatized glucagon-like peptide-1 receptor agonist. *Science Translational Medicine* 10 (468): eaar7047.

2 Maher, S. and Brayden, D.J. (2021). Formulation strategies to improve the efficacy of intestinal permeation enhancers. *Advanced Drug Delivery Reviews* 177: 113925.

3 Rønholt, S., Kristensen, M., and Nielsen, H.M. (2015). Cell-penetrating peptides as carriers for transepithelial delivery in vitro. In: *Cell-Penetrating Peptides: Methods and Protocols*, 2e (ed. Ü. Langel), 261–277. New York: Humana Press.

4 Daugherty, A.L. and Mrsny, R.J. (1999). Transcellular uptake mechanisms of the intestinal epithelial barrier – Part one. *Pharmaceutical Science & Technology Today* 2 (4): 144–151.

5 Boegh, M. and Nielsen, H.M. (2015). Mucus as a barrier to drug delivery – understanding and mimicking the barrier properties. *Basic & Clinical Pharmacoly & Toxicology* 116 (3): 179–186.

6 Zhu, Q., Chen, Z., Paul, P.K. et al. (2021). Oral delivery of proteins and peptides: challenges, status quo and future perspectives. *Acta Pharmaceutica Sinica B* 11 (8): 2416–2448.

7 Kristensen, M. and Nielsen, H.M. (2015). Cell-penetrating peptides as carriers for oral delivery of biopharmaceuticals. *Basic & Clinical Pharmacoly & Toxicology* 118: 99–106.

8 Kristensen, M. and Nielsen, H.M. (2016). Cell-penetrating peptides as tools to enhance non-injectable delivery of biopharmaceuticals. *Tissue Barriers* 4 (2): e1178369.

9 Drucker, D.J. (2020). Advances in oral peptide therapeutics. *Nature Reviews Drug Discovery* 19 (4): 277–289.

10 Kamei, N., Morishita, M., Ehara, J. et al. (2008). Permeation characteristics of oligoarginine through intestinal epithelium and its usefulness for intestinal peptide drug delivery. *Journal of Controlled Release* 131 (2): 94–99.

11 Kamei, N., Morishita, M., Eda, Y. et al. (2008). Usefulness of cell-penetrating peptides to improve intestinal insulin absorption. *Journal of Controlled Release* 132 (1): 21–25.

12 Kamei, N., Morishita, M., and Takayama, K. (2009). Importance of intermolecular interaction on the improvement of intestinal therapeutic peptide/protein absorption using cell-penetrating peptides. *Journal of Controlled Release* 136 (3): 179–186.

13 Kamei, N., Morishita, M., Kanayama, Y. et al. (2010). Molecular imaging analysis of intestinal insulin absorption boosted by cell-penetrating peptides by using positron emission tomography. *Journal of Controlled Release* 146 (1): 16–22.

14 Kamei, N., Kikuchi, S., Takeda-Morishita, M. et al. (2013). Determination of the optimal cell-penetrating peptide sequence for intestinal insulin delivery based on molecular orbital analysis with self-organizing maps. *Journal of Pharmaceutical Sciences* 102 (2): 469–479.

15 Zhu, X., Shan, W., Zhang, P. et al. (2014). Penetratin derivative-based nanocomplexes for enhanced intestinal insulin delivery. *Molecular Pharmaceutics* 11 (1): 317–328.

16 Nielsen, E.J.B., Yoshida, S., Kamei, N. et al. (2014). in vivo proof of concept of oral insulin delivery based on a co-administration strategy with the cell-penetrating peptide penetratin. *Journal of Controlled Release* 189: 19–24.

17 Kristensen, M., Franzyk, H., Klausen, M.T. et al. (2015). Penetratin-mediated transepithelial insulin permeation: importance of cationic residues and pH for complexation and permeation. *AAPS Journal* 17 (5): 1200–1209.

18 Nielsen, E.J.B., Kamei, N., and Takeda-Morishita, M. (2015). Safety of the cell-penetrating peptide penetratin as an oral absorption enhancer. *Biological and Pharmaceutical Bulletin* 38 (1): 144–146.

19 Khafagy, E.S., Iwamae, R., Kamei, N. et al. (2015). Region-dependent role of cell-penetrating peptides in insulin absorption across the rat small intestinal membrane. *AAPS Journal* 17 (6): 1427–1437.

20 Shan, W., Zhu, X., Liu, M. et al. (2015). Overcoming the diffusion barrier of mucus and absorption barrier of epithelium by self-assembled nanoparticles for oral delivery of insulin. *ACS Nano* 9 (3): 2345–2356.

21 Kamei, N., Bech Nielsen, E.J., Nakakubo, T. et al. (2016). Applicability and limitations of cell-penetrating peptides in noncovalent mucosal drug or carrier delivery systems. *Journal of Pharmaceutical Sciences* 105 (2): 747–753.

22 Kamei, N., Shigei, C., Hasegawa, R. et al. (2018). Exploration of the key factors for optimizing the in vivo oral delivery of insulin by using a noncovalent strategy with cell-penetrating peptides. *Biological and Pharmaceutical Bulletin* 41 (2): 239–246.

23 Kristensen, M., Diedrichsen, R.G., Vetri, V. et al. (2020). Increased carrier peptide stability through ph adjustment improves insulin and PTH(1–34) delivery in vitro and in vivo rather than by enforced carrier peptide-cargo complexation. *Pharmaceutics* 12 (10): 1–18.

24 Diedrichsen, R.G., Harloff-Helleberg, S., Werner, U. et al. (2021). Revealing the importance of carrier-cargo association in delivery of insulin and lipidated insulin. *Journal of Controlled Release* 338: 8–21.

25 Khafagy, E.S., Morishita, M., Kamei, N. et al. (2009). Efficiency of cell-penetrating peptides on the nasal and intestinal absorption of therapeutic peptides and proteins. *International Journal of Pharmaceutics* 381 (1): 49–55.

26 Morishita, M., Kamei, N., Ehara, J. et al. (2007). A novel approach using functional peptides for efficient intestinal absorption of insulin. *Journal of Controlled Release* 118 (2): 177–184.

27 Zhang, Y., Li, L., Han, M. et al. (2015). Amphiphilic lipopeptide-mediated transport of insulin and cell membrane penetration mechanism. *Molecules* 20 (12): 21569–21583.

28 Kristensen, M., de Groot, A.M., Berthelsen, J. et al. (2015). Conjugation of cell-penetrating peptides to parathyroid hormone affects its structure, potency, and transepithelial permeation. *Bioconjugate Chemistry* 26: 477–488.

29 Kristensen, M., Birch, D., and Nielsen, H.M. (2016). Applications and challenges for use of cell-penetrating peptides as delivery vectors for peptide and protein cargos. *International Journal of Molecular Sciences* 17 (185): 1–17.

30 Khafagy, E.S., Morishita, M., Ida, N. et al. (2010). Structural requirements of penetratin absorption enhancement efficiency for insulin delivery. *Journal of Controlled Release* 143 (3): 302–310.

31 Liang, J.F. and Yang, V.C. (2005). Insulin-cell penetrating peptide hybrids with improved intestinal absorption efficiency. *Biochemical and Biophysical Research Communication* 335 (3): 734–738.

32 Kamei, N., Aoyama, Y., Khafagy, E.S. et al. (2015). Effect of different intestinal conditions on the intermolecular interaction between insulin and cell-penetrating peptide penetratin and on its contribution to stimulation of permeation through intestinal epithelium. *European Journal of Pharmaceutics and Biopharmaceutics* 94: 42–51.

33 Iqbal, J., Shahnaz, G., Perera, G. et al. (2012). Thiolated chitosan: development and in vivo evaluation of an oral delivery system for leuprolide. *European Journal of Pharmaceutics and Biopharmaceutics* 80 (1): 95–102.

34 Pekow, C.A. and Baumans, V. (2003). Common nonsurgical techniques and procedures. In: *Handbook of Laboratory Animal Science – Volume I Essential Principles and Practices*, 2e (ed. J. Hau and G.L. Van Hoosier), 351–390. London: CRC Press.

35 Abbott, C.R., Small, C.J., Sajedi, A. et al. (2006). The importance of acclimatisation and habituation to experimental conditions when investigating the anorectic effects of gastrointestinal hormones in the rat. *International Journal of Obesity* 30: 288–292.

36 Meijer, M.K., Spruijt, B.M., Van Zutphen, L.F.M. et al. (2006). Effect of restraint and injection methods on heart rate and body temperature in mice. *Laboratory Animals* 40 (4): 382–391.

37 Harloff-Helleberg, S., Nielsen, L.H., and Nielsen, H.M. (2017). Animal models for evaluation of oral delivery of biopharmaceuticals. *Journal of Controlled Release* 28 (268): 57–71.

38 Lozoya-Agullo, I., Zur, M., Beig, A. et al. (2016). Segmental-dependent permeability throughout the small intestine following oral drug administration: Single-pass vs. Doluisio approach to in-situ rat perfusion. *International Journal of Pharmaceutics* 515: 201–208.

39 Lozoya-Agullo, I., Zur, M., Wolk, O. et al. (2015). In-situ intestinal rat perfusions for human Fabs prediction and BCS permeability class determination: Investigation of the single-pass vs. the Doluisio experimental approaches. *International Journal of Pharamceutics* 480: 1–7.

40 van Witteloostuijn, S.B., Pedersen, S.L., and Jensen, K.J. (2016). Half-life extension of biopharmaceuticals using chemical methods: alternatives to PEGylation. *ChemMedChem* 11 (22): 2474–2495.

41 Pénicaud, L., Ferré, P., Kande, J. et al. (1987). Effect of anesthesia on glucose production and utilization in rats. *American Journal of Physiology – Endocrinology and Metabolism* 252 (3): E365–E369.

42 Tabata, H., Kitamura, T., and Nagamatsu, N. (1998). Comparison of effects of restraint, cage transportation, anaesthesia and repeated bleeding on plasma glucose levels between mice and rats. *Laboratory Animals* 32 (2): 143–148.

43 McCartney, F., Gleeson, J.P., and Brayden, D.J. (2016). Safety concerns over the use of intestinal permeation enhancers: a mini-review. *Tissue Barriers* 4 (2): e1176822.

44 Smith, A.J., Clutton, R.E., Lilley, E. et al. (2018). PREPARE: guidelines for planning animal research and testing. *Laboratory Animals* 52 (2): 135–141.

45 du Sert, N.P., Hurst, V., Ahluwalia, A. et al. (2020). The ARRIVE guidelines 2.0: updated guidelines for reporting animal research. *PLOS Biology* 18 (7): e3000410.

46 du Sert, N.P., Ahluwalia, A., Alam, S. et al. (2020). Reporting animal research: explanation and elaboration for the arrive guidelines 2.0. *PLOS Biology* 18 (7): e3000411.

47 Festing, M.F.W., Overend, P., Borja, M.C. et al. (2016). The determination of sample size. In: *The Design of Animal Experiments*, 2e, 102–104. Dorchester: SAGE Publications.

48 Boegh, M., Baldursdóttir, S.G., Müllertz, A. et al. (2014). Property profiling of biosimilar mucus in a novel mucus-containing in vitro model for assessment of intestinal drug absorption. *European Journal of Pharmaceutics and Biopharmaceutics* 87 (2): 227–235.

49 Boegh, M., García-Díaz, M., Müllertz, A. et al. (2015). Steric and interactive barrier properties of intestinal mucus elucidated by particle diffusion and peptide permeation. *European Journal of Pharmaceutics and Biopharmaceutics* 95: 136–143.

50 Gestin, M., Dowaidar, M., and Langel, Ü. (2017). Uptake mechanisms of cell-penetrating peptides. In: *Peptides and Peptide-based Biomaterials and their Biomedical Applications. Advances in Experimental Medicine and Biology* (ed. Ü. Langel), 255–264. Cham: Springer.

51 Kamei, N., Onuki, Y., Takayama, K. et al. (2013). Mechanistic study of the uptake/permeation of cell-penetrating peptides across a caco-2 monolayer and their stimulatory effect on epithelial insulin transport. *Journal of Pharmaceutical Sciebnces* 102 (11): 3998–4008.

52 Kristensen, M., Kucharz, K., Fernandes, E.F.A. et al. (2020). Conjugation of therapeutic psd-95 inhibitors to the cell-penetrating peptide Tat affects blood–brain barrier adherence, uptake, and permeation. *Pharmaceutics* 12 (7): 1–24.

53 Danielsen, E.M. and Hansen, G.H. (2018). Impact of cell-penetrating peptides (CPPs) melittin and Hiv-1 Tat on the enterocyte brush border using a mucosal explant system. *Biochimica et Biophysica Acta – Biomembranes* 1860 (8): 1589–1599.

54 Panou, D.A., Diedrichsen, R.G., Kristensen, M. et al. Cell-penetrating peptides as carriers for transepithelial drug delivery. In: *Cell Penetrating Peptides. Methods in Molecular Biology*, vol. *2383* (ed. Ü. Langel), 371–384. New York: Humana Press.

55 Escribano, E., Sala, X.G., Salamanca, J. et al. (2012). Single-pass intestinal perfusion to establish the intestinal permeability of model drugs in mouse. *International Journal of Pharmaceutics* 436: 472–477.

56 Cao, X., Gibbs, S.T., Fang, L. et al. (2006). Why is it challenging to predict intestinal drug absorption and oral bioavailability in human using rat model. *Pharmaceutical Research* 23 (8): 1675–1686.

19

"Intranasal Delivery"

Noriyasu Kamei and Mariko Takeda-Morishita

Laboratory of Drug Delivery Systems, Faculty of Pharmaceutical Sciences, Kobe Gakuin University, Kobe, Japan

19.1 Enhanced Systemic Absorption of Biologics by Coadministration with CPPs

Intranasal administration is being considered a potentially noninvasive route for the systemic absorption of biologics such as peptides and proteins, as well as drug delivery. Many drugs can easily permeate through the thin layer of the nasal epithelium; additionally, the underlying lamina propria is rich with capillary blood vessels, allowing for faster absorption [1, 2]. In addition to the advantage of easier permeation, the nasal mucosa houses fewer degradation enzymes, such as peptidases and proteases, than intestinal mucosa [3, 4]. For these reasons, several peptide-based medicines have been approved and clinically used intranasally. Some examples of these drugs include small peptides, such as desmopressin and buserelin; all are around 1000 Da in size. The bioavailability (BA) of nasally administered desmopressin has been reported [5] and although its BA is low, the drug has been shown to be effective due to its high pharmacological potency. However, larger peptides and proteins (>2000 Da) cannot efficiently permeate through the nasal epithelium [2, 4]. Therefore, macromolecular peptides and proteins such as insulin must be administered with absorption enhancers or a different drug delivery system in order to achieve sufficient systemic absorption.

Researchers have tried to enhance the nasal epithelial permeation of macromolecular drugs by opening tight junctions [6]; however, the enhancement of paracellular transport has the potential to increase the nonspecific influx of harmful molecules, including exogenous pathogens. Therefore, we can consider enhancing the transcellular transport via the basic biological functions of epithelial cells themselves as a safer alternative. The internalization into epithelial cells is the most significant limitation step for improving transcellular transport and absorption of macromolecular drugs through the nasal epithelium. As described in previous chapters, CPPs can transport various types of drugs and molecules, including peptides and proteins, into numerous cell types [7–9]. Therefore, CPPs have the potential to efficiently improve the nasal absorption of therapeutic peptides and proteins. We found that the systemic absorption of peptides and proteins, such as insulin, Exendin-4, and interferon β, could be significantly increased by intranasally coadministering with the cationic CPP, octaarginine (R8), and the amphipathic CPP, penetratin, in rats (Figure 19.1) [10, 11]. Based on the calculation of the area under the plasma concentration curves (AUCs), penetratin could increase the BA of insulin after nasal administration up to 100%. The coadministration strategy with CPPs could also be applicable to enhancing the nasal absorption of polyethylene glycol (PEG)-decorated protein (PEGylated interferon), which is approximately 60 kDa in size [12]. The formation of the complex between drugs and CPPs via electrostatic and/or hydrophobic interaction is essential to facilitate the epithelial cellular internalization of drugs [13, 14]. As the significant enhancement effect on nasal absorption can be achieved by only administering the mixture of drugs and CPPs, this strategy can be easily used without considering any modification or conjugation to the drugs or carrier molecules. Repeated nasal administration of the mixture of insulin and penetratin for one month did not change the effectiveness of

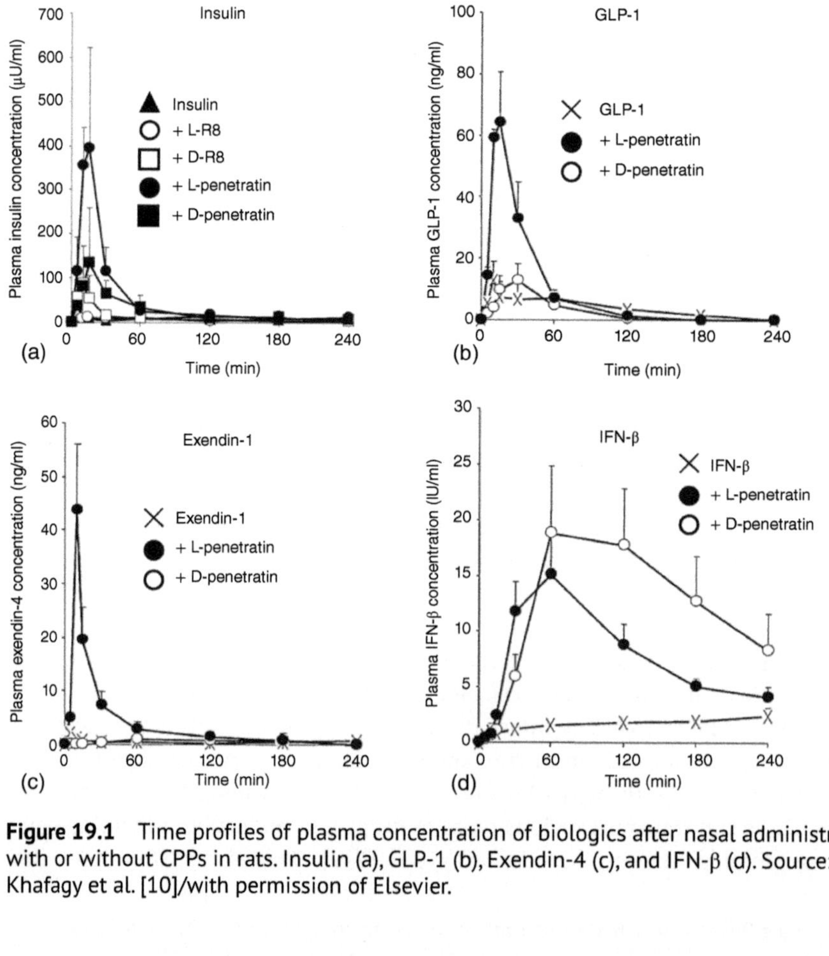

Figure 19.1 Time profiles of plasma concentration of biologics after nasal administration with or without CPPs in rats. Insulin (a), GLP-1 (b), Exendin-4 (c), and IFN-β (d). Source: Khafagy et al. [10]/with permission of Elsevier.

the absorption enhancement and did not show toxic outcomes in the nasal mucosal tissue specimens [15]. Thus, we can consider the coadministration method with CPPs as a useful and safe strategy for the efficient systemic absorption of macromolecular biologics, which can be applied to enhancing the absorption of biologics.

19.2 Enhanced Nose-to-Brain Delivery of Biologics by Coadministration with CPPs

The coadministration of CPPs has the potential to effectively deliver biologics to the brain and to improve their therapeutic effect on the central nervous system (CNS)-related disorders such as Alzheimer's disease (AD). Since not only conventional drugs but also macromolecular biologics systemically administered cannot permeate through the blood–brain barrier (BBB), many researchers have attempted to enhance drug permeation through the BBB by using permeation enhancers, tight junction opening agents, brain-targeting ligands, and nanoparticulate carrier systems [16–18]. Some of these strategies can be effective for facilitating drug permeation through the BBB, but it is difficult to avoid significant dilution and undesirable exposure in the body after administration. In contrast, intranasal administration has been considered a potential route for direct drug transport from the administration site to the brain, as the nasal cavity is closely connected to the brain via nasal epithelium [19, 20]. Direct drug delivery to the brain via this route is called "nose-to-brain drug delivery" (**Figure 19.2**). Most published research reports that drugs can be

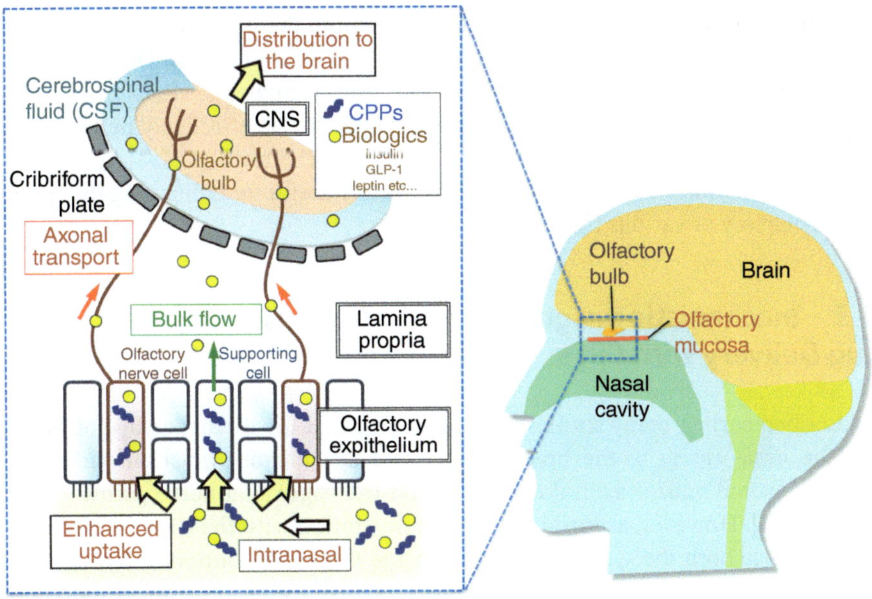

Figure 19.2 Possible mechanisms of nose-to-brain delivery of biologics and the concept of its enhancement by coadministration with CPPs. Source: The illustration was created by Noriyasu Kamei.

transported from the nasal cavity to the olfactory bulb, which is an anterior part of the brain, through the olfactory epithelium. It has been suggested that drugs move along the olfactory neurons or in the perineuronal spaces to get access to the olfactory bulb and/or cerebrospinal fluid (CSF) [19, 21–24]. Other research showed that drugs can be transported along trigeminal nerves to be distributed to the brainstem, the posterior part of the brain [22, 25, 26]. Thus, intranasal administration is likely to be the superior method for drug delivery to the brain, and some works suggested that peptides and proteins, such as insulin-like growth factor (IGF) and interferon, could be detected in the brain after intranasal administration to rats and monkeys [22, 27]. However, our quantitative data, as well as others demonstrated that macromolecular drugs cannot be effectively delivered to the brain by intranasal administration alone [28–30]. Therefore, it is needed to establish a strategy to maximize the potential of the intranasally administered drugs.

The most significant barrier to nose-to-brain delivery of biologics is the nasal epithelial layer, which is the entrance for olfactory and trigeminal nerves, and the lamina propria, which connects to the perineuronal space surrounding the axon of neurons. Therefore, we examined the efficiency of nose-to-brain delivery of biologics after intranasal coadministration with CPPs to mice and rats. We found that the hydrophobic CPP penetratin could significantly improve the transport of peptides and proteins from the nasal cavity to the brain [28]. Figure 19.3 shows the autoradiography of the brain slices isolated after intranasal administration of ^{64}Cu-NODAGA-labeled insulin with or without penetratin in rats [31]. The radioactivity in the brain, especially the olfactory bulb and the bottom part of the brain, was significantly increased after coadministration with the L-form of penetratin (L-penetratin). The increased distribution of insulin to the brain was observed and quantitatively validated by measurement via enzyme immunoassay. Other peptide drugs, such as glucagon-like peptide-1 (GLP-1) and its analogue (Exendin-4), and protein drugs, such as leptin and antibodies, could also be effectively delivered to the brain after intranasal coadministration with L-penetratin [32–34]. Thus, the noncovalent coadministration strategy with CPPs is useful for enhancing the nose-to-brain delivery of various biologics.

19.3 Mechanistic Insight for Enhanced Nose-to-Brain Drug Delivery with CPPs

As described earlier, many studies have suggested that the direct passage of drugs from the nasal cavity to the brain is mediated by olfactory mucosal pathways, including axonal and/or axonal transport, and the trigeminal nerve pathway [19, 21–23, 25, 26]. However, as intranasal coadministration with the CPP L-penetratin could enhance both the systemic absorption and brain distribution of biologics, it is possible that the increased systemic absorption of drugs via coadministration with L-penetratin can contribute to their enhanced brain delivery. We therefore examined whether the efficiency of the brain distribution of biologics could be enhanced by increasing the plasma concentration. The excessive increase in plasma

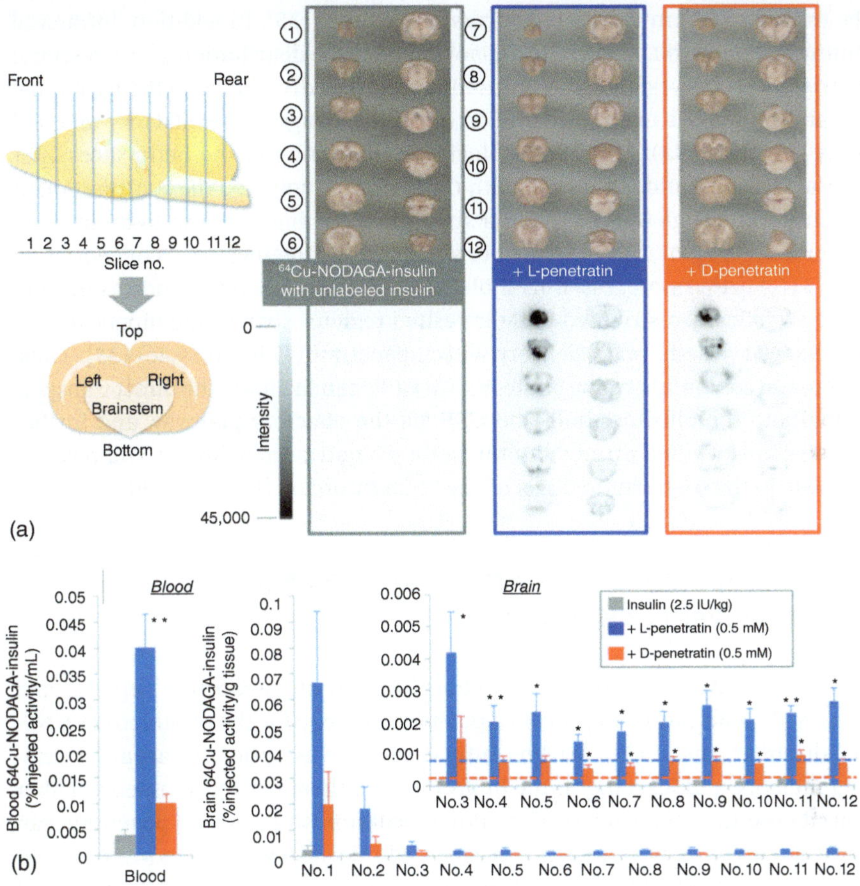

Figure 19.3 Brain distribution of radioactivity after intranasal administration of ^{64}Cu-NODAGA-insulin with or without penetratin to rats. The upper images show pictures of brain coronal slices, and the bottom images show autoradiography obtained after two hours exposure. Source: Kamei et al. [31]/with permission of American Chemical Society.

Exendin-4 concentration by subcutaneous injection to mice did not increase the brain concentration of Exendin-4, suggesting that coadministration with L-penetratin can increase the brain concentration of biologics by facilitating the direct nose-to-brain transport [35].

The experimental data further suggests that peptide drugs are mainly transported to the brain via the olfactory mucosal route, as the concentration of insulin and Exendin-4 after intranasal coadministration with L-penetratin was highest at the olfactory bulb (**Figure 19.3**). This was again emphasized when measuring the CSF concentration of drugs, as the concentration of insulin was increased only in the anterior fraction of the CSF surrounding the brain [31]. In contrast, fluorescence imaging analysis of the trigeminal nerves isolated after intranasal administration of Cy7-labeled insulin or Exendin-4 demonstrated that peptide drugs can be effectively distributed to the trigeminal nerves even without penetratin, but the drugs could not be transferred from trigeminal nerves to the brain [35]. In addition, intranasal coadministration with L-penetratin did not enhance the distribution of fluorescence to the trigeminal nerves; rather, it reduced the fraction distributed to the trigeminal nerves and shifted the major distribution of peptide drugs to the olfactory mucosal pathway (Figure 19.4a). Immunohistological examination of the brain slices after intranasal administration of insulin with L-penetratin suggested the contribution of drug transport through the perivascular spaces to the enhanced nose-to-brain delivery (Figure 19.4b) [35]. This was further supported by the increased concentration of CSF [31]. Furthermore, essential molecules for brain functions, such as insulin, could be specifically distributed to their desired regions, such as the hippocampus, after enhanced nose-to-brain delivery with L-penetratin (Figure 19.4c) [35]. Thus, the intranasal coadministration strategy with CPPs can facilitate the direct transport of biologics to the olfactory bulb and CSF via the olfactory pathway, and further drugs can be distributed throughout the brain via perivascular spaces and possibly accumulate to the therapeutic target of the brain disorders (Figure 19.4d).

19.4 Therapeutic Effects After Enhanced Nose-to-Brain Delivery of Biologics

As described above, the intranasal coadministration of therapeutic peptides and proteins with penetratin can provide a potential strategy for the treatment of CNS-related disorders. Insulin has been recently recognized as a potential agent for stimulating memory and learning in the brain [36, 37], therefore we assessed whether enhanced nose-to-brain delivery of insulin by coadministration with penetratin can improve the learning in senescence-accelerated mice (SAMP8). The Morris water maze (MWM) test used on the SAMP8 mice (16–24 weeks old) showed that repeated intranasal administration of insulin with L-penetratin could hamper the delay in escape of the SAMP8 mice to the platform (Figure 19.5a), suggesting that this strategy could be effective for preventing the progression of memory loss [38]. This may be related to the continuous stimulation of insulin signaling in the brain created by the enhanced nose-to-brain delivery of insulin. In contrast, the intranasal administration of insulin with L-penetratin did not improve the severe cognitive dysfunction developed in SAMP8 mice (24–29 weeks old) (Figure 19.5b), but rather increased the amyloid β (Aβ)-positive area and decreased the NeuN-positive neuronal density [38]. As insulin-degrading enzyme (IDE), which is a major enzyme for the clearance of Aβ from the brain, can be competed between Aβ and highly delivered insulin by intranasal administration with L-penetratin, the results suggested that insulin delivery to the brain is not always effective for the treatment of dementia such as AD.

To overcome this pitfall with insulin for AD therapy in progressive stages, we tried to deliver Exendin-4, which has the potential to stimulate the downstream of insulin signaling in the brain [39], to the brain via repeated intranasal administration. The phosphorylation of Akt in the hippocampus was facilitated after intranasal administration of a mixed solution of Exendin-4, L-penetratin, and supplemental

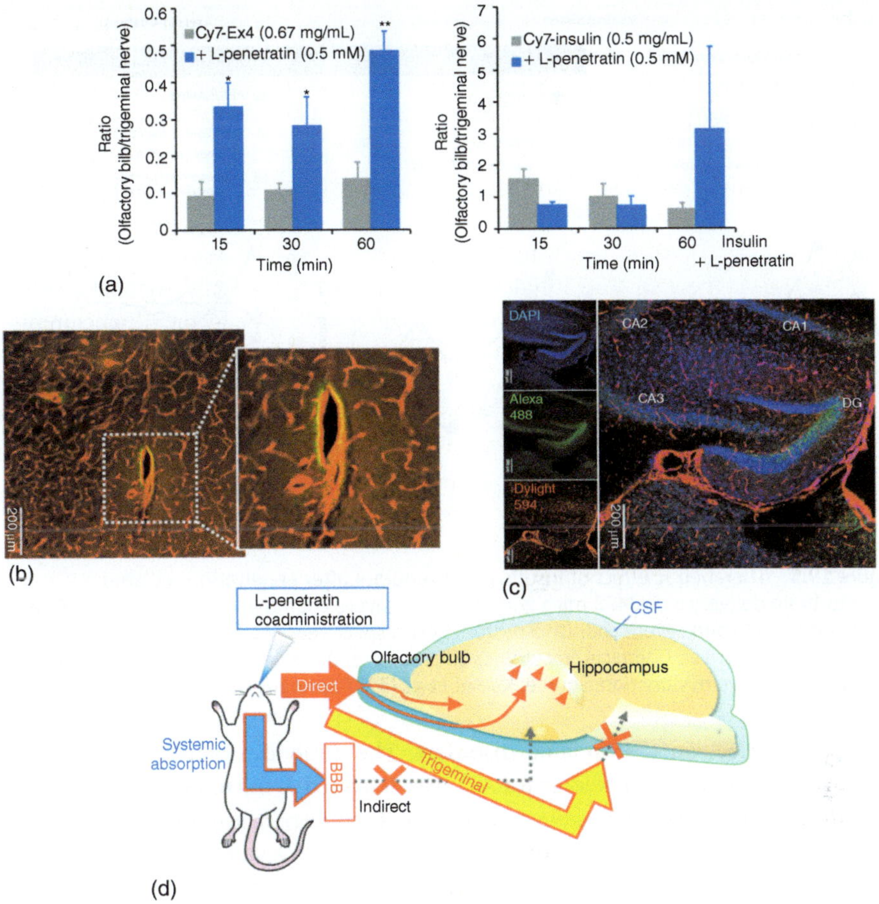

Figure 19.4 Investigation of transport pathways associated with enhanced brain delivery of peptide drugs by intranasal coadministration with penetratin. (a) Fluorescence ratios of olfactory bulbs per trigeminal nerves after intranasal administration of Cy7-Exendin-4 with or without L-penetratin to rats. (b and c) Immunohistological staining of the brain section (0.85 and −1.91 mm from bregma, respectively) after intranasal administration of insulin with L-penetratin to mice. (d) Schematic of the mechanisms associated with the transport of peptide drugs to the brain after intranasal coadministration with penetratin. Source: Kamei et al. [35]/MDPI / CC BY-4.0.

insulin in mice [32]. Furthermore, the MWM test clarified that repeated intranasal administration of Exendin-4 with L-penetratin and low doses of insulin significantly improved the severe cognitive dysfunction of SAMP8 mice in the progressive stages of dementia (Figure 19.5c) [32]. Thus, intranasal coadministration of Exendin-4 and supplemental insulin with L-penetratin can synergistically facilitate insulin signaling in the brain and contribute to the therapeutic effects on the severe cognitive dysfunction seen in the SAMP8 mice.

In addition to the peptide drugs, such as insulin and Exendin-4, the protein drugs delivered to the brain via intranasal administration with penetratin could show a

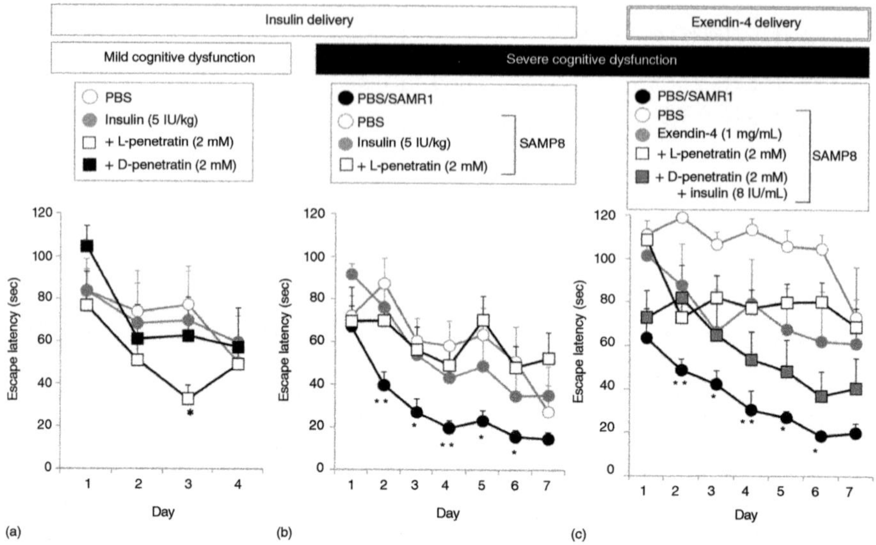

Figure 19.5 Therapeutic effect of insulin and Exendin-4 after penetratin-mediated nose-to-brain delivery in SAMP8 mice with mild and severe cognitive dysfunction. (a, b) Effect of insulin on mild and severe cognitive dysfunction, respectively. *Source:* Kamei et al. [38]/with permission of American Chemical Society. (c) Effect of Exendin-4 with supplemental insulin on severe cognitive dysfunction. Source: Kamei et al. [32] / Springer Nature / CC BY-4.0.

therapeutic effect on CNS-related disorders. The antibody against Aβ, which is known as a pathological molecule for AD, was delivered to the brain after intranasal administration with L-penetratin, which then significantly improve the cognitive dysfunction in the amyloid precursor protein (APP) knock-in mice [34]. This effect on memory was likely generated by the prevention of aggregation of Aβ. In addition, intranasal coadministration with L-penetratin facilitated the delivery of leptin, which is an appetite-regulating hormone, to the brain, and repeated delivery of leptin induced a decrease in food intake and reduced the body weight of rats [33].

19.5 Perspectives

In this chapter, we described the applicability of CPPs to enhance nasal absorption to the systemic circulation and the nose-to-brain delivery of biologics. Drugs such as peptides and large proteins modified with PEG moieties can be delivered effectively by coadministration with CPPs; penetratin has the strongest ability for enhancing both systemic absorption and brain delivery. The coadministration strategy with CPPs may have the potential to deliver nucleic acids such as siRNA and mRNA, which are currently recognized as medical modalities with high therapeutic potential for various diseases, including CNS disorders. Moreover, other researchers showed that CPPs have the capacity to transport nanoparticulate carriers, such as polymer micelles, via the intranasal route [40]. On the other hand, the sorting of biologics to blood circulation or the brain after intranasal administration is not controllable at present. To avoid untoward off-target effects in the brain or systemic body, we need to control the direction of delivery, dependent on the target site. Based on our recent works, in which larger molecules such as antibodies could be preferentially delivered to the brain [34], we may be able to consider the molecular weight of drugs as the determining factor for sorting to the blood or brain. Most importantly, noninvasive intranasal administration can be a patient-friendly route compared to invasive administration routes such as intravenous and subcutaneous injections. Thus, CPPs can be useful tools for enhancing the nasal permeation of various therapeutic modalities, thereby increasing their clinical value. The ideal medication for systemic and CNS disorders with potential biologics has yet to be established; further optimizing the pharmaceutical formulation and dosing devices for nasal administration is required.

References

1 Arora, P., Sharma, S., and Garg, S. (2002). Permeability issues in nasal drug delivery. *Drug Discovery Today* 7 (18): 967–975.

2 Ozsoy, Y., Gungor, S., and Cevher, E. (2009). Nasal delivery of high molecular weight drugs. *Molecules* 14 (9): 3754–3779.

3 Sarkar, M.A. (1992). Drug metabolism in the nasal mucosa. *Pharmaceutical Research* 9 (1): 1–9.

4 Illum, L. (2003). Nasal drug delivery – possibilities, problems and solutions. *Journal of Controlled Release* 87 (1–3): 187–198.

5 Harris, A.S., Ohlin, M., Lethagen, S. et al. (1988). Effects of concentration and volume on nasal bioavailability and biological response to desmopressin. *Journal of Pharmaceutical Science* 77 (4): 337–339.

6 Illum, L. (2012). Nasal drug delivery – recent developments and future prospects. *Journal of Controlled Release* 161 (2): 254–263.

7 Bechara, C. and Sagan, S. (2013). Cell penetrating peptides: 20 years later, where do we stand? *FEBS Letters* 587 (12): 1693–1702.

8 Copolovici, D.M., Langel, K., Eriste, E. et al. (2014). Cell-penetrating peptides: design, synthesis, and applications. *ACS Nano* 8 (3): 1972–1994.

9 Lönn, P. and Dowdy, S.F. (2015). Cationic PTD/CPP-mediated macromolecular delivery: charging into the cell. *Expert Opinion on Drug Delivery* 12 (10): 1627–1636.

10 Khafagy, E.S., Morishita, M., Kamei, N. et al. (2009). Efficiency of cell-penetrating peptides on the nasal and intestinal absorption of therapeutic peptides and proteins. *International Journal of Pharmaceutics* 381 (1): 49–55.

11 Khafagy, E.S., Morishita, M., Isowa, K. et al. (2009). Effect of cell-penetrating peptides on the nasal absorption of insulin. *Journal of Controlled Release* 133 (2): 103–108.

12 Iwase, Y., Kamei, N., Khafagy, E.S. et al. (2016). Use of a non-covalent cell-penetrating peptide strategy to enhance the nasal delivery of interferon beta and its PEGylated form. *Intenational Journal of Pharmaceutics.* 510 (1): 304–310.

13 Kamei, N., Morishita, M., and Takayama, K. (2009). Importance of intermolecular interaction on the improvement of intestinal therapeutic peptide/protein absorption using cell-penetrating peptides. *Journal of Controlled Release* 136 (3): 179–186.

14 Khafagy, E.S., Morishita, M., and Takayama, K. (2010). The role of intermolecular interactions with penetratin and its analogue on the enhancement of absorption of nasal therapeutic peptides. *International Journal of Pharmaceutics* 388 (1–2): 209–212.

15 Khafagy, E.S., Kamei, N., Nielsen, E.J. et al. (2013). One-month subchronic toxicity study of cell-penetrating peptides for insulin nasal delivery in rats. *European Journal of Pharmaceutics and Biopharmaceutics* 85 (3 Pt A): 736–743.

16 Banks, W.A. (2016). From blood-brain barrier to blood-brain interface: new opportunities for CNS drug delivery. *Nature Reviews Drug Discovery* 15 (4): 275–292.

17 Terstappen, G.C., Meyer, A.H., Bell, R.D. et al. (2021). Strategies for delivering therapeutics across the blood-brain barrier. *Nature Reviews Drug Discovery* 20 (5): 362–383.

18 Anraku, Y., Kuwahara, H., Fukusato, Y. et al. (2017). Glycaemic control boosts glucosylated nanocarrier crossing the BBB into the brain. *Nature Communications* 8 (1): 1001.

19 Lochhead, J.J. and Thorne, R.G. (2012). Intranasal delivery of biologics to the central nervous system. *Advanced Drug Delivery Reviews* 64 (7): 614–628.

20 Samaridou, E. and Alonso, M. (2018). Nose-to-brain peptide delivery – the potential of nanotechnology. *Bioorganic & Medicinal Chemistry* 26 (10): 2888–2905.

21 Dhuria, S.V., Hanson, L.R., and Frey, W.H. 2nd. (2010). Intranasal delivery to the central nervous system: mechanisms and experimental considerations. *Journal of Pharmaceutical Science* 99 (4): 1654–1673.

22 Thorne, R.G., Pronk, G.J., Padmanabhan, V. et al. (2004). Delivery of insulin-like growth factor-I to the rat brain and spinal cord along olfactory and trigeminal pathways following intranasal administration. *Neuroscience* 127 (2): 481–496.

23 Lochhead, J.J., Kellohen, K.L., Ronaldson, P.T. et al. (2019). Distribution of insulin in trigeminal nerve and brain after intranasal administration. *Scientific Reports* 9 (1): 2621.

24 Kumar, N.N., Lochhead, J.J., Pizzo, M.E. et al. (2018). Delivery of immunoglobulin G antibodies to the rat nervous system following intranasal administration: distribution, dose-response, and mechanisms of delivery. *Journal of Controlled Release* 286: 467–484.

25 Johnson, N.J., Hanson, L.R., and Frey, W.H. 2nd (2010). Trigeminal pathways deliver a low molecular weight drug from the nose to the brain and orofacial structures. *Molecular Pharmaceutics* 7 (3): 884–893.

26 Kanazawa, T., Kaneko, M., Niide, T. et al. (2017). Enhancement of nose-to-brain delivery of hydrophilic macromolecules with stearate- or polyethylene glycol-modified arginine-rich peptide. *International Journal of Pharmaceutics* 530 (1-2): 195–200.

27 Thorne, R.G., Hanson, L.R., Ross, T.M. et al. (2008). Delivery of interferon-beta to the monkey nervous system following intranasal administration. *Neuroscience* 152 (3): 785–797.

28 Kamei, N. and Takeda-Morishita, M. (2015). Brain delivery of insulin boosted by intranasal coadministration with cell-penetrating peptides. *Journal of Controlled Release* 197: 105–110.

29 Yadav, S., Gattacceca, F., Panicucci, R. et al. (2015). Comparative biodistribution and pharmacokinetic analysis of cyclosporine-A in the brain upon intranasal or intravenous administration in an oil-in-water nanoemulsion formulation. *Molecular Pharmaceutics* 12 (5): 1523–1533.

30 Nedelcovych, M.T., Gadiano, A.J., Wu, Y. et al. (2018). Pharmacokinetics of intranasal versus subcutaneous insulin in the mouse. *ACS Chemical Neuroscience* 9 (4): 809–816.

31 Kamei, N., Shingaki, T., Kanayama, Y. et al. (2016). Visualization and quantitative assessment of the brain distribution of insulin through nose-to-brain delivery based on the cell-penetrating peptide noncovalent strategy. *Molecular Pharmaceutics* 13 (3): 1004–1011.

32 Kamei, N., Okada, N., Ikeda, T. et al. (2018). Effective nose-to-brain delivery of exendin-4 via coadministration with cell-penetrating peptides for improving progressive cognitive dysfunction. *Scientific Reports* 8 (1): 17641.

33 Khafagy, E.S., Kamei, N., Fujiwara, Y. et al. (2020). Systemic and brain delivery of leptin via intranasal coadministration with cell-penetrating peptides and its therapeutic potential for obesity. *Journal of Controlled Release* 319: 397–406.

34 Kamei, N., Hashimoto, A., Tanaka, E. et al. (2022). Therapeutic effects of anti-amyloid beta antibody after intravenous injection and efficient nose-to-brain delivery in Alzheimer's disease mouse model. *Drug Delivery and Translational Research* https://doi.org/10.1007/s13346-022-01117-6.

35 Kamei, N., Suwabe, S., Arime, K. et al. (2021). Investigation of the transport pathways associated with enhanced brain delivery of peptide drugs by intranasal coadministration with penetratin. *Pharmaceutics* 13 (11): 1745.

36 Craft, S., Baker, L.D., Montine, T.J. et al. (2012). Intranasal insulin therapy for Alzheimer disease and amnestic mild cognitive impairment: a pilot clinical trial. *Archives of Neurology* 69 (1): 29–38.

37 Claxton, A., Baker, L.D., Hanson, A. et al. (2015). Long-acting intranasal insulin detemir improves cognition for adults with mild cognitive impairment or early-stage Alzheimer's disease dementia. *Journal of Alzheimer's Disease* 44 (3): 897–906.

38 Kamei, N., Tanaka, M., Choi, H. et al. (2017). Effect of an enhanced nose-to-brain delivery of insulin on mild and progressive memory loss in the senescence-accelerated mouse. *Molecular Pharmaceutics* 14 (3): 916–927.

39 Bomfim, T.R., Forny-Germano, L., Sathler, L.B. et al. An anti-diabetes agent protects the mouse brain from defective insulin signaling caused by Alzheimer's disease- associated Abeta oligomers. *Journal of Clinical Investigation* 122 (4): 1339–1353.

40 Kanazawa, T., Morisaki, K., Suzuki, S. et al. (2014). Prolongation of life in rats with malignant glioma by intranasal siRNA/drug codelivery to the brain with cell-penetrating peptide-modified micelles. *Molecular Pharmaceutics* 11 (5): 1471–1478.

20

Clinical Trials

Makoto Oba

Kyoto Prefectural University of Medicine, Graduate School of Medical Science, 1-5 Shimogamohangi-cho, Sakyo-ku, Kyoto, 606-0823, Japan

20.1 Introduction

Although novel cell-penetrating peptides (CPPs) are still under development, several CPPs for the delivery of therapeutic cargos are in clinical trials or have been clinically tested [1–8]. Unfortunately, there are no CPPs available on the market as yet. In this chapter, such CPPs that have completed or are under development in clinical trials are presented. However, there are a few notes of caution. The term "cell-penetrating peptide" is sometimes used for therapeutic peptides that have the cell-penetrating ability in addition to other functions. Only CPPs possessing a cell-penetrating ability and used for the delivery of other functional cargos are presented here. Another caution is that access to information from clinical trials is often limited, and information from them is sometimes omitted after the completion of the clinical trial. Based on the currently accessible information, several examples of CPPs in clinical trials are presented here (Table 20.1). CPPs are shown according to the delivered cargos, peptides, proteins, low-molecular-weight (LMW) drugs, nucleic acids, and fluorescent dyes.

20.2 Peptides

Peptides are the most common cargos in clinical trials using CPPs. In general, peptides are membrane-impermeable compounds, and, thus, delivery tools such as CPPs are needed to target peptides to intracellular compartments. Therapeutic peptides can be easily conjugated to CPPs based on peptide chemistry.

The first clinical trial utilizing a CPP was PsorBan initiated by CellGate. PsorBan was composed of heptaarginine (R7) as a CPP and cyclosporine A as a peptide drug [9]. Systemically administered cyclosporine A is effective against a broad range

Cell-Penetrating Peptides: Design, Development and Applications, First Edition. Makoto Oba and Yosuke Demizu.

Table 20.1 Examples of CPP-related therapeutics in clinical development.

Compound	Company	CPP	Cargo	Therapeutic application	Status	ClinicalTrials.gov. ID
PsorBan	CellGate	R7 RRRRRRR	Cyclosporine A	Psoriasis	Phase II	N/A
KAI-9803 (Delcasertib)	KAI Pharmaceuticals (Amgen)	Tat_{47-57} YGRKKRRQRRR	δ-PKC inhibitor peptide	Myocardial infarction	Phase II	NCT00093197 NCT00785954
KAI-1678	KAI Pharmaceuticals (Amgen)	Tat_{47-57} YGRKKRRQRRR	ε-PKC inhibitor peptide	Pain: postherpetic neuralgia, spinal cord injury, postoperative	Phase II	NCT01015235 NCT01135108 NCT01106716
KAI-1455	KAI Pharmaceuticals (Amgen)	Tat_{47-57} YGRKKRRQRRR	ε-PKC activator peptide	Ischemic injury	Phase I	N/A
AM-111 (Sonsuvi)	Auris Medical	D-Tat_{48-57} grkkrrqrrr	JNK binding peptide	Hearing loss	Phase III Received orphan drug designation from FDA and EMA	NCT00802425 NCT02561091 NCT02809118
XG-102 (Brimapitide)	Xigen SA	D-Tat_{48-57} grkkrrqrrr	JNK binding peptide	Inflammation, pain, cataract	Phase III	NCT01570205 NCT02235272 NCT02508337
AZX100	Capstone Therapeutics	PTD4 YARAAARQARA	Phosphorylated peptide analog of HSP20	Scar prevention/reduction	Phase II	NCT00811577 NCT00892723

(Continued)

Table 20.1 (Continued)

Compound	Company	CPP	Cargo	Therapeutic application	Status	ClinicalTrials. gov. ID
ATX-101	Therapim Pty Ltd (fully owned subsidiary of APIM Therapeutics)	R11 RRRRRRRRRRR	PCNA interacting peptide	Advanced solid tumor	Phase I/II Estimated study completion in 2025	NCT04814875
PEP-010	PEP-Therapy	N/A	Interfering peptide between caspase-9 and PP2A	Metastatic solid cancer	Phase Ia/Ib Estimated study completion in 2023	NCT04733027
RT001 **(DaxibotulinumtoxinA topical gel)**	Revance Therapeutics	Tat_{49-57} RKKRRQRRR	Botulinum toxin type A	Lateral canthal lines, crow's feet, facial wrinkles	Phase II	NCT00884234 NCT00888914 NCT00907387 NCT00968825 NCT00968942 NCT01064518 NCT01124552 NCT01124565 NCT01940991
RT002 **(DaxibotulinumtoxinA for injection: DAXI)**	Revance Therapeutics	Tat_{49-57} RKKRRQRRR	Botulinum toxin type A	Glabellar frown line, spasmodic torticollis, plantar fasciitis, cervical dystonia	Phase III	NCT02303002 NCT02706795 NCT02973269 NCT03014622 NCT03014635 NCT03608397

(Continued)

Table 20.1 (Continued)

Compound	Company	CPP	Cargo	Therapeutic application	Status	ClinicalTrials.gov. ID
ATP128 (KISIMA cancer vaccine)	AMAL Therapeutics (acquired by Boehringer-Ingelheim)	N/A	Toll-like receptor 2/4 agonist sequence, multi-antigenic sequence	Colorectal cancer	Phase Ib Estimated study completion in 2023	NCT04046445
DTS-108	Diaton SA, Drais Pharmaceuticals	DPV peptide CVKRGLKLRH VRPRVTRMDV	SN-38	Advanced/ metastatic solid tumor	Phase I	N/A
AVI-5126	Sarepta Therapeutics	$(R\text{-}Ahx\text{-}R)_4$ (Ahx: 6-aminohexanoic acid)	c-Myc antisense PMO	Cardiovascular disease, coronary artery bypass	Phase II	NCT00451256
SRP-5051	Sarepta Therapeutics	N/A	Exon 51 skipping PMO	Duchenne muscular dystrophy	Phase II Estimated study completion in 2022 Phase I/II Estimated study completion in 2024	NCT03375255 NCT04004065 NCT03675126
AVB-620 (Pegloprastide)	Avelas Biosciences	R8 rrrrrrrr	Fluorescent dyes (Cy5 and Cy7)	Breast cancer	Phase II	NCT02391194 NCT03113825

of inflammatory skin diseases but it can cause severe side effects. On the other hand, topical application of cyclosporine A to the skin can suppress systemic side effects but is not effective enough because of low absorption through the skin. To solve these problems, a CPP was used in PsorBan. This CPP-conjugated peptide drug PsorBan commenced a phase II clinical trial in 2003 for the topical treatment of psoriasis by transdermal delivery of cyclosporine A. R7 was coupled with cyclosporine A via a linker, which was stable at acidic pH but labile at physiological pH, in order to release cyclosporine A as an active drug within the tissues. In vitro experiments using a human T-cell line revealed that the releasable R7 conjugate of cyclosporine A (PsorBan) was more biologically active than cyclosporine A alone, and the non-releasable R7 conjugate showed no activity. Furthermore, R7 conjugation facilitated the penetration of cyclosporine A into a mouse and human skin, delivering it to dermal T lymphocytes, and inhibited cutaneous inflammation in vivo. However, despite the efficient uptake of PsorBan, the release of free cyclosporine A was not rapid enough to compete with clearance in a phase IIb clinical trial, so the subsequent clinical trial was interrupted [10].

KAI Pharmaceuticals (acquired by Amgen in 2012) developed a class of therapeutics targeting protein kinase C (PKC), KAI-9803, KAI-1678, and KAI-1455 utilizing transactivator of transcription (Tat) peptide as a CPP. KAI-9803 (delcasertib) is a 23-amino acid peptide drug composed of a potent and selective δ-PKC inhibitor peptide as a therapeutic peptide and Tat_{47-57} as a CPP. Tat_{47-57} was conjugated to the δ-PKC inhibitor peptide via a disulfide bond between cysteine residues. In preclinical experiments, KAI-9803 ameliorated injury associated with ischemia and reperfusion in animal models of acute myocardial infarction [11, 12]. Thus, KAI-9803 moved into a first-in-human dose-escalation study (NCT00093197) that demonstrated signs of potential drug activity using a biomarker [13]. However, a phase II clinical trial (NCT00785954) revealed selective inhibition of δ-PKC with an intravenous fusion of KAI-9803 during primary percutaneous coronary intervention for acute ST-elevation myocardial infarction in a population of patients treated according to the contemporary standard of care did not reduce a biomarker of myocardial injury [14]. In KAI-1678, which is a 21-amino acid peptide, an ε-PKC inhibitor peptide was used instead of the δ-PKC inhibitor peptide of KAI-9803. ε-PKC has been identified in multiple neuronal systems related to pain processing. KAI-1678 competitively and selectively inhibited the interaction of ε-PKC and its receptor for activated C kinase, resulting in a reduction in pain. Phase II clinical trials of KAI-1678 were carried out for the treatment of pain associated with postherpetic neuralgia, spinal cord injury, and postoperative pain relief (NCT01015235, NCT01106716, NCT01135108). However, KAI Pharmaceuticals concluded that KAI-1678 was not efficient as an acute analgesic for chronic neuropathic pain [15]. KAI-1455 was a selective ε-PKC activator designed to reduce ischemic organ injury. In preclinical experiments, KAI-1455 showed a drastic reduction in infarct size when delivered prior to the ischemic insult. A phase I clinical trial of KAI-1455 was initiated for ischemic injury in 2007; however, no study results were posted on ClinicalTrials.gov or are available yet [16].

AM-111 (sonsuvi), which is a D-stereoisomer of c-Jun N-terminal kinase inhibitor 1 (D-JNK-1) gel for intratympanic injection, was developed by Auris Medical.

D-JNKI-1 is a cell-permeable 31-amino acid peptide, in which D-stereoisomer of Tat_{48-57} (D-Tat_{48-57}) is coupled with a 20-amino acid JNK-binding peptide, and acts as an inhibitor of JNK-mediated apoptosis [17]. Note that the peptide stability of the D-stereoisomer against enzymatic degradation was much higher than that of the L-stereoisomer. AM-111 is formulated in a biocompatible and fully biodegradable hyaluronic acid gel and administered in a single dose intratympanic injection into the middle ear after acute hearing loss [18]. Phase III clinical trials on the efficacy and safety of AM-111 revealed that it provided effective otoprotection in cases of profound idiopathic sudden sensorineural hearing loss, and activation of the JNK stress kinase seemed to set in only following pronounced acute cochlear injury associated with large hearing threshold shifts (NCT02561091, NCT02809118) [19]. AM-111 received orphan drug designation from both Food and Drug Administration (FDA) and European Medicines Agency (EMA) for the treatment of acute sensorineural hearing loss. However, further clinical testing is required.

XG-102 (brimapitide), developed by Xigen SA, is also D-JNK-1, a selective inhibitor of JNK activity in a non-ATP competitive manner [20, 21]. In preclinical experiments against the rat intraocular model of inflammation, XG-102 significantly reduced the ocular inflammation score, inflammatory cell filtration, and inducible oxide synthase [22]. A phase Ib clinical trial revealed that XG-102 administered as a single subconjunctival injection as adjunct therapy was safe and well-tolerated (NCT01570205) [23]. Phase III clinical trials for the efficacy and safety of XG-102 in the reduction of post-cataract surgery intraocular inflammation and pain were carried out after subconjunctival injection of XG-102 (NCT02235272, NCT02508337). Furthermore, in another phase II clinical trial, a single subconjunctival injection of XG-102 at the end of ocular surgery was noninferior to dexamethasone eye drops in the treatment of postoperative ocular inflammation [24]. However, we cannot obtain further information on XG-102 currently.

Capstone Therapeutics developed AZX100, in which a CPP sequence (PTD4: YARAAARQARA) was linked to a phosphorylated region from heat shock-related protein HSP20 (WLRRAsAPLPGLK; where s denotes phosphorylated serine) [25]. The CPP sequence was optimized by alanine substitution of Tat_{47-57} (YGRKKRRQRRR). The phosphorylated peptide competes for binding with phosphor-cofilin to 14-3-3 protein, and it induces the subsequent dephosphorylation of cofilin, resulting in actin depolymerization. AZX100 decreased the expression of connective tissue growth factor and type I collagen induced by transforming growth factor-β1, endothelin, and lysophosphatidic acid. AZX100 revealed its potential to prevent excessive scarring and fibrotic disorders [26]. In preclinical experiments, a single injection of AZX100 immediately following wound closure significantly improved both collagen fiber density and collagen fiber maturity [27]. Capstone Therapeutics started phase II clinical trials to evaluate the safety and efficacy of AZX100 in preventing or reducing scars from trocars and re-growth of surgically removed keloid scars (NCT00811577, NCT00892723). Human clinical data was positive; however, the intellectual property license was returned to Arizona State University, where AZX100 was first developed.

ATX-101 (MDRWLVKWKKKRKIRRRRRRRRRRR) is a 25-amino acid peptide composed of three parts; a AlkB homologue 2 proliferating cell nuclear antigen (PCNA)-interacting motif (APIM; RWLVK), a SV40 nuclear localization signal (KKKRK), and the CPP undeca-arginine (R11) [28]. APIM peptide inhibits the interaction between PCNA and APIM-containing proteins, including DNA repair proteins, and thus, ATX-101 potentiates the action of several anti-cancer drugs such as DNA damaging drugs, microtubulin targeting drugs, molecular targeting agents, and γ-irradiation. Preclinical experiments revealed that ATX-101 rapidly penetrated cells and targeted PCNA/APIM-containing protein complexes, and had single agent activity in certain cancers and combinatorial activity with more than 25 different anti-cancer drugs in different indications [29–31]. Therapim Pty Ltd (fully owned subsidiary of APIM Therapeutics) initiated a first-in-human study of ATX-101 for its safety profile and disease stabilization in patients with late-stage, heavily pretreated, solid tumors and reported positive results in 2021 [32]. Accordingly, a phase Ib/IIa clinical trial to evaluate the combination of ATX-101 and platinum-based chemotherapy is being initiated and will be completed in 2025 (NCT04814875).

PEP-010, developed by PEP-Therapy, comprises two domains in a single peptide, a CPP to deliver the active peptide into the cellular cytosol and an interfering peptide to disrupt the interaction between caspase-9 and protein phosphatase 2A (PP2A). The peptide sequences of the CPP and interfering peptide have not been disclosed so far. In multiple preclinical experiments, PEP-010 demonstrated a good safety profile and anti-cancer efficacy and was shown to inhibit growth in breast and ovary cancer models. Furthermore, the combinatorial activity of PEP-010 with chemotherapies was observed. Accordingly, a first-in-human phase I clinical trial to evaluate PEP-010 as a single agent and in combination with paclitaxel in patients with metastatic solid cancers is being investigated and is estimated for completion in 2023 (NCT04733027).

20.3 Proteins

Protein pharmaceuticals have gained a large market share among drugs during the past two decades. Almost all the targets of current protein pharmaceuticals are extracellular components because of the low cell membrane permeability of proteins. However, there are a lot of potential drug targets in the intracellular milieu. Therefore, therapeutic protein delivery tools into the cells are in high demand, and CPPs have great potential as tools.

Revance Therapeutics developed RT001 (daxibotulinumtoxinA topical gel) to treat patients with lateral canthal lines, crow's feet, and facial wrinkles. RT001 is a non-injectable topical gel botulinum toxin type A drug product candidate, which is composed of a TransMTS peptide and a 150 kDa fragment of botulinum toxin type A [33, 34]. TransMTS peptide has a positively charged lysine-rich central peptidic domain located between two Tat_{49-57} domains. The positively charged domain interacts with the anionic portion of botulinum toxin type A via electrostatic interaction and the Tat_{49-57} domains worked as a CPP. The result of phase II clinical trial provided strong clinical evidence establishing the ability of RT001 to safely deliver botulinum toxin type A (NCT00884234, NCT00888914, NCT00907387, NCT00968825, NCT00968942, NCT01064518, NCT01124552, NCT01124565, NCT01940991). However, in phase III clinical trial, RT001 did not achieve its co-primary and other endpoints. Based on these results, Revance Therapeutics decided not to pursue the clinical development plan for RT001 in 2016. On the other hand, a similar but different drug candidate RT002 (daxibotulinumtoxinA for injection: DAXI) has been also developed by Revance Therapeutics [35–37]. RT002 is an injectable botulinum toxin type A drug product candidate utilizing a TransMTS peptide and limits the spread of botulinum toxin type A, which contributes to safe administration of larger doses and extension of its duration of the function. There have been several clinical trials (NCT02303002, NCT02706795, NCT02973269) for glabellar frown lines, spasmodic torticollis, plantar fasciitis, and cervical dystonia. One of phase III clinical trials (NCT03014622, NCT03014635, NCT03608397) was completed for the treatment of glabellar frown lines, and the FDA accepted for review the Biologics License Application in 2020.

AMAL Therapeutics (acquired by Boehringer-Ingelheim in 2019) developed ATP128 (KISIMA cancer vaccine), a self-adjuvanted chimeric recombinant protein vaccine. Three functional components were assembled into one fusion protein: (i) a CPP for antigen delivery; (ii) a Toll-like receptor 2/4 agonist domain as an adjuvant; and (iii) a multi-antigenic cargo that is tailored to raise an immune response against colorectal cancers. Currently, the sequence of ATP128 is not available. Preclinical experiments indicated that ATP128 had great potential to be a safe and potent immunopreventive intervention for individuals at high risk of developing colorectal cancer [38]. ATP128 is currently being investigated as a monotherapy and in combination with ezabenlimab, which is a humanized programmed cell death protein-1 targeting monoclonal antibody, in patients with stage IV colorectal cancer. The phase Ib clinical trial is planned for completion in 2023 (NCT04046445).

20.4 Low-Molecular-Weight (LMW) Drugs

To the best of my knowledge, one clinical trial has been performed for a chemotherapeutic LMW drug with a CPP. DTS-108, which comprises the SN-38 moiety covalently linked to a DPV1047 peptide using a specific esterase-sensitive cross-linker, is a CPP-LMW drug conjugate. This technology (Vectocell Technology) was also demonstrated in a CPP-doxorubicin drug conjugate to enhance both the in vitro and in vivo efficacy of doxorubicin [39]. SN-38 is the active metabolite of the widely used cancer drug irinotecan and acts as a topoisomerase I inhibitor. Although SN-38 has an activity a thousand times higher than irinotecan, the poor aqueous solubility of SN-38 has hampered its clinical development. The DPV1047 peptide is a highly positive charged 20-amino acid CPP (CVKRGLKLRHVRPRVTRMDV) [40]. The conjugation of SN-38 to a DPV1047 rendered it soluble for intravenous injection in purely aqueous vehicles at high dose levels. The specific esterase-sensitive cross-linker allowed the release of SN-38 from DTS-108 through esterase bond cleavage. In preclinical experiments, DTS-108 delivered significantly higher levels of SN-38 than irinotecan with negligible toxicity, and it showed a significant dose-dependent anti-cancer efficacy [41]. Accordingly, Diatos SA and Drais Pharmaceuticals initiated a first-in-human phase I clinical study on the pharmacokinetics and safety of DTS-108 in patients with advanced/metastatic carcinomas. The phase I clinical study concluded that based on the positive results of its study the recommended dose in phase II clinical trial for intravenous DTS-108 was 313 mg/m^2 every two weeks in patients with advanced/metastatic solid tumor [42]. However, no novel information has been updated from then.

20.5 Nucleic Acids

There have been many preclinical studies on the delivery of nucleic acids and genes utilizing CPPs. DNA and RNA have negative charges, so they can form complexes with positively charged CPPs by electrostatic interaction. Furthermore, CPP-DNA and CPP-RNA conjugates have also been developed. Such ion complexes and conjugates can be used for the intracellular delivery tools of plasmid DNA, small interfering RNAs, mRNAs, antisense oligonucleotides, and so on. However, therapeutic DNA and RNA complexed or conjugated with CPPs have not yet entered clinical trials.

Phosphorodiamidate morpholino oligomers (PMOs), which are uncharged analogues of nucleic acids with an unnatural backbone, are becoming one of the promising targets for using CPPs. In 2016, the innovative PMO-drug Exondys 51 (eteplirsen) was approved by the FDA for use in patients with Duchenne muscular dystrophy [43]. Exondys 51 uses exon-skipping technology to allow the body to make a shorter form of the dystrophin protein by skipping over a specific exon in the dystrophin gene. Thus, PMOs have great potential for the treatment of intractable diseases, but they have a problem of low cell-membrane permeability. CPPs may be able to solve this problem.

The first clinical trial of a CPP-PMO conjugate was AVI 5126 planned by Sarepta Therapeutics, which was a company involved in the development of Exondys 51. AVI-5126 is composed of an Arg-rich CPP $(R\text{-}Ahx\text{-}R)_4$ (Ahx: 6-aminohexanoic acid) and PMO to inhibit the c-Myc gene [44]. Proto-oncogene c-myc is a key factor in transplant rejection. Preclinical experiments, in which AVI-5126 was evaluated in a rat cornea transplant model, demonstrated that graft rejection was significantly delayed after pre-transplantation storage of graft corneas in Optisol containing AVI-5126 followed by topical application post-transplantation [45]. Thus, a phase II clinical trial was initiated in patients undergoing coronary artery bypass to evaluate whether immersion of the excised saphenous vein in an AVI-5126 solution before transplantation in heart bypass graft surgery was able to prevent subsequent graft failure at 1 year (NCT00451256). Unfortunately, this clinical trial was discontinued due to insufficient efficacy in 2009 [3].

Sarepta Therapeutics also initiated clinical trials of a CPP-PMO conjugate, SRP-5051, in which a CPP is covalently linked to Exondys 51 [46]. SRP-5051 is a next-generation Exondys 51 and uses a next-generation technology for enhanced PMO tissue penetration leading to greater exon skipping and dystrophin production.

The CPP sequence in SRP-5051 is currently not disclosed. Preclinical experiments in a mouse model revealed an increase in muscle uptake of PMO, prolonged pharmacodynamics effects of SRP-5051, no adverse safety findings, and higher exon skipping with a steep dose-response. A phase I clinical trial was carried out to evaluate the safety, tolerability, and pharmacokinetics of SRP-5051 in patients with Duchenne muscular dystrophy, and the trial was completed successfully in 2019 (NCT03375255). Recently, SRP-5051 has entered phase I/II clinical trials, which are estimated to reach completion in 2022 and 2024 (NCT03675126, NCT04004065).

20.6 Fluorescent Dye

Avelas Biosciences has developed AVB-620 (pegloprastide) designed to deliver a fluorescent marker specifically into cancer cells to enable surgeons to identify cancer tissue in real-time. AVB-620 has the peptide backbone and fluorescent moiety [47]. Two fluorescent dyes, Cy5 and Cy7, are linked to the C-terminus and N-terminus of the peptide backbone, respectively, and the efficient Förster resonance energy transfer (FRET) is observed between Cy5 and Cy7 in intact AVB-620. The peptide backbone is composed of D-pentaglutamic acid (e5), PLGC(Me)AG sequence, and D-octaarginine (r8) from N-terminus to C-terminus. Cationic r8 acts as a CPP, but anionic e5 masks the cationic nature of r8, resulting in inhibition of the non-specific interaction of AVB-620 with cells. PLGC(Me)AG sequence is a substance for matrix metalloproteases (MMPs), which are overexpressed in cancer tissue. Intravenous injected AVB-620 can circulate in the blood stream in an intact form with FRET, and then, the PLGC(Me)AG sequence is cleaved by MMPs after reaching the cancer tissue. Released r8-Cy5 from AVB-620 can enter cancer cells selectively with disrupted FRET. We can detect Cy5 fluorescence signals in cancer tissue only. AVB-620 has completed a phase II clinical trial (NCT03113825) for fluorescence detection of cancer during breast cancer surgery as well as phase Ib clinical trial (NCT02391194) in women undergoing breast cancer surgery [48]. These clinical trials revealed that AVB-620 has the potential to significantly improve cancer detection in real-time during breast cancer surgery.

References

1 Vasconselos, L., Pärn, K., and Langel, Ü. (2013). Therapeutic potential of cell-penetrating peptides. *Therapeutic Delivery* 4 (5): 573–591.

2 Papadopoulou, L.C. and Tsiftsoglou, A.S. (2013). The potential role of cell penetrating peptides in the intracellular delivery of proteins for therapy of erythroid related disorders. *Pharmaceuticals* 6: 32–53.

3 Guidotti, G., Brambilla, L., and Rossi1, D. (2017). Cell-penetrating peptides: from basic research to clinics. *Trends in Pharmacological Sciences* 38 (4): 406–424.

4 Tsoumpra, M.K., Fukumoto, S., Matsumoto, T. et al. (2019). Peptide-conjugate antisense based splice-correction for Duchenne muscular dystrophy and other neuromuscular diseases. *EBioMedicine* 45: 630–645.

5 Young, C.C., Vedadghavami, A., and Bajpayee, A.G. (2020). Bioelectricity for drug delivery: the promise of cationic therapeutics. *Bioelectricity* 2 (2): 68–82.

6 Xie, J., Bi, Y., Zhang, H. et al. (2020). Cell-penetrating peptides in diagnosis and treatment of human diseases: from preclinical research to clinical application. *Frontiers in Pharmacology* 11: 697.

7 Matijass, M. and Neundorf, I. (2021). Cell-penetrating peptides as part of therapeutics used in cancer research. *Medicine in Drug Discovery* 10: 100092.

8 Stiltner, J., McCandless, K., and Zahid, M. (2021). Cell-penetrating peptides: application in tumor diagnosis and therapeutics. *Pharmaceutics* 13: 890.

9 Rothbard, J.B., Garlington, S., Lin, Q. et al. (2000). Conjugation of arginine oligomers to cyclosporin A facilitates topical delivery and inhibition of inflammation. *Nature Medicine* 6 (11): 1253–1257.

10 Vivès, E., Schmidt, J., and Pèlegrin, A. (2008). Cell-penetrating and cell-targeting peptides in drug delivery. *Biochimica et Biophysica Acta* 1786: 126–138.

11 Inagaki, K., Chen, L., Ikeno, F. et al. (2003). Inhibition of δ-protein kinase C protects against reperfusion injury of the ischemic heart in vivo. *Circulation* 108: 2304–2307.

12 Miyaji, Y., Walter, S., Chen, L. et al. (2011). Distribution of KAI-9803, a novel δ-protein kinase C inhibitor, after intravenous administration to rats. *Drug Metabolism and Disposition* 39 (10): 1946–1953.

13 Bates, E., Bode, C., Costa, M. et al. (2008). Intracoronary KAI-9803 as an adjunct to primary percutaneous coronary intervention for acute ST-segment elevation myocardial infarction. *Circulation* 117 (7): 886–896.

14 Lincoff, A.M., Roe, M., Aylward, P. et al. (2014). Inhibition of delta-protein kinase C by delcasertib as an adjunct to primary percutaneous coronary intervention for acute anterior ST-segment elevation myocardial infarction: results of the PROTECTION AMI Randomized Controlled Trial. *European Heart Journal* 35: 2516–2523.

15 Cousins, M.J., Pickthorn, K., Huang, S. et al. (2013). The safety and efficacy of KAI-1678 – an inhibitor of epsilon protein kinase C (εPKC) – versus lidocaine and placebo for the treatment of postherpetic neuralgia: a crossover study design. *Pain Medicine* 14: 533–540.

16 Rizzuti, M., Nizzardo, M., Zanetta, C. et al. (2015). Therapeutic applications of the cell-penetrating HIV-1 Tat peptide. *Drug Discovery Today* 20 (1): 76–85.

17 Wang, J., De Water, T.R., Bonny, C. et al. (2003). A peptide inhibitor of c-Jun N-terminal kinase protects against both aminoglycoside and acoustic trauma-induced auditory hair cell death and hearing loss. *The Journal of Neuroscience* 23 (24): 8596–8607.

18 Suckfuell, M., Lisowska, G., Domka, W. et al. (2014). Efficacy and safety of AM-111 in the treatment of acute sensorineural hearing loss: a double-blind, randomized, placebo-controlled phase II study. *Otology & Neurotology* 35 (8): 1317–1326.

19 Staecker, H., Jokovic, G., Karpishchenko, S. et al. (2019). Efficacy and safety of AM-111 in the treatment of acute unilateral sudden deafness – a double-blind, randomized, placebo-controlled phase 3 study. *Otology & Neurology* 40 (5): 584–594.

20 Michel-Monigadon, D., Bonny, C., and Hirt, L. (2010). c-Jun N-terminal kinase pathway inhibition in intracerebral hemorrhage. *Cerebrovascular Diseases* 29: 564–570.

21 Reinecke, K., Eminel, S., Dierck, F. et al. (2012). The JNK inhibitor XG-102 protects against TNBS-induced colitis. *PLoS ONE* 7 (3): e30985.

22 Toucbard, E., Omri, S., Naud, M.C. et al. (2010). A peptide inhibitor of c-Jun N-terminal kinase for the treatment of endotoxin-induced uveitis. *Investigative Ophthalmology & Visual Science* 51 (9): 4683–4693.

23 Chiquet, C., Aptel, F., Creuzot-Garcher, C. et al. (2016). Postoperative ocular inflammation: a single subconjunctival injection of XG-102 compared to dexamethasone drops in a randomized trial. *American Journal of Ophthalmology* 174: 76–84.

24 Beydoun, T., Deloche, C., Perino, J. et al. (2015). Subconjunctival injection of XG-102, a JNK inhibitor peptide, in patients with intraocular inflammation: a safety and tolerability study. *Journal of Ocular Pharmacology and Therapeutics* 31 (2): 93–99.

25 Komalavilas, P., Penn, R.B., Flynn, C.R. et al. (2008). The small heat shock-related protein, HSP20, is a cAMP dependent protein kinase substrate that is involved in airway smooth muscle relaxation. *American Journal of Physiology Lung Cellular and Molecular Physiology* 294 (1): L69–L78.

26 Flynn, C.R., Cheung-Flynn, J., Smoke, C.C. et al. (2010). Internalization and intracellular trafficking of a PTD-conjugated anti-fibrotic peptide, AZX100, in human dermal keloid fibroblasts. *Journal of Pharmaceutical Sciences* 99 (7): 3100–3121.

27 Lopes, L.B., Furnish, E.J., Komalavilas, P. et al. (2009). Cell permeant peptide analogues of the small heat shock protein, HSP20, reduce TGF-β1-induced CTGF expression in keloid fibroblasts. *Journal of Investigative Dermatology* 129 (3): 590–598.

28 Müller, R., Misund, K., Holien, T. et al. (2013). Targeting proliferating cell nuclear antigen and its protein interactions induces apoptosis in multiple myeloma cells. *PLoS ONE* 8 (7): e70430.

29 Søgaard, C.K., Moestue, S.A., Rye, M.B. et al. (2018). APIM-peptide targeting PCNA improves the efficacy of docetaxel treatment in the TRAMP mouse model of prostate cancer. *Oncotarget* 9 (14): 11752–11766.

30 Søgaard, C.K., Blindheim, A., Røst, L.M. et al. (2018). "Two hits - one stone"; increased efficacy of cisplatin-based therapies by targeting PCNA's role in both DNA repair and cellular signaling. *Oncotarget* 9 (65): 32448–32465.

31 Søgaard, C.K., Nepal, A., Petrovic, V. et al. (2019). Targeting the non-canonical roles of PCNA modifies and increases the response to targeted anti-cancer therapy. *Oncotarget* 10 (68): 7185–7197.

32 Lemech, C.R., Kichenadasse, G., Marschner, J.P. et al. (2021). Safety profile and disease stabilization in late stage, heavily pretreated, solid tumor patients in a first-in-human (FIH) study of ATX-101, a drug targeting proliferating cell nuclear antigen (PCNA). *Journal of Clinical Oncology* 39 (suppl 15): abstr 3067.

33 Fredric, B., Christopher, O., Alex, C. et al. (2010). Efficacy and safety evaluation of a novel botulinum toxin topical gel for the treatment of moderate to severe lateral canthal lines. *Dermatologic Surgery* 36: 2111–2118.

34 Glogau, R., Blitzer, A., Brandt, F. et al. (2012). Study to evaluate the efficacy and safety of a botulinum toxin type a topical gel for the treatment of moderate-to-severe lateral canthal lines. *Journal of Drugs in Dermatology* 11 (1): 38–45.

35 Carruthers, J., Fagien, S., Joseph, J.H. et al. (2020). DaxibotulinumtoxinA for injection for the treatment of glabellar lines: results from each of two multicenter, randomized, double-blind, placebo-controlled, phase 3 studies (SAKURA 1 and SAKURA 2). *Plastic and Reconstructive Surgery* 145 (1): 45–58.

36 Bertucci, V., Solish, N., Kaufman-Janette, J. et al. (2020). DaxibotulinumtoxinA for injection has a prolonged duration of response in the treatment of glabellar lines: pooled data from two multicenter, randomized, double-blind, placebo-controlled, phase 3 studies (SAKURA 1 and SAKURA 2). *Journal of the American Academy of Dermatology* 82 (4): 838–845.

37 Fabi, S.G., Cohen, J.L., Green, L.J. et al. (2021). DaxibotulinumtoxinA for injection for the treatment of glabellar lines: efficacy results from SAKURA 3, a large, open-label, phase 3 safety study. *Dermatologic Surgery* 47 (1): 48–54.

38 Belnoue, E., Leystra, A.A., Carboni, S. et al. (2021). Novel protein-based vaccine against self-antigen reduces the formation of sporadic colon adenomas in mice. *Cancers* 13: 845.

39 Meyer-Losic, F., Quinonero, J., Dubois, V. et al. (2006). Improved therapeutic efficacy of doxorubicin through conjugation with a novel peptide drug delivery technology (Vectocell). *Journal of Medicinal Chemistry* 49: 6908–6916.

40 De Coupade, C., Fittipaldi, A., Chagnas, V. et al. (2005). Novel human-derived cell-penetrating peptides for specific subcellular delivery of therapeutic biomolecules. *Biochemical Journal* 390: 407–418.

41 Meyer-Losic, F., Nicolazzi, C., Quinonero, J. et al. (2008). DTS-108, a novel peptidic prodrug of SN38: in vivo efficacy and toxicokinetic studies. *Clinical Cancer Research* 14 (7): 2145–2153.

42 Coriat, R., Faivre, S.J., Dreyer, C. et al. (2016). Pharmacokinetics and safety of DTS-108, a human oligopeptide bound to SN-38 with an esterase-sensitive cross-linker in patients with advanced malignancies: a phase I study. *International Journal of Nanomedicine* 11: 6207–6216.

43 Mendell, J.R., Goemans, N., Lowes, L.P. et al. (2016). Longitudinal effect of eteplirsen versus historical control on ambulation in Duchenne muscular dystrophy. *Annals of Neurology* 79: 257–271.

44 Lebleu, D.A., Moulton, H.M., Abes, R. et al. (2008). Cell penetrating peptide conjugates of steric block oligonucleotides. *Advanced Drug Delivery Review* 60: 517–529.

45 Hosseini, A., Lattanzio, F.A. Jr., Samudre, S.S. et al. (2012). Efficacy of a phosphorodiamidate morpholino oligomer antisense compound in the inhibition of corneal transplant rejection in a rat cornea transplant model. *Journal of Ocular Pharmacology and Therapeutics* 28 (2): 194–201.

46 Fortunato, F., Rossi, R., and Ferlini, A. (2021). Innovative therapeutic approaches for Duchenne muscular dystrophy. *Journal of Clinical Medicine* 10: 820.

47 Miampamba, M., Liu, J., Harootunian, A. et al. (2017). Sensitive in vivo visualization of breast cancer using ratiometric protease-activatable fluorescent imaging agent, AVB-620. *Theranostics* 7 (13): 3369–3386.

48 Unkart, J.T., Chen, S.L., Wapnir, I.L. et al. (2017). Intraoperative tumor detection using a ratiometric activatable fluorescent peptide: a first-in-human phase 1 study. *Annuals of Surgical Oncology* 24: 3167–3173.

21

Plant

Kousuke Tsuchiya and Keiji Numata

Department of Material Chemistry, Graduate School of Engineering, Kyoto University, Kyoto, 615-8510, Japan

21.1 Introduction

Cell-penetrating peptides (CPPs) have been developed and engineered to be applied to biomedical tools to deliver impermeable molecules into living cells. Rational designs of CPPs for efficient penetration into dynamic lipid membranes enable the internalization of various cargos accommodated with a specific CPP into cells. Extensive efforts to elucidate the mechanism for internalization of CPP have expanded the scope of available situations for CPP-mediated material delivery. Besides mammalian cell systems, plant cells are also attractive targets of material delivery in agronomic fields in order not only to produce useful metabolites but also to enhance crop yields. To date, there are very few innovative approaches to utilize CPP for plant cells. Several research groups have attempted to apply CPP for material delivery into plants. Gene modification of plants via delivery of an exogenous gene is a massive method to provide the plants with novel useful traits. Traditionally, such genetic transformation has been performed by agrobacterium-mediated or biolistic introduction of exogenous genes into plant cells. The agrobacterium-mediated delivery of genes relies on the infectious ability of *Agrobacterium* (*Rhizobium radiobacter* is most frequently used) which is able to transfer its own DNA into the nucleus of the infected plant cell. However, the agrobacterium-mediated gene transfer is only applicable to a limited number of plants which are susceptible to agrobacterium, and the material delivery into recalcitrant plants still remains challenging. On the other hand, the biolistic method utilizes costly, specialized particle bombardment apparatus to directly deliver inorganic particles coated with exogenous DNA into plant cells. The bombarded particles physically penetrate the plant cell membrane but deliver the cargo into the cytoplasm or lumen of organelles randomly. Furthermore, there is a risk that the biolistic method may cause severe damage to target tissues/genes. Compared with these methods, the material delivery mediated by peptide carriers consisting of CPPs potentially has a broader range of applicable plant species, plant

Cell-Penetrating Peptides: Design, Development and Applications, First Edition. Makoto Oba and Yosuke Demizu.

tissues, and organelles. Representatively, DNAs/RNAs with abundant anionic charges spontaneously form a nanoassembly with cationic peptides containing a CPP sequence via electrostatic interaction. Such DNA/peptide complex assemblies are delivered into plant cells by mild infiltration/immersion techniques [1]. Various types of CPPs have been utilized for gene/protein delivery into plants to date. Nevertheless, both the availability of CPPs and their internalization mechanism have not been thoroughly investigated for material delivery into plant cells. In this section, the CPPs reported for material delivery into plants are reviewed for a comprehensive understanding of the scope of CPP designs, cargo materials, and host plants.

21.2 CPP Design and Availability for Plant Cells

As per the first report for systematic investigation on the permeability of CPPs into plant cells, Numata et al. demonstrated that a variety of CPPs were subjected to internalization assay for model plants [2]. To assess the availability of conventional CPPs for material delivery into plants, the cell-penetrating ability of known CPPs into plant cells was investigated using 55 libraries consisting of cationic, anionic, and amphiphilic types of CPPs, which are originally developed for cell penetration in animal cells. Initial screening of fluorescent probe-labeled CPPs using tobacco (*Nicotiana tabacum*) suspension-cultured cells (BY-2 cells) by confocal laser scanning microscopy (CLSM) revealed that the cell uptake efficiency was weakly correlated with the cytotoxicity causing plasmolysis, whereas no obvious similarity in the secondary structures was obtained among the CPPs showing high cell-penetrating efficiencies. The top three CPPs, D-R9 (rrrrrrrrr, D-form), R12 (RRRRRRRRRRRR), and KLA10 (KALKKLLAKWLAAAKALL) exhibited excellent cell-penetrating efficiency in BY-2 cells without lethal toxicity. The CPPs with high cell-penetrating properties tend to contain some cationic amino acid residues such as arginine and lysine, but no obvious correlation between the efficiency and the cationic amino acids, indicating that the specific cationic residues are not essential for increasing cell-penetrating efficiency in cultured plant cells.

The cell penetrating ability into plant leaf cells was also evaluated using the CPP library. The real leaf of tobacco (*N. benthamiana*) was infiltrated with fluorescent probe-labeled CPPs to investigate the cell-penetrating efficiency using CLSM. The high penetrating efficiency CPPs for the tobacco leaf differed from those for BY-2 cells. The top three CPPs with high penetrating efficiency were BP100 (KKLFKKILKYL), K9 (KKKKKKKKK), and DPV3 (RKKRRRESRKKRRRES). The remarkable feature is that BP100 and DPV3 are categorized in amphiphilic CPPs and tend to adopt an amphiphilic helical conformation. Another point is that these CPPs contain multiple lysine residues as cationic amino acids. The lysine-rich CPPs relatively tend to be effective for the cell penetration into plant tissue cells whereas the arginine-rich CPPs generally show high cell-penetrating ability for mammalian cells by inducing negative Gaussian membrane curvature, do not exhibit effective penetration properties except cultured plant cells. The versatility of the CPPs for plant types was also investigated using other types of plants such as *Arabidopsis thaliana* (thale cress, a model plant),

Solanum lycopersicum (tomato), hybrid aspen *Populus tremula* × *tremuloides* line T89 (poplar), and *Oryza sativa* (rice). The CPPs with high cell-penetrating efficiencies for BY-2 cells showed low cell-penetrating efficiency for the cells of these plant species and D-R9 especially did not penetrate the plant leaf cells. Some of the CPPs were plant type-specifically internalized in the plant leaf cells and the variation in the cell-penetrating efficiency was obviously found among the plant types. In particular, the leaf of rice was not penetrated by any CPPs examined in this study, whereas almost all the CPPs in the library showed medium to high penetration efficiencies into a rice callus, a cultured tissue of unorganized, undifferentiated plant parenchyma cells. The substantial discrepancy in the cell-penetrating efficiency between leaf and callus of rice is assumedly explained by the anatomical difference between these tissues. The monocot rice leaves are covered with a rigid silica layer, which possibly hampers the CPPs from effectively contacting cell membranes for internalization.

Collectively, some amphiphilic CPPs such as BP100 are possible candidates as a versatile tool to translocate cargo materials in plant cells for a wide range of plant species, although no CPP was found to be omnipotent for the universal use across all plant species/tissues. The optimal CPP will need to be selected for each plant type and target tissue to maximize internalization efficiency.

Peptide chain length is also a determinant factor of cell-penetrating ability. Kimura et al. treated plant-cultured cells with a series of polyhistidine peptides (PHPs) with different numbers of histidine residues (6–24) [3]. The cellular uptake of tetramethylrhodamine (TAMRA)-labelled PHPs in BY-2 cells was evaluated by CLSM. Shorter PHPs with 6–10 histidine residues efficiently internalized in BY-2 cells, whereas PHPs with longer histidine residues than 16, which showed high cellular uptake by mammalian cells, did not enter BY-2 cells. The cellular uptake was inhibited by a low-temperature treatment (4 °C), indicating that the internalization process of PHPs predominantly proceeds through an energy-dependent endocytic pathway. Long PHPs were not internalized and trapped in the cell wall of BY-2 cells, and yeast cells possessing cell walls also showed similar cellular uptake behavior. Therefore, the existence of cell walls might be a major reason of the cellular uptake difference between mammalian and plant cells.

21.3 Artificial CPP for Plant Cells

Current CPPs include various types of functional sequences, some of which are originated from natural proteins/polypeptides such as Tat peptide (GRKKRRQ RRRPPQ) derived from human immunodeficiency virus type 1 (HIV-1), penetratin (RQIKIWFQNRRMKWKK) derived from antennapedia protein discovered in *Drosophila*, and transportan (GWTLNSAGYLLGKINLKALAALAKKIL), a chimeric peptide constructed from galanin and mastoparan. To improve the cell-penetrating property of CPPs, sophisticated engineering of known CPPs has been demonstrated by optimizing the effective sequences and introducing unnatural structures. By sequence modification, conventional CPPs originally known for internalization into mammalian cells are further engineered to improve the cell-penetrating efficiency.

A facile, useful method to improve the internalization is the dimerization of known CPPs. Chugh et al. evaluated the cell-penetrating properties of several types of CPPs delivered into the freshly isolated triticale microspores by fluorometric analysis [4]. Among the various types of CPPs, Tat peptide, dimeric Tat peptide (Tat2), and mutated Tat peptide (M-Tat) were compared in their uptake efficiency. The dimeric Tat2 peptide showed 1.3 times higher uptake efficiency than the monomeric Tat peptide. Notably, the M-Tat peptide, where the first arginine residue was replaced by hydrophobic alanine residue, showed a significantly reduced fluorescence level than the Tat peptide. This indicates the significance of the amino acid sequence, which is crucial for assembling an active secondary structure, on the permeation property. Another example was demonstrated by Ng et al. in the intracellular delivery of proteins into intact plants [5]. In this study, two types of CPPs were designed: BP100-KH9 and 2BP100-K8 which possess a monomeric or dimeric amphiphilic CPP sequence, respectively, fused with a cationic domain to complex with negatively charged proteins. The fluorescently labeled bovine serum albumin (BSA) was complexed with these CPPs and delivered into the *A. thaliana* leaf via infiltration. The efficiency of BSA delivery into the leaves using the dimeric 2BP100-K8 was significantly higher than that using the monomeric BP100-KH9. The high efficiency was explained by that the second CPP may act as a linker preventing the polycationic domain from affecting the helical conformation of the first CPP. The trimer of a known cationic CPP, R9, was also designed by Wang et al. [6] The new peptide R9Z contains three copies of nona-arginine spaced by a cysteine residue, which may form a zinc-finger structure similar to zinc finger proteins in the presence of Zn^{2+} ion. The R9Z peptide was well internalized into both mammalian and plant cells, although a detailed comparable study was not reported.

Among molecular strategies for designing artificial CPPs, a modification with unnatural amino acid residues is a powerful way to enhance cell-penetrating ability by means of optimizing the conformation of peptides suitable for efficient interaction with the cell membrane. A rational sequence design consisting of unnatural amino acid residues leads to a fine control in the secondary structure of peptides, which will offer an appropriate peptide conformation for cell membrane penetration. An illustrative work was done by Terada et al. for designing a novel artificial CPP which shows an effective cell-penetrating ability in plant cells [7]. Amphiphilic CPPs are known to exhibit cell membrane-penetrating ability by adopting an amphiphilic helical conformation. In this study, 2-aminoisobutyric acid (Aib) residue, the simplest α,α-disubstituted amino acid, was introduced into periodic peptide sequences to induce stable helical structures for efficient internalization into plant cells. The Aib residue has two methyl groups on the α carbon, which affords a bulky peptide backbone to effectively induce and stabilize a helical conformation. Periodic sequences consisting of LysAibGly, LysAibAla, or LysAibLys motifs were designed as amphiphilic peptides and synthesized via enzyme-catalyzed polymerization method (Figure 21.1). Among them, poly(LysAibGly) and poly(LysAibAla) (designated as KAibG and KAibA, respectively) showed higher cell internalization efficiency not only in mammalian cells (HEK293 cells) but also

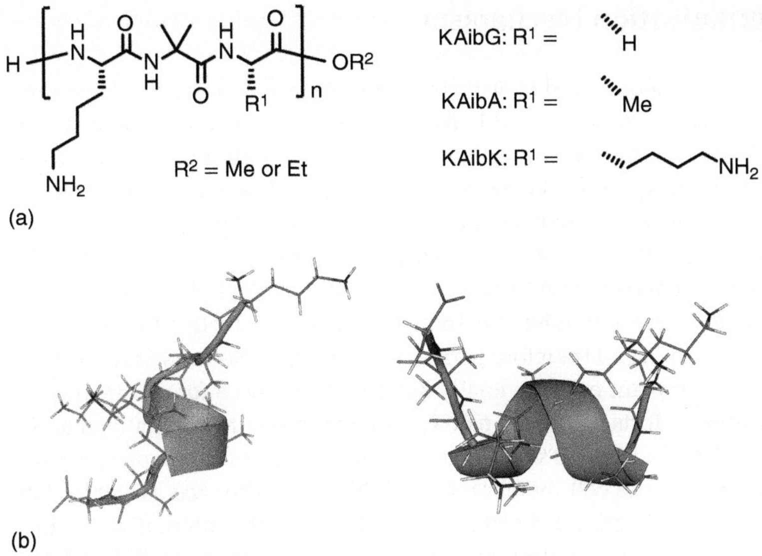

Figure 21.1 (a) Chemical structures of artificial CPPs designed for plant cells. (b) The most stable structures obtained by KAibA in GaMD simulations. Source: Terada et al. [7] / American Chemical Society / CC BY-4.0.

in plant leaf cells compared with conventional CPPs such as R9 and Tat peptides. In particular, KAibA peptide functioned as the most effective CPP in plant leaf cells regardless of plant species. The leaves of both *A. thaliana* and *Hibiscus cannabinus* (kenaf) were found to be internalized with KAibA peptide. Kenaf leaves have relatively rigid anatomical structures compared with model plants such as *A. thaliana* and tend to extremely resist the internalization of the conventional CPPs. In contrast, the KAibA peptide was clearly penetrated into the cytoplasm and vacuole of kenaf leaf cells. The result strongly supports that the KAibA peptide can be used as a versatile CPP to deliver cargo materials into plant cells. The novel Aib-containing CPP with a LysAibAla repetitive sequence predominantly adopted a helical conformation in a cell-membrane-mimicking medium (0.1 wt% sodium dodecyl sulfate solution), indicating that the helical structure stabilized by Aib residues probably improve the internalization process across the cell membrane. The helical conformation of KAibA peptide was also confirmed by Gaussian accelerated molecular dynamics (GaMD) simulation (Figure 21.1). In addition, the high internalization efficiency of these novel artificial CPPs was prolonged for up to 24 hours. This remarkable feature is attributed to their significant proteolytic stability in the presence of proteases (trypsin) without notable cytotoxicity, which is due to the existence of unnatural Aib residues. The high stability against proteolysis for the KAibA peptide leads to long-term internalization, which is assumed to be another key factor for the high cell-penetrating ability in plant cells because the material delivery into plant tissue cells tends to take longer incubation time than that into mammalian cells.

21.4 Internalization Mechanism

Plant cells possess a unique hierarchical structure which is distinct from mammalian cells. The most different feature is the existence of cell walls. The cell wall consists of a dense cellulose microfibrillar network filled with multiple biomacromolecules such as amorphous polysaccharides (pectin and hemicellulose) and cell wall relating enzymes mostly engaged in biosynthesis/degradation of cellulose. In most of the cells in plant tissues, the cell wall can be a physical barrier for material delivery rather than a dynamic cell membrane. In addition, the lipid components in the cell membrane differ from those of mammalian cells, which offer different rigidity, membrane potential, and therefore permeability. From these facts, it is no doubt that suitable CPPs for plant cells are totally different from those for mammalian or bacterial cells. Some efforts have been paid for elucidation of the mechanism of the cellular uptake of CPPs in plant cells by approaching from multiple perspectives. For the penetration of the cell membrane, an inhibitory assay was performed by several researchers to investigate the plausible pathway for the internalization process using cationic CPPs such as oligoarginine (R9) and Tat dimer (Tat2) [4, 8]. The peptides accompanied with cargo molecules were subjected to internalization into both cultured plant cells (wheat immature embryos) and plant tissue cells (root cells of mung bean) in the presence of various endocytosis inhibitors. Cytochalasin D, an endocytosis inhibitor by preventing polymerization of actin monomers, and 5-(*N*-ethyl-*N*-isopropyl)amiloride (EIPA), a micropinocytosis inhibitor which inhibits Na^+/H^+ exchanger, effectively suppressed the internalization of cationic CPPs into these plant cells. In the case of R9 peptide for the root cells of mung bean, some endocytosis inhibitors (fusicoccin and nigericin, H^+-ATPase activators) did not sufficiently interrupt the internalization, revealing that the actin-mediated endocytosis (presumed as macropinocytosis) is most probable for the major pathway. Low-temperature treatment at 4 °C also dramatically decreased the cell penetrating efficiency of these cationic peptides in both cultured and tissue plant cells. Therefore, the cationic CPPs could be dominantly internalized in plant cells via energy-dependent endocytic pathway(s). The CPP-mediated internalization of cargo molecules via vesicular formation was visually confirmed by CLSM observation by Thagun et al. [9] A peptide/DNA complex was formed by mixing a fluorescently labeled plasmid DNA and a cationic peptide containing BP100 sequence (KHKHKHKHKHKHKHKHKHKKLFKKILKYL) and internalized into tobacco leaf cells (*N. benthamiana*) with the cell membrane stained with FM4-64. The CLSM observation revealed that the colocalization of intercellular vesicles with the peptide/DNA complex was detected on the plasma membrane of plant cells after 2 hours post transfection, indicating that the peptide/DNA complex was translocated by the vesicle-forming endocytic pathway. The delivery process of the peptide/DNA complex into the nuclear membrane via vesicle trafficking was also detected. Most recently, a novel peptide which internalizes into plant cells via the macropinocytosis pathway was developed by Miyamoto et al. [10, 11] The peptide, dTat-Sar-EED4, consists of a cationic D-arginine-rich CPP domain dTat (rrrqrrkkr) and a hydrophobic endosomolytic domain EED4 (GWWG) linked with a hexameric sarcosine linker, and effectively transduce various functional proteins into plant cells. According to transmission electron microscopy (TEM) observation of *A. thaliana* root cells treated with dTat-Sar-EED4, the peptide induced the formation of plasma membrane protrusions and a large intracellular vesicle with a diameter >200 nm in plant cells (Figure 21.2). The internalization of dTat-Sar-EED4 was not inhibited by Endosidin 9-17, an inhibitor for clathrin-mediated endocytosis in plants, but significantly suppressed by the treatment with wortmannin (an inhibitor of phosphatidylinositol-3 kinase) and cytochalasin D. Collectively, the mechanism of dTat-Sar-EED4 internalization is assumed to be a clathrin-independent macropinocytosis-like pathway, which is a first report to directly visualize how the peptide induces an endocytic process characterized as macropinocytosis in plants.

Amphiphilic CPPs generally show a high affinity for lipid membranes by adopting a specific amphiphilic helical conformation. Some of the amphiphilic CPPs such as BP100 were found to be effective for internalization into plant cells as revealed by the screening assay discussed above, although the internalization mechanism is yet to be fully unraveled. Recent attempts to clarify the internalization process of the amphiphilic CPPs exploited atomic force microscopy (AFM) observation and molecular dynamics (MD) simulation. Yilmaz et al. investigated the interaction of BP100 with the plasma membrane-mimicking lipid membrane composed of 1,2-dioleoyl-*sn*-glycero-3-phosphocholine (DOPC)/1,2-dipalmitoyl-*sn*-glycero-3-phosphocholine (DPPC)/

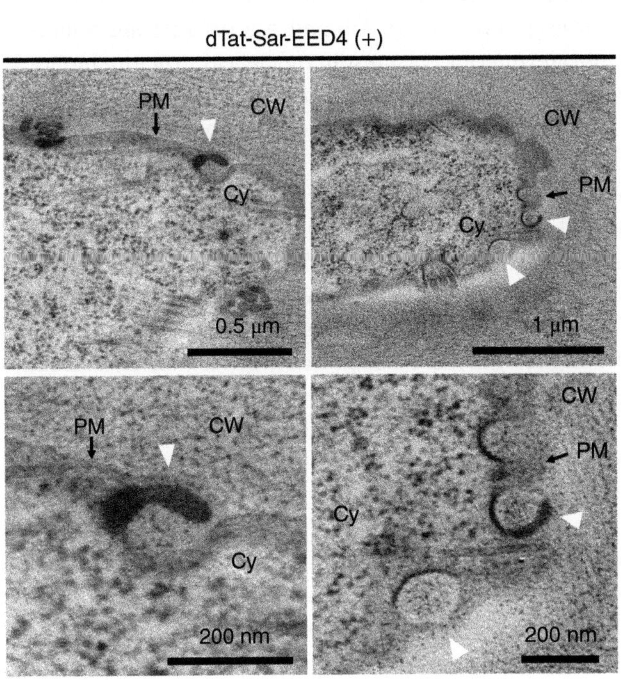

Figure 21.2 Plasma membrane protrusions (arrowhead) observed in TEM images of *A. thaliana* root cells treated with dTat-Sar-EED4. CW: cell wall; PM: plasma membrane; Cy: cytoplasm. Source: Miyamoto et al. [10], Reproduced with permission from American Chemical Society.

cholesterol (1/1/1) by quartz crystal microbalance (QCM) and high-speed AFM techniques [12]. The binding affinity of BP100 for the lipid membrane was evaluated by QCM, whereas the dynamic behavior of BP100 during the internalization process was monitored by high-speed AFM observation. In the presence of BP100, the cell-membrane mimicking lipid membrane showed thinning of the lipid layer. On the other hand, the fused CPP with a cationic peptide, BP100-KH9, disrupted the lipid membrane via pore formation. The different actions between BP100 and BP100-KH9 on the lipid membrane were attributed to their amphiphilic and cationic characteristics, respectively. The more cationic BP100-KH9 induced the toroidal pores causing the lipid headgroups to bend inward, whereas the amphiphilic BP100 adopted parallel alignment to interact with the lipid membrane interface, leading to a reduction in the lipid acyl chain order and thereby the membrane thinning. The dynamics of the internalization process of the amphiphilic CPP into the lipid membrane were also demonstrated using MD simulations by Gimenez-Dejoz et al. [13] Adaptively biased MD (ABMD) simulation with various CPPs, KAibA, KAibG, BP100, and R9, was performed in the lipid membrane that is composed of DPPC/DOPC/cholesterol. The amphiphilic peptides, KAibA, KAibG, and BP100 showed lower internalization energies compared with the cationic peptide R9, especially for KAibA with the lowest energy. The KAibA peptide had a higher propensity toward α- and 3_{10}-helices than for turns and bends in the lipid membrane. During the simulation, the KAibA peptide adopted the amphiphilic helical conformation where its Lys residues are in contact with the phosphatidyl colin head groups and its Aib and Ala residues are oriented toward the hydrophobic membrane interior. As a result, the helical structure induced the membrane distortion, leading to the efficient penetration of the peptide. These molecular insights on the internalization process of the amphiphilic peptides will help better design novel artificial CPPs for improved cell-penetrating efficiency.

To date, no direct evidence was reported on how CPPs and their complex with cargos penetrate the cell wall of plant cells. The cargo molecules accompanied by CPPs are eventually internalized into the cytoplasm of plant cells across the cell wall and membrane, although the conventional CPPs are basically designed to interact with cell membranes. The high-speed AFM observation of the cultured plant cells (BY-2 cells) revealed that the aligned cellulose fibrillar network is densely and randomly packed in the layered structure of cell walls [14]. The cellulose network has interfibrillar spaces with a size of less than 100 nm, where the CPP-mediated complex with an optimized size is assumed to pass through the cell walls [11]. Additionally, amorphous polysaccharides such as pectin filled in cell walls are rich in anionic charges, which interact with cationic residues of CPPs and hinder the traverse of CPPs through the cell wall. This is probably the reason for the remarkable difference in the cell-penetrating efficiency of some cationic CPPs into mammalian and plant cells. From a more macroscopic viewpoint, the structural difference of plant tissues also destinates the uptake pathway of CPP-mediated complex when the cargo materials are delivered into various plant tissues. Midorikawa et al. discovered that the preferable pathway of the CPP-fused peptide/plasmid DNA (pDNA) complex into plant cells is different among the plant species [15]. In this study, the delivery of peptide/pDNA complex into the seedlings of *A. thaliana* and

N. benthamiana was performed by infiltration using the vacuum/compression method. The uptake of the complex mainly occurred via the hydathodes of leaves in *A. thaliana* whereas the uptake in *N. benthamiana* preferably occurred via the root hairs. The peptide/pDNA complex featured by its specific size and electrostatic nature putatively penetrates the plant cells through these weak regions that are also targeted by the bacterial invasion pathway.

21.5 Polypeptide-Based Carriers for Material Delivery

Material delivery into the plant cells has been achieved by utilizing peptide-based carriers. Various types of genetic cargos such as pDNAs, double-strand DNAs (dsDNAs), and RNAs are electrostatically condensed with cationic CPPs or fusion peptides containing cationic and CPP sequences to form a peptide/gene nanoassembly. By selecting appropriate types of peptide carriers, the target cargo molecules are not only limited to genes but also expanded to proteins. Table 21.1 summarizes recent reports regarding the material delivery mediated by CPP-containing peptides into plants reported within the last two decades. As seen in the list, various types of plant species including model plants, dicots and monocots, and crops have been transduced by peptide-mediated delivery systems. In addition, a wide range of tissue types from cultured cells to real leaf and root cells were used for material delivery.

Anecdotally, arginine-rich CPPs rather than lysine-rich CPPs are an excellent tool to transfect mammalian cells with less cytotoxicity. A number of studies regarding the transfection into mammalian cells exploited arginine-rich CPPs. Indeed, the early studies employed arginine-based CPPs such as R9, R12 and Tat for the material delivery into plants, especially cultured plant cells and small size tissues. Synthetic oligoarginine (R12) was used to deliver a double-strand RNA (dsRNA) into the cultured tobacco suspension cells [16]. The Tat-based peptides containing multiple arginine residues were also used for the DNA and protein delivery into wheat zygotic immature embryos, which are an important tissue for the tissue culture of wheat crops [4]. As discussed in the previous section, the amphiphilic CPPs containing multiple lysine residues have been found to effectively internalize into plant tissues. Therefore, the lysine-based amphiphilic CPPs such as transportan, BP100, and most recently KAibA were applied to material internalization into various intact plant tissues including leaves and roots [15, 20, 22, 31].

Various types of cargo materials have been delivered into plant cells. By delivering exogenous DNA into the plant nucleus or organelles (chloroplasts and mitochondria carrying each distinct genomic DNA), desired functional proteins can be expressed in the plant cells to provide additional traits for plants. The pDNAs combined with CPPs or CPP-fused peptides have been most frequently used for the delivery into plants to transiently express the functional proteins of interest. The expression of the marker genes encoded in pDNAs such as β-glucuronidase (GUS) or a green fluorescent protein (GFP) was examined by GUS reporter gene assay or fluorescence observation in vivo, respectively, indicating that the pDNAs were successfully internalized into plant cells by CPP-mediated delivery [4, 20].

Table 21.1 Material delivery into various plants using CPPs.

Authors	Year	CPPs	Cargos	Plant species	Plant types	Ref.
Unnamalai et al.	2004	R12	dsRNA	*N. tabacum* (tobacco)	Cultured suspension cells	[16]
Wang et al.	2006	Tat, R9, R9Z	protein (GFP)	*Allium cepa* L. (onion)	Root tip cells	[6]
Chen et al.	2007	R9	pDNA	*Vigna radiata* L. (mung bean), *Glycine max* L. (soybean)	Root cells	[8]
Chang et al.	2007	Tat, R9	protein (GFP, RFP)	*Zea mays* L. (Corn) *A. cepa* L. (Onion)	Root tip cells	[17]
Chugh et al.	2008	Tat, Tat2, pVEC, transportan	enzyme (GUS), pDNA	*Triticum aestivum* (wheat)	Immature embryos	[4]
Chugh et al.	2009	Tat, Tat2, transportan, Pep-1	enzyme (GUS), pDNA	triticale	Microspores (mid-late uninuclear stage)	[18]
Ziemienowicz et al.	2012	Tat2	T-DNA, protein (VirD2)	*Triticosecale spp.* (triticale)	microspores	[19]
Lakshmanan et al.	2013	KH9-BP100, R9-Tat2	pDNA	*A. thaliana*, *N. benthamiana* (tobacco)	Leaf cells	[20]
Shim et al.	2013	Tat2, Pep-1	dsDNA, protein (RecA)	triticale	Microspores	[21]
Chuah et al.	2015	BP100	pDNA	*A. thaliana*	Leaf cells	[22]
Jain et al.	2015	tachyplesin	enzymes	*T. aestivum* (wheat), *Brassica napus* (rapeseed)	protoplasts, root cells, coleoptile, hypocotyl	[23]
Lakshmanan et al.	2015	KH9-BP100	dsDNA	*N. benthamiana* (tobacco)	Leaf cells	[24]
Ng et al.	2016	2BP100-K8 BP100-KH9	proteins (BSA, ADH, YFP)	*A. thaliana*	Leaf cells	[5]

(Continued)

Table 21.1 (Continued)

Authors	Year	CPPs	Cargos	Plant species	Plant types	Ref.
Numata et al.	2016	2BP100-K8	proteins (NPTII)	*A. thaliana*, *Malus prunifolia* (Willd.) Borkh.264 × *M. pumila* Mill. var. *paradisiaca* Schneid. (apple)	Leaf cells	[25]
Chuah et al.	2018	BP100-CH7	pDNA	*A. thaliana*	Leaf cells	[26]
Guo et al.	2019	BP100-KH9, BP100-CH7	protein	*O. sativa* (rice)	Callus cells	[27]
Midorikawa et al.	2019	BP100-KH9	pDNA	*A. thaliana*, *N. benthamiana* (tobacco)	Leaf cells, Root cells	[15]
Miyamoto et al.	2019	BP100	pDNA	*N. benthamiana* (tobacco)	Leaf cells	[28]
Thagun et al.	2019	BP100	pDNA	*A. thaliana*, *N. benthamiana* (tobacco), *S. lycopersicum* (tomato), potato	Leaf cells, fruit (tomato), tuber (potato)	[29]
Bilichak et al.	2020	D-R9, BP100-KH9	protein (ZFN)	*T. aestivum* L. (wheat)	Microspores	[30]
Miyamoto et al.	2020	BP100-KH9, dTat-Sar-EED4	pDNA	*A. thaliana*, *O. sativa* (rice)	Protoplasts, callus cells	[11]
Thagun et al.	2020	KH9-BP100	pDNA, RNA, protein	*N. benthamiana* (tobacco)	Leaf cells	[9]
Miyamoto et al.	2021	KAibA	pDNA	*A. thaliana*	Leaf cells	[31]
Fujita et al.	2021	KAibA	proteins (YFP, NPTII)	*A. thaliana*	Seedlings	[32]
Fujita et al.	2021	KAibA	protein (RNase)	*A. thaliana*	Seedlings	[33]
Odahara et al.	2021	KAibA	RNA, protein (Cas9)	*A. thaliana*	Callus cells	[34]

Fusing the CPP with the other peptides improves the internalization process. The cationic domains such as KH9 and R9 were ligated with CPPs to effectively condense pDNAs into a nano-sized complex. A cyclic DNA-binding domain CH7 (HHCRGHTVHSHHHCIR), in which the two cysteine residues are cross-linked by an S-S bond to generate a cyclic structure, was fused with a CPP (BP100) for a stimuli-triggered release of cargo DNA [26]. The cyclic structure in the CH7-fused peptide undergoes the cleavage of the S-S bond in the presence of reducing agents such as intracellular glutathione. The cargo pDNA complexed with the CH7-fused peptide was successfully released in plant cells, resulting in higher expression efficiency than the case using a linear analogue peptide without cross-link by the S-S bond. A combination of CPPs and organellar targeting peptides destinates the cargo materials to desired organelles after internalization into plant cells. The CPP-fused peptides containing mitochondria or plastid (chloroplast) targeting peptide were used for the DNA delivery into plant mitochondria or plastid, respectively [22, 29]. Miyamoto et al. demonstrated that the dual functionalization of peptide/pDNA complex with an artificial CPP (KAibA) and an endosome disrupting peptide (EDP) was performed by post-modification of the complex using click chemistry [31]. Additional functionalization with the EDP enabled the internalized peptide/pDNA complex to effectively escape from an endosome to the cytosol, which significantly improved the gene expression efficiency compared with the peptide/pDNA complex modified only with KAibA.

The delivery of dsDNA/dsRNA mediated by the peptides was also examined. Unlike pDNA requiring construction onto a suitable vector, dsDNA can be readily prepared as PCR products and easily handled in plant science fields. Using a BP100-based fusion peptide (KH9-BP100), the dsDNA coding luciferase as a reporter gene was successfully delivered into intact *N. benthamiana* leaves [24]. On the other hand, the post-transcriptional gene silencing of the target genes was achieved by oligoarginine-mediated dsRNA delivery into the genetically modified tobacco suspension cells. The gene silencing was confirmed by the degradation of the corresponding mRNA that resulted from a small interfering RNA (siRNA) generated from the delivered dsRNA [16].

Proteins are another attractive cargo molecule that directly introduces its functionality into plants without gene transfection process. Conventionally, functional proteins have been delivered into mammalian cells by employing an engineered protein which is covalently fused with a CPP, and a similar protocol has been applied to plant cells [35]. Recently, various types of functional proteins including GFP, yellow fluorescent protein (YFP), and alcohol dehydrogenase (ADH) were delivered into plant cells in a form of a complex with CPPs [5, 17]. A fusion peptide containing BP100 was able to deliver a functional protein neomycin phosphotransferase II (NPTII) into intact apple leaves without any genetic transfection [25]. The leaf infiltrated with NPTII showed transient resistance to kanamycin, which inhibits plant growth, for two days. Proteins regarding the genetic modification of plants such as RecA and a zinc finger nuclease (ZNF) were also attempted to be delivered using CPPs into the triticale or wheat microspores, respectively [21, 30].

Since native proteins are susceptible to proteolysis in vivo, the functionalization of plants by the exogenous proteins is transiently achieved in a fairly short period. Recently, vesicle-type carriers based on peptides have been utilized for the delivery of proteins into living cells. The equimolar combination of cationic and anionic peptides provides a vesicular assembly via electrostatic interactions. The polyion complex vesicle, known as PICsome, can stably encapsulate various intact proteins without sacrificing their native functionality, promising a longer duration of novel traits in plants conferred by exogenous proteins. Fujita et al. demonstrated the delivery of proteins using peptide-based PICsome in intact plants [32, 33]. As a functional protein, NPTII was encapsulated in PICsome modified with an artificial CPP KAibA and delivered into the seedlings of *A. thaliana*. The seedling infiltrated with NPTII exhibited kanamycin resistance up to 10 days, indicating that the PICsome-protected exogenous protein imparted the plant with a long-term transient trait. Most recently, CPP-modified PICsome was also used for the delivery of the clustered regularly interspaced short palindromic repeat (CRISPR)-associated nuclease Cas9 into plant callus cells [34]. The complex of Cas9 with a ribonucleoprotein (RNP) was encapsulated in the CPP-modified PICsome and delivered into the callus cells of *A. thaliana*. The DNA extracted from the calli transfected by the RNP complex encapsulated in PICsome was subjected to amplicon sequencing, and multiple mutations such as a single nucleotide deletion and a 14-base indel were detected in the target region. This technique using CPP-modified PICsome can offer a novel plant genome editing platform that requires no unfavorable permanent genetic alterations of the host genome.

21.6 Perspective

The CPPs are fascinating tools to modify plants for the current biotechnology as is for the material delivery systems in mammalian cells. In the fields of plant science and agriculture, the CPP-mediated material delivery into plants is making an increasingly large impact from both academic and practical viewpoints. Nevertheless, the delivery systems utilizing CPPs still have some limitations, which are attributed to insufficient insights into the internalization mechanisms and correlation between the function of CPPs and plant types. A continuous effort to elucidate how the CPPs interact with the cell membrane, adopt a suitable secondary structure in the membrane, and penetrate the cytoplasm will lead to further improvement of the sequence design to achieve higher internalization efficiency into plant cells. The CPP-mediated material delivery system in plants will open the way to an innovative plant technology for material and crop production.

References

1 Watanabe, K., Odahara, M., Miyamoto, T. et al. (2021). Fusion peptide-based biomacromolecule delivery system for plant cells. *ACS Biomaterials Science & Engineering* 7 (6): 2246–2254.

2 Numata, K., Horii, Y., Oikawa, K. et al. (2018). Library screening of cell-penetrating peptide for BY-2 cells, leaves of Arabidopsis, tobacco, tomato, poplar, and rice callus. *Scientific Reports* 8 (1): 10966.

3 Kimura, S., Kawano, T., and Iwasaki, T. (2017). Short polyhistidine peptides penetrate effectively into Nicotiana tabacum-cultured cells and Saccharomyces cerevisiae cells. *Bioscience, Biotechnology, and Biochemistry* 81 (1): 112–118.

4 Chugh, A. and Eudes, F. (2008). Study of uptake of cell penetrating peptides and their cargoes in permeabilized wheat immature embryos. *The FEBS Journal* 275 (10): 2403–2414.

5 Ng, K.K., Motoda, Y., Watanabe, S. et al. (2016). Intracellular delivery of proteins via fusion peptides in intact plants. *PLOS ONE* 11 (4): e0154081.

6 Wang, Y.-H., Chen, C.-P., Chan, M.-H. et al. (2006). Arginine-rich intracellular delivery peptides noncovalently transport protein into living cells. *Biochemical and Biophysical Research Communications* 346 (3): 758–767.

7 Terada, K., Gimenez-Dejoz, J., Miyagi, Y. et al. (2020). Artificial cell-penetrating peptide containing periodic α-aminoisobutyric acid with long-term internalization efficiency in human and plant cells. *ACS Biomaterials Science & Engineering* 6 (6): 3287–3298.

8 Chen, C.-P., Chou, J.-C., Liu, B.R. et al. (2007). Transfection and expression of plasmid DNA in plant cells by an arginine-rich intracellular delivery peptide without protoplast preparation. *FEBS Letters* 581 (9): 1891–1897.

9 Thagun, C., Motoda, Y., Kigawa, T. et al. (2020). Simultaneous introduction of multiple biomacromolecules into plant cells using a cell-penetrating peptide nanocarrier. *Nanoscale* 12 (36): 18844–18856.

10 Miyamoto, T., Toyooka, K., Chuah, J.-A. et al. (2022). A synthetic multi-domain peptide that drives a macropinocytosis-like mechanism for cytosolic transport of exogenous proteins into plants. *ACS JACS Au.*

11 Miyamoto, T., Tsuchiya, K., and Numata, K. (2020). Dual peptide-based gene delivery system for the efficient transfection of plant callus cells. *Biomacromolecules* 21 (7): 2735–2744.

12 Yilmaz, N., Kodama, Y., and Numata, K. (2021). Lipid membrane interaction of peptide/DNA complexes designed for gene delivery. *Langmuir* 37 (5): 1882–1893.

13 Gimenez-Dejoz, J. and Numata, K. (2022). Molecular dynamics study of the internalization of cell-penetrating peptides containing unnatural amino acids across membranes. *Nanoscale Advances* 4: 397–407.

14 Yilmaz, N., Kodama, Y., and Numata, K. (2020). Revealing the architecture of the cell wall in living plant cells by bioimaging and enzymatic degradation. *Biomacromolecules* 21 (1): 95–103.

15 Midorikawa, K., Kodama, Y., and Numata, K. (2019). Vacuum/compression infiltration-mediated permeation pathway of a peptide-pDNA complex as a non-viral carrier for gene delivery in planta. *Scientific Reports* 9 (1): 271.

16 Unnamalai, N., Kang, B.G., and Lee, W.S. (2004). Cationic oligopeptide-mediated delivery of dsRNA for post-transcriptional gene silencing in plant cells. *FEBS Letters* 566 (1-3): 307–310.

17 Chang, M., Chou, J.-C., Chen, C.-P. et al. (2007). Noncovalent protein transduction in plant cells by macropinocytosis. *New Phytologist* 174 (1): 46–56.

18 Chugh, A., Amundsen, E., and Eudes, F. (2009). Translocation of cell-penetrating peptides and delivery of their cargoes in triticale microspores. *Plant Cell Reports* 28 (5): 801–810.

19 Ziemienowicz, A., Shim, Y.-S., Matsuoka, A. et al. (2012). A novel method of transgene delivery into triticale plants using the agrobacterium transferred DNA-derived nano-complex. *Plant Physiology* 158 (4): 1503–1513.

20 Lakshmanan, M., Kodama, Y., Yoshizumi, T. et al. (2013). Rapid and efficient gene delivery into plant cells using designed peptide carriers. *Biomacromolecules* 14 (1): 10–16.

21 Shim, Y.-S., Eudes, F., and Kovalchuk, I. (2013). dsDNA and protein co-delivery in triticale microspores. *in vitro Cellular & Developmental Biology – Plant* 49 (2): 156–165.

22 Chuah, J.-A., Yoshizumi, T., Kodama, Y. et al. (2015). Gene introduction into the mitochondria of *Arabidopsis thaliana* via peptide-based carriers. *Scientific Reports* 5 (1): 7751.

23 Jain, A., Yadav, B.K., and Chugh, A. (2015). Marine antimicrobial peptide tachyplesin as an efficient nanocarrier for macromolecule delivery in plant and mammalian cells. *The FEBS Journal* 282 (4): 732–745.

24 Lakshmanan, M., Yoshizumi, T., Sudesh, K. et al. (2015). Double-stranded DNA introduction into intact plants using peptide–DNA complexes. *Plant Biotechnology* 32 (1): 39–45.

25 Numata, K., Horii, Y., Motoda, Y. et al. (2016). Direct introduction of neomycin phosphotransferase II protein into apple leaves to confer kanamycin resistance. *Plant Biotechnology* 33 (5): 403–407.

26 Chuah, J.-A. and Numata, K. (2018). Stimulus-responsive peptide for effective delivery and release of DNA in plants. *Biomacromolecules* 19 (4): 1154–1163.

27 Guo, B., Itami, J., Oikawa, K. et al. (2019). Native protein delivery into rice callus using ionic complexes of protein and cell-penetrating peptides. *PLOS ONE* 14 (7): e0214033.

28 Miyamoto, T., Tsuchiya, K., and Numata, K. (2019). Block copolymer/plasmid DNA micelles postmodified with functional peptides via Thiol–Maleimide conjugation for efficient gene delivery into plants. *Biomacromolecules* 20: 653–661.

29 Thagun, C., Chuah, J.-A., and Numata, K. (2019). Targeted gene delivery into various plastids mediated by clustered cell-penetrating and chloroplast-targeting peptides. *Advanced Science* 6 (23): 1902064.

30 Bilichak, A., Sastry-Dent, L., Sriram, S. et al. (2020). Genome editing in wheat microspores and haploid embryos mediated by delivery of ZFN proteins and cell-penetrating peptide complexes. *Plant Biotechnology Journal* 18 (5): 1307–1316.

31 Miyamoto, T., Tsuchiya, K., and Numata, K. (2021). Endosome-escaping micelle complexes dually equipped with cell-penetrating and endosome-disrupting peptides for efficient DNA delivery into intact plants. *Nanoscale* 13 (11): 5679–5692.

32 Fujita, S., Motoda, Y., Kigawa, T. et al. (2021). Peptide-based polyion complex vesicles that deliver enzymes into intact plants to provide antibiotic resistance without genetic modification. *Biomacromolecules* 22 (3): 1080–1090.

33 Fujita, S., Tsuchiya, K., and Numata, K. (2021). All-peptide-based polyion complex vesicles: facile preparation and encapsulation of the protein in active form. *ACS Polymers Au* 1 (1): 30–38.

34 Odahara, M., Watanabe, K., Kawasaki, R. et al. (2021). Nanoscale polyion complex vesicles for delivery of cargo proteins and Cas9 ribonucleoprotein complexes to plant cells. *ACS Applied Nano Materials* 4 (6): 5630–5635.

35 Chang, M., Chou, J.-C., and Lee, H.-J. (2005). Cellular internalization of fluorescent proteins via arginine-rich intracellular delivery peptide in plant cells. *Plant and Cell Physiology* 46 (3): 482–488.

Index

r

s